T0413537

LEARNING AND INTERTEMPORAL INCENTIVES

World Scientific Series in Economic Theory
(ISSN: 2251-2071)

Series Editor: Eric Maskin *(Harvard University, USA)*

Published

Forthcoming

World Scientific Series in Economic Theory – Vol. 8

LEARNING AND INTERTEMPORAL INCENTIVES

Dirk Bergemann

Yale University, USA

Juuso Välimäki

Aalto University School of Business, Finland

W₲ World Scientific

NEW JERSEY · LONDON · SINGAPORE · BEIJING · SHANGHAI · HONG KONG · TAIPEI · CHENNAI · TOKYO

Published by

World Scientific Publishing Co. Pte. Ltd.
5 Toh Tuck Link, Singapore 596224
USA office: 27 Warren Street, Suite 401-402, Hackensack, NJ 07601
UK office: 57 Shelton Street, Covent Garden, London WC2H 9HE

British Library Cataloguing-in-Publication Data
A catalogue record for this book is available from the British Library.

World Scientific Series in Economic Theory — Vol. 8
LEARNING AND INTERTEMPORAL INCENTIVES

Copyright © 2020 by World Scientific Publishing Co. Pte. Ltd.

ISBN 978-981-121-441-7

For any available supplementary material, please visit
https://www.worldscientific.com/worldscibooks/10.1142/11669#t=suppl

Desk Editors: Aanand Jayaraman/Shreya Gopi

Typeset by Stallion Press
Email: enquiries@stallionpress.com

To our families, Kishwar, Reza, and Yasmin, and Anu, Esko, Vilja, and Aapo, without whose love and company this material would not have been written, who cheered us on when the economics wore us out.

— Dirk Bergemann and Juuso Välimäki

Foreword

Economists have long been interested in markets where time plays an important role. Indeed, Stackelberg's model of a leader–follower duopoly has been a standard part of the economics curriculum for half a century.

But until recently, little attention was paid to the intemporal interaction between markets and information — how much should a firm invest today to learn about demand for a good it will be selling tomorrow?

Dirk Bergemann and Juuso Välimäki — both outstanding and internationally recognized economic theorists — have been leading figures in the development of this exciting and important area. This is why I am grateful and delighted that they have assembled in this volume a collection of sixteen of their papers (both jointly and with other collaborators), together with a helpful introduction showing how the different papers fit together. The volume will prove to be essential reading.

— Professor Eric S. Maskin
American Economist

About the Authors

 **Dirk Bergemann** is Douglass and Marion Campbell Professor of Economics at Yale University. He was Chair of the Department of Economics from 2013 to 2019 and Co-Editor of *Econometrica* from 2014 to 2018. He has been affiliated with the Cowles Foundation for Research in Economics at Yale since 1996 and has been a fellow of the Econometric Society since 2007. His research is in the area of game theory, contract theory and venture capital, and mechanism design. His most recent work is in the area of dynamic mechanism design and dynamic pricing, robust mechanism design, and information design. His research has been supported by grants from the National Science Foundation, the Alfred P. Sloan Research Fellowship, Google Faculty Fellow, and the German National Science Foundation.

Juuso Välimäki is a microeconomic theorist, whose research focuses on dynamic incentive problems and mechanism design. After receiving his Ph.D. from the University of Pennsylvania, he has taught at leading universities such as Northwestern University and Yale University and has been a fellow of the Econometric Society since 2007. He is currently a Professor of Economics and the Head of Department of Economics at Aalto University School of Business. He has published numerous articles in leading journals, and until the end of 2018, he was the Managing Editor of the *Journal of the European Economic Association.*

Contents

Chapter 1

Introduction

Dirk Bergemann[*] and Juuso Välimäki[†]

*Department of Economics, and Cowles Foundation of Research in Economics,
Yale University, New Haven, CT 06520-8268, USA
dirk.bergemann@yale.edu
†Department of Economics, Aalto University School of Business,
02150 Espoo, Finland
juuso.valimaki@aalto.fi

This book collects our articles (sometimes with different co-authors) on the effects of dynamic resolution of uncertainty in models of industrial organization and contracting. The main objective behind this strand of literature is to understand how market institutions affect the intertemporal incentives of various parties to engage in costly production of information and to assess the equilibrium performance of different institutions. If future prices depend on information produced in current purchases, firms and buyers have differential stakes in the outcomes. If a principal has to incentivize an agent to collect costly information over time, new dynamic costs of information emerge in games with symmetric information and with informational asymmetries.

For example, when a firm decides to launch a new product in the marketplace its eventual success depends on many uncertain factors. For example, it is often difficult to predict how highly buyers value new features in a product. The market position of each product and the profitability of all firms depend on the resolution of this uncertainty. Buyers as well as all sellers in the market have a stake in the launch and their market actions reflect this. This gives rise to a fundamentally dynamic view of the market place where investment and pricing decisions transform the competitive landscape in an endogenous manner.

The first seven papers in this volume consider settings where information between the players in the game is symmetric and transfers between the players are possible. In the last eight papers, various informational asymmetries are introduced and their effects on the dynamic equilibrium are analyzed.

In "Learning and Strategic Pricing" (Bergemann and Välimäki (1996)), we analyze a simple game where two firms with differentiated products of initially unknown qualities compete in prices over a single customer. New information arrives only for the firm serving the buyer in any period and a new imperfect quality signal for the selling firm's product is drawn. We show that spot prices are sufficient to guarantee the efficiency of truthful Markov Perfect Equilibria in this game (where truthfulness is defined in the sense of common agency games in Bernheim and Whinston (1986)). It should be noted that this result requires forward-looking behavior on part of all the firms as well as the buyer and depends on the full internalization of future price effects in the buyer's decisions.

When buyers exert informational externalities on each other, the situation changes. If the value of the product is the same for all buyers, then information resulting from individual purchases has an effect on all future sales due to changes in the competitive position of the firms. We analyze this possibility in "Experimentation in Markets" (Bergemann and Välimäki (2000)) in the context of an entrant firm with uncertain quality in price competition against an incumbent of known quality. We show that whenever both firms are forward-looking in their decisions, but the buyers only care about their own individual purchases (and in particular, not about future purchases of other buyers), the relevant Markov Perfect Equilibria feature socially excessive experimentation. If the firms were myopic, equilibria would feature too little experimentation. Hence, we see that details of the market organization have somewhat unexpected effects on dynamic market performance.

In "Market Diffusion with Two-Sided Learning", (Bergemann and Välimäki (1997)) and "Entry and Vertical Differentiation" (Bergemann and Välimäki (2002a)), we extend the analysis to cover differentiated markets. The first paper covers learning about the common quality component in a model with horizontally differentiated buyers while the second assumes vertical differentiation. By using the strong long-run average payoff criterion introduced in Dutta (1991) and a diffusion model for belief dynamics, we derive detailed qualitative predictions on the sales paths for new products depending on the mode of competition (prices vs. quantities) and the type of differentiation.

"Dynamic Pricing of New Experience Goods" (Bergemann and Välimäki (2006b)) changes focus to learning about private values. While still focusing on the launch of a new product in the market, we assume away all aggregate uncertainty. The distribution of valuations within the population of buyers is known, but no single buyer knows her own valuation before trying the product: even if after clinical tests, it is known that a pharmaceutical works for 40% of the population and causes side effects in 10%, each patient learns her own reaction to the drug only by trying it. We assume that the firm still must set a single market price so that price discrimination between new and existing buyers is ruled out and we derive different diffusion and pricing patterns for niche and mass products.

In "Dynamic Common Agency" (Bergemann and Välimäki (2003)) and "Dynamic Price Competition" (Bergemann and Välimäki (2006a)), we turn to a more abstract learning set-up that includes the previous product markets as special cases. In "Dynamic Common Agency", we allow a set of principals (corresponding to the selling firms in the previous papers) to propose transfers conditional on a spot action to an agent (the buyer). In the dynamic common agency game, all information is symmetric in the sense that all parties observe all the payoff relevant variables publicly and all parties are forward-looking. We show that truthful equilibria of the dynamic game are efficient and that they are unique if and only if the payoff of each principal coincides with her marginal contribution to the social value of the game.

"Dynamic Price Competition" restricts the transfers between the principals and the agent to be bilateral. Hence, the model provides a bridge between the general common agency contracting environment and the pricing games of our earlier papers. If the choice that the buyer makes in any period is payoff-relevant only for the future rewards from purchases from the same seller, then the truthful Markovian equilibria of any stationary dynamic pricing game are efficient. This result offers us a new look at the reasons for the efficiency of equilibria in "Learning and Strategic Pricing" and other similar models.

Informational asymmetries open entirely new questions in game theoretic models of learning. "Dynamic Venture Capital Financing, Learning and Moral Hazard" (Bergemann and Hege (1998)) and "The Financing of Innovation: Learning and Stopping" (Bergemann and Hege (2005)) introduce imperfect monitoring in a model of venture capital financing. A principal and an agent learn about a common payoff-relevant state variable and the agent makes an unobservable (and privately costly) effort choice. After deviations from equilibrium effort, this gives rise to divergent beliefs on the state. These papers make a methodological contribution in showing that a tractable equilibrium analysis is possible even in this quite complicated setting. The interplay between learning about the state and the moral hazard inherent in the agent's actions gives rise to a variety of optimal equilibrium patterns of financing.

"Learning about the Arrival of Sales" (by Mason and Välimäki (2011)) analyzes the optimal dynamic pricing pattern in markets for a single indivisible good where the buyers' actions are not perfectly monitored. Learning is about the rate at which potential buyers arrive in a market and only decisions to purchase are observable. By choosing the spot prices, the seller also determines the speed of learning conditional on not selling in the current period.

"Learning and Information Aggregation in an Exit Game" (Murto and Välimäki (2011)) analyzes social learning in a model of common and idiosyncratic learning about market conditions. The key informational imperfection here is that individual market experiences are unobservable and only summary market actions (exit decisions in this case) transmit information between players.

The final four papers come closer to dynamic mechanism design problems where the focus is on the dynamic evolution of private information. "Information Acquisition and Efficient Mechanism Design" (Bergemann and Välimäki (2002b)) considers private incentives to learn prior to participating in a mechanism. These incentives are perfectly aligned in the case of independent private information as long as final allocation is determined through the pivot mechanism and individual agents get their marginal contribution to the social surplus. With interdependent (but independent) types, generalized VCG-mechanisms sometimes result in efficient allocation decisions. Unfortunately, they do not result in marginal contribution payoffs and hence the private and social incentives for information production are not aligned. In auction-type settings, individual incentives for information acquisition are shown to be excessive.

"The Dynamic Pivot Mechanism" (Bergemann and Välimäki (2010)) develops the notion of dynamic marginal contribution payoffs in the independent quasi-linear setting with quite general dynamic private information. We connect the symmetric information model in "Learning and

Strategic Pricing" and "Dynamic Common Agency" to their private information counterparts and in doing this we demonstrate the close analogy between dynamic pricing games, common agency models, and VCG-mechanisms. "Dynamic Revenue Maximization: A Continuous Time Approach" (Bergemann and Strack (2015)) considers a profit-maximizing seller in a dynamic mechanism design model. Finally, "Dynamic Mechanism Design: An Introduction" (Bergemann and Välimäki (2019)) surveys recent developments in dynamic mechanism design with quasi-linear values and independent types in more detail.

Many interesting questions remain open in this area. Information is of value to many market participants, but its creation often involves a private cost. Different market structures distribute the future benefits from current investments differently between the market participants. Different assumptions on privacy of the information generated give rise to different degrees of market power to the sellers. We look forward to seeing more research on the overall efficiency of competition with different modes of information transmission between the market participants.

References

BERGEMANN, D., AND U. HEGE (1998): "Dynamic Venture Capital Financing, Learning and Moral Hazard," *Journal of Banking and Finance* 22, 703–735.

———— (2005): "The Financing of Innovation: Learning and Stopping," *RAND Journal of Economics*, 36, 719–752.

BERGEMANN, D., AND P. STRACK (2015): "Dynamic Revenue Maximization: A Continuous Time Approach," *Journal of Economic Theory*, 159, 819–853.

BERGEMANN, D., AND J. VÄLIMÄKI (1996): "Learning and Strategic Pricing," *Econometrica* 64, 1125–49.

———— (1997): "Market Diffusion with Two-Sided Learning," *Rand Journal of Economics* 28, 773–795.

———— (2000): "Experimentation in Markets," *Review of Economic Studies* 67, 213–234.

———— (2002a): "Entry and Vertical Differentiation," *Journal of Economic Theory* 106, 91–125.

———— (2002b): "Information Acquisition and Efficient Mechanism Design," *Econometrica* 70, 1007–1033.

———— (2003): "Dynamic Common Agency," *Journal of Economic Theory* 111, 23–48.

———— (2006a): "Dynamic Price Competition," *Journal of Economic Theory* 127, 232–263.

———— (2006b): "Dynamic Pricing of New Experience Goods," *Journal of Political Economy* 114, 713–743.

———— (2010): "The Dynamic Pivot Mechanism," *Econometrica* 78, 771–790.

———— (2019): "Dynamic Mechanism Design: An Introduction," *Journal of Economic Literature* 57, 235–274.

BERNHEIM, D., AND M. WHINSTON (1986): "Common Agency," *Econometrica*, 54, 923–942.

DUTTA, P. (1991): "What do Discounted Optima Converge to?" *Journal of Economic Theory*, 55, 64–94.

MASON, R., AND J. VÄLIMÄKI (2011): "Learning about the Arrival of Sales," *Journal of Economic Theory*, 146, 1699–1711.

MURTO, P., AND J. VÄLIMÄKI (2011): "Learning and Information Aggregation in an Exit Game," *Review of Economic Studies*, 78, 1426–1461.

Chapter 2

Learning and Strategic Pricing*

Dirk Bergemann[†] and Juuso Välimäki[‡]
†*Department of Economics, and Cowles Foundation of Research in Economics,
Yale University, New Haven, CT 06520-8268, USA*
dirk.bergemann@yale.edu
‡*Department of Economics, Aalto University School of Business,
02150 Espoo, Finland*
juuso.valimaki@aalto.fi

We consider the situation where a single consumer buys a stream of goods from different sellers over time. The true value of each seller's product to the buyer is initially unknown. Additional information can be gained only by experimentation. For exogenously given prices the buyer's problem is a multi-armed bandit problem. The innovation in this paper is to endogenize the cost of experimentation to the consumer by allowing for price competition between the sellers. The role of prices is then to allocate intertemporally the costs and benefits of learning between buyer and sellers. We examine how strategic aspects of the oligopoly model interact with the learning process.

All Markov perfect equilibria (MPE) are efficient. We identify an equilibrium which besides its unique robustness properties has a strikingly simple, seemingly myopic pricing rule. Prices below marginal cost emerge naturally to sustain experimentation. Intertemporal exchange of the gains of learning is necessary to support efficient experimentation. We analyze the asymptotic behavior of the equilibria.

Keywords: Learning, experimentation, dynamic oligopoly, Markov perfect equilibrium, infinite stochastic game, multi-armed bandit.

1. Introduction

Much of the existing literature on dynamic choice under uncertainty has focused on the case where a single decision-maker chooses sequentially

*This chapter is reproduced from *Econometrica*, **64**, 1125–1150, 1996.

among a fixed set of alternatives. In many economic situations, the alternatives are supplied by a separate economic agent (or a group of economic agents) and the decision theoretic analysis then provides only a description of the demand side of the market.

We develop a simple dynamic equilibrium model of price formation under learning and uncertainty. In an infinite horizon model with price competition, a buyer chooses sequentially between products whose qualities are initially unknown to all parties in the model; the buyer does not know the underlying characteristics of the products while the producers are uncertain about the tastes of the buyer. Each purchase yields additional information about the true product quality to all parties in the model.

In the decision theoretic situation where prices are exogenously fixed, it is well known that the optimal purchasing strategy by the buyer may involve experimentation. That is, in some periods the buyer is willing to sacrifice some of her *current* payoff in order to gain additional information which is valuable for *future* decisions. This temporal separation of costs and benefits causes no ex-ante efficiency losses in the single player case since the costs of experimentation have to be born by the same agent who enjoys any gains from successful experiments. However, if prices are set by profit maximizing producers, a nontrivial problem of intertemporal allocation arises as current and future prices determine the costs and benefits of experimentation to all the parties in the model.

To illustrate the point, consider a factory manager (the buyer) choosing between two alternative technologies supplied by outside contractors (the sellers). In each period, she signs a one period lease with one of the contractors. The output from each technology is a random variable depending on both the true productivity of the technology in the factory (the value of the match between the factory and the technology) and some outside random effects. If output is publicly observable, then all parties receive a common noisy signal about the true productivity of the technology chosen in each period. Beliefs are updated in a Bayesian fashion and posterior beliefs determine the relative competitive positions of the two contractors. We want to compare the incentives of the factory manager in the equilibrium model, where the contractors react optimally to changes in beliefs, to the decision theoretic model, where prices are exogenously fixed. In particular, the incentives to undertake experimentation may be fundamentally different under the two scenarios.

Suppose that the beliefs about product qualities are such that the optimal strategy in the decision theoretic case suggests experimentation (i.e.

choosing the product with lower expected quality) in the current period. Is the buyer still willing to pay for an experiment in the equilibrium model if the outcomes of experiments are publicly observed? A bad outcome in the experiment results in more pessimistic beliefs on the quality of the product purchased. If the buyer decides to switch suppliers, the bad outcome results in a higher price to be paid in the next period since the relative competitive position of the nonselling firm has improved. On the other hand, if the outcome of the experiment is good, the buyer becomes more optimistic about the product quality and as a consequence, her willingness to pay for the product is increased. Since the outcome is publicly observable, the seller observes an increase in her monopoly power relative to the competitors and has an incentive to raise the price in future periods to appropriate the maximal amount of consumer surplus from the buyer. Since neither a good nor a bad outcome leads to an increase in consumer surplus, the willingness of the buyer to experiment is reduced. If it is in the firms' interest to sustain experimentation, the consumer will have to be compensated for experimentation costs in the current period.

For the case of two sellers and one buyer, we characterize the set of Markov perfect equilibria. The central result of the paper states that in spite of future rent seeking by the firms, all Markov perfect equilibria in the model are efficient, i.e. the discounted expected sum of consumer surplus and the two firms' profits is maximized along any equilibrium path.[1] In particular, an efficient amount of experimentation is undertaken on any Markov perfect equilibrium path. Using this fact, we can deduce the sequencing of consumer purchases immediately, since the efficient paths coincide with the solution paths of the buyer's decision problem when prices are fixed to be identically zero. A solution for this decision problem is available in the statistical literature on multi-armed bandits. The remaining task is thus to calculate the prices that support efficient experimentation in equilibrium and determine the division of surplus between the buyer and the sellers along the efficient path.

In analogy with the one-shot pricing game with heterogeneous product quality, we need a refinement similar in nature to trembling hand perfection

---

[1] With multiple buyers and publicly observed signals, the externalities involved in the experimentation process cannot be fully internalized by the firms and hence the equilibrium path will differ from the efficient path, in general. The restriction to two sellers is made purely for expositional convenience. We discuss various extensions to the model in Section 3.

to select a unique equilibrium. In this equilibrium, which we call the cautious equilibrium, current prices provide the buyer with insurance against future rent seeking resulting from successful experiments. Note, however, that a bad outcome for the currently selling firm is a good outcome for the nonselling firm. As a consequence, prices do not provide insurance against bad outcomes in the experimentation. The buyer is left worse off since the nonselling firm acts as an outside option for the buyer in the determination of selling prices. Rent seeking by the nonselling firm in these contingencies allows the selling firm to charge higher prices. The equilibrium pricing rule is quite simple. In each period, the selling price is equal to the difference in expected qualities and is hence similar to the equilibrium price in the myopic Bertrand game. The identity of the seller does not, however, coincide with the myopic game since the efficient path involves experimentation at some nodes.

Recent papers by Smith (1992) and Bolton and Harris (1993) also introduce dynamic learning models with many agents. Smith considers a sequence of sellers entering the market. Each seller can individually observe the fraction of incumbents charging a low price, before solving his own bandit problem. He shows that the most profitable pricing option is eventually chosen by the market with probability one, as opposed to the case of an individual seller, who might charge the less profitable price forever even under optimal learning, as Rothschild (1974) showed. Bolton and Harris introduce strategic interaction in a learning model in which N players face simultaneously the same experimentation problem. Although the alternatives are still exogenously given, the informational externality, which arises through the public good aspect of experimentation, transforms the bandit problem into a game of strategic experimentation. The idea of an informational externality arising in a sequential learning model is already central to Rob (1991), who studies a dynamic model of entry when the size of the market is uncertain. In our work, public observability of the signals creates no informational externalities since the product qualities are assumed to be statistically independent. With multiple buyers and publicly observed signals, free riding on other buyers' experimentation becomes a problem. The firms are, however, able to internalize this externality to a large extent and this may reverse the typical results on underinvestment in information as discussed in Section 3.

While we consider the situation where a single consumer makes purchases over time in an oligopolistic market, several other economic applications could be analyzed within our framework. The bandit framework is often used as a matching model in the analysis of the labor market as in Jovanovic (1979) and Miller (1984), which ignore however the aspect of strategic interaction between employee and employer.[2] The choice and financing of new and uncertain technologies and R&D projects also fits exactly into the framework we develop in this paper.

A brief outline of the paper follows. The duopoly model is introduced in Section 2. In Section 3 we investigate the efficiency of the Markov perfect equilibria for the infinite game. In Section 4 we single out a particular equilibrium (we call it *cautious equilibrium*), which besides its unique robustness properties, has very appealing economic features. Subsequently we analyze the asymptotic properties of the equilibria and characterize the entire set of Markov perfect equilibria by the lower and upper bounds of the payoffs. We conclude in Section 5 with a discussion of some variants of the basic model.

2. The Model

In this section, we describe the players, the learning environment, and the strategies. Then we compare the strategic pricing model briefly with the multi-armed bandit model. The comparison will prove useful for the welfare analysis of the equilibrium model in Section 3.

Price competition between two firms, indexed by $i = 1, 2$, takes place in discrete time with an infinite horizon, $t = 0, 1, 2, \ldots$. The firms announce in each period their prices, p_t^i, simultaneously. The goods produced by the two firms differ only with respect to their (expected) quality. Firms have the same unit costs normalized to zero. The buyer has unit demand in each period. At time t, the buyer's expected valuation of a purchase is a linear function of the expected quality and the price:

$$E_t X_t^i - p_t^i = x_t^i - p_t^i,$$

where the random realization of the quality of product i in period t is denoted by X_t^i.[3] Each X_t^i is a nonnegative real valued random variable with finite expectations on some probability space $(\Omega, \mathscr{F}, \mathscr{P})$. The expected value of the quality realization, X_t^i, conditional on the history until period t,

---

[2]Recently, Felli and Harris (1994) introduced a matching model in continuous time where wages are renegotiated at every instant of time. The equilibrium in their basic model is the continuous time equivalent of the cautious equilibrium in our model.

[3]Any quasilinear utility function $U(X_t^i) - p_t^i$ could be used alternately.

is given by $x_t^i = E_t X_t^i$. All parties have common priors about the reward processes $X^i = \{X_t^i\}_{t=0}^\infty$ at the beginning of the game. Moreover, the sample realization X_t^i is publicly observable.

We concentrate our attention for simplicity on *sampling processes*. A sampling process is a sequence $X^i = \{X_t^i\}_{t=0}^\infty$ of independent, identically distributed random variables $X_0^i, X_1^i, \ldots$, drawn from a distribution with an unknown (vector-valued) parameter θ^i belonging to a family of distributions $\mathscr{D}$. The associated density functions are denoted by $f^i(\cdot|\theta^i)$. The *prior density* for the parameter $\theta^i \in \mathbb{R}^n$ is given by $\pi_0^i(\cdot)$. The posterior beliefs are represented by $\pi_t = (\pi_t^1, \pi_t^2)$. After observing the random variable X_t^i in period t, π_t^i is converted by Bayes rule into π_{t+1}^i:

$$\pi_{t+1}^i(\theta^i|X_t^i) = \frac{\pi_t^i(\theta^i) \cdot f^i(X_t^i|\theta^i)}{\int \pi_t^i(\phi) \cdot f^i(X_t^i|\phi)d\phi}. \tag{2.1}$$

Starting with prior beliefs and applying (2.1) recursively, we obtain a sequence of beliefs $\{\pi_t\}_{t=0}^\infty$.[4]

The consumer and the firms discount the future with the same discount factor β, with $0 \le \beta < 1$. Past quality realizations together with past prices and past consumer decisions constitute the history of the game. We denote with H_t the set of all possible histories up to, but not including period t. An element $h_t \in H_t$ includes all past prices, $p_s = (p_s^1, p_s^2)$, $0 \le s < t$, the consumers decision variable, $d_s = (d_s^1, d_s^2)$, where

$$d_s^i = \begin{cases} 1 & \text{if the consumer } accepts \text{ the offer of firm } i \text{ in period } s, \\ 0 & \text{if the consumer } rejects \text{ the offer of firm } i \text{ in period } s, \end{cases}$$

and the random realizations X_s^i of the purchased product i, $0 \le s < t$. Hence h_t is

$$h_t = (p_0, d_0, X_0^s, \ldots, p_{t-1}, d_{t-1}, X_{t-1}^s),$$

where the upper index $s = 1, 2$ indicates the identity of the selling firm.

---

[4]While our exposition will be restricted to sampling processes, all our results remain true for a general non-i.i.d. filtration set-up. Similarly all our results extend naturally from a duopoly to a general N-seller oligopoly model.

A *pricing strategy* of seller i at any time t is a function from the history into a distribution on the real numbers,

$$p_t^i : H_t \to \Delta(\mathbb{R}).$$

The buyer makes her purchase decision knowing the past play and the prices currently offered. Her *acceptance strategy* is a function from the history and the current prices into her decision space $\Delta(\{0,1\} \times \{0,1\}\backslash(1,1))$:

$$d_t : H_t \times \mathbb{R} \times \mathbb{R} \to \Delta(\{0,1\} \times \{0,1\}\backslash(1,1)).$$

Notice that unit demand imposes the constraint $d_t^1 + d_t^2 \le 1$, which allows the buyer not to purchase at all, if she prefers to do so. We denote by $d_s = \{d_t\}_{t=s}^\infty$ the sequence of decision functions starting in period s. Similarly $p_s^i = \{p_t^i\}_{t=s}^\infty$ is the sequence of future pricing strategies of firm i starting in period s.

The discounted expected profit for firm i under a given strategy triple (d_s, p_s^1, p_s^2) at time s is

$$E_s \left[\sum_{t=s}^\infty \beta^{t-s} d_t^i p_t^i \right], \tag{2.2}$$

and the expected present value for the consumer in period s is

$$E_s \left[\sum_{t=s}^\infty \beta^{t-s} [d_t^1 (X_t^1 - p_t^1) + d_t^2 (X_t^2 - p_t^2)] \right]. \tag{2.3}$$

Each player acts so as to maximize the expected discounted return given the beliefs over the return processes and the strategies of the other players. To facilitate the equilibrium analysis in the next section, we compare our model with the multi-armed bandit problem.

An n-armed bandit consists of n statistically independent alternatives (arms) which may be chosen in any order and one at a time. We concentrate our attention without loss of generality to $n = 2$. The maximization problem of the decision maker is to find an allocation strategy d^* which solves

$$\max_d E_t \left[\sum_{t=0}^\infty \beta^t d_t^1 X_t^1 + \sum_{t=0}^\infty \beta^t d_t^2 X_t^2 \right], \quad \text{subject to}$$

$$d_t^1 + d_t^2 \le 1. \tag{2.4}$$

The solution to (2.4) is the celebrated index policy. Gittins and Jones (1974) showed that it is possible to assign to each alternative i an *index function* $M^i(\pi_t^i)$ which depends only on the state π_t^i of project i. The optimal policy based on the index function is simple: Compute at any given time the indices of the different alternatives and select a project with maximal index. The index of project i is defined in terms of the following optimization problem involving only project i. Suppose the decision maker is facing in each period only the choice between continuing with the random sequence $\boldsymbol{X}^i$ or stopping the sequence to obtain a terminal reward z. The value $G^i(\pi_t^i, z)$ of this problem is defined by the dynamic programming equation

$$G^i(\pi_t^i, z) = \max\{z, E_{\pi_t^i}[X_t^i + \beta E G^i(\pi_{t+1}^i, z)]\}. \tag{2.5}$$

The *dynamic allocation* or *Gittins index* $M^i(\pi_t^i)$ is given through the equation (2.5).

Definition 1:   The *dynamic allocation index of alternative i* is defined as

$$M^i(\pi_t^i) = \sup\{z \in \mathbb{R}|G^i(\pi_t^i, z) > z\},$$
$$= \inf\{z \in \mathbb{R}|G^i(\pi_t^i, z) = z\}.$$

In words, the index $M^i(\pi_t^i)$ of alternative i in state π_t^i, is the supremum over all terminal rewards, such that the decision-maker still prefers to continue with the random stream; or alternatively, it is the infimum over all terminal rewards such that the decision maker is indifferent between continuing with the random sequence and retiring with the stopping reward $M^i(\pi_t^i)$. The *index process* $M_t^i = \{M^i(\pi_t^i), t \in N\}$ reduces the n-dimensional problem to a comparison of n 1-dimensional problems.[5]

The buyer's decision problem in our model differs from the multi-armed bandit problem in two important aspects. First, in the strategic model the return stream of the buyer is affected by the pricing policies of the sellers, where each pricing policy is in turn the solution to the firm's profit maximization problem (2.2). The second difference is central to the intertemporal aspect of the game. In the multi-armed bandit problem, the value of the random sequence i depends only on the information acquired along the sequence i, but not on the history of the other random sequences. In the duopoly game, however, we naturally expect any pricing strategy of seller i to react not only to its own quality realizations, but also to

---

[5]Whittle (1982) and Gittins (1989) are excellent references for more details on the multi-armed bandit theory.

those of the competing alternative. Hence, the *current* expected reward, $x_t^1 - p_t^1$ or $x_t^2 - p_t^2$, and the expectation over *future* rewards of the competing alternatives are naturally dependent.

3. Equilibrium Price Competition and Efficiency

The conceptual distinction between our strategic model and the decision theoretic learning model is that in our model the alternatives are owned by separate economic agents. The pricing of alternative i is now a strategic decision made by seller i in each period. By introducing the separation in ownership we can examine how the costs and benefits of learning are allocated intertemporally between the buyer and sellers in equilibrium. First, we investigate the efficiency of the learning process in the presence of strategic interaction. In the next section, we analyze the dynamic allocation of the payoffs needed to sustain the equilibrium learning process.

We are interested in Markov perfect equilibria of the game for which π_t is the state variable.[6] By requiring that players base their decisions in equilibrium only on payoff relevant variables, current prices, and current information (summarized in the densities $\pi_t = (\pi_t^1, \pi_t^2)$), we focus the equilibrium analysis on the strategic effects of the learning process. The Markov property will also limit the set of equilibria significantly and we will discuss the main differences between Markovian and non-Markovian equilibria briefly in the next section.

An equilibrium in the game is defined as follows.

Definition 2: A *subgame perfect equilibrium* (*SPE*) is a triple of decision rules $\{d, p^1, p^2\}$ which form a Nash equilibrium in every subgame.[7]

Player i's strategy is Markov if it depends only on the payoff relevant history.

---

[6]See Maskin and Tirole (1994) for a detailed account of the Markov perfect equilibrium concept.

[7]There are two alternative ways to look at this game. The first is an incomplete information game where nature moves at the first node to select the types of sellers. Information is partially revealed at subsequent nodes. In this representation the game has no proper subgames. An alternative representation, adopted for this paper, is a complete information game with a unique starting node given by the priors and a perfectly observed move by nature in each period determining the transition on the state variables of all players. In this representation, each choice by the buyer starts a new subgame.

Definition 3: A *Markov perfect equilibrium* *(MPE)* is an SPE, where

$$p_t^i(h_t) = p^i(\pi), \quad \text{and}$$

$$d_t^i(\pi_t, p_t^1, p_t^2) = d^i(\pi, p^1, p^2) \quad \text{for } i = 1, 2.$$

The definition explicitly states that the Markov strategies should be time-invariant or stationary in the sense that whenever $\pi_t = \pi_{t+s}$, then $p^i(\pi_t) = p^i(\pi_{t+s})$ and also $d_t^i(\pi_t, p_t^1, p_t^2) = d_{t+s}^i(\pi_{t+s}, p_{t+s}^1, p_{t+s}^2)$ have to hold.[8]

For the characterization of the Markov equilibria we cast the players' decision problems in a stochastic dynamic programming framework. We define the value function of each player i as

$$V^i(\pi_t) \equiv v^i(\pi_t|\cdot, \cdot),$$

where we take the strategies of the other players as given and π_t as the state variable of the system. The buyer has to choose simultaneously between the current returns, $X_t^i - p^i(\pi_t)$ or $X_t^j - p^j(\pi_t)$ and their associated learning opportunities as indicated by (π_t, X_t^i) or (π_t, X_t^j). We shall write $V^B(\pi_t, X_t^i)$ rather than $V^B(\pi_{t+1})$ to indicate which sample, X_t^i or X_t^j, conditions the transition from π_t to π_{t+1}. The Bellman equation for the buyer is given by

$$V^B(\pi_t) = \max E_t\{X_t^1 - p_t^1 + \beta V^B(\pi_t, X_t^1),$$

$$X_t^2 - p_t^2 + \beta V^B(\pi_t, X_t^2), 0 + \beta V^B(\pi_t)\}. \tag{3.1}$$

The consumer can always refuse to make a purchase and receive a reservation value of zero. If the best decision for the buyer should be to accept neither offer, then under the Markov assumption, this will remain her best possible decision forever. Consider next the dynamic programming problem for the firms. Each seller, when choosing his price, has to consider the benefits of realizing a sale today, or foregoing that possibility today and instead betting on future sales in a possibly changed environment. The value function for the first seller is

$$V^1(\pi_t) = \max_{p^1} E_t\{d_t^1[p^1(\pi_t) + \beta V^1(\pi_t, X_t^1)]$$

$$+ d_t^2 \beta V^1(\pi_t, X_t^2) + (1 - d_t^1)(1 - d_t^2)\beta V^1(\pi_t)\}, \tag{3.2}$$

---

[8]We index π with t henceforth only as a matter of recording time.

and for the second seller it is symmetrically

$$V^2(\pi_t) = \max_{p^2} E_t\{d_t^2[p^2(\pi_t) + \beta V^2(\pi_t, X_t^2)]$$

$$+ d_t^1\beta V^2(\pi_t, X_t^1) + (1 - d_t^1)(1 - d_t^2)\beta V^2(\pi_t)\}. \qquad (3.3)$$

If the buyer decides not to accept any offer in π_t, then we have of course $\pi_{t+1} = \pi_t$.

We concentrate our attention for the moment on MPE in pure strategies. It will be shown in Proposition 1, that this involves no loss of generality.

Lemma 1: *In any pure MPE the buyer makes a purchase in every period and*

$$V^B(\pi_t) = E_t[X_t^s - p^s(\pi_t) + \beta V^B(\pi_t, X_t^s)] \geq 0,$$

where $s = 1, 2$ is the accepted seller.

Proof: Notice first that $V^B(\pi_t) \geq 0$, $\forall \pi_t$, by the no purchase option and $V^i(\pi_t) \geq 0$, $\forall \pi_t$, since the seller can always ask for positive prices, which can at most be refused. Suppose that no purchase is made in period t. By Markov assumption no sales are made in future periods either and hence all agents' value is zero. Since $E_t X_t^i > 0$ by nonnegativity of X^i, one of the firms can offer a strictly positive price $0 < p_t^i < E_t X_t^i$ that yields a strictly positive value to the firm as well as the buyer. Hence a sale has to be made in all periods in equilibrium. $\qquad\square$

For pure strategy equilibria, price competition implies that the consumer is indifferent between the choices offered by the two firms (or between one firm and the no-purchase option) at all points in time. At equilibrium prices, the selling firm, s, must (weakly) prefer to make the sale, whereas the nonselling firm, n, must (weakly) prefer to concede in the current period. The following (in-) equalities for the buyer (B), the selling firm (S^s), and the nonselling firm (S^n) are equilibrium conditions and central to the following analysis. We state them independently in the following lemma.

Lemma 2: *Assume $E_t[X_t^s - p^s(\pi_t) + \beta V^B(\pi_t, X_t^s)] > 0$. The strategy triple $\{d, p^1, p^2\}$ is a pure MPE if and only if the conditions (B), (S^s),* *and (S^n) are met for all t and π_t:*

$$E_t[X_t^s - p^s(\pi_t) + \beta V^B(\pi_t, X_t^s)] = E_t[X_t^n - p^n(\pi_t) + \beta V^B(\pi_t, X_t^n)], \quad (B)$$

$$p^s(\pi_t) + \beta E_t V^s(\pi_t, X_t^s) \geq \beta E_t V^S(\pi_t, X_t^n), \quad (S^s)$$

$$\beta E_t V^n(\pi_t, X_t^s) \geq p^n(\pi_t) + \beta E_t V^n(\pi_t, X_t^n), \quad (S^n)$$

where $n \neq s$.

Proof: ($\Rightarrow$) Consider (B) first. Suppose not. Then the two terms in the Bellman equation (3.1) of the consumer differ by a strictly positive number, which implies that the firm offering the higher value to the consumer could raise her price by $\epsilon > 0$ and, by $E_t[X_t^s - p^s(\pi_t) + \beta V^B(\pi_t, X_t^s)] > 0$, this would not affect the consumer's decision, which contradicts the equilibrium assumption. Consider now (S^s) and (S^n). If (S^s) does not hold, then by (B) we know that a deviation to a higher price induces the buyer to switch sellers and consequently is a profitable deviation. Similarly, if (S^n) does not hold, a downward deviation by ε is profitable for the nonselling firm by (B).

($\Leftarrow$) Recall that the value function of each player is defined given the opponents' strategies. Conditions (B), (S^s), and (S^n) are then the appropriate conditions to make sure that no one-shot deviations are profitable for any of the players. Since the payoffs are bounded from below, we refer to the principle of optimality to conclude that strategies satisfying conditions (B), (S^s), and (S^n) are optimal given the other players' strategies and hence form an equilibrium. $\square$

Remark: Lemma 2 is true for all pure MPE for which $E_t[X_t^s - p^s(\pi_t) + \beta V^B(\pi_t, X_t^s)] > 0$ holds along the equilibrium path for all π_t. If the assumption doesn't hold, an equilibrium, and some π_t, could conceivably exist such that $E_t[X_t^s - p^s(\pi_t) + \beta V^B(\pi_t, X_t^s)] > 0$. The equilibrium price $p^n(\pi_t)$ would then not necessarily satisfy (S^n) and this could possibly break the equality (B). It is easy to verify that whenever such an equilibrium exists, an outcome equivalent MPE with price $\bar{p}^n(\pi_t) < p^n(\pi_t)$ also exists, which contrary to the latter satisfies both conditions (B) and (S^n). We will see in Proposition 3 that all pure MPE for which (B), (S^s), and (S^n) are satisfied have $E_t[X_t^s - p^s(\pi_t) + \beta V^B(\pi_t, X_t^s)] > 0$, which allows the conclusion that all pure MPE are characterized by (B), (S^s), and (S^n), and that the qualification made for the moment is only a temporary one.

Price competition between the sellers in each period makes the stage game similar to a static Bertrand pricing game in which firms have different costs. To illustrate this point, we take the continuation values of each subgame for the moment as given. Upon entering the competition each firm has to consider the benefits as well as the costs of making a sale today. The benefits for firm i of selling today come from the realized current price and the future sales following (π_t, X_t^i), but by doing so firm i foregoes all possible sales along the continuation game of (π_t, X_t^j):

$$\overbrace{p^i(\pi_t) + \beta E_t V^i(\pi_t, X_t^i)}^{\text{benefit is of making a sale}} - \overbrace{\beta E_t V^i(\pi_t, X_t^j)}^{\text{costs of making a sale}} \;.$$

If we take the difference in the future payoffs of the two paths, $\beta E_t V^i(\pi_t, X_t^j)$ and $\beta E_t V^i(\pi_t, X_t^i)$, to be the net costs $c^i(\pi_t)$ of making a sale today, which are of course endogenous in equilibrium:

$$c^i(\pi_t) \equiv \beta E_t V^i(\pi_t, X_t^i) - \beta E_t V^i(\pi_1, X_t^i), \tag{3.4}$$

then we can read conditions (S^s) and (S^n) simply as

$$p^s(\pi_t) \geq c^n(\pi_t), \tag{$S^{s\prime}$}$$

$$p^n(\pi_t) \leq c^n(\pi_t). \tag{$S^{n\prime}$}$$

The equilibrium price for the selling firm, s, must exceed the costs of making a sale, whereas the price of the conceding seller, n, must not exceed the costs of making a sale, for otherwise he could lower his price slightly and attract the consumer. We may note at this point that the dynamic duopoly model inherits the multiplicity of equilibria present in the static Bertrand game with different costs.[9] Since the buyer is indifferent between the sellers in equilibrium, choosing n instead of s is without costs for her. By quoting a high enough price, seller n makes sure that a deviation by the buyer is not to his disadvantage. In other words, he is *cautious* enough to ask for a price which he does not regret should he be chosen against all expectations. For future reference we shall call the equilibrium in which the conceding seller bids exactly his intertemporal costs of competition, *cautious equilibrium*.

---

[9]The reader may recall that in a static Bertrand game where firms have different costs $c_1 < c_2$, the "standard" equilibrium is $p_1 = p_2 = c_2$ and the low cost firm is making the sale. However any price combination $p_1 = p_2 \in (c_1, c_2)$ can also be sustained as an equilibrium if the consumer chooses the low cost producers with probability one.

Definition 4: An *MPE is cautious* if seller n is indifferent between selling and conceding to the competitor:

$$p^n(\pi_t) + \beta E_t V^n(\pi_t, X_t^n) = \beta E_t V^n(\pi_t, X_t^s). \qquad (3.5)$$

Before we attack the question of how efficient the learning process is under competition, we clarify a more technical issue concerning the *similarity* of the mixed strategy equilibria with the pure strategy equilibria.

Proposition 1: *Every Markov perfect equilibrium is outcome equivalent to some Markov perfect equilibrium in pure strategies.*

Proof: We notice first that there is no MPE in which all sellers use mixed strategies in any one period simultaneously. For given continuation payoffs the price setting game in any period is just like a static Bertrand duopoly game with different costs for each seller and unit demand. By a standard, but tedious, argument which we omit here, one can show that both sellers do not simultaneously engage in mixed strategies. We now show that only the seller who is not chosen by the consumer in equilibrium can ever use mixed strategies. Consider a pure strategy $p^i(\pi_t)$ by seller i when seller j is employing a mixed strategy. Take the price $\hat{p}^j(\pi_t)$ at which the buyer is just willing to buy from j:

$$E_t[X_t^i - p^i(\pi_t) + \beta V^B(\pi_t, X_t^i)] = E_t[X_t^j - \hat{p}^j(\pi_t) + \beta V^B(\pi_t, X_t^j)].$$

Should the inequality

$$\hat{p}^j(\pi_t) + \beta E_t V^j(\pi, X_t^j) > \beta E_t V^j(\pi_t, X_t^i).$$

hold, then seller j would never use a price lower than $\hat{p}^j(\pi_t)$ in his mixed price strategy. But at any price higher than $\hat{p}^j(\pi_t)$, he would be rejected by the buyer; thus seller j offers a unique price $\hat{p}^j(\pi_t)$. If seller j is then using a mixed strategy with $\hat{p}^j(\pi_t)$ in its support, it must satisfy

$$\hat{p}^j(\pi_t) + \beta E_t V^j(\pi, X_t^j) \le \beta E_t V^j(\pi_t, X_t^i).$$

The prices $p^j(\pi_t)$ which are in the support of the mixed strategy, cannot be lower than $\hat{p}^j(\pi_t)$, because otherwise j would be chosen by the buyer, although he prefers not to. Thus the support can only contain $\hat{p}^j(\pi_t)$ and higher prices. But at higher prices than $\hat{p}^j(\pi_t)$, j will never be chosen by the buyer. Thus this equilibrium is outcome equivalent to the one in which the seller, who plays the mixed strategy, simply charges the lower bound in the support of his mixed strategies, namely $\hat{p}^j(\pi_t)$. Finally, if the buyer is randomizing, then both sellers have to be indifferent between selling and nonselling; otherwise one of them would deviate. And if both sellers are indifferent, then choosing one of them with probability one is again a pure strategy equilibrium. □

It will prove instructive to express the value function of each player for given (equilibrium) policies explicitly by the entire sequence of payoffs. The buyer's alternating between the sellers as she acquires experience can be represented by two sequences of switching times, which are in fact stopping times: $\{\sigma_n\}_{n=1}^{\infty}$ and $\{\tau_n\}_{n=1}^{\infty}$. The switching times are random times which depend on the sample path and the prices offered along the sample path. We define τ_n as a (stochastic) time at which the consumer stops buying from the first seller and switches to the second, and σ_n as a (stochastic) time at which the reverse switching behavior occurs. In other words, σ_n is a time period in which the buyer begins her nth round of purchases from the first seller and τ_n is the period in which the buyer begins her nth round of purchases from the second seller. The value function of the buyer then admits the following representation

$$
V^B(\pi_t) = E_t \left\{ \sum_{n=1}^{\infty} \sum_{s=\sigma_n}^{\tau_n - 1} \beta^{s-t} [X_s^1 - p^1(\pi_s)] \right.
$$
$$
\left. + \sum_{n=1}^{\infty} \sum_{s=\tau_n}^{\sigma_{n+1} - 1} \beta^{s-t} [X_s^2 - p^2(\pi_s)] \right\}. \tag{3.6}
$$

The value of the buyer is the discounted expected sum of the per period net gains, $X_s^1 - p^1(\pi_s)$ or $X_s^2 - p^2(\pi_s)$, along all sample paths.[10] For seller i it is the expected discounted sum of all realized sales. The value of the game for the first seller is then

$$
V^1(\pi_t) = E_t \left\{ \sum_{n=1}^{\infty} \sum_{s=\sigma_n}^{\tau_n - 1} \beta^{s-t} p^1(\pi_s) \right\},
$$

and for the second seller

$$
V^2(\pi_t) = E_t \left\{ \sum_{n=1}^{\infty} \sum_{s=\tau_n}^{\sigma_{n+1} - 1} \beta^{s-t} p^2(\pi_s) \right\}.
$$

---

[10] By Lemma 1, there is no time period where she does not buy at all, so that the purchasing behavior is completely described by $\{\sigma_n\}_{n=1}^{\infty}$ and $\{\tau_n\}_{n=1}^{\infty}$.

The social value $W(\pi_t)$ of the game in any state π_t is simply $W(\pi_t) \equiv V^B(\pi_t) + V^1(\pi_t) + V^2(\pi_t)$ and can be explicitly expressed through

$$W(\pi_t) = E_t \left\{ \sum_{n=1}^{\infty} \sum_{s=\sigma_n}^{\tau_n - 1} \beta^{s-t} X_s^1 + \sum_{n=1}^{\infty} \sum_{s=\tau_n}^{\sigma_{n+1} - 1} \beta^{s-t} X_s^2 \right\}. \qquad (3.7)$$

We define an efficient equilibrium.

Definition 5: An equilibrium is *efficient* if it maximizes the social value $W(\pi_t)$ for all π_t.

The notion of efficiency should, of course, be understood as a notion of (informationally) constrained or ex-ante Pareto efficiency.

The problem of maximizing the social value of the game as depicted in (3.7) is in fact identical to the multi-armed bandit problem given in (2.4). An equilibrium is then efficient if and only if the stopping times $\{\sigma_n\}_{n=1}^{\infty}$ and $\{\tau_n\}_{n=1}^{\infty}$ coincide with the stopping times prescribed by the dynamic allocation index policy. With this identification in place no ambiguity should arise when we refer to the *efficient path* or the *efficient (inefficient)* or *superior (inferior)* *alternative* $i(j)$, by which we simply mean that $M^i(\pi_t) > M^j(\pi_t)$.[11]

The explicit representation of the payoff sequence of the buyer in (3.7) suggests also that the buyer should be willing to support the efficient allocation path through her choices if she is guaranteed to share sufficiently, *today* and in the *future*, in the gains from learning, which are limited only by the prices she has to pay. Ultimately, the question of efficient experimentation then hinges on the price path induced through the duopoly game.

We may distinguish two different situations. When long-run efficiency as indicated by the allocation index and high current return coincide in alternative i, i.e. $M^i(\pi_t) \geq M^j(\pi_t)$ and $E_t X_t^i \geq E_t X_t^j$, then firm i should be able to attract the buyer and yet obtain a relatively high price for its product since the competing product is inferior on both accounts. The inferior firm's best strategy is to wait, since a price low enough to attract the consumer today would not be justified on the current expectation for future profits.

---

[11]By the index theorem efficiency follows already by $M^i(\pi_t^i) > M^j(\pi_t^j)$, i.e. the index $M^i(\pi_t^i)$ is independent of π_t^j. To save on notation we shall neglect this distinction.

The intertemporal incentives are more complicated when long-run efficiency and current high returns do not coincide, i.e. $M^i(\pi_t) \geq M^j(\pi_t)$ but $E_t X_t^i < E_t X_t^j$. Suppose for simplicity of the argument that the quality of firm j's product is known with certainty. We may ask how long firm i is willing to make sales. In this simplified case, we know by the Markovian assumption that once the buyer selects firm j, she will buy from firm j forever and consequently firm i's profit is zero from then on. As long as the total surplus along paths beginning with a sale by firm i exceed the total value of paths switching immediately, firm i can offer low enough prices today to attract the consumer while making a positive expected profit in future periods. But this is exactly the condition stated earlier in the form of the dynamic allocation indices: $M^i(\pi_t) \geq M^j(\pi_t)$. The following theorem shows that this intuition extends to the case of two uncertain products. In the following section, we determine equilibrium prices needed to support the appropriate intertemporal division of gains from trade.

Theorem 1 (Efficiency): *All Markov perfect equilibria are efficient: If seller i is chosen in period t, then*

$$M^i(\pi_t) \geq M^j(\pi_t).$$

Proof: [12]Suppose that firm i is chosen in period t. By Lemma 2 the following (in-)equalities have to hold:

$$E_t[X_t^i - p^i(\pi_t) + \beta V^B(\pi_t, X_t^i)] = E_t[X_t^j - p^j(\pi_t) + \beta V^B(\pi_t, X_t^j)], \quad (B)$$

$$p^i(\pi_t) + \beta E_t V^i(\pi_t, X_t^i) \geq \beta E_t V^i(\pi_t, X_t^j), \quad \text{and} \quad (S^1)$$

$$\beta E_t V^j(\pi_t, X_t^i) \geq p^j(\pi_t) + \beta E_t V^j(\pi_t, X_t^j). \quad (S^2)$$

By summing both sides of the three (in-)equalities and recalling the definition of $W(\pi)$, we get

$$E_t X_t^i + \beta E_t W(\pi_t, X_t^i) \geq E_t X_t^j + \beta E_t W(\pi_t, X_t^j).$$

Hence the social value of the game, $W(\pi_t)$, calculated along the equilibrium path, satisfies the following functional equation:

$$W(\pi_t) = \max\{E_t X_t^i + \beta E_t W(\pi_t, X_t^i), E_t X_t^j + \beta W(\pi_t, X_t^j)\}. \quad (3.8)$$

---

[12]We thank the editor for suggesting a different proof strategy, which led to a shorter and more transparent argument.

But (3.8) characterizes the value function of the planner's problem as well. An easy application of the contraction mapping theorem establishes the uniqueness of solutions to (3.8). Consequently, firm i is selected in equilibrium only if firm i is selected along some optimal path in the planner's problem. By the Gittins index theorem, this is equivalent to $M^i(\pi_t) \geq M^j(\pi_t)$. $\square$

The message of Theorem 1 is unambiguous. The fact that all MPE are efficient demonstrates that no firm has an interest in stopping the efficient learning process, since the costs involved in doing so are too high at each stage. In particular, the conceding seller, rather than forcing a sale through a very low price, prefers to postpone any sales in the expectation of a more favorable competitive context in the future.

The efficiency result of Theorem 1 continues to be valid in more general settings. If the buyer is not restricted to a single experiment, but can allocate up to N experiments among the sellers in each period, the resulting equilibria are still efficient. The necessary modification to establish efficiency in this framework is to allow firms the use of nonlinear pricing schemes.

The sellers are offering nonlinear pricing schedules to the buyer:

$$p_t^j = \{p_t^{j1}, p_t^{j2}, \ldots, p_t^{jN}\},$$

where p_t^{jk} denotes the unit price of firm j's product if the buyer purchases k units. The purchasing decision of the buyer is then represented by two numbers:

$$d_t = \{n_t^1, n_t^2\},$$

where n_t^j denotes the number of units demanded from firm j.

The optimization problem of the consumer in the value function form is then given by

$$V(\pi_t) = \max_{0 \leq n_t^1 + n_t^2 \leq N} E_t\{n_t^1(X_t^1 - p_t^{1n_t^1})$$

$$+ n_t^2(X_t^2 - p_t^{2n_t^2}) + \beta V(\pi_t, n_t^1, n_t^2)\}.$$

The sellers' problems are described similarly. As in the case of a single unit, the consumer will always buy N units and will never use her no-purchase option. Analogues to Lemmas 1 and 2 continue to hold in this setting and using the same equilibrium concepts and welfare criteria as above we are able to prove the following theorem.

Theorem 2: *All MPE in the multi-unit game are efficient.*

Furthermore, all the results pertaining to the price path of the cautious equilibrium discussed in Section 4 continue to hold in the multi-unit case. We also point out that the efficiency result would continue to hold if the quality realizations of the products were mutually dependent.[13]

Extending these results to a game with multiple buyers proves to be substantially more difficult. Since the experiments are publicly observed, each purchase creates an informational externality on the other buyers. In the case of fixed prices this leads to well-known free-rider problems as in Bolton and Harris (1993). In our model, the firms are able to internalize some of these effects since a successful experiment by one consumer leads to an improved competitive position with respect to all consumers in the next period. It turns out, however, that we cannot expect efficient experimentation in general, even with nonlinear prices. A consumer has to be compensated for her experimentation costs only, while the gains of experimentation are collected from all consumers through higher prices. As a consequence, experimentation tends to be *too* cheap from the firm's point of view in the multiple buyer case and ex-ante optimal experimentation is not achieved.[14] This has to be contrasted to the single buyer *with* multiple unit demand, who perfectly internalizes the price increase on *all* units in future periods.

4. Characterization of the Markov Perfect Equilibria

The equilibrium choice path of the buyer has been established by the efficiency property of the MPE and we focus now on the equilibrium price path. Prices determine the intertemporal allocation of gains from experimentation between buyer and sellers. We focus on the *cautious equilibrium* where the pricing path provides the intertemporal incentives to experiment in a surprisingly simple and intuitive way. Finally, we

---

[13] All model extensions as mentioned above have, however, the drawback that efficient policies cannot be characterized by index policies anymore, since they are either not known or simply don't exist as in the case of mutually dependent alternatives.

[14] A well known fact on ex-ante efficient experimentation in multi-armed bandit models states that ex-post efficiency fails with positive probability. Since experimentation is relatively cheap in the multiple buyer case, an interesting conjecture to be checked in future research is that equilibrium in the pricing game is closer to the ex-post efficient path than the ex-ante efficient path.

characterize the *entire set* of MPE by giving upper and lower bounds on the payoffs for the players in Proposition 3.

4.1. The cautious equilibrium

We recall that the equilibrium condition as given in Definition 4 made the conceding seller (S^n) indifferent between realizing a sale or foregoing the sale in the current period,

$$p^n(\pi_t) + \beta E_t V^n(\pi_t, X_t^n) \geq \beta E_t V^n(\pi_t, X_t^s). \qquad (4.1)$$

In other words, in the cautious equilibrium the conceding seller always sets his price equal to his net costs of competing $c^n(\pi_t) = \beta E_t V^n(\pi_t, X_t^s) - \beta E_t V^n(\pi_t, X_t^n)$.

$$p^n(\pi_t) = c^n(\pi_t).$$

The intertemporal allocation of the costs and benefits of the learning process in the *cautious* MPE are described completely in the following theorem.

Theorem 3: *The cautious equilibrium is unique, efficient, and*

$$p^s(\pi_t) = E_t X_t^s - E_t X_t^n = x_t^s - x_t^n, \qquad (4.2)$$

$$p^n(\pi_t) = \beta E_t V^n(\pi_t, X_t^s) - \beta E_t V^n(\pi_t, X_t^n). \qquad (4.3)$$

The pricing rule of the conceding seller is a submartingale:

$$p^n(\pi_t) \leq \beta E_t \{p^n(\pi_t, X_t^s)\}. \qquad (4.4)$$

Proof: The nonselling firm in period t, say j, is indifferent between selling and not selling by the definition of the cautious equilibrium:

$$p^j(\pi_t) + \beta E_t V^j(\pi_t, X_t^j) = \beta E_t V^j(\pi_t, X_t^i). \qquad (4.5)$$

By the consumer's indifference,

$$E_t[X_t^i - p^i(\pi_t) + \beta V^B(\pi_t, X_t^i) = E_t[X_t^j - p^j(\pi_t) + \beta V^B(\pi_t, X_t^j)], \qquad (4.6)$$

we can express $V^B(\pi_t)$ for a given equilibrium either as

$$V^B(\pi_t) = E_t[X_t^i - p^i(\pi_t) + \beta V^B(\pi_t, X_t^i)], \qquad (4.7)$$

or as

$$V^B(\pi_t) = E_t[X_t^j - p^j(\pi_t) + \beta V^B(\pi_t, X_t^j)]. \qquad (4.8)$$

We can extend (4.7) and (4.8) in this way for any number of periods. Since the equality (4.6) has to hold in each period, we are free to choose which alternative, i or j, to use in any period in the particular extensions. For now we extend (4.7) and (4.8) by the continuation game in which j is accepted forever. Extending (4.7) we get

$$V^B(\pi_t) = E_t \left\{ \sum_{s=t}^{\infty} \beta^{s-t}[X_s^j - p^j(\pi_s)] \right\}, \qquad (4.9)$$

and extending (4.8) we have

$$V^B(\pi_t) = E_t \left\{ X_t^i - p^i(\pi_t) + \sum_{s=t+1}^{\infty} \beta^{s-t}[X_s^j - p^j(\hat{\pi}_s)] \right\}. \qquad (4.10)$$

We decorated the state variable $\hat{\pi}_s$ in (4.10) to distinguish it from the state variables π_s in (4.9), since their experimenting paths are different. We solve equation (4.6) for the price of the superior seller i in period t, when the inferior seller j adheres to his pricing policy (4.5). By extending (4.5) in the same manner as (4.9) or (4.10) we obtain

$$E_t \left\{ \sum_{s=1}^{\infty} \beta^{s-t} p^j(\pi_s) \right\} = E_t \left\{ \sum_{s=t+1}^{\infty} \beta^{s-t} p^j(\hat{\pi}_s) \right\}.$$

We can finally use equality (4.5) to simplify equality (4.6) and obtain

$$p^i(\pi_t) = E_t X_t^i - E_t X_t^j = x_t^i - x_t^j,$$

describing the price of the successful seller in period t. To obtain an expression of $p^j(\pi_t)$ for the nonselling firm, we start again with equation (4.5) and extend both sides by one period. Since the Gittins index of seller j might rise above the one of seller i in $t+1$ after an observation of X_t^j in t, equality (4.5) might turn into an inequality in $t+1$, conditional on X_t^j:

$$p^j(\pi_t, X_t^j) + \beta E_t V^j(\pi_t, X_t^j, X_{t+1}^i) \geq \beta E_t V^j(\pi_t, X_t^j, X_{t+1}^i). \qquad (4.11)$$

As a consequence we obtain from (4.5):

$$p^j(\pi_t) + \beta^2 E_t V^j(\pi_t, X_t^j, X_{t+1}^i) \leq \beta E_t p^j(\pi_t, X_t^j)$$
$$+ \beta^2 E_t V^j(\pi_t, X_t^i, X_{t+1}^j),$$

resulting in

$$p^j(\pi_t) \leq \beta E_t p^j(\pi_t, X_t^i),$$

which concludes the proof. $\qquad\qquad\qquad\qquad\qquad\qquad\qquad\qquad\qquad\square$

A few aspects of the pricing strategies in the cautious equilibrium deserve discussion. Experimentation is efficient and all sales are made at $p^s(\pi_t) = E_t X_t^s - E_t X_t^n$. Notice that efficiency, $M^s(\pi_t) \geq M^n(\pi_t)$, does not imply $E_t X_t^s \geq E_t X_t^n$. It may be that $M^s(\pi_t) \geq M^n(\pi_t)$, although $E_t X_t^s < E_t X_t^n$, in which case $p^s(\pi_t) = E_t X_t^s - E_t X_t^n < 0$. Negative (or below cost) prices then appear in equilibrium as a natural instrument to support the learning process of the consumer. Negative prices are associated with states π_t, where $M^s(\pi_t) \geq M^n(\pi_t)$ *but* $E_t X_t^s < E_t X_t^n$. In these states seller s is willing to offer negative prices because the superiority of his dynamic allocation index indicates that he will be able to recover the initial losses through higher future prices. On the other hand, the consumer expects negative prices as an advance payment because the gains from learning will eventually be diluted through higher prices.

The net value of the sale to the buyer is the current expected quality of the dynamically inefficient firm: $E_t X_t^s - p^s(\pi_t) = E_t X_t^n$. As long as the buyer does not switch from seller i to seller j, her net value of a purchase remains thus constant at $E_t X_t^j$ and the price of the successful seller i forms a martingale. Marginal gains and costs of experimentation are reflected in the price set by seller i in response to new information. The buyer is thus insured during any single *round* of experimentation with a particular seller and realizes intertemporal costs or benefits only when she is switching from one seller to the other.

The conceding seller hence represents, in equilibrium, the insurance or outside option to the buyer. If the buyer continues to experiment with i although the estimate $E_t X_t^i$ decreases, she weakens her future insurance position vis-à-vis seller j. Or, put differently, the future switching costs from i to j are increasing when the expected quality of seller i decreases. Consequently, if seller i still intends to make the sale, i has to compensate the buyer for the induced future risk of higher switching costs. Conversely, if the buyer expects to switch from i to j because the value of learning is high, but not because the current return of j dominates i, then i provides the buyer with a comparatively high insurance level in the future. In consequence seller i can ask today for a price higher than that justified by current quality differences. The discrepancy between the intertemporal pricing rule and the static pricing rule can be systematically linked to this insurance effect.

In a myopic environment the optimal pricing rule $p_m^i(\pi_t)$ of the successful seller i is given by

$$p_m^i(\pi_t) - p^j(\pi_t) = x_t^i - x_t^j. \tag{4.12}$$

The price $p_m^i(\pi_t)$ at which i can make a sale is such that the price difference between i and j equates the estimated quality difference between the competing products. We contrast this with the incentives provided by the intertemporal pricing rules. Recall from the definition of the cautious equilibrium that

$$p^j(\pi_t) = \beta E_t[V^j(\pi_t, X_t^i) - V^j(\pi_t, X_t^j)], \qquad (4.13)$$

and from Theorem 4.1 that

$$p^i(\pi_t) = x_t^i - x_t^j. \qquad (4.14)$$

Comparing equations (4.12) and (4.14) we observe immediately that the price differential in the cautious equilibrium is smaller than in the myopic case if $p^j(\pi_t) > 0$. By (4.13) we can relate this condition to the future competitive position of the currently conceding seller j. When $p^j(\pi_t) > 0$, then the expected continuation payoff for firm j is higher if the buyer experiments today with firm i's product rather than with j's product:

$$p^j(\pi_t) > 0 \Leftrightarrow E_t V^j(\pi_t, X_t^i) > E_t V^j(\pi_t, X_t^j).$$

Experimentation with seller i must hence provide better prospects for an improvement in j's competitive position than a direct experiment with j himself. The improvement can only come through a decrease in the expected quality of firm i's product along the path of the play. But the decrease in $E_t X_t^i$ along the equilibrium path implies that the outside option seller i provides to the buyer, conditional on switching from seller i to seller j, is expected to decrease. Since the buyer realizes the negative future impact of experimentation with seller i today, she is not willing to pay the myopic price $p_m^i(\pi_t)$ and seller i has to settle for less than the myopic price:

$$p^i(\pi_t) < p_m^i(\pi_t).$$

The potential of future rent-seeking by seller j, after current experimentation with i, is consequently expressed in the fact that only a strictly positive price determined by (4.13) makes seller j indifferent between a sale today and the improvement in the competitive position induced by experimenting with seller i. A similar argument can be given for $p^j(\pi_t) < 0$, in which case seller i can extract a higher than myopic price since he provides the buyer with a relatively stable insurance value and consequently:

$$p^i(\pi_t) > p^i_m(\pi_t).$$

With more accurate estimates of the product qualities, the changes in the competitive environment become smaller over time. In turn, we would expect the deviation from the myopic pricing policy to become less significant as more information accumulates. In the next subsection, we describe the asymptotic properties of the cautious equilibrium and show that in general, the long-run prices are different from the myopic Bertrand prices under perfect information.

4.2. *The asymptotic behavior of the cautious equilibrium*

As time goes by, the buyer will learn more about the true value of the purchased products. By Lemma 1 a purchase is made in every period and at least one seller will be chosen infinitely often. The value of learning, $(1 - \beta)M^i(\pi_t) - x^i_t$, as distinct from the value of the current return, diminishes as the dynamic allocation index (and the posterior mean) converges to the true mean of the reward process. Consequently the value of sampling decreases and the ranking of the alternatives with respect to their current payoff tends to coincide with the ranking of the indices. Below cost prices associated with states π_t, where the learning effects dominate the current return effect, should therefore gradually disappear. We may ask whether the pricing and acceptance policies will approach those of the static Bertrand competition in the limit.

Proposition 2 (Asymptotic Behavior): *The asymptotic behavior of the cautious MPE is given by*

(i) $\lim_{t\to\infty} p^s(\pi_t) = x^s - \lim_{t\to\infty} E_t X^n_t \geq 0,$
(ii) $\lim_{t\to\infty} V^n(\pi_t) = 0.$

Proof: (i) By Definition 1 and the Martingale convergence theorem,

$$\lim_{t\to\infty} M^\infty(\pi_t) = x^\infty/(1 - \beta) \quad \text{a.s.}$$

where M^∞ and x^∞ are the dynamic allocation index and the true mean of any alternative which is chosen infinitely often. Since

$$M^s(\pi_t) \geq M^n(\pi_t) \geq E_t X^n_t/(1 - \beta)$$

has to hold by the efficiency of the MPE, we have

$$\lim_{t\to\infty} p^s(\pi_t) = \lim_{t\to\infty} [E_t X_t^s - E_t X_t^n] = x^s - \lim_{t\to\infty} E_t X_t^n \geq 0.$$

Since the claim applies to the prices of all sellers who are chosen infinitely often, we don't exclude the case where switching between the sellers occurs infinitely often. $\text{Lim}_{t\to\infty} E_t X_t^n$ may not converge to x^n, since if a seller j is abandoned after a finite time, convergence to x^n is by no means guaranteed.

(ii) Clearly $\liminf_{t\to\infty} V^{n_t}(\pi_t) \geq 0$ by the no sale option and we want to show that in fact $\lim_{t\to\infty} V^{n_t}(\pi_t) = 0$, where n_t is the conceding seller in period t, and s_u is the successful seller in period u. We proceed by contradiction. Assume that $\limsup_{t\to\infty} V^{n_t}(\pi_t) > 0$, which implies that there exists $\in > 0$ such that

$$\lim_{t\to\infty} \Pr[s_u = n_1, u > t | \pi_t] \geq \epsilon,$$

where $\Pr[\cdot]$ is the probability assessment of the players along the path of the play. Standard convergence arguments imply that

$$\lim_{t\to\infty} \Pr[s_u = n_t, u > t | x^i \neq x^j] = 0,$$

and consequently if

$$\lim_{t\to\infty} \Pr[\pi_u, s_u = n_t, u > t | \pi_t] \geq \epsilon,$$

then

$$\lim_{t\to\infty} \Pr[x^i = x^j | \pi_t] = 1,$$

in which case it has to be that

$$\lim_{t\to\infty} p^{s_u}(\pi_u, s_u = n_t) = \lim_{t\to\infty} [E_t X_t^s - E_t X_t^n] = 0,$$

which implies that $\limsup_{t\to\infty} V^{n_t}(\pi_t) = 0$, concluding the proof. □

The equilibrium converges exactly to the myopic Bertrand equilibrium if both sellers are chosen infinitely often and all learning possibilities are exhausted, in which case even $\lim_{t\to\infty} E_t X_t^n) = x^n$ holds. The asymptotic behavior is somewhat different when only one seller is chosen infinitely often in equilibrium. The price $p^n(\pi_t) < 0$ is determined by Theorem 4.1 as

$$p^n(\pi_t) = \beta E_t V^n(\pi_t, X_t^s) - \beta E_t V^n(\pi_t, X_t^n).$$

By Proposition 2 we also have

$$\lim_{t \to \infty} V^n(\pi_t) = \lim_{t \to \infty} \beta E_t V^n(\pi_t, X_t^s) > 0,$$

so that

$$\lim_{t \to \infty} p^n(\pi_t) = -\beta E_t V^n(\pi_t, X_t^n).$$

Now, if a single experiment with n at π_t could change the ranking of the indices, then we have $\beta E_t V^n(\pi_t, X_t^n) > 0$ and therefore $p^n(\pi_t) < 0$. The fact that $V^n(\pi_t)$ converges nevertheless to zero is only confirming that in equilibrium it is optimal not to explore the remaining learning opportunities. This has to be contrasted to statistical decision problems which generally converge in the limit to the short-run, myopic decision problems as in Aghion *et al.* (1991, Proposition 2.2–2.4). Here it is the strategy of the conceding seller which indicates that if some uncertainty remains unresolved in the game then the convergence will not be complete.

4.3. *The set of MPE*

We come to the characterization of the entire set of MPE. By Theorem 1, all MPE are efficient and hence have the same social value $W^*(\pi_t)$. The multiplicity of equilibria then pertains only to different allocations of the surplus, $W^*(\pi_t)$, among the players. The equilibrium which maximizes the buyer's payoff is therefore simultaneously minimizing the sellers' payoff. Conversely, the equilibrium which is maximizing the seller's payoff is simultaneously minimizing the buyer's payoff. The characterization of these two extremal equilibria is then sufficient to describe the lower and upper bounds on the payoffs of the players.

Proposition 3 (Characterization of MPE): *The set of Markov perfect equilibria is characterized by the lower and upper bounds on the payoffs. The lower bounds are given by:*

$$V^B(\pi_t) = E_t \left\{ \sum_{n=1}^{\infty} \sum_{s=\sigma_n}^{\tau_n - 1} \beta^{s-t} x_{\sigma_n}^2 + \sum_{n=1}^{\infty} \sum_{s=\tau_n}^{\sigma_{n+1} - 1} \beta^{s-t} x_{\tau_n}^1 \right\}, \tag{b}$$

$$V^1(\pi_t) = 0, \qquad\qquad (s^1)$$

$$V^2(\pi_t) = 0. \qquad\qquad (s^2)$$

The upper bounds are given by:

$$V^B(\pi_t) = W^*(\pi_t), \qquad\qquad (B)$$

$$V^1(\pi_t) = E_t \left\{ \sum_{n=1}^{\infty} \sum_{s=\sigma_n}^{\tau_n - 1} \beta^{s-t}[x_s^1 - x_{\sigma_n}^2] \right\}, \qquad\qquad (S^1)$$

$$V^2(\pi_t) = E_t \left\{ \sum_{n=1}^{\infty} \sum_{s=\tau_n}^{\sigma_{n+1} - 1} \beta^{s-t}[x_s^1 - x_{\tau_n}^2] \right\}. \qquad\qquad (S^2)$$

We omit the proof, which can be found in Bergemann and Välimäki (1995), since the construction of the extremal equilibria, while entirely straightforward, is long and tedious.

The equilibrium which generates payoffs $\{(B), (s^1), (s^2)\}$ reduces the payoffs of the sellers permanently to zero, which is their individual participation constraint. The buyer receives the entire value of the game. The second equilibrium, which generates $\{(b), (S^1), (S^2)\}$ is in fact the *cautious equilibrium* of Theorem 4.1. Let us here just recall the payoff structure of this equilibrium. The consumer is buying the product of the superior seller i at the price $p^i(\pi_t) = E_t X_t^i - E_t X_t^j = x_t^i - x_t^j$. Let $i = 1$ and $j = 2$. Starting at τ_n, when the buyer switches from j to i and until σ_n when she switches back to j, the estimate of $E_t X_t^j$ does not change since no new information on j's product becomes available. We can consequently write $E_t X_t^j = E_{\tau_n} X_{\tau_n}^j = x_{\tau_n}^j$ for t between $\pi_n \leq t \leq \sigma_n - 1$. During this time span the buyer is experimenting with i, and only the estimate of X_t^i, $E_t X_t^i$ is changing over time. For the same time interval, $\pi_n \leq t \leq \sigma_n - 1$, the buyer's periodic return is constant and given by $E_t X_t^i - p^i(\pi_t) = E_t X_t^i - (x_1^i - x_{\tau_n}^j) = x_{\tau_n}^j$, which is represented in (b).

The symmetry in the extremal equilibria is apparent. The equilibrium which maximizes the buyer's payoff makes the successful seller s always indifferent between selling and not selling: the equilibrium condition (S^s) holds as an equality. In the equilibrium which minimizes the buyer's payoff it is, on the contrary, always the conceding seller n who is indifferent between selling and not selling and in turn (S^n) holds as an equality. Since the "successful" prices, $p^s(\pi_t)$, and the "threat" prices, $p^n(\pi_t)$, that each seller is employing are strategically almost independent devices, it is not difficult to show that the entire convex hull spanned by the payoffs of the extremal equilibria constitute the set of MPE payoffs.

Proposition 3 tells us that the main difference between the set of Markovian and the set of non-Markovian equilibria lies in the possibility of collusion among the sellers. This may imply efficiency losses as the following example demonstrates. Consider the following collusive equilibrium, in which the sellers alternate in selling at prices $p^s = E_t X_t^s$ and $p^n \leq E_t X_t^n$, and use the trigger strategy to convert to the equilibrium $\{(B), (s^1), (s^2)\}$ should a price deviation by one of the sellers occur. The buyer's payoff is now reduced forever to zero and the allocation path is clearly inefficient, since the alternating is independent of the actual learning experience.

5. Conclusion

We presented a simple dynamic equilibrium pricing model under uncertainty where the players take into account the costs and benefits of learning. All MPE are efficient. The cautious MPE, which was the focus of our analysis implements the efficient learning solution by a simple and intuitive equilibrium pricing policy of the firms. The restriction to Markovian equilibria allowed us to focus on the interaction of pricing and learning policies.

Since there was only one large consumer and the qualities of the firms were statistically independent, experimentation did not give rise to any externalities. The extension to statistically dependent alternatives is straightforward and would yield exactly the same conclusions in terms of efficiency and equilibrium prices as the statistically independent case. The reader may verify that the independence assumption was only used for the characterization of the efficient policy in terms of the dynamic allocation index. This result underlines the basic mechanism at work in the dynamic pricing model. Strategic competition can sustain efficient learning outcomes if the exchange of the costs and benefits is frictionless both intertemporally *and* interpersonally. The necessity of intertemporal exchange was a major theme throughout the paper. The extension of our model to a multiple buyer market illustrates the necessity of frictionless interpersonal exchange to sustain efficient experimentation. While our results for multiple buyers are only preliminary, they suggest some interesting possibilities. In the simplest case of one known and one unknown product with many buyers, market experimentation will continue beyond the social optimum.

As in the case with a single buyer, the seller of the unknown product has to offer negative prices to compensate for the current quality differential.

The essential difference arises as the seller with the unknown product needs to compensate *his* buyers only for their own future expected losses. Since experimentation is public, he will eventually appropriate the benefits of a positive sample path from all consumers. Conversely, the firm with the known product would need to offer to each and every consumer a price low enough in order to completely prevent experimentation with the unknown product. This policy is very costly and, rather than trying to exclude his competitor entirely, he prefers to let experimentation continue beyond the social optimum to further weaken his opponent's position. In Bergemann and Välimäki (1996), we show that the underinvestment in learning result as described in Bolton and Harris (1993) can be reversed and the equilibrium may involve socially excessive experimentation.

Acknowledgments

We would like to express our gratitude to George J. Mailath for his encouraging support since the very beginning of this project. We thank Dieter Balkenborg, Pierpaolo Battigalli, Bruno Biais, In-Koo Cho, Matthias Kahl, Richard Kihlstrom, Roger Lagunoff, Stephen Morris, Andrew Postlewaite, Rafael Rob, and Jean Tirole for many helpful comments. We are grateful to an editor and two anonymous referees for very helpful comments and suggestions. The first author would like to thank Avner Shaked for his hospitality during a stay at the SFB 303, University of Bonn. Financial support by the German Science Foundation and Yrjö Jahnsson Foundation is gratefully acknowledged.

References

Aghion, P., P. Bolton, C. Harris, and B. Jullien (1991): "Optimal Learning by Experimentation," *Review of Economic Studies*, 58, 621–654.

Banks, J. S., and R. K. Sundaram (1992): "Denumerable Armed Bandits," *Econometrica*, 60, 1071–1096.

Bergemann, D., and J. Välimäki (1995): "Learning and Strategic Pricing: Further Results," Mimeo, Northwestern University and Yale University.

———— (1996): "Market Experimentation and Pricing," Mimeo, Northwestern University and Yale University.

Bolton, P., and C. Harris (1993): "Strategic Experimentation," Discussion Paper TE/93/261, LSE, London.

Easley, D., and N. M. Kiefer (1988): "Controlling a Stochastic Process with Unknown Parameters," *Econometrica*, 56, 1045–1064.

Felli, L., and C. Harris (1994): "Job Matching, Learning and the Distribution of Surplus," Mimeo, LSE.

Gittins, J. C. (1989): *Multi-Armed Bandit Allocation Indices*. Chichester: Wiley.

Gittins, J. C., and D. M. Jones (1974): "A Dynamic Allocation Index for the Sequential Design of Experiments," in *Progress in Statistics*, ed. by J. Gani *et al.* Amsterdam: North-Holland, 241–266.

Jovanovic, B. (1979): "Job Search and the Theory of Turnover," *Journal of Political Economy*, 87, 972–990.

Kihlstrom, R., L. Mirman, and A. Postlewaite (1984): "Experimental Consumption and the 'Rothschild Effect,' " in *Bayesian Models of Economic Theory*, ed. by M. Boyer and R. Kihlstrom. Amsterdam: Elsevier.

Maskin, E., and J. Tirole (1988): "A Theory of Dynamic Oligolpoly, II," *Econometrica*, 56, 571–599.

——— (1994): "Markov Perfect Equilibrium," mimeo.

McLennan, A. (1984): "Price Dispersion and Incomplete Learning in the Long-Run," *Journal of Economic Dynamics and Control*, 7, 331–347.

Miller, R. A. (1984): "Job Matching and Occupational Choice," *Journal of Political Economy*, 92, 1086–1120.

Rob, R. (1991): "Learning and Capacity Expansion under Demand Uncertainty," *Review of Economic Studies*, 58, 655–675.

Rothschild, M. (1974): "A Two-Armed Bandit Theory of Market Pricing," *Journal of Economic Theory*, 9, 185–202.

Smith, L. (1992): "Error Persistence, and Experimental versus Observational Learning," Mimeo, MIT.

Whittle, P. (1982): *Optimization over Time*, Vols. 1&2. Chicester: Wiley.

Chapter 3

Experimentation in Markets[*]

Dirk Bergemann[†] and Juuso Välimäki[‡]

[†] *Department of Economics, and Cowles Foundation of Research in Economics,*
Yale University, New Haven, CT 06520-8268, USA
dirk.bergemann@yale.edu
[‡] *Department of Economics, Aalto University School of Business,*
02150 Espoo, Finland
juuso.valimaki@aalto.fi

We present a model of entry and exit with Bayesian learning and price competition. A new product of initially unknown quality is introduced in the market, and purchases of the product yield information on its true quality. We assume that the performance of the new product is publicly observable. As agents learn from the experiments of others, informational externalities arise.

We determine the Markov Perfect Equilibrium prices and allocations. In a single market, the combination of the informational externalities among the buyers and the strategic pricing by the sellers results in excessive experimentation. If the new product is launched in many distinct markets, the path of sales converges to the efficient path in the limit as the number of markets grows.

1. Introduction

In multi-agent learning situations, informational externalities may reduce the number of experiments undertaken below the socially efficient level. As buyers choose among new experience goods or firms decide whether to adopt a new technology, the availability of information from others' decisions gives rise to a free rider problem. Rather than perform a costly experiment herself, a buyer may opt to wait and see how the market evaluates the new product. In this paper, we develop a simple market model of experimentation and analyze the informational effects in a model with

---

[*]This chapter is reproduced from *Review of Economic Studies*, **67**, 213–234, 2000.

buyers and sellers. In contrast to the one-sided experimentation problems, we find that equilibrium experimentation often exceeds the Pareto optimal level in two-sided models.

The first model we consider is a dynamic duopoly in continuous time with homogeneous buyers. Two firms with differentiated products engage in price competition over time. One product has a known value while the other is new and its true value is initially uncertain to all the parties in the model. The value of the product is determined by the quality of the match between consumer preferences and product characteristics. Additional information is acquired only through repeat purchases. In each period, buyers observe a noisy signal of the true value of the product. We assume that all signals are publicly observable and that all uncertainty is about a common value component. With this assumption, all buyers and sellers condition their behavior on the same information and we can abstract from individual differences in past observations. This is justified in economic situations where public data on other buyers' choices is available and informative. For example, when choosing among providers of communication or transportation services, such as mail delivery firms or airlines, it is reasonable to rely on consumer reports and other published data in addition to one's private experience. We also show that our model has an alternative interpretation as one where a flow of buyers make a once and for all purchase of a durable good. Under that scenario, signals from earlier purchases are the only source of information for a given buyer. In both interpretations of the model, the public observability of signals gives rise to an informational externality in the market.

We determine the paths of sales and prices in Markov Perfect Equilibrium and compare them to the Pareto optimal paths.[1] In a model without sellers (or equivalently with fixed prices), informational externalities among the buyers result in too little investment in information acquisition as in Bolton and Harris (1999). When we allow the firms to set their prices optimally, a new source of potential inefficiency is discovered: experimentation determines the future competitive positions of the two firms. The new firm extracts benefits from successful experiments through higher future prices while buyers bear the risk of unsuccessful experiments.

---

[1]There is a multiplicity of equilibrium prices. The situation is similar to the static Bertrand model with differentiated products, We concentrate on equilibria in trembling hand perfect prices.

When buyers become more pessimistic about the new firm's product, the established firm is able to charge higher prices on its product in equilibrium. The new seller has to compensate each buyer for the costs resulting from higher future prices charged by the established seller. The individual buyer fails, however, to take into account the effect of her own purchases on all other buyers. This makes sales relatively inexpensive for the new firm and the equilibrium displays excessive experimentation. A well known result on optimal learning states that the *ex post* efficient alternative is not always selected in the long run along the *ex ante* efficient path (*e.g.* Rothschild (1974)). Our results therefore imply that introducing price competition on the supply side of the market may achieve *ex post* optimality with a larger probability.

The second model analysis price competition in multiple markets. Each market has a separate established producer, but the new product is a competitor in all markets. One may think of each single market as a different geographic location or, alternatively, as the customer base of a small firm producing under a capacity constraint. Multiple markets introduce a new externality. Sales of the new product in any given market provide information to all of the small producers. As a result, they are more aggressive in their own pricing decisions as undercutting the new firm no longer results in a complete stop in the flow of information. As the number of markets increases, the equilibrium experimentation path converges to the efficient path.

The continuous time techniques we use allow us to derive the equilibria in closed form. In particular, we can analyze the prices in a more detailed manner than would be possible in a discrete time model. By examining the price paths, we gain new insights into the strength of the rivalry between the firms at various points in the game.

There are a number of papers examining learning and experimentation in multi-agent settings. The most relevant papers for our purposes are Bolton and Harris (1999), Bergemann and Välimäki (1996) and Felli and Harris (1996). Bolton and Harris analyze a continuous-time game of strategic experimentation and our methodology follows theirs closely. They focus on the pure informational externality between a set of identical agents and show that all the equilibria involve too little experimentation. Bergemann and Välimäki present a model with a single buyer, and consequently the issue of informational externalities between the buyers does not arise. Finally, Felli and Harris study the wage dynamics in a continuous time learning model about an employee's firm-specific productivity. Again, the model considers only a single employee and hence informational externalities do not arise.

The first paper to address the potential free rider problems in the presence of informational externalities is the Hendricks and Kovenock (1989) analysis of oil exploration games. They showed that experimentation can be insufficient or excessive due to the externalities. Rob (1991) studies the informational impact of entry decisions into an industry with uncertain profitability. Successful entry attracts more entry which reduces the long-run profits of early successful entrants. As a result the amount of entry in each period is inefficiently low. Chamley and Gale (1994) present the free-riding aspect in a similar timing game in which each agent has to make a single investment decision. Vettas (1998) allows for two-sided learning in a market of new products. The firms learn about the market size while the buyers make inferences about product quality. Bergemann and Välimäki (1997) consider the diffusion of a new product in a model of horizontal differentiation. Keller and Rady (1999) consider experimentation in a changing environment by a monopolist. The public observability of the utility signals, or a subset of those signals, is central to some recent models of world-of-mouth communication and social learning such as McFadden and Train (1996) and Banerjee and Fudenberg (1997).

Section 2 introduces the elements of the dynamic game in a two-period example. Its structure is kept as simple as possible to convey the basic aspects of the two-sided market environment. The continuous-time model is introduced in Section 3. The single market model is analyzed in Section 4. In Section 5 we present the model with many markets. Section 6 concludes.

2. A Two-Period Example

Two sellers provide quality differentiated products to a unit mass of identical buyers with unit demand in each period.[2] The incumbent supplies a product with known quality, while the quality supplied by the entrant is initially unknown. The value of the established product is s per period and the new product has a value of either μ_L or μ_H with $\mu_L < s < \mu_H$. Let α be the common prior probability that the product has value μ_H and denote the expected quality by $\mu(\alpha) \triangleq \alpha\mu_H + (1 - \alpha)\mu_L$. The marginal costs of production are identical and normalized to zero.

---

[2]We are grateful to Eric Maskin for suggesting this example.

Firm j chooses price p_j^t in period $t \in \{1,2\}$, where $j = 0$ indexes the entrant, and $j = 1$ the incumbent. The net utility of a purchase to the buyer is the (expected) quality of the product minus its current price. Buyers and sellers maximize the sum of their per-period payoffs.

The revelation of uncertainty takes an extremely simple form. If a fraction x of the buyers experiment with the new product, then its true quality is revealed to all agents in the second period with probability x. With the complementary probability, no new information arrives.

Consider first the continuation games after zero or complete experimentation: $x \in \{0,1\}$. Under complete experimentation in the first period $(x = 1)$, the new product is worth μ_H with probability α in the second period. The second period prices are given by Bertrand competition: $p_0^2 = \mu_H - s$, and $p_1^2 = 0$, and all buyers purchase from the entrant. With probability $1 - \alpha$, the quality is low and the second-period prices are $p_0^2 = 0$ and $p_1^2 = s - \mu_L$, and all buyers purchase from the incumbent. Conditional on complete experimentation in the first period, the expected second-period profits for the two firms are $\pi_0^2 = \alpha(\mu_H - s)$ and $\pi_1^2 = (1 - \alpha)(s - \mu_L)$. If there is zero experimentation in the first period, then second-period prices are given by $p_0^2 = \max\{\mu(\alpha) - s, 0\}$ and $p_1^2 = \max\{s - \mu(\alpha), 0\}$, and the firm with a positive price sells to the entire market. We assume for the rest of this section that $\mu(\alpha) < s$. This implies that from a myopic point of view, experiments are costly.

The first-period equilibrium prices, p_0^1 and p_1^1, are found by backward induction. Since each consumer is of measure zero, the future payoff of an individual buyer is independent of her current product choice. The equilibrium condition under Bertrand pricing then requires the buyer to be indifferent between the two offers

$$\mu(\alpha) - p_0^1 = s - p_1^1 \tag{2.1}$$

and hence the price differential has to be equal to the (expected) quality difference. Moreover, we require that the non-selling firm be indifferent between selling and not selling at equilibrium prices. Prices satisfying this requirement are called *cautions*. With the linearity of the payoffs in x, either all buyers or none buy from the new firm in equilibrium. The values of α at which experimentation occurs in equilibrium are characterized by two conditions. First, the incumbent must prefer to concede the market in the first period and to make sales in the second period if the new good fails in the first period

$$p_1^1 + s - \mu(\alpha) \leqq (1 - \alpha)(s - \mu_L).$$

With cautious pricing, this holds as an equality and

$$p_1^1 = \alpha(\mu_H - s). \tag{2.2}$$

Second, the entrant has to make nonnegative expected profits by selling today and betting on a favorable resolution of uncertainty tomorrow

$$p_0^1 + \alpha(\mu_H - s) \geqq 0. \tag{2.3}$$

The values of α that satisfy (2.1)–(2.3) induce experimentation in the first period. The conditions (2.1)–(2.3) imply that

$$p_0^1 = \alpha(\mu_H - s) + \mu(\alpha) - s \geqq -\alpha(\mu_H - s).$$

Hence experimentation occurs in equilibrium whenever

$$\alpha \geqq \alpha^* = \frac{s - \mu_L}{(\mu_H - \mu_L) + 2(\mu_H - s)}.$$

On the other hand, the socially efficient policy requires experimentation whenever current costs of experimentation are outweighed by future gains

$$\alpha(\mu_H - s) \geqq s - \mu(\alpha),$$

or

$$\alpha \geqq \hat{\alpha} = \frac{s - \mu_L}{(\mu_H - \mu_L) + (\mu_H - s)}.$$

As $\alpha^* < \hat{\alpha}$, we conclude that the cautious equilibrium exhibits excessive experimentation.

This inefficiency can be traced to the divergence of the private cost from the social cost of experiments in equilibrium. The social benefit, $\alpha(\mu_H - s)$, coincides with the entrant's private benefit. The social cost is given by the myopic losses, $s - \mu(\alpha)$. The private cost of supporting the experiment, *i.e.* the negative price that the entrant has to quote, is $p_0^1 = \mu(\alpha) - s + \alpha(\mu_H - s)$. The additional term $\alpha(\mu_H - s)$ is the price of the incumbent, and thus reflects his informational gain through cautious pricing. The failure of the buyers to take the future surplus extraction into account reduces the private cost to finance experimentation. In contrast to the duopoly, where the identity of the benefiting seller depends on the outcome of the experiment, a monopoly would extract the social surplus at every stage and the equilibrium would be efficient. More insight into the discrepancy between the efficient and the equilibrium allocation may be obtained by considering the case where all buyers act collectively and make purchases as a cooperative.[3]

In equilibrium, the cooperative is indifferent between the two products at current prices. Hence the price differential is equal to the sum of the quality differential *and* the change in the continuation payoff resulting from experimentation,

$$p_0^1 - p_1^1 = \mu(\alpha) - s + \alpha s + (1 - \alpha)\mu_L - \mu(\alpha)$$
$$= \mu(\alpha) - s - \alpha(s - \mu_H). \tag{2.4}$$

By cautious pricing, p_1^1 equals the expected gain from experimentation for the incumbent when the entrant is selling in the first period. Notice that with the cooperative, the expected losses from experimentation for the buyer equal exactly the incumbent's expected gain. The equilibrium condition (2.4) then shows that the private cost of experimentation for the entrant coincides with the social cost, or $p_0^1 = \mu(\alpha) - s$, and efficiency follows. The dynamic implications of the informational externality are considered next in a more gradual process of information revelation and with a finite number of buyers and sellers.

3. The Model

The market consists of buyers and sellers. The buyers are indexed by $i \in \{1, \ldots, N\}$. Two sellers, $j = 0, 1$, offer differentiated products and compete in prices in a continuous time model with an infinite horizon. Firm 0 is called the new firm or the entrant. The value, μ, of its product is initially unknown to *all* parties in the market and we refer to it as the new or uncertain product. It can be either low or high

$$\mu \in \{\mu_L, \mu_H\},$$

with $0 < \mu_L < s < \mu_H$. The value of firm 1's product is s, and the firm is called the established firm or the incumbent. All players share a common prior $\Pr(\mu = \mu_H) = \alpha_0$. Let $n_j(t)$ denote the number of buyers that purchase firm j's product in period t.

---

[3]This is the case analyzed in Bergemann and Välimäki (1996).

3.1. *Bayesian learning*

The uncertainty about the value of the new product can be resolved only by experimenting. The performance of the new product is, however, subject to random disturbances. The information resulting from any single purchase provides a noisy signal of the true underlying quality. The flow utility from the uncertain alternative is

$$du_i(t) = \mu dt + \sigma dW_i(t),$$

where $dW_i(t)$ is the increment of the one-dimensional standard Wiener process. We assume that dW_i and $dW_{i'}$ are independent for $i \neq i'$. The flow utility of the established product is given by

$$du_i(t) = sdt.$$

The aggregate performance of the product over all buyers is the sum of flow utility realizations, which we assume to be publicly observable. Let $n_0(t)$ be the number of the new firm's customers and w.l.o.g. assume that for a given $n_0(t)$, the buyers with the smallest indices buy from the new firm. The aggregate performance of the new product is given by

$$\sum_{i=1}^{n_0(t)} du_i(t) = n_0(t)\mu dt + \sigma \sum_{i=1}^{n_0(t)} dW_i(t). \tag{3.1}$$

All relevant information is contained in the aggregate outcome.[4] As the value of μ is either μ_L or μ_H, the posterior beliefs are given by $\alpha(t)$, with

$$\alpha(t) = \Pr(\mu = \mu_H | \mathscr{F}(t)),$$

where $\mathscr{F}(t)$ is the history (or more accurately, the filtration) generated by the past observations. The conditional expected quality $\mu(\alpha(t))$ of the uncertain product is

$$\mu(\alpha(t)) = (1 - \alpha(t))\mu_L + \alpha(t)\mu_H.$$

The players extract the information contained in the noisy market outcome (3.1) to update their beliefs. The game is thus one of incomplete but symmetric information, and no issues of asymmetric information arise. As the beliefs are characterized by $\alpha(t)$, the inference problem reduces to the

---

[4]Recall that in Bayesian learning models with normal distributions, the sample mean and the number of observations constitute a sufficient statistic for the model observation of n i.i.d. draws from a distribution with known variance and unknown mean.

description of the law of motion of $\alpha(t)$.[5] It can be shown that $\alpha(t)$ is a diffusion process with zero drift and instantaneous variance $n_0(t)\Sigma(\alpha(t))$, where

$$\Sigma(\alpha(t)) = \left(\frac{\alpha(t)(1-\alpha(t))(\mu_H - \mu_L)}{\sigma} \right)^2.$$

The process of the posterior $\alpha(t)$ has zero drift since posterior beliefs form a martingale and any change in $\alpha(t)$ has zero expectation. The variance $\Sigma(\alpha(t))$ measures the additional information obtained through an experiment. An increase in the variance causes a more rapid change in the posterior. The variance of $\alpha(t)$ is linear in the size of the aggregate experiment, $n_0(t)$, and in the "signal to noise" ratio $((\mu_H - \mu_L)^2)/\sigma^2$. A large market share for the new firm results in a faster change of the posterior $\alpha(t)$ as more information is generated.

The key assumption in our model is the public observability of the utility signals. This allows us to abstract from the issue of idiosyncratic differences in posterior beliefs that might arise from different experiences. While this is clearly a strong assumption, we feel that it approximates well two important economic situations. In the first, a large number of buyers are faced with the same dynamic decision problem. Even an imperfect aggregate measure of performance by the new product is valuable to the buyers' choices. This situation is common in the provision of services, and examples include the data collected by consumer agencies on the percentage of flights arriving on time on a new airline or average waiting times for internet services on a new provider. The second class of problems involves once and for all purchases of durable goods. In Section 4, we argue that our model can be reinterpreted as one of durable goods sales to an inflow of new buyers. In that case, no buyer has individual information on the products at the moment when the choice is made and public information is the only basis for assessing the products.

3.2. *Strategies and equilibrium*

The dynamic game consists of two components: the pricing strategies of the sellers and the acceptance decisions of the buyers. The strategies depend on information available to the agents at the instant of decision. Since we

---

[5]See Liptser and Shirayayev (1977), Chapter 9, for the filtering equations of the Brownian motion in continuous time.

want to analyze how the resolution of uncertainty affects the pricing game between the sellers, we concentrate on equilibria in Markovian strategies. This allows us to rule out collusive equilibria with continuation strategies that depend on information that is not payoff relevant.

We view the model as a continuous-time analogue of the repeated extensive form game where the sellers set prices at the beginning of each stage, and buyers then choose where to buy. With this in mind, the natural state variable in period t is $\alpha(t)$ for the sellers, and for the buyers it is $\alpha(t)$ together with the prevailing prices, $p_0(t)$ and $p_1(t)$. A *pricing policy* p_j is a measurable function $p_j : [0,1] \to \mathbb{R}$. In the equilibrium analysis we show that the restriction to pure strategies is without loss of generality. Each buyer has unit demand at every instant and her *acceptance policy*, a_i, determines where to purchase the product: $a_i : [0,1] \times \mathbb{R} \times \mathbb{R} \to \{0,1,R\}$. The acceptance policy specifies whether the buyer accepts seller 0 or 1 or rejects both (R).

The firms offer prices $p_j(t)$ at each instant of time and their instantaneous revenues evolve as a function of market share $n_j(t)$ and price $p_j(t)$

$$d\pi_j(t) = n_j(t)p_j(t)dt.$$

The marginal cost of production is constant and normalized to zero. The flow utility of a buyer is determined by her choice among the competing products. It is her flow utility net of the current price

$$du_i(t) - p_j(t)dt.$$

The intertemporal profits of each firm j are a function of the pricing and acceptance policies of the market players

$$V_j(a, p_j, p_{-j}; \alpha) = \mathbb{E}_\alpha \int_0^\infty e^{-rt} d\pi_j(t),$$

where $r > 0$ is the discount rate and $a = (a_1, \ldots, a_N)$. The intertemporal utility functional of each buyer is the discounted flow of net utilities

$$V_i(a_i, a_{-i}, p; \alpha) = \mathbb{E}_\alpha \int_0^\infty e^{-rt}(du_i(t) - p_j(t)dt),$$

where $a_i(\alpha, p_0(t), p_1(t)) = j$ and $p = (p_0, p_1)$.

Definition 1 (Markov Perfect Equilibrium, MPE): A collection of strategies, $\{a^*, p^*\}$ is a Markov Perfect Equilibrium if

(i) $V_j(a^*, p_j^*, p_{-j}^*; \alpha) \geq V_j(a^*, p_j, p_{-j}^*; \alpha), \forall j, \forall p_j, \forall \alpha,$

(ii) $V_i(a_i^*, a_{-i}^*, p^*; \alpha) \geq V_i(a_i, a_{-i}^*, p^*; \alpha), \forall i, \forall a_i, \forall \alpha.$

Notice that the deviation strategies a_i and p_j are not required to be Markovian. The equilibrium analysis presented in Sections 4 and 5 involves the solution of stopping problems derived from the equilibrium conditions. The basic technique is most clearly illustrated in the solution of the socially efficient allocation.

3.3. *Efficiency*

The efficient allocation policy in the market model can be obtained by solving a specific multi-armed bandit problem. Here we provide only the essentials of the solution technique and refer the reader to Karatzas (1984) for details. Since we have assumed quasilinear utilities for all of the players in the game, finding the set of Pareto-efficient allocations is equivalent to solving the total surplus maximization problem.

An allocation policy n is a measurable and adapted process with values $n_j(t) \in \{1, \ldots, N\}$. Clearly, an optimal policy allocates all buyers to one of the firms. Let $n(t)$ be the number of the new firm's customers and w.l.o.g. assume that for a given $n(t)$, the buyers with the smallest indices buy from the new firm. The expected value of an allocation policy n is given by

$$V(n; \alpha) = \mathbb{E}_\alpha \int_0^\infty e^{-rt} \left(\sum_{i=1}^{n(t)} (\mu dt + \sigma dW_i(t)) + \sum_{i=n(t)+1}^{N} s dt \right). \quad (3.2)$$

The problem is then to find a policy n^* so as to maximize expected value

$$V(\alpha) \triangleq \max_n V(n; \alpha), \text{ with } V(\alpha) = V(n^*; \alpha), \forall \alpha \in [0, 1].$$

The controls in the allocation problem are continuous, but due to the convexity of the value function, it can be shown that (3.2) is equivalent to the optimal stopping problem[6]

$$V(\alpha) = \max_\tau V(\tau; \alpha) = \mathbb{E}_\alpha \left[\int_0^\tau e^{-rt} \left(\sum_{i=1}^{N} \mu dt + \sigma dW_i(t) \right) + e^{-r\tau} \frac{Ns}{r} \right],$$

where τ is a stopping time.[7] As the instantaneous payoff $\mu(\alpha)$ is increasing in α, one might expect a solution that chooses the uncertain alternative in

---

[6]The assumed convexity is shown to hold in the solution of the stopping problem and the uniqueness of the value function then justifies the initial hypothesis.

[7]Recall that a stopping time is a real valued, $\mathscr{F}(t)$-measurable random variable. In other words, stopping at t has to be decided based upon history at lime t.

a half-open interval $(\hat{\alpha}, 1]$ and stops the process at $\hat{\alpha}$. Using Itô's lemma and the stochastic differential equation governing the posterior process, the Hamilton–Jacobi–Bellman (HJB) equation of this problem for $\alpha > \hat{\alpha}$ is the following differential equation[8]

$$rV(\alpha) = N(\mu(\alpha) + \frac{1}{2}\Sigma(\alpha)V''(\alpha)), \qquad (3.3)$$

with the initial boundary conditions given by the value matching and the smooth pasting conditions

$$rV(\hat{\alpha}) = Ns,$$

$$V'(\hat{\alpha}) = 0.$$

The differential equation (3.3) represents the flow benefit during the experimentation phase. It consists of the expected payoff $N\mu(\alpha)$ and the informational gains $(N/2)\Sigma(\alpha)V''(\alpha)$ which improve the intertemporal policy. The instantaneous variance $\Sigma(\alpha)$ indicates the quantity of information released by a unit of experimentation and the curvature of the value function $V''(\alpha)$ is the shadow price of information for the planner. It is then optimal to experiment as long as the flow payoff and the flow value of information exceed the value of the safe alternative.

Theorem 1 (Efficient Stopping): *The Pareto efficient stopping point $\hat{\alpha}$ is*

$$\hat{\alpha} = \frac{(s - \mu_L)(\gamma - 1)}{(\mu_H - \mu_L)(\gamma - 1) + 2(\mu_H - s)},$$

with

$$\gamma = \sqrt{1 + \frac{8r\sigma^2}{N(\mu_H - \mu_L)^2}}.$$

Proof: See Appendix.   □

For notational brevity, we define ρ by

$$\rho \triangleq \frac{8r\sigma^2}{(\mu_H - \mu_L)^2}. \qquad (3.4)$$

---

[8]The HJB equation is the dynamic programming equation in continuous time. It can be shown that the HJB equation admits a unique solution that is piecewise twice continuously differentiable. See Dixit and Pindyck (1994) or Harrison (1985) for the details.

The stopping point $\hat{\alpha}$ is strictly increasing and $V(.)$ is strictly decreasing in ρ/N.

4. Single Market

The basic model with a finite number of buyers in a single market is presented in Subsection 4.1. The limiting case with a continuum of small buyers is analyzed in Subsection 4.2. The limiting model is also interpreted as a model of durable goods sales with a flow of incoming buyers.

4.1. *Finitely many buyers*

Since we are looking for equilibria in Markovian strategies, we can use dynamic programming techniques directly. The value functions of the sellers are functions of the current posterior $\alpha(t) \in [0, 1]$. Using Itôs lemma and the equation governing the posterior process, we get the Hamilton–Jacobi–Bellman equations

$$rV_0(\alpha) = \max_{p_0} \left\{ np_0 + \frac{n}{2}\Sigma(\alpha)V_0''(\alpha) \right\}, \tag{4.1}$$

and

$$rV_1(\alpha) = \max_{p_1} \left\{ (N - n)p_1 + \frac{n}{2}\Sigma(\alpha)V_1''(\alpha) \right\}, \tag{4.2}$$

where n is the number of buyers who purchase from the entrant. Observe that the objective functions are linear in sales for both firms.

Each $V_j(.)$ can be decomposed into the flow revenue resulting from sales, $n_j P_j$, and the expected change in the competitive position generated by the sales of the new product, captured by $(n/2)\Sigma(\alpha)V_j''(\alpha)$. As the posterior belief $\alpha(t)$ forms a martingale, $\mathbb{E}[d\alpha(t)] = 0$, only the second-order term in the expected change remains. With Markovian strategies, each firm's future competitive position is influenced only by the arrival of new information. Hence $\frac{1}{2}\Sigma(\alpha)V_j''(\alpha)$ can be interpreted as the value of information to firm j. Experiments lead to more differentiation (in *ex post* terms) between the two competitors as information is accumulated. In a more differentiated environment, competitive pressure between the firms is reduced and they are able to extract more surplus from the buyers. As a result, one would expect that $V''(\alpha) > 0$.

The value function of buyer i is

$$rU_i(\alpha) = \max \left\{ s - p_1 + \frac{n}{2}\Sigma(\alpha)U_i''(\alpha), \mu(\alpha) - p_0 \right.$$

$$\left. + \frac{n+1}{2}\Sigma(\alpha)U_i''(\alpha) \right\}, \tag{4.3}$$

when n other buyers choose the uncertain object. The choice of the buyer is determined by the expected flow payoff and the flow value of information released through the choice. By selecting the uncertain firm, her future payoffs are changed by the amount of $\frac{1}{2}\Sigma(\alpha)U_i''(\alpha)$. Equation (4.3) of buyer i is representative of all buyers as they have identical preferences and they have access to the same information. We can therefore omit the index i entirely and consider the representative buyer. Differentiation leads to less competitive prices and hence enables the sellers to extract more surplus from the buyers. As a consequence, it is to be expected that the buyers have a negative value of information, or $U''(\alpha) < 0$.

Due to price competition, each buyer has to be indifferent between the alternatives, or formally

$$s - p_1 = \mu(\alpha) - p_0 + \frac{1}{2}\Sigma(\alpha)U''(\alpha). \tag{4.4}$$

It is never optimal to leave any buyer with more net surplus than she would obtain from the alternative seller, nor could any seller ever expect to make sales if he would offer strictly less utility than the competitor.

As the objective functions of the sellers are linear in market shares and (4.4) holds, we conclude that nontrivial market sharing can happen only if both sellers are indifferent between selling any amount $n_j > 0$ or not selling at all. But at any such state α, stopping the experiments must also be an equilibrium outcome. We then conjecture that the equilibrium allocation has the following simple structure: Since the payoff of the second seller is increasing in α, there is a half-open interval $(\alpha^*, 1]$, called the continuation region, where experiments occur at the maximal rate $n_0 = N$, and a closed interval $[0, \alpha^*]$, called the stopping region, in which no experiments take place and $n_1 = N$.

Before we verify this conjecture, a qualification for the pricing policies is made. We require that any price quoted by a firm which is not selling in a given period would make the firm at least weakly better off if accepted. In an earlier paper, Bergemann and Välimäki (1996), we called prices which satisfy this property "cautious".[9] the value functions in (4.1) and (4.2) imply that cautious prices satisfy

$$p_0(\alpha) \geq -\frac{1}{2}\Sigma(\alpha)V_0''(\alpha), \tag{4.5}$$

and

$$p_t(\alpha) \geq \frac{1}{2}\Sigma(\alpha)V_1''(\alpha). \tag{4.6}$$

Condition (4.5) states that the entrant is willing to sell only if the price is at least offset by the value of the information flow. In contrast, condition (4.6) states that the incumbent is willing to sell only if he receives at least enough revenue to compensate for the foregone informational gains. A simple undercutting argument establishes that the prices p_j need to satisfy the appropriate inequality as an equality $n_j = 0$. With positive sales, the prices are obtained by using (4.4)

$$p_0(\alpha) = \mu(\alpha) - s + \frac{1}{2}\Sigma(\alpha)(U''(\alpha) + V_1''(\alpha)), \; if \; n_0 > 0, \tag{4.7}$$

and

$$p_1(\alpha) = s - \mu(\alpha) - \frac{1}{2}\Sigma(\alpha)(U''(\alpha) + V_0''(\alpha)), \; if \; n_1 > 0. \tag{4.8}$$

The price of each seller has two components. First, the price extracts or subsidizes the difference in the current expected value of the alternatives. The second component reflects the intertemporal incentives of the competitor and the individual buyer. The pricing policies (4.7) and (4.8) display an important asymmetry in the influence the value of information has on the pricing policies. If the new firm sells its product, then the experiments generate information in the market. In contrast, if the established firm intends to make a sale, it has to recognize that it reduces the information flow in the market, which is reflected in the sales prices in (4.8).

Since any single experiment provides relevant information not only to the buyer who purchases the new good, but to all buyers, the experiment of an individual buyer generates an externality among all buyers. This effect is not properly reflected in the equilibrium prices. If, as was previously

---

[9]This requirement captures the logic behind trembling hand perfection in this infinite lime horizon framework. Notice that prices p_j, at which sales, $n_j > 0$, occur always satisfy the cautious property. Without cautiousness, any switching point between $\hat{\alpha}$ and the cautious equilibrium switching point α^* can be supported.

argued, the value of information, $U''(\alpha)$, is negative to the buyers, then the equilibrium price overstates the value of an experiment. In fact, if the new seller were to absorb the cost of the negative externality imposed on all buyers by a single experiment, then the price would have to be

$$p_0(\alpha) = \mu(\alpha) - s + \frac{1}{2}\Sigma(\alpha)(NU''(\alpha) + V_1''(\alpha)). \qquad (4.9)$$

The difference between (4.7) and (4.9) indicates the divergence between the market price and the social price of the experiment, which increases in the number of buyers.

Using (4.7) and (4.8), the sellers' optimality conditions can be written as

$$rV_0(\alpha) = N \max\left\{\mu(\alpha) - s + \frac{1}{2}\Sigma(\alpha)(V_0''(\alpha) + V_1''(\alpha) + U''(\alpha)), 0\right\},$$
$$(4.10)$$

and

$$rV_1(\alpha) = N \max\left\{\frac{1}{2}\Sigma(\alpha)(V_1''(\alpha),\right.$$
$$\left. s - \mu(\alpha) - \frac{1}{2}\Sigma(\alpha)(V_0''(\alpha) + U''(\alpha))\right\}. \qquad (4.11)$$

It is easily verified that the two value functions represent the same stopping problem. The indifference condition (4.4) and cautious pricing reduce the two-dimensional control problem in $p_0(\alpha)$ and $p_1(\alpha)$ into a one-dimensional stopping problem in α which can be stated as: How long can the entrant afford a pricing policy that captures the entire market? The value function $V_0(\alpha)$ indicates that the extent of experimentation depends on the benefits to the sellers and the costs to the buyers. The equilibrium stopping point is obtained by continuity conditions on the value functions of all players and a smoothness condition associated with the stopping problem of the new seller.[10]

Theorem 2 (Equilibrium Stopping): *There is a unique MPE in cautious strategies. The equilibrium path displays excessive experimentation.*

---

[10]The explicit derivation of the value functions is presented in Lemma 1 in the Appendix.

The stopping point is given by

$$\alpha^* = \frac{(s - \mu_L)(\gamma N - \lambda(N-1) - 1)}{(\mu_H - \mu_L)(\gamma N - \lambda(N-1) - 1) + 2(\mu_H - s)} < \hat{\alpha},$$

with $\lambda = \sqrt{1 + \rho/(N-1)}$, *and* γ *and* ρ *as in Theorem 1.*

Proof: See Appendix.   □

The threshold α^* at which stopping occurs is decreasing in the number of buyers. An increase in the market size N increases the signal to noise ratio of the outcome when all buyers experiment and thus increases the return from experimentation. For $N = 1$, the equilibrium coincides with efficient stopping: $\alpha^* = \hat{\alpha}$, as shown in Bergemann and Välimäki (1996). The equilibrium price policies $p_j(\alpha)$ and the curvatures of the value functions $V_j(\alpha)$ and $U(\alpha)$ follow directly from the solution of the equilibrium stopping problem.

Corollary 1 (Submartingale and Convexity):

1. *The pricing policies $p_j(\alpha)$ are submartingales.*
2. *The value functions of the sellers are convex.*
3. *The value functions of the buyers are concave.*

The submartingale characterization of the prices illustrates the dilemma facing the buyers. As the expected quality of each product follows a martingale, the submartingale prices imply that buyers expect decreasing net utilities over time. In fact, the instantaneous utilities, $du_i(\alpha) = \mu(\alpha) - p_0(\alpha)$, of the buyers form a strict supermartingale in the continuation region. As the established seller is indifferent between selling and not selling in the experimental phase, we have

$$p_1(\alpha) = \frac{1}{2}\Sigma(\alpha)V_1''(\alpha),$$

and in consequence

$$p_1(\alpha) = \frac{r}{N}V_1(\alpha).$$

The foregone revenue at any instant of time must be equal to the expected increase in discounted future revenue, or

$$p_1(\alpha)dt = \frac{\mathbb{E}[dp_1(\alpha)]}{r} \Leftrightarrow \mathbb{E}[dp_1(\alpha)] = rp_1(\alpha)dt,$$

from which it follows that the price of the incumbent has a positive drift.

The selling price $p_0(\alpha)$ has two interesting features. It is negative between α^* and some α with $\mu(\alpha) - s < 0$. The negative price compensates the buyers for their purchases of the (myopically) lower quality product. But $p_0(\alpha)$ is not monotone increasing in α as one might have expected. As α approaches α^* from the right, the value function of the incumbent firm increases as the likelihood of stopping increases and the expected time to stopping (conditional on eventual stopping) decreases. Through cautious pricing, this leads to higher prices posted by the incumbent. The competitive pressure on the entrant is thus relieved and he can charge higher prices. Thus a segment of decreasing prices (as a function of α) is observed. Less aggressive prices by the incumbent allow the entrant to shift the cost of the experiments to the buyers as beliefs become more pessimistic.

The curvatures of the value functions reflect the attitudes towards information. As the posterior belief α approaches the stopping point α^*, the value of information for the entrant declines and at α^*, we observe $V_0''(\alpha^*) = 0$ and by implication $p_0(\alpha)$ converges to 0 from the right. For posterior beliefs close to α^*, stopping becomes almost certain and the buyers become less averse to experimentation, and at α^*, we find $U''(\alpha^*) = 0$. A typical pair of equilibrium price policies $p_j(\alpha)$ and flow utilities $du(\alpha) - p_j(\alpha)dt$ as a function of α are presented in Figure 3.1.

4.2. *Infinitely many buyers*

The previous analysis suggests that the inefficiency increases as the size of a single buyer becomes small relative to the size of the market. The

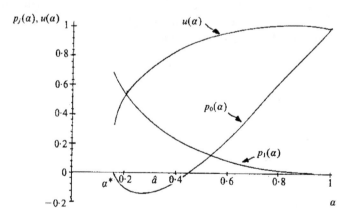

Figure 3.1.   Prices $p_j(\alpha)$ and flow utilities $u(\alpha)$ in a single market: $\mu_H = 2$, $\mu_L = 0$, $s = 1$, $N = 2$, $\sigma = 4$, $r = \frac{1}{2}$.

experiments should then extend to threshold levels $\alpha^*(N)$ which decline in the number N of buyers. To make this intuition precise, we have to be careful when changing N. If we were to simply increase the number of buyers, we would also change the informativeness of the aggregate outcome in the continuation region. In fact, as the number of buyers increases, the instantaneous aggregate outcome would become completely informative by the law of large numbers, and even the efficient stopping point $\hat{\alpha}$ would converge to 0.

It is therefore necessary to keep the informativeness of the aggregate outcome independent of the number of buyers if we want to analyze the effect of making the buyers small without eliminating the aggregate uncertainty. This can be accomplished by decreasing the size of the individual experiment as we increase the total number of experiments. By normalizing the flow utility from the experiment to

$$du_i(t) = \frac{\mu}{N}dt + \frac{\sigma}{\sqrt{N}}dW_i(t),$$

the size of the aggregate experiment remains unchanged as we increase N. Intuitively, in a market with fixed aggregate variance σ^2, the effect of increasing N then reflects solely the increase in the externality between the buyers. Note in particular that the socially-efficient allocation policy is independent of N.

The inefficiency due to the free-riding aspect increases as each buyer has an ever smaller influence on the release of information. In consequence, the entrant has to compensate each individual buyer less and less for the participation in the experiment. It then becomes less costly for the entrant to finance the experiments and the stopping point $\alpha^*(N)$ decreases as a function of N. The equilibrium stopping point $\alpha^*(N)$ is as in Theorem 2 given by

$$\alpha^*(N) = \frac{(s - \mu_L)(\tilde{\gamma}N - \tilde{\lambda}(N-1) - 1)}{(\mu_H - \mu_L)(\tilde{\gamma}N - \tilde{\lambda}(N-1) - 1) + 2(\mu_H - s)}, \tag{4.12}$$

where the only changes are introduced through the parameters $\tilde{\gamma}$ and $\tilde{\lambda}$ which reflect the normalization of the market size, with

$$\tilde{\gamma} = \sqrt{1 + \rho}, \tag{4.13}$$

and

$$\tilde{\lambda} = \sqrt{1 + \rho\frac{N}{N-1}}. \tag{4.14}$$

The parameter $\tilde{\gamma}$ is now independent of N as the aggregate market size is constant and equal to 1. The parameter $\tilde{\lambda}$ represents the share of the variance in the posterior belief for which the individual consumer is not compensated, namely $(N-1)/N$. The social value of the game is decreasing in N due to the excessive information acquisition. The monotonicity of the allocation in N is associated with the monotonicity of the value functions of the agents. As N increases, the seller of the unknown product receives a successively larger share of the social surplus to the detriment of the known seller and the aggregate value of the buyers. The comparative static results follow directly from the equilibrium value functions.

Corollary 2 (Convergence and Monotonicity):

1. *The equilibrium stopping point $\alpha^*(N)$ is strictly decreasing in N and converges to the stopping point with infinitely many buyers*

$$\alpha_\infty^* = \frac{(s - \mu_L)(\tilde{\gamma} - 1)^2}{(\mu_H - \mu_L)(\tilde{\gamma} - 1)^2 + 4\tilde{\gamma}(\mu_H - s)}.$$

2. *The value function $V_0(\cdot)$ is strictly increasing in N for all $\alpha > \alpha^*(N)$.*
3. *The value functions $NU(\cdot)$ and $V_1(\cdot)$ are strictly decreasing in N for all $\alpha > \alpha^*(N)$.*

Proof: See Appendix. $\square$

The extent of the inefficiency as N increases is depicted in Figure 3.2, which shows the price path of $p_0(\alpha)$ as N varies. As N increases the new firm succeeds in shifting the costs of the experiments to the buyers. Moreover the plateau with $p_0(\alpha) = 0$ recedes as N increases.

In the limit as $N \to \infty$, any single experiment carries no information about the value of the uncertain alternative. The single buyer is now infinitesimally small relative to the market and her purchase decision has no impact on the informational content of the market outcome. Consequently, neither the individual buyer nor the sellers attach any strategic importance to her current decision. In the limit the representative buyer then behaves as if she were completely myopic. The indifference condition of each buyer simplifies to

$$\mu(\alpha) - p_0 = s - p_1, \tag{4.15}$$

as the size of the individual experiment is infinitesimal. We can therefore strengthen Corollary 1 and obtain symmetry in the expected rates of change of the prices quoted by the sellers.

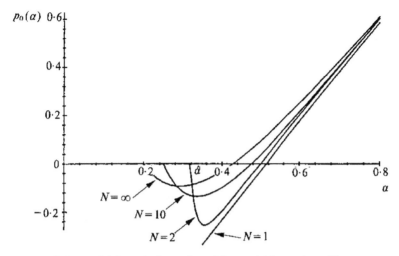

Figure 3.2.   Prices $p_0(\alpha)$ in a single market with a variable number of buyers: $\mu_H = 2$, $\mu_L = 0$, $s = 1$, $\sigma = 4$, $r = \frac{1}{2}$.

Corollary 3: *In the continuation region, the pricing policies of both firms are submartingales with identical and positive drift*

$$\mathbb{E}[dp_j(\alpha)] = rp_1(\alpha)dt, \quad \text{for } j = 0, 1.$$

Interestingly, the model with a continuum of buyers can be reinterpreted as one with a constant inflow of buyers that make a once and for all purchase of a durable good. With perfect durables, it is natural that the experiences of past buyers have a large impact on current buyers' beliefs on product quality. In the absence of repeat purchases, it is also incentive compatible for all buyers to report their experiences truthfully *even* in the case of private observability. The only new constraint that appears in the durable goods case is that we need to make sure that no buyer wants to delay her purchase. As prices are always below the expected value of the product and the expected change in prices is nonnegative, waiting entails a loss to the buyers in all periods. Notice also that it is never in the firms' interest to delay sales. Hence the equilibrium as described above yields an equilibrium in this durable goods model as well. Instead of having the same buyers switch from one product to the other, different buyers choose different products in the durable goods interpretation. As each buyer purchases only once, the informational effects on future prices are not internalized.

5. Many Markets

In the single market model, the sellers internalize the effect from all information flows in the model while the buyers are not able to do this. In this section, we consider the case where the informational externality affects both sellers and buyers.

We start by considering M distinct markets, each having a separate incumbent offering an established product of value s to the buyers in his market. The M markets may be thought of as local markets and the entrant as a global competitor, who can introduce his new product into all local markets. An alternative interpretation of this model would be that the incumbents are using a technology with a capacity constraint while the entrant has a technology with unlimited capacity available. The true value of the new product, $\mu \in \{\mu_L, \mu_H\}$, is the same across all markets. The main difference compared to the previous sections is that there is now an informational externality across the markets in addition to the one within the markets.

We assume that the buyers in all markets are identical and that each incumbent is serving the same number of buyers, N/M with $N \geq M$. We normalize the size of the sum of all markets to 1 and as a consequence, each incumbent sells to a market of size $1/M$.[11] The main result in this section is a characterization of the equilibrium stopping for different market structures in terms of M and N. We consider only the symmetric equilibria of the model. The representation of the equilibrium conditions is similar to the previous section and we focus on the differences. By symmetry, it is sufficient to consider a representative incumbent and a representative buyer.

Denote the representative incumbent by j and the entrant by 0. The equilibrium in this model can again be characterized as an optimal stopping problem. The indifference condition of the buyer is, as before

$$s - p_j(\alpha) = \mu(\alpha) - p_0(\alpha) + \frac{1}{2}\Sigma(\alpha)U''(\alpha). \tag{5.1}$$

The optimality condition of the incumbent in the continuation region reflects the trade-off between sales and information

$$rV_j(\alpha) = \max\left\{\frac{1}{M}p_j(\alpha) + \frac{1}{2}\frac{M-1}{M}\Sigma(\alpha)V_j''(\alpha), \frac{1}{2}\Sigma(\alpha)V_j''(\alpha)\right\}. \tag{5.2}$$

---

[11]This normalization allows us to keep the efficient stopping point constant and independent of the size of the individual buyers and sellers. However, the convergence results to be presented are independent of the normalization.

The cautious price per unit of sale is again given by setting the price equal to the value of information generated by a sale of the new product

$$p_j(\alpha) = \frac{1}{2}\Sigma(\alpha)V_j''(\alpha).$$

Despite the similarity to the single-market behavior, note how M appears in the decision of the individual incumbent in the Bellman equation (5.2). The trade-off between sales and information is now relaxed by the fact that even if the incumbent makes a sale in his local market, information may still be generated in the remaining markets. This leads the incumbent to adopt a more aggressive pricing policy which in turn makes it more expensive for the entrant to sell in that particular local market.

The failure of the individual incumbent to take the value of information to the other local incumbents into account leads to less experimentation in equilibrium. This is perhaps most clearly reflected in the equilibrium stopping problem for the entrant. After inserting (5.1) and (5.2) into the value function of the entrant, we obtain

$$rV_0(\alpha) = \max\left\{\mu(\alpha) - s + \frac{1}{2}\Sigma(\alpha)(V_0''(\alpha) + V_j''(\alpha) + U''(\alpha)), 0\right\}. \quad (5.3)$$

In contrast, the stopping problem which reflects the social value of all the players and hence would lead to the efficient stopping point is given by

$$\max\left\{\mu(\alpha) - s + \frac{1}{2}\Sigma(\alpha)(V_0''(\alpha) + MV_j''(\alpha) + NU''(\alpha)), 0\right\}.$$

The symmetry in the way that the externalities among buyers and incumbents effect the equilibrium stopping is now apparent. The qualitative difference is that these externalities work in opposite directions in terms of determining α as $V_j''(\alpha) > 0$ and $U''(\alpha) < 0$. This symmetry is also evident in the solution to the equilibrium stopping problem in (5.3).

Theorem 3 (Many Market Equilibrium):

1. *The equilibrium stopping point in the unique cautious equilibrium is*

$$\alpha^*(M,N) = \frac{(N\tilde{\gamma} - \tilde{\lambda}(N-M) - M)(s - \mu_L)}{(N\tilde{\gamma} - \tilde{\lambda}(N-M) - M)(\mu_H - \mu_L) + 2M(\mu_H - s)},$$

   with $\tilde{\gamma}$ and $\tilde{\lambda}$ as in (4.13) and (4.14).
2. *If $M = N$, the equilibrium stopping is efficient: $\alpha^*(M,N) = \hat{\alpha}$.*

3. *The stopping point* $\alpha^*(M, N)$ *is decreasing in* N *and increasing in* M; *and*

$$\lim_{M,N\to\infty} \alpha^*(M, N) = \tilde{\alpha}.$$

Proof: See Appendix.   □

The case of $M = N$ has a few interesting properties. If there is a single buyer in each market, then the local incumbent and buyer can trade intertemporal payoffs on a one-to-one basis. They both have control of $1/M$ of the total market. Hence the future gains from experiments for the incumbent equal exactly the future losses of the buyer. Cautious pricing then allows the new firm to make sales at the myopic quality differential $\mu(\alpha) - s$.

As the number of markets M increases, the strategic aspect in the acceptance and pricing policies vanishes. In consequence, the sales prices in the experimental phase converge to the myopic quality differential

$$\lim_{M,N\to\infty} p_0(\alpha) = \mu(\alpha) - s, \quad \lim_{M,N\to\infty} p_j(\alpha) = 0, \quad \text{for all } \alpha > \alpha^*(M, N),$$

and α^* (M, N) converges to the efficient stopping point. The convergence result holds also if instead of the single large entrant, each market had a small entrant with access to the same new product as all other local entrants. This may represent a situation where local franchisees share a common franchising product.

With many markets, the prices at the switching point display a discontinuity which is linked to the externality among the incumbents. For $\alpha > \alpha^*(M, N)$, the value $V_j(\alpha)$ of each incumbent is given by the value of information generated in all markets, or

$$rV_j(\alpha) = \frac{1}{2}\Sigma(\alpha)V_j''(\alpha), \quad \text{for } \alpha > \alpha^*.$$

After the switch the value is determined by sales in the local market, or

$$rV_j(\alpha^*) = \frac{1}{M}p_j(\alpha^*).$$

We can immediately infer the size of the jump in prices at the switching point by the continuity of the value functions. We denote the limiting point from the right by

$$p_j^+(\alpha^*) \triangleq \lim_{\alpha\downarrow\alpha^*} p_j(\alpha).$$

Corollary 4: *The equilibrium prices at the stopping point display an upward jump for $M > 1$*

$$Mp_j^+(\alpha^*) = p_j(\alpha^*) = s - \mu(\alpha^*),$$

and

$$p_0^+(\alpha^*) = \frac{N(M-1)}{(N-1)M}(\mu(\alpha^*) - s) < 0 = p_0(\alpha^*).$$

The size of the jump is positively related to the number of markets M and disappears for $M = 1$ and $N > 1$. The discontinuity of the prices implies that the flow utility of the buyers decreases discontinuously when the new firm stops selling in the markets.

6. Conclusion

This paper shows that the conventional wisdom that informational externalities lead to inefficiently low levels of experimentation may be reversed in a two-sided learning model. The introduction of sellers into the multi-agent learning model creates a market where experiments are priced. The new seller *sponsors* the uncertain alternative and rewards buyers for experiments through low prices. In contrast to one-sided learning models, the seller provides direct incentives for the buyers to experiment. The ownership of the product allows the seller to extract the future benefits of current experimentation which would have evaporated without the assignment of property rights.

The main theme of the paper is the importance of the market structure for efficiency conclusions in a model of informational externalities. To abstract from other forms of distortions we analyze a stage game without static distortions. The cost of this modeling choice is that it is very difficult to generate genuine market sharing between firms. In an earlier working paper (Bergemann and Välimäki (1998)), we considered a variation of the current model where buyers in two markets differ in terms of the volume of their purchases. For simplicity, the first market segment consists of a continuum of identical small buyers and the second is formed by a single large buyer. Buyers in both segments have the same flow valuation for the products and the sellers can price discriminate between the segments. In contrast to the small buyers, the large buyer controls a sizable part of the current information flow and thus internalizes the impact of her current purchases on her future utilities. The willingness to pay for the new good is then different across the market segments. As a result, there is an intermediate range of the values of α where the new firm sells to the small buyers in equilibrium while the established firm caters to the large buyer.

When interpreting the basic model as a durable goods model, an interesting question arises. Since the buyers have a strict preference to purchase the product at the moment of their arrival rather than wait for a period, it might be useful to analyze a model where the entry of new buyers is endogenous. For example, one could imagine a model with a constant population of buyers that own a physically depreciating durable good. The possibility of early purchases at an endogenously determined cost might have important implications on the shape of the price process and the speed of information transmission.

Appendix

The appendix contains the proofs of all propositions and theorems presented in the main body of the paper.

Proof of Theorem 1.

We describe the solution procedure for the inhomogeneous second-order differential equation in some detail as it reappears in the construction of the various equilibrium value functions. All solutions of the inhomogeneous equation (3.3) permit the following representation

$$V(\alpha) = c_1 H_1(\alpha) + c_2 H_2(\alpha) + \psi(\alpha),$$

where $H_1(\alpha)$ and $H_2(\alpha)$ are two linearly independent solutions of the corresponding homogeneous equation and $\psi(\alpha)$ is a particular solution of the inhomogeneous equation. The complete solution to (3.3) is established by the variation of parameters method, see Chapter 2 in Birkhoff and Rota (1978). A particular solution to the inhomogeneous differential equation is given by

$$\psi(\alpha) = u_1(\alpha)H_1(\alpha) + u_2(\alpha)H_2(\alpha),$$

where the parameters $u_1(\alpha)$ and $u_2(\alpha)$ are determined by

$$u_1'(\alpha) = -\frac{G(\alpha)H_2(\alpha)}{W(H_1(\alpha),\, H_2(\alpha))},$$

and

$$u_2'(\alpha) = -\frac{G(\alpha)H_1(\alpha)}{W(H_1(\alpha),\, H_2(\alpha))},$$

and $G(\alpha)$ is the forcing term of the inhomogeneous differential equation. The Wronskian determinant, $W(H_1(\alpha), H_2(\alpha))$, is given by

$$W(H_1(\alpha), H_2(\alpha)) = H_1(\alpha)H_2'(\alpha) - H_1'(\alpha)H_2(\alpha).$$

The solution to the homogeneous version of the differential equation (3.3) is

$$H(\alpha) = b_1\alpha^{(\gamma+1)/2}(1-\alpha)^{(\gamma-1)/2} + b_2\alpha^{-(\gamma-1)/2}(1-\alpha)^{(\gamma+1)/2}, \quad (A.1)$$

with $\gamma = \sqrt{1+\rho/N}$. The general solution for the value function $V(\alpha)$ is

$$V(\alpha) = \frac{N\mu(\alpha)}{r} + b_1\alpha^{(\gamma+1)/2}(1-\alpha)^{-(\gamma-1)/2} + b_2\alpha^{-(\gamma-1)/2}(1-\alpha)^{(\gamma+1)/2}.$$

The efficient switching point $\hat\alpha$ is obtained by requiring that the value matching and the smooth pasting conditions are satisfied[12]:

$$rV(\hat\alpha) = Ns,$$
$$V'(\hat\alpha) = 0. \quad (A.2)$$

Boundedness of the value function as $\alpha \to 1$ implies $b_1 = 0$. The conditions in (A.2) determine $\hat\alpha$ and b_2

$$\hat\alpha = \frac{(s-\mu_L)(\gamma-1)}{2(\mu_H - s) + (\mu_H - \mu_L)(\gamma-1)},$$

$$b_2 = \frac{2N(\mu_H - s)}{(\gamma-1)r}\left(\frac{\hat\alpha}{1-\hat\alpha}\right)^{(\gamma+1)/2}.$$

In particular $b_2 > 0$ implies the convexity of the value function. Further computation shows that the efficient cut-off $\hat\alpha$ is strictly increasing in ρ/N and $V(\cdot)$ is strictly decreasing in ρ/N.

To characterize the equilibrium stopping point, we need the solutions to the differential equations describing the value functions, which is obtained after inserting the equilibrium prices (4.7)–(4.8) in the value functions.

---

[12]See Shiryayev (1978), Chapter 3, for a formal statement of this principle.

Lemma 1 (Value functions): *The value functions in the experimentation region display the general form*

$$V_0(\alpha) = \frac{N}{r}(\mu(\alpha)-s)-cN\alpha^{-(\lambda-1)/2}(1-\alpha)^{(\lambda+1)/2}+d\alpha^{-(\gamma-1)/2}(1-\alpha)^{(\gamma+1)/2},$$

$$V_1(\alpha) = b\alpha^{-(\gamma-1)/2}(1-\alpha)^{(\gamma+1)/2},$$

and

$$U(\alpha) = \frac{s}{r} + c\alpha^{-(\lambda-1)/2}(1-\alpha)^{(\lambda+1)/2} - b\alpha^{-(\gamma-1)/2}(1-\alpha)^{(\gamma+1)/2},$$

with $\lambda = \sqrt{1+\rho/(N-1)}$, and γ as in Theorem 1.

Proof: The construction of the value functions proceeds from the homogenous part of the differential equation (4.11) of the incumbent to the inhomogeneous differential equation of the individual buyer (4.3) and the entrant (4.10). The solution to (4.11) is as in (A.1), where the second term vanishes as $V_1(\alpha) \to 0$ as $\alpha \to 1$

$$V_1(\alpha) = b\alpha^{(\gamma-1)/2}(1 - \alpha)^{(\gamma+1)/2}.$$

The general solution to the homogeneous version $U_h(\alpha)$

$$rU_h(\alpha) = \frac{1}{2}(N - 1)\Sigma(\alpha)U_h''(\alpha),$$

of the differential equation $U(\alpha)$ is

$$U_h(\alpha) = c_1\alpha^{(\lambda+1)/2}(1 - \alpha)^{-(\lambda-1)/2} + c_2\alpha^{-(\lambda-1)/2}(1 - \alpha)^{(\lambda+1)/2}.$$

A particular solution of the equation (4.3) is

$$\psi(\alpha) = \frac{s}{r} - b\alpha^{(\gamma-1)/2}(1 - \alpha)^{(\gamma+1)/2}.$$

The complete solution to the inhomogeneous equation (4.3) is

$$U(\alpha) = \frac{s}{r} + c\alpha^{-(\lambda-1)/2}(1 - \alpha)^{(\lambda+1)/2} - b\alpha^{(\gamma-1)/2}(1 - \alpha)^{(\gamma+1)/2}.$$

As before we set $c_1 = 0$, since the value function of the buyer has to be bounded as α goes to one. For simplicity set $c_2 = c$. The final step is to construct $V_0(\alpha)$. A particular solution to the inhomogeneous equation is

$$\psi(\alpha) = \frac{N}{r}(\mu(\alpha) - s) - cN\alpha^{-(\lambda-1)/2}(1 - \alpha)^{(\lambda+1)/2}. \qquad (A.3)$$

The fundamental solutions together with the particular solution $\psi(\alpha)$ gives us the format of all possible solutions of the value function $V_0(\alpha)$. Boundedness of the value function forces one term of the homogenous solution to vanish

$$V_0(\alpha) = \frac{N}{r}(\mu(\alpha)-s)+d\alpha^{-(\gamma-1)/2}(1-\alpha)^{(\gamma+1)/2}-cN\alpha^{-(\lambda-1)/2}(1-\alpha)^{(\lambda+1)/2},$$

which completes the construction of the value functions. □

Proof of Theorem 2.

The existence of an equilibrium is established by construction. Given the guess on the shape of the continuation and stopping regions, we construct the value functions. Following the derivation of the value functions, it is then verified that the initial guess is satisfied.

The equilibrium conditions for the stopping point α^* is the smooth-pasting condition of the optimal stopping problem for the new seller and the continuity of the value functions for the sellers and the buyers at the stopping point α^*

$$V_0(\alpha^*) = 0,$$
$$V_0'(\alpha^*) = 0,$$
$$rV_1(\alpha^*) = N(s - \mu(\alpha^*)),$$
$$rU(\alpha^*) = \mu(\alpha^*). \qquad (A.4)$$

The conditions in (A.4) yield the stopping point α^* and the values of the parameters b,c, and d, which determine the curvature of the value functions

$$\alpha^* = \frac{(s - \mu_L)(\gamma N - \lambda(N - 1) - 1)}{(\mu_H - \mu_L)(\gamma N - \lambda(N - 1) - 1) + 2(\mu_H - s)} < \hat{\alpha},$$

and

$$b = \frac{2N}{(N\gamma - (N - 1)\lambda - 1)}\frac{(\mu_H - s)}{r}\left(\frac{\alpha^*}{1 - \alpha^*}\right)^{(\gamma+1)/2},$$

$$c = \frac{2(N - 1)}{(N\gamma - (N - 1)\lambda - 1)}\frac{(\mu_H - s)}{r}\left(\frac{\alpha^*}{1 - \alpha^*}\right)^{(\gamma+1)/2},$$

$$d = \frac{2N^2}{(N\gamma - (N-1)\lambda - 1)} \frac{(\mu_H - s)}{r} \left(\frac{\alpha^*}{1 - \alpha^*} \right)^{(\gamma+1)/2}. \qquad (A.5)$$

To show uniqueness, it is sufficient to prove that no other shape of the stopping region is possible. To see this, let $C \subset [0,1]$ denote the continuation region and $S \subset [0,1]$ the stopping region of an arbitrary equilibrium. By cautiousness and Markovian strategies, it is immediate that sales are made at all $\alpha \in [0,1]$, *i.e.* $S \cup C = [0,1]$.

We need to show that there is an $\alpha^* \in [0,1]$ such that $S = [0, \alpha^*]$, and $C - (\alpha^*, 1]$. First note that $p_1(\alpha) \geq 0$ for all α and therefore, $\alpha \in S \Rightarrow \alpha \leq \hat{\alpha}$. By cautiousness, $p_0(\alpha) = 0$ for all $\alpha \in S$. By the continuity of the value functions in α we may take S to be closed. Therefore C is a union of pairwise disjoint (relatively) open intervals. We need to show that if $(\alpha_1, \alpha_2) \subset C$, then $\alpha_2 > \hat{\alpha}$. But this follows immediately from cautiousness as $p_0(\alpha_1) = p_0(\alpha_2) = 0$ and there must be an $\alpha \in (\alpha_1, \alpha_2)$ with $p_0(\alpha) \geq 0$. This yields the desired conclusion.

Proof of Corollary 2.

The equilibrium value functions are as in Theorem 2, with the exception of the normalization as transparent in the stopping point $\alpha^*(N)$ through the parameters $\hat{\gamma}$ and $\hat{\lambda}$. The parameters b, c, d are as in (A.5) only to be divided through N and inserting $\hat{\gamma}$ and $\hat{\lambda}$. The comparative statics follow after straightforward algebra.

Proof of Theorem 3.

(1) The value functions are derived as in Lemma 1 with the obvious modifications due to the normalization of the size of the market. The solutions to the differential equations are given by

$$V_0(\alpha) = \frac{1}{r}(\mu(\alpha) - s) + d\alpha^{-(\gamma-1)/2}(1-\alpha)^{(\gamma+1)/2}$$
$$- cN\alpha^{-(\lambda-1)/2}(1-\alpha)^{(\lambda+1)/2},$$
$$V_j(\alpha) = b\alpha^{-(\gamma-1)/2}(1-\alpha)^{(\gamma+1)/2},$$

and

$$U(\alpha) = \frac{s}{rN} + c\alpha^{-(\lambda-1)/2}(1-\alpha)^{(\lambda+1)/2} - b\alpha^{-(\gamma-1)/2}(1-\alpha)^{(\gamma+1)/2},$$

with $\gamma = \sqrt{1+\rho}$ and $\lambda = \sqrt{1 + \rho N/(N-1)}$. The equilibrium stopping conditions are modified only by the normalization in the market size to

$$V_0(\alpha^*) = 0,$$
$$V_0'(\alpha^*) = 0,$$
$$rMV_j(\alpha^*) = (s - \mu(\alpha^*)),$$
$$rNU(\alpha*) = \mu(\alpha^*). \tag{A.6}$$

The stopping point is given as a solution to the equations in (A.6)

$$\alpha^* = \frac{(N\gamma - \lambda(N-M) - M)(s - \mu_L)}{(N\gamma - \lambda(N-M) - M)(\mu_H - \mu_L) + 2M(\mu_H - s)},$$

and the parameters of the value functions are given by

$$b = \frac{2(\mu_H - s)}{(N\gamma - \lambda(N-M) - M)r} \left(\frac{\alpha^*}{1 + \alpha^*} \right)^{(\gamma+1)/2},$$

$$c = \frac{2(N-M)(\mu_H - s)}{(N\gamma - \lambda(N-M) - M)Nr} \left(\frac{\alpha^*}{1 - \alpha^*} \right)^{(\lambda+1)/2},$$

$$d = \frac{2N(\mu_H - s)}{(N\gamma - \lambda(N-M) - M)r} \left(\frac{\alpha^*}{1 - \alpha^*} \right)^{(\gamma+1)/2}.$$

Uniqueness is proved as in Theorem 2. (2.2) and (2.3) follow directly. ‖

Acknowledgements

The authors express their gratitude to Drew Fudenberg, Bengt Holmström, Al Klevorick, George Mailath, Eric Maskin, Georg Nöldeke, Andy Postlewaite, Mike Riordan, Larry Samuelson, Karl Schlag and Avner Shaked for many helpful discussions. Detailed comments by two anonymous referees and the editor, Patrick Bolton, greatly improved the paper. We benefited from seminar participants at Bonn, Harvard, Mannheim, MIT, Northwestern and Paris. The first author would like to thank Avner Shaked for his hospitality during a stay at the SFB 303 at Bonn University. The authors acknowledge support from NSF Grant SBR 9709887 and 9709340 respectively.

References

BANERJEE, A. and FUDENBERG, D. (1997), "A Simple Model or Word-of-Mouth Learning" (Mimeo).

BERGEMANN, D. and VÄLIMÄKI, J. (1996), "Learning and Strategic Pricing", *Econometrica*, **64**, 1125–1149.

BERGEMANN, D. and VÄLIMÄKI, J. (1997), "Market Diffusion with Two-Sided Learning", *Rand Journal of Economics*, **28**, 773–795.

BERGEMANN, D. and VÄLIMÄKI, J. (1998), "Experimentation in Markets" (Northwestern University CMSEMS DP. 1220).

BIRKHOFF, G. and ROTA, G. C. (1978), *Ordinary Differential Equations*. 3rd ed. (New York: Wiley).

BOLTON, P. and HARRIS, C. (1999), "Strategic Experimentation", *Econometrica*, **67**, 349–374.

CHAMLEY, C. and GALE, D. (1994), "Information Revelation and Strategic Delay in a Model of Investment", *Econometrica*, **62**, 1065–1086.

DIXIT, A. and PINDYCK, R. (1994) *Investment under Uncertainty* (Princeton: Princeton University Press).

FELLI, L. and HARRIS, C. (1996), "Job Matching, Learning and Firm-Specific Human Capital", *Journal of Political Economy*, **104**, 838–868.

HARRISON, J. (1985), *Brownian Motion and Stochastic Flow Systems* (Malabar: Krieger).

HENDRICKS, K. and KOVENOCK, D. (1989), "Asymmetric Information, Information Externalities, and Efficiency: the Case of Oil Exploration", *Rand Journal of Economics*, **20**, 164–182.

KARATZAS, I. (1984), "Gittins Indices in the Dynamic Allocation Problem for Diffusion Processes", *Annals of Probability*, **12**, 173–192.

KELLER, G. and RADY, S. (1999), "Optimal Experimentation in a Changing Environment", *Review of Economic Studies*, **66**, 475–509.

LIPTSER, R. S. and SHIRYAYEV, A. N. (1977) *Statistics of Random Process I* (New York: Springer-Verlag).

McFADDEN, D. and TRAIN, K. (1996), "Consumers' Evaluation of New Products: Learning from Self and Others", *Journal of Political Economy*, **104**, 683–703.

ROB, R. (1991), "Learning and Capacity Expansion under Demand Uncertainty", *Review of Economic Studies*, **58**, 655–675.

ROTHSCHILD, M. (1974), "A Two-Armed Bandit Theory of Market Pricing", *Journal of Economic Theory*, **9**, 185–202.

SHIRYAYEV, A. N. (1978) *Optimal Stopping Rules* (Berlin: Springer-Verlag).

VETTAS, N. (1998), "Demand and Supply in New Markets: Diffusion with Bilateral Learning", *Rand Journal of Economics*, **29**, 215–233.

Chapter 4

Market Diffusion with Two-Sided Learning*

Dirk Bergemann[†] and Juuso Välimäki[‡]

[†]Department of Economics, and Cowles Foundation of Research in Economics,
Yale University, New Haven, CT 06520-8268, USA
dirk.bergemann@yale.edu
[‡]Department of Economics, Aalto University School of Business,
02150 Espoo, Finland
juuso.valimaki@aalto.fi

We analyze the diffusion of a new product of uncertain value in a duopolistic market. Both sides of the market, buyers and sellers, learn the true value of the new product from experiments with it. Buyers have heterogeneous preferences over the products and sellers compete in prices. The pricing policies and market shares in the unique Markov-perfect equilibrium are obtained explicitly. The dynamics of the equilibrium market shares display excessive sales of the new product relative to the social optimum in early stages and too-low sales later on. The diffusion path of a successful product is S-shaped.

1. Introduction

The design of a new product or the improvement of an existing product is only the necessary first step in launching a product in the market. Commercial success depends critically on the speed and cost at which buyers learn the relevant characteristics of the product. An eventually

The authors would like to thank Colin Campbell, Ron Goettler, Phillip Leslie, Ariel Pakes, Jim Peck, Mike Riordan, and seminar participants at the Institut d'Anàlisi Econòmica, Barcelona, Ohio State University, and the University of Wisconsin for many helpful comments. We are grateful to three anonvmous referees
*This chapter is reproduced from RAND Journal of Economics, 28, 773–795, 1997.

successful product may go through a phase of sluggish sales in the beginning of its life cycle simply because buyers are not aware of its true quality. It may then be in the firm's interest to engage in strategies that sacrifice current revenue to generate more information about the product, for example through aggressive penetration pricing, to capture a larger clientele.

In this article we model dynamic competition in a duopolistic market for experience goods. An established firm and a firm with a new product compete in prices in an infinite-horizon, continuous-time model. Buyers have to try the new product to learn how well it suits their needs. We assume that the product incorporates both a common and a private-value component to the buyers. To keep the model analytically tractable, we assume that the private-value component of every buyer is common knowledge and may reflect idiosyncratic taste, location, or the like. In contrast, the common component is learned gradually over time as more experience is accumulated. The information obtained in any single trial with the new product is a noisy signal of product quality or more generally of the utility it provides to the consumer. The total available information on the common-value component then depends on the cumulative sales. We assume that a statistic on the aggregate performance of the new product (obtained via either consumer reports or an unmodeled process of word-of-mouth communication) is available to all parties in the model. This introduces an informational externality into the model. Each individual purchase has effects on all buyers in this model through changes in the market statistic.

The simplest economic example for our story is a linear city with, for example, two fish markets located at the ends of the city. Consumers are uniformly distributed along the segment between the fish markets and incur a transportation cost linear in distance to the stores. One of the stores has been in operation for a long time, whereas the ownership of the other has recently been changed. Buyers choose the store at the beginning of each period depending on their beliefs on the quality of the fish in the new store, their transportation cost, and current prices. The quality of the storekeeping has to be learned through experience, while the idiosyncratic transportation cost to every buyer is common knowledge at the outset. Every purchase in the new store yields an imperfect signal on the quality (taste, freshness, etc.) of the fish in the new market. Based on the buyers' experiences, the two firms and the consumers update their belief on the value the new fish market provides to the buyers in the city.

We characterize the diffusion path of the new product (or the time path of the clientele of the new seller in the fish market example) and derive the equilibrium price path in the unique Markov-perfect equilibrium. Our continuous time specification allows us to derive analytical solutions to both the price paths of the firms and the associated path of market share evolution. The pricing policies display an interesting asymmetry. Both firms prefer *ex post* differentiation, as is typical in models of price competition, and hence the value of information is positive for both firms. Only sales of the new product, however, produce more information. Hence, if the firms want to speed up the information transmission, the new firm must make relatively large sales in early periods. Since both firms benefit from sales by the new firm, competitive pressures in the stage game are reduced, and uncertain (and vertical) product differentiation relaxes competition in a similar sense as deterministic product differentiation in Shaked and Sutton (1982). In equilibrium, this results in the new firm's collecting higher revenues due to increased market share early on. As buyers and sellers become more convinced about the true quality of the new product, the market shares converge to those of the full-information game. The value of accumulating more information thus has two components for the established seller: Due to product differentiation, he expects to get higher profits in expected terms, and as the information becomes more accurate, the need to distort sales in favor of the new firm diminishes as well.

The asymmetry is also apparent in the intertemporal behavior of the market share of the new firm. In the myopic case, i.e., the case where the stage game is played repeatedly as a one-shot game and all players ignore the informational effects, the expected market share of the new firm is constant. In the Markov-perfect equilibrium, however, the expected market share of the new firm is decreasing (its market share is a supermartingale). High initial sales reflect the value of information to both sellers. Furthermore, the expected revenues of both sellers are increasing over time. Thus firms are sacrificing current profits early on to enhance the accumulation of information.

In equilibrium, the pricing strategies induce sales paths that differ from the socially optimal path. As long as the beliefs about the type of the new product are sufficiently pessimistic, the level of experimentation and hence purchases of the new product exceed socially optimal levels. The intuition behind this result is straightforward: At pessimistic beliefs, the market share of the new firm is small and the cost of attracting prices, is small.

The established firm has a large market share and concedes some market share to protect inframarginal profits. On the other hand, if beliefs about the new product are optimistic, then the old firm prices more aggressively and experimentation falls below the social optimum.

The time path of product adoption, or the diffusion curve of a successful new product, displays the S-shape documented in empirical work such as Mansfield (1968) and Gort and Klepper (1982). At relatively pessimistic beliefs, increases in the number of adopting consumers increase the inflow of new information. As beliefs about the product quality become sufficiently optimistic, the growth in the market share of the new firm eventually slows down.

Our model of dynamic competition involves simultaneous determination of two value functions representing the sellers' intertemporal problems in an infinite-horizon model. The dynamic programming equations under positive discounting, which describe the equilibrium of the model, are, unfortunately, nonlinear differential equations and can only be solved numerically. To avoid these complications we use a technique recently employed by Bolton and Harris (1993), considering the limiting model as discounting in the model becomes small. The limiting model preserves all the desirable features of the original discounted dynamic program, and in particular, the optimal policies in the limiting model are the unique limits of optimal policies with discounting.

There are a number of related articles in which issues of strategic pricing are analyzed in a learning environment. Aghion, Espinosa, and Jullien (1993) and Harrington (1995) consider product-differentiated duopolies in which firms learn about the substitutability of their products. In their models learning is one-sided. More precisely, the firms try to learn the degree of substitutability between their products. The goods are not experience goods, as consumers have perfect information about the products at the outset and it is observed demand conditional on the price differential between the products, rather than cumulative sales of a new product, that generates the learning. Caminal and Vives (1996) analyze a two-period duopoly model where a sequence of short-lived buyers are uncertain about the quality differential. As consumers are restricted to observe only quantities of past sales, the long-lived firms attempt to signal-jam the learning process of the buyers.

Our techniques are similar to those used by Bolton and Harris (1993), who were the first to consider a continuous-time model of strategic experimentation. Bergemann and Välimäki (1996) analyze strategic experimentation in a duopoly with a continuum of identical consumers. The homogeneity excludes market sharing, and the analysis there concentrates on how the informational externalities affect the efficiency of the market. Issues of informational externalities in a market for new products were already present in the competitive-entry and capacity-expansion model of Rob (1991) and its extension to a two-sided learning model by Vettas (forthcoming). Our focus on strategic interaction between the firms results in very different price dynamics and welfare conclusions.

Judd and Riordan (1994) analyze a two-sided learning model in a monopoly with two periods. Their information structure differs from ours, as the consumption of the experience good yields a private signal to the individual buyer and the aggregate signal of sales is a private signal to the monopolist. Finally, Schlee (1996) has analyzed the incentives of a monopolist to engage in introductory pricing of a new experience good in a two-period model where the signals are publicly observed.

The model and the Bayesian learning process are introduced in Section 2. The socially efficient allocation policy is described in Section 3. The definition of the Markov-perfect equilibrium is given in Section 4, where we completely describe the pricing policies and market shares in the unique equilibrium. The diffusion path of a successful product is presented in some detail in Section 5. We conclude in Section 6.

2. The model

In a dynamic duopoly, firms with differentiated products compete in prices in an infinite-horizon, continuous-time setting. The first firm is well established in the market, and its product characteristics are common knowledge at the beginning of the game. The second firm has a new product whose value has to be learned over time.

The preferences of the buyers are described by a Hotelling location model. The buyers are uniformly distributed on the interval $[0, 1]$ and they have unit demand at each instant of time. The value of the certain product for individual n is given by s_n, with

$$s_n = s + nh, \quad n \in [0, 1]. \tag{2.1}$$

The parameter $h > 0$ represents the horizontal differentiation between the products, and as such h is a measure of the heterogeneity among the buyers. Symmetrically, the value of the uncertain product for individual n is given by μ_n, with

$$\mu_n = \mu + (1 - n)h, \quad n \in [0, 1]. \tag{2.2}$$

The value, μ, of the new product is initially unknown to all parties. It can be either low or high:

$$\mu \in \{\mu_L, \mu_H\}, \tag{2.3}$$

with

$$0 < s - h < \mu_L < s < \mu_H < s + h. \tag{2.4}$$

The inner inequalities in (2.4) imply that the new product can be of either lower or higher value than the established one. The outer inequalities assert that in either case, the efficient allocation would assign a positive measure of buyers to both products.[1] The marginal cost of production for both products is normalized to zero.[2]

The size of h determines how much the value of the product to the buyer and ultimately the choice behavior of the buyer is influenced by her location. The model in (2.1)–(2.3) is one of horizontal and vertical differentiation, where the horizontal differentiation is common knowledge at the outset but the extent of vertical differentiation is uncertain.

The uncertainty about the value of the second product can be resolved only by experimentation, i.e., through purchases of the new product. The performance of the new product is, however, subject to random disturbances, and any single experiment with the new product provides only a noisy signal about the true underlying value. The information conveyed by an experiment depends on the size of the experiment. As each buyer is of measure zero, the size of her purchase is negligible and hence the information generated by an individual experiment is also negligible. In consequence, all relevant information is contained in the aggregate outcome. The aggregate or market outcome is the performance of the product over all buyers, which is assumed to be publicly observable.

To derive the law of motion for the market outcome, it is most intuitive to start with the discrete approximation of the model. The approximation is discrete in time as well as in the number of buyers. In an economy with N buyers, each individual experiment, X_i^N, is an independent draw

---

[1] In other words, the innovation is not drastic.

[2] The parameter μ is to be interpreted as an unknown mean governing the random payoff realizations from the uncertain alternative. The new product is of random quality and each additional observation yields information on the true value of μ.

from a normal distribution with an unknown mean, $\mu_N = \mu/N$, where $\mu \in \{\mu_L, \mu_H\}$, and a known variance, $\sigma_N^2 = \sigma^2/N$, for fixed μ and σ^2. Since the mean and the variance of each individual draw are normalized by the number of buyers in the market, the aggregate mean and the aggregate variance of the market experiment,

$$\sum_{i=1}^{N} X_i^N,$$

remain constant at (μ, σ^2). For $k \leq N$, we may compute the expected value and variance of the k buyer experiment,

$$\sum_{i=1}^{k} X_i^N \sim N\left(\frac{k}{N\mu}, \frac{k}{N\sigma^2}\right).$$

Taking the limit as $N \to \infty$, we can express the distribution of the aggregate experiment of a fraction

$$n = \frac{k}{N}$$

of the buyers' experiment as[3]

$$X(n) \sim N(n\mu, n\sigma^2). \tag{2.5}$$

In the continuous-time formulation, the market outcome process, $dX(n(t))$, becomes a stochastic differential equation:

$$dX(n(t)) = n(t)\mu dt + \sigma\sqrt{n(t)}dB(t), \quad t \in [0, \infty), \tag{2.6}$$

where the instantaneous or flow payoffs $dX(n(t))$ in continuous time take the same form as (2.5) in discrete time. As before, $n(t)$ is the fraction of buyers who use the new product at time t. The expected flow payoff from the aggregate experiment is $n(t)\mu$. Its variance is $\sigma^2 n(t)$, and $dB(t)$ is the increment of a standard Brownian motion. At each instant, $dX(n(t))$ provides a noisy signal of the true value of the uncertain alternative, $n(t)\mu$, subject to a random perturbation $\sigma\sqrt{n(t)}dB(t)$.

It is immediately verified that the instantaneous mean and the variance of the market outcome are linear in the market share of the new seller. As

---

[3]This construction does not really require the individual experiments to be normal, since we can use the central limit theorem in the limiting procedure, the number of buyers become large, to derive normality of the sums in the limit.

the value of μ can only be μ_L or μ_H, posterior beliefs about the quality are completely characterized by $\alpha(t)$, with

$$\alpha(t) = \Pr(\mu = \mu_H | \mathcal{F}(t)), \qquad (2.7)$$

where $\mathcal{F}(t)$ is the history generated by $X(n(t))$. The conditional expected quality $\mu(\alpha(t))$ of the uncertain product is

$$\mu(\alpha(t)) = (1 - \alpha(t))\mu_L + \alpha(t)\mu_H. \qquad (2.8)$$

The market players extract the information provided by the noisy market outcome (2.6) to improve their common prior beliefs $\alpha(0) = \alpha_0$ over time. The game is thus one of incomplete but symmetric information, and no issues of asymmetric information arise.[4]

The learning process of the market represents a signal-extraction problem: Given the information generated by $X(n(t))$, what is the posterior belief about the value of the uncertain alternative? As the beliefs are completely characterized by $\alpha(t)$, the signal-extraction problem reduces to the description of the law of motion of the posterior belief $\alpha(t)$.

Proposition 1 (Posterior Belief): *The process $\alpha(t)$ is a Brownian motion with zero drift and variance $n(t) \sum^2(\alpha(t))$:*

$$n(t) \sum{}^2 (\alpha(t)) = n(t) \left[\frac{\alpha(t)(1 - \alpha(t))(\mu_H - \mu_L)}{\sigma} \right]^2. \qquad (2.9)$$

Proof: See Liptser and Shiryaev (1977), Theorem 9.1.   □

The posterior belief $\alpha(t)$ is a martingale, as the posterior belief incorporates all predictable information. The variance of $\alpha(t)$ can be interpreted as the amount of information generated by the market outcome, as it indicates how rapidly $\alpha(t)$ can change. The variance increases linearly in the market share of the new firm and depends on the signal-to-noise ratio $(\mu_H - \mu_L)/\sigma$ and the diffuseness of the prior information $\alpha(t)(1 - \alpha(t))$.

The only payoff-relevant source of uncertainty in this model is the value of the new product. All available information in period t is incorporated in the common posterior belief $\alpha(t)$, and the dynamics in the model are driven by the process of belief change. The focus in subsequent sections is the analysis of Markovian policies where the posterior belief α is a natural

---

[4]Because of this simplifying assumption, we do not have to address issues of prices signaling the qualities as in Judd and Riordan (1994). We shall come back to this issue in the conclusion.

state variable. We start by solving for the socially efficient experimentation policy to get a benchmark for the equilibrium model.

3. Efficient experimentation

This section has two objectives. First, we want to introduce the basic technique of dynamic programming with zero discounting to describe intertemporal policies in our model. Second, we apply the technique to characterize the socially efficient experimentation policy, which maximizes the social surplus. The technique is most clearly illustrated here, since the efficient allocation is the solution to a single optimization problem without any strategic interaction.

The reader mainly interested in the duopoly may want to go directly to Section 4 and refer back to this section for details only when needed.

An experimentation policy prescribes for every posterior belief α the shares of buyers allocated to the sellers. Denote by $n(\alpha)$ the market share of the new product and by $1 - n(\alpha)$ the share of the established product. The average flow value from the established product when the $1 - n(\alpha)$ buyers receive the product is with preferences as in (2.1) and (2.2):

$$s(n(\alpha)) \equiv s + \frac{(1 + n(\alpha))h}{2}, \tag{3.1}$$

and for the new product it is similarly

$$\mu(n(\alpha)) \equiv \mu + \frac{(2 - n(\alpha))h}{2}. \tag{3.2}$$

The expected flow value of consumption from the two alternatives under posterior belief α and allocation policy $n(\alpha)$ is then given by

$$n(\alpha)\mu(n(\alpha)) + (1 - n(\alpha))s(n(\alpha)).$$

The optimal allocation problem with discounting can be written as a dynamic programming problem in continuous time:

$$rV(\alpha) = \max_{n(\alpha)} \left\{ n(\alpha)\mu(n(\alpha)) + (1 - n(\alpha))s(n(\alpha)) \right.$$

$$\left. + \frac{1}{2}n(\alpha) \sum{}^2(\alpha)V''(\alpha) \right\}, \tag{3.3}$$

where $V(\alpha)$ is the value function of the allocation problem and $r > 0$ is the discount rate.[5]

The expression (3.3) has a very simple interpretation. The flow benefit $rV(\alpha)$ from an experimentation policy $n(\alpha)$ consists of two parts. The first part is the aggregated flow payoff from the two alternatives, and the second part is the flow value of experimentation $\frac{1}{2}n(\alpha)\sum^2(\alpha)V''(\alpha)$. Notice that since $\alpha(t)$ is a posterior belief and hence a martingale, terms involving the first derivative $V'(\alpha)$ disappear in the dynamic programming equation. The last term captures the impact that oscillations in α have on the flow value. The instantaneous variance $\sum^2(\alpha)$ of the belief process indicates the quantity of information released through a unit of experimentation, and $n(\alpha)$ is the current size of the experiment. The curvature $V''(\alpha)$ of the value function is the shadow price of information. If $V''(\alpha)$ is positive, then additional information is valuable for better future allocation decisions. The flow value of the optimal experimentation policy then maximizes the sum of the flow payoff and the flow value of information.

The optimal allocation $n(\alpha)$ can be obtained in principle from the first-order conditions of the right-hand-side term of (3.3). But the differential equation that results when $n(\alpha)$ is replaced by its solution is quadratic in the second-order term $V''(\alpha)$. As a consequence, solutions to the differential equation implied by (3.3) can be obtained only through numerical methods, and no analytical solutions are available. This feature is present in all specifications with idiosyncratic preferences as long as the value of information, $\frac{1}{2}\sum^2(\alpha)V''(\alpha)$, is nonzero, since the optimal market share, $n(\alpha)$, for the new firm is a function of $V''(\alpha)$ and the Bellman equation is

$$rV(\alpha) = f(V''(\alpha)),$$

where f is a nonlinear function of $V''(\alpha)$.

We sidestep these problems by analyzing the optimal allocation problem under zero discounting. With the appropriate optimality criterion, called the "strong long-run average criterion" in Dutta (1991), we preserve the recursive representation of the dynamic programming equation (3.3) as $r \to 0$. Most importantly for our purposes, the optimal policies under this criterion are the unique limits to the associated policies under discounting and as such maintain the intertemporal aspect of the experimentation

---

[5]See Dixit and Pindyck (1994) or Harrison (1985) for a detailed derivation of the dynamic programming equation in continuous time when uncertainty is represented by a Brownian motion.

policies. In other words, all the qualitative properties of the equilibrium we derive in the following will hold also for small, but positive, discount rates $r > 0$.

The strong long-run average criterion refines the long-run average criterion, which discriminates insufficiently between alternative intertemporal policies.[6] The long-run average under the initial belief α_0 is given by

$$v(\alpha_0) \equiv \sup_{n(\alpha)} \lim_{T \to \infty} \frac{1}{T} E \left[\int_0^T (n(\alpha)\mu(n(\alpha)) + (1 - n(\alpha))s(n(\alpha))) \, dt | \alpha_0 \right],$$
$$(3.4)$$

where we suppress dependence on t. The long-run average $v(\alpha)$ is equal to the expected full-information payoff

$$v(\alpha) = \alpha v(2.1) + (1 - \alpha)v(0),$$
$$(3.5)$$

if $\alpha(t)$ converges almost surely to zero or one, as is the case here. Hence almost any allocation policy in finite time is consistent with long-run payoff maximization. The full-information payoffs $v(0)$ and $v(1)$ are the solutions to the static allocation problems, when μ is known to be either μ_L or μ_H. These values can be computed immediately and hence so can all long-run average values $v(\alpha)$.[7] The strong long-run average is defined by the following optimization problem:

$$V(\alpha_0) \equiv \sup_{n(\alpha)} \lim_{T \to \infty} E$$

$$\left[\int_0^T (n(\alpha)\mu(n(\alpha)) + (1 - n(\alpha))s(n(\alpha)) - v(\alpha)) \, dt | \alpha_0 \right], \quad (3.6)$$

where $v(\alpha)$ is as defined in (3.4). The strong long-run average maximizes the expected returns net the long-run average. As the flow value $n(\alpha)\mu(n(\alpha)) + (1 - n(\alpha))s(n(\alpha))$ of the optimal allocation policy under imperfect information is necessarily less than the expected full-information payoff $v(\alpha)$, a different interpretation of the strong long-run average criterion is that it minimizes the losses due to imperfect information compared to the maximum achievable under full information. The limit as $T \to \infty$ is well-defined

---

[6]See Dutta (1991) for a very careful and detailed analysis on the connection between optimality criteria under discounting and under no discounting.
[7]The long-run average values for the efficient and the equilibrium program are recorded in Lemma A1.

and finite, and hence this criterion discriminates between policies based on their performance on finite time intervals as well. The infinite-horizon problem (3.6) can be presented via the dynamic programming equation as

$$\max_{n(\alpha)} \left\{ n(\alpha)\mu(n(\alpha)) + (1 - n(\alpha))s(n(\alpha)) - v(\alpha) \right.$$

$$\left. + \frac{1}{2}n(\alpha) \sum{}^2(\alpha)V''(\alpha) \right\} = 0, \tag{3.7}$$

under the condition that $n(\alpha)$ is bounded away from zero for all α. The latter condition means that even for low values of α, it is optimal to allocate some buyers to the new product. The condition is satisfied with the inequalities (2.4) introduced in the previous section.

The simplicity of the Bellman equation (3.7) in the case of no discounting is transparent, as it contains only the second-order term but no lower-order terms of the value function. The flow value $rV(\alpha)$ that appeared in the discounted case is replaced by the long-run average $v(\alpha)$ in the undiscounted case. The interpretation of the value-of-information term, $1/2n(\alpha) \sum{}^2(\alpha)V''(\alpha)$, remains the same as in the discounted case.

We proceed to solve for the optimal experimentation policy with the assistance of the Bellman equation (3.7). As the optimal allocation $n^*(\alpha)$ achieves a maximum value of zero for the left-hand term of (3.7), and as $n^*(\alpha)$ is bounded away from zero, we may divide the term inside the maximum operator through $n(\alpha)$ and claim that the maximizer of the modification,

$$\max_{n(\alpha)} \left\{ \frac{s + \frac{h}{2} - v(\alpha)}{n(\alpha)} - hn(\alpha) \right\} + \mu(\alpha) - s + h + \frac{1}{2}\sum{}^2(\alpha)V''(\alpha) = 0,$$

$$\tag{3.8}$$

is identical to the maximizer of the original expression (3.7). The optimal allocation is found by the first-order condition of (3.8), which is independent of the second-order term of the value function.

Proposition 2 (Efficient Experimentation Policy): *The efficient allocation $n^*(\alpha)$ is given by*

$$n^*(\alpha) = \sqrt{\frac{v(\alpha) - s - \frac{h}{2}}{h}}, \tag{3.9}$$

with $n^*(\alpha) > 0$ for all $\alpha \in [0, 1]$.

Proof: See the Appendix. $\square$

The optimal allocation $n^*(\alpha)$ is naturally an increasing function of α. The comparative statics are as expected, and $n^*(\alpha)$ is increasing in μ_H, μ_L, and decreasing in s and h. As we compare $n^*(\alpha)$ with the myopically optimal allocation $m^*(\alpha)$ that solves the static problem

$$m^*(\alpha) \in \arg\max_{n(\alpha)} n(\alpha)\mu(n(\alpha)) + (1 - n(\alpha))s(n(\alpha)),$$

it is verified after some computations that $n^*(\alpha) > m^*(\alpha)$ for all $\alpha \in (0,1)$. This provides additional evidence that the optimal policy under the strong long-run average preserves the intertemporal aspect of the experimentation policy as the optimal size of the experiment $n^*(\alpha)$ is larger than the myopically efficient allocation. The difference is of course attributable to the additional incentive to use the new product to generate information for future decisions. This intertemporal benefit is absent in the static decision.

4. Dynamic equilibrium

In this section we first define our solution concept, the Markov-perfect equilibrium. Next we characterize the unique equilibrium and the associated equilibrium policies. Finally we consider the efficiency of the equilibrium and examine its dynamic properties.

4.1. *Equilibrium and policies*

In this model of dynamic competition, buyers and sellers learn over time more about the true value of the new product. We focus on Markovian strategies with the posterior belief $\alpha(t)$ as the state variable to emphasize.

The sellers choose at each instant of time their prices noncooperatively. We consider only pricing policies, $p_i(\alpha)$, of the sellers that are measurable with respect to the state variable α. The strategic considerations of the sellers involve both current revenues and the influence that experimentation has on future revenues. Again, we are interested in the limiting case of no discounting, and the value functions of the sellers can be constructed in the same way as the value function of the efficient program in (3.7). The dynamic programming equation for the established seller is

$$0 = \max_{p_1(\alpha)} \left\{ (1 - n(\alpha))p_1(\alpha) - v_1(\alpha) + \frac{1}{2}n(\alpha)\sum{}^2(\alpha)V_1''(\alpha) \right\} \qquad (4.1)$$

and for the new seller it is

$$0 = \max_{p_2(\alpha)} \left\{ n(\alpha)p_2(\alpha) - v_2(\alpha) + \frac{1}{2}n(\alpha)\sum{}^2(\alpha)V_2''(\alpha) \right\}, \qquad (4.2)$$

where $v_1(\alpha)$ and $v_2(\alpha)$ are the long-run average payoffs of the sellers.[8] The equilibrium market share,

$$n(\alpha) \equiv n(\alpha, p_1(\alpha), p_2(\alpha)),$$

naturally depends on the prices offered by the sellers.

The buyers, in turn, make their purchase decisions as a function of the current estimate $\mu(\alpha)$ of the new product, the current prices, and their individual preferences. As a single buyer is negligible in the market, the informativeness of the market outcome is independent of any individual purchase. In consequence, the buyer's decision is based exclusively on the current (expected) values and prices, as she cannot influence the informativeness of the market outcome. The equilibrium market shares are then determined by the critical consumer $n(\alpha)$ who is indifferent between the two alternatives:

$$s + n(\alpha)h - p_1(\alpha) = \mu(\alpha) + (1 - n(\alpha))h - p_2(\alpha).$$

After rearranging, the equilibrium share of the new seller $n(\alpha)$ emerges as a function of the current values and prices and the degree of horizontal differentiation:

$$n(\alpha) = \frac{(\mu(\alpha) - p_2(\alpha)) - (s - p_1(\alpha)) + h}{2h}. \qquad (4.3)$$

As the buyers decide as they would under myopia, the equilibrium share condition (given the prices) is as in the static pricing game. We now formally define the Markov-perfect equilibrium (Maskin and Tirole, 1995).

Definition 1 (Markov-perfect equilibrium): A Markov-perfect equilibrium is a triple $\{p_1(\alpha), p_2(\alpha), n(\alpha)\}$ such that equations (4.1)–(4.3) are satisfied for all $\alpha \in [0, 1]$.

The process of experimentation gradually establishes whether the new product is of low or high value. Experimentation then leads ultimately to an increase in the (vertical) differentiation between the products. The

---

[8]The long-run average payoffs are again the expected full-information payoffs, which are the static profits when the value of the product is known. The values are recorded in Lemma A1 in the Appendix.

expected joint profits for the sellers are higher after differentiation, as the superior seller will be able to extract a larger part of the social surplus from the buyers via higher prices. *Ex ante*, both sellers could become the sellers with the superior product, and thus both sellers attach positive value to additional information. Consequently, the shadow price of information, represented by $V_i''(\alpha)$, is positive to both of them.[9] This has important but different implications for the strategies of the sellers.

By inserting the equilibrium share condition (4.3) in the optimization problems of the competing sellers, the strategic considerations of the sellers become more transparent:

$$0 = \max_{p_1(\alpha)} \left\{ \begin{array}{l} \left(1 - \dfrac{\mu(\alpha) - p_2(\alpha) - s + p_1(\alpha) + h}{2h}\right) p_1(\alpha) - v_1(\alpha) \\[3mm] + \dfrac{\mu(\alpha) - p_2(\alpha) - s + p_1(\alpha) + h}{4h} \sum{}^2(\alpha) V_1''(\alpha) \end{array} \right\} \quad (4.4)$$

and

$$0 = \max_{p_2(\alpha)} \left\{ \begin{array}{l} \dfrac{\mu(\alpha) - p_2(\alpha) - s + p_1(\alpha) + h}{2h} p_2(\alpha) - v_2(\alpha) \\[3mm] + \dfrac{\mu(\alpha) - p_2(\alpha) - s + p_1(\alpha) + h}{4h} \sum{}^2(\alpha) V_2''(\alpha) \end{array} \right\}. \quad (4.5)$$

The equilibrium conditions (4.4) and (4.5) illustrate that for both firms, an increase in their respective market shares requires a decrease in the price at which their product is offered. But the equilibrium conditions also show that the incentives to acquire a larger market share are quite different for the two sellers. The value of information, which is generated by sales of the new product only, operates like an additional revenue source. And while it is associated with sales of the new product, new information actually benefits *both* firms. The additional value from sales will prompt the new seller to price more aggressively and seek a larger market share than he would in a static world. In contrast, the incentives for the established seller to increase his market share are relatively weaker because a larger market share would imply less experimentation and hence reduce the informational gains. This will lead the established firm to adopt a less aggressive pricing policy in equilibrium.

By the same argument we presented for the efficient program, we can divide (4.4) and (4.5) by $n(\alpha)$, or its equivalent in (4.3). The resulting

---

[9]The formal result will be stated in Corollary 2.

equilibrium conditions for the optimal pricing strategies do not involve the second derivatives of the value functions:

$$0 = \frac{1}{2}\sum^{2}(\alpha)V_1''(\alpha) + \max_{p_1(\alpha)}\left\{\frac{2h(p_1(\alpha) - v_1(\alpha))}{p_1(\alpha) - p_2(\alpha) + \mu(\alpha) - s + h} - p_1(\alpha)\right\},$$

(4.6)

and similarly for the second seller,

$$0 = \frac{1}{2}\sum^{2}(\alpha)V_2''(\alpha) + \max_{p_2(\alpha)}\left\{\frac{-2hv_2(\alpha)}{p_1(\alpha) - p_2(\alpha) + \mu(\alpha) - s + h} + p_2(\alpha)\right\}.$$

(4.7)

The advantage of this approach is rather obvious. The optimal pricing policies of the sellers can now be obtained without explicit reference to the second derivatives of the value functions. The optimal pricing policies are the solutions to the first-order conditions of (4.6) and (4.7).

Proposition 3 (Equilibrium Experimentation): *There is a unique Markov-perfect equilibrium. The equilibrium prices are*

$$p_1(\alpha) = \frac{2}{3}(s - \mu(\alpha)) + \sqrt{2hv_2(\alpha)}$$

(4.8)

and

$$p_2(\alpha) = \frac{1}{3}(\mu(\alpha) - s) + h.$$

(4.9)

The market share of the new seller is given by

$$n(\alpha) = \sqrt{\frac{v_2(\alpha)}{2h}}.$$

(4.10)

Proof: See the Appendix.   □

The uniqueness property of the equilibrium is naturally due to the Markovian restriction of the equilibrium strategies, and the set of non-Markovian (perfect) equilibria is obviously much larger.

As the market is always completely shared by the sellers, the market share of the established firm is $1 - n(\alpha)$. The properties of the dynamic pricing equilibrium are most easily illustrated by contrasting the dynamic with the static policies. In the corresponding static equilibrium, buyers and sellers take the current posterior belief as given and make decisions as if they were in a one-shot game without any intertemporal considerations. We denote the static (or myopic) unique equilibrium prices by $p_1^m(\alpha)$ and

$p_2^m(\alpha)$, and the market share of the new firm by $n^m(\alpha)$. The following corollary is immediate.

Corollary 1 (Static and Dynamic Prices):

(i) *The level of prices is* $p_1(\alpha) > p_1^m(\alpha)$ *and* $p_2(\alpha) = p_2^m(\alpha)$.
(ii) *The level of sales is* $n(\alpha) > n^m(\alpha)$.

The static prices and market shares are recorded in Lemma A1 in the Appendix. The equality of the new firm's dynamic price and the myopic price is a joint consequence of the linear preference structure and no discounting. In the linear model with discounting, the price of the new seller would in fact be below the myopic price, as one might have expected.

The equality in the case of no discounting can be explained as follows. The revenue from a marginal buyer to the new firm is the sum of the current price, p_2, and the marginal value of information, $v_2(\alpha)/n - p_2$. Notice that when summing up these effects, current prices cancel out. The losses resulting from lower prices on the inframarginal buyers are given by $2hn$. On the other hand, the marginal losses from losing market share for the established seller are given by $[p_1 - v_1(\alpha)]/n$, and the inframarginal gains (from higher prices associated with lower market shares) are $2h\,(1-n)$. Summing the marginal revenues across the firms, we get $[p_1 - v_1(\alpha)]/n + v_2(\alpha)/n = 2h$, where $[p_1 - v_1(\alpha)/n$ and $v_2(\alpha)/n$ can be interpreted as the flow gains from the marginal buyer.

To recover from here the current prices, we solve for $p_1 = v_1(\alpha) - v_2(\alpha) + 2hn$ by observing that in our linear specification, $v_1(\alpha) - v_2(\alpha) = p_1^m(\alpha) - p_2^m(\alpha)$. But then

$$p_1 = p_1^m(\alpha) - p_2^m(\alpha) + 2hn \qquad (4.11)$$

and the marginal buyer is indifferent between the two firms only if $p_2 = p_2^m$. The equilibrium price of the established firm is then linear in the equilibrium market share by (4.11), while the equilibrium price of the new firm is constant.

The experience with the new product generates information that is valuable for both firms. The equilibrium value of information is given by the dynamic programming equations (4.1) and (4.2) as

$$\frac{1}{2}n(\alpha)\sum^2(\alpha)V_1''(\alpha) = v_1(\alpha) - (1 - n(\alpha))p_1(\alpha)$$

and

$$\frac{1}{2}n(\alpha) \sum^{2}(\alpha)V_2''(\alpha) = v_2(\alpha) - n(\alpha)p_2(\alpha),$$

respectively. For each firm, the equilibrium value of information $1/2n(\alpha) \sum^{2}(\alpha)V_i''(\alpha)$ is thus precisely the difference between its expected full-information revenue and its current revenue. As the equilibrium prices and allocations are established in Proposition 3, it can be verified that the value of information is twice as large for the established firm as for the new firm. Thus, the established firm values information about the true quality of the new product even higher than the new seller does himself. This initially puzzling result can be traced back to the impact the information flow has on the players' strategies. To see this better, consider for the moment the value of information if the sellers would follow myopic policies. In this case, the value of information for the two sellers is in fact equal:

$$v_1(\alpha) - (1 - n^m(\alpha))p_1^m(\alpha) = v_2(\alpha) - n^m(\alpha)p_2^m(\alpha),$$

as the symmetry of the model and the updating process would suggest. In the dynamic equilibrium, the less aggressive position of the established firm lowers its own revenue flow and raises the revenue flow for the new firm. Thus the current revenue shortfall from the expected full-information revenues increases for the established seller relative to the new seller. In consequence, the resolution of uncertainty is valued more highly by the established firm.

The influence of the information flow on the pricing strategies has a surprising consequence for the optimal choice of innovations. Suppose for the moment that the new seller was currently offering a product of low quality μ_L. Suppose further that he can choose between two forms of innovation: either he gets a product with value $\mu(\alpha)$ with certainty, or he gets an innovation of superior quality, μ_H, with probability α, and with probability $1 - \alpha$ it presents no improvement beyond μ_L. Corollary 1 then implies that the second seller will strictly prefer the uncertain innovation.[10] Moreover, the stochastic innovation increases the prices the consumers have to pay on average and thus relaxes the price competition relative to the

---

[10]The established seller also prefers the uncertain innovation. To see this, observe that he can always deviate to the myopic price in all periods. This gives him the myopic profit in all periods. As the value function is convex, his payoff along this deviation must exceed the myopic payoff.

certain innovation. Next we analyze the efficiency and dynamic features of the equilibrium.

4.2. *Inefficiency of the equilibrium*

The pricing policies of the sellers change systematically in the size of their market shares. As the value functions of the sellers indicate, the marginal benefit from persuading an additional buyer to experiment at any instant of time is constant at $1/2 \sum^2_i(\alpha)V_i''(\alpha)$ for seller i and independent of the market share. But as the marginal buyer has always the lowest valuation for the product among the current buyers, to convince her, each seller would have to decrease the price on all inframarginal buyers. The static revenue loss associated with the acquisition of an additional buyer is therefore increasing in the market share, since the price decrease is granted to a larger measure of inframarginal buyers. For a seller with a small market share, the revenue loss is small and his price policy is mostly determined by intertemporal considerations. Conversely, a seller with a large market share will not be very responsive to a lower price from the competitor as he attempts to maintain the current price level even at the expense of losing some market share.

This suggests that for low values of α, the new firm acquires market share aggressively to generate information and the established firm responds only slowly with lower prices. In conjunction, these two strategies will generate excessive experimentation compared to the socially optimal level. As α increases, the market share of the new firm increases, and this weakens its aggressive stance. Simultaneously, the established firm becomes more responsive in its pricing policy as its market share shrinks. For high values of α, the new seller captures the market to a large extent and his incentives to support additional experimentation become very weak.

Proposition 4 (Inefficiency): *Equilibrium experimentation is excessive for low values of α and insufficient for high values of α. The difference*

$$n^*(\alpha) - n(\alpha)$$

is increasing in α and crosses zero once.

Proof: See the Appendix. □

The dynamic response of the sellers to changes in the market condition as represented in the posterior belief α has several implications for the expected changes in the market shares and revenues of the sellers over time. The following properties of the equilibrium prices and market shares are established by Proposition 3 and the fact that the posterior belief $\alpha(t)$ is a martingale (see Proposition 1).

Corollary 2 (Martingale and convexity):

(i) *The price $p_1(\alpha)$ is a supermartingale; the price $p_2(\alpha)$ is a martingale.*
(ii) *The market share $n(\alpha)$ of the new seller is a supermartingale.*
(iii) *The revenues $(1 - n(\alpha))p_1(\alpha)$ and $n(\alpha)p_2(\alpha)$ are submartingales.*
(iv) *The value functions $V_i(\alpha)$ of the sellers are convex.*

For the new seller, the incentives to acquire a larger market share weaken with the size of his current market share. In consequence, marginal changes in the market share of the new seller in response to changes in the posterior belief α are more accentuated for low market shares $n(\alpha)$. The concavity of $n(\alpha)$, together with the martingale property of the posterior belief $a(t)$, turns $n(\alpha)$ into a strict supermartingale. A symmetric argument leads to a strict submartingale property for the market share of the established seller and thus to the resilience of the established seller, whose market shares are expected to increase over time.

The response of the established seller to a decline in his market share mirrors the behavior of the new firm. His pricing behavior becomes more aggressive and more $p_1(\alpha)$ is a decreasing and concave function of α and thus the prices of the established seller are expected to decrease over time. In consequence, the market share of the established seller is expected to increase, while its price is expected to decrease over time. The market share of the established firm responds initially, i.e., for low values of $\alpha(t)$, very fast to changes in the market condition, whereas his price is initially very slow in response to changes in the posterior belief $\alpha(t)$.

We argued earlier that the joint profits of the sellers are increasing in the extent of (vertical) differentiation. As experimentation induces the process of differentiation, the revenues of the sellers are expected to increase over time. In probabilistic terms, the revenues of both sellers are strict submartingales with respect to the posterior belief $\alpha(t)$. Finally, the positive value of the experiment for the firms, as it generates *ex post* differentiation, is documented by the convexity of the value functions.

5. Market diffusion

So far we have described the equilibrium policies as functions of the state variable $\alpha(t)$. But as the changes in the posterior belief $\alpha(t)$ are endogenously determined by the equilibrium policies, one would like to know more about the evolution of the posterior belief $\alpha(t)$ and the associated policies over real time t. We therefore take the analysis one final step further and examine how market shares and prices typically develop over time for a successful or unsuccessful new product. This will give us a more detailed description of the diffusion paths of new products in real time rather than in the state variable $\alpha(t)$ which, after all, only represents the information available at time t.

In the following, we focus on the case when the true value of the product is high, or $\mu = \mu_H$, as only successful products will display on average increasing market shares over time. While the sample path of the market shares and prices is of course stochastic, we start by analyzing the expected changes in the posterior belief $\alpha(t)$ under the equilibrium policies.

The stochastic differential equation that describes the market outcome for the high-value product is given by

$$dX(n(\alpha(t))) = n(\alpha(t))\mu_H dt + \sqrt{n(\alpha(t))}\sigma dB(t) \qquad (5.1)$$

and is similar to the outcome process as defined earlier in (2.6). Here, $n(\alpha(t))$ is the equilibrium market share when the market participants hold the belief $\alpha(t)$. The stochastic differential equation (5.1) combines two different elements: the true data-generating process and the equilibrium policies of the imperfectly informed traders, which were obtained in the previous section. More precisely, in the case of $dX(n(\alpha(t)))$ the signal is generated by the true value μ_H, but as the market participants possess only imperfect information in the form of $\alpha(t)$, the size $n(\alpha(t))$ of the experiment is determined by the equilibrium policies based on $\alpha(t)$. The evolution of the posterior belief $\alpha(t)$ conditional on $\mu = \mu_H$ is given by

$$d\alpha(t) = (\mu_H - \mu_L)\alpha(t)(1 - \alpha(t))[dX(n(\alpha(t))) - n(\alpha(t))\mu(\alpha(t))dt], \quad (5.2)$$

as an application of the general filtering equation; see again Liptser and Shiryaev (1977). The change in the posterior belief $\alpha(t)$ is then determined by the difference inside the square bracket, which is simply the difference between the true signal and the expected signal, where the expectation is based on the current imperfect information of the market traders. By inserting the differential equation (5.1) into (5.2), we obtain

$$d\alpha(t) = (\mu_H - \mu_L)^2\alpha(t)(1 - \alpha(t))^2 n(\alpha(t))dt$$
$$+ (\mu_H - \mu_L)\alpha(t)(1 - \alpha(t))\sqrt{n(\alpha(t))}dB(t). \qquad (5.3)$$

The difference between the evolution of the conditional posterior belief $\alpha(t)$ here and the posterior belief in Proposition 1 is that the former receives a continuous push upward from the conditioning on μ_H. On average, the signal, which is the outcome realization, exceeds the expectations of the market, which still puts some positive probability, namely $1 - \alpha(t)$, on the event that the product is of low quality, while in fact the product is of high quality. Hence, $\alpha(t)$ is not a martingale anymore, but a submartingale. The drift of the process $\alpha(t)$ is then given by

$$E[d\alpha(t)] = n(\alpha(t))(\mu_H - \mu_L)^2\alpha(t)(1 - \alpha(t))^2 dt. \tag{5.4}$$

Equation (5.4) presents the drift rate of the conditional posterior belief as a function of time. Our interest rests with the description of the posterior belief when its evolution is governed by the expected or mean changes. In other words, we analyze the behavior of the market when the evolution of $\alpha(t)$ is determined only by the drift of the process (5.3). We denote the new and deterministic process by $\hat{\alpha}(t)$, and the evolution of the mean posterior $\hat{\alpha}(t)$ is given by

$$d\hat{\alpha}(t) = n(\hat{\alpha}(t))(\mu_H - \mu_L)^2\hat{\alpha}(t)(1 - \hat{\alpha}(t))^2 dt. \tag{5.5}$$

As the differential equation (5.5) is deterministic, $\hat{\alpha}(t)$ is a function of the initial condition $\hat{\alpha}_0 = \alpha_0$ and time t only.[11] Next we analyze the behavior of the market shares and the prices under the deterministic process $\hat{\alpha}(t)$ rather than under the stochastic process $\alpha(t)$. The market share $n(\hat{\alpha}(t))$ is a composite function of the state variable $\hat{\alpha}(t)$ and time t. The differential equation that describes the changes in the market share of the new seller $dn(\hat{\alpha}(t))$ is given by

$$dn(\hat{\alpha}(t)) = n'(\hat{\alpha}(t))n(\hat{\alpha}(t))(\mu_H - \mu_L)^2\hat{\alpha}(t)(1 - \hat{\alpha}(t))^2 dt. \tag{5.6}$$

As $n(\cdot)$ is given in Proposition 3, we can infer immediately from (5.6) that the market share of the new firm increases over time. The market share $n(\hat{\alpha}(t))$, as $\hat{\alpha}(t)$ before, is a deterministic process, and as such it is a function of the initial condition $\hat{\alpha}_0 = \alpha_0$ and time t only. For transparency

---

[11] An alternative approach would derive the distribution of the state variable $\alpha(t)$ by the Kolmogorov forward equation. The nonlinearities present in this model don't allow us to solve the partial differential equation explicitly. But numerical simulations indicate the same as the one we derive in the following.

we relabel

$$\hat{n}(t) \equiv n(\hat{\alpha}(t))$$

for a given $\hat{\alpha}_0 = \alpha_0$. The question we would like to address, then, is the speed by which the new firm gains market shares. In particular we would like to know whether the acquisition of the new buyers is performed more aggressively over time or whether the acquisition speed slows down with the passage of time.

Proposition 5 (S-shaped Diffusion Path):

(i) *The mean posterior belief $\hat{\alpha}(t)$ is increasing over time. There exists $\bar{\alpha} \in (1/3, 2/3)$ such that if $\hat{\alpha}(t) \leq \bar{\alpha}$, then the rate of increase is increasing in time; if $\hat{\alpha}(t) > \bar{\alpha}$, then it is decreasing in time.*

(ii) *The mean market share $\hat{n}(t)$ is increasing over time. The rate of increase is increasing if $\hat{\alpha}(t) \leq 1/3$, and decreasing if $\hat{\alpha}(t) > 1/3$.*

Proof: See the Appendix. □

If the initial beliefs about the new product are pessimistic enough, then its expected market share, conditional on being of high value, is an S-shaped function, increasing, and initially convex and then concave. A symmetric result holds for the expected market share of a low-value product and optimistic initial beliefs. In that case, the market share of the new firm is decreasing, initially concave and then convex.

The evolution of the market shares over time is a composition of the behavior of the equilibrium market share $n(\hat{\alpha}(t))$ as a function of $\hat{\alpha}$ and the evolution of the posterior belief $\hat{\alpha}(t)$ as a function of time t. The drift in $\hat{\alpha}(t)$ is increasing when starting from low values of $\hat{\alpha}(t)$. This in turn accelerates the growth in market shares and increasingly increases the drift of $n(\hat{\alpha}(t))$ as a function of t. But as $\hat{\alpha}$ approaches one, the learning process slows down and the drift in $\hat{\alpha}(t)$, while remaining positive, approaches zero. In conjunction with the concavity of $n(\cdot)$, this decreases the drift $dn(\hat{\alpha}(t))$ and hence the speed of market acquisition by the new seller.

The evolution of the mean market share $\hat{n}(t)$ and an actual sample path of the stochastic share process $n(\alpha(t))$ are displayed in Figure 4.1. The evolution of the market shares is accompanied by a similar but somewhat shifted movement in the prices. The differences in the drift rates of market share and price processes reflect again the strategic trade-offs for the sellers. We continue the discussion here for the case when the true value of the new

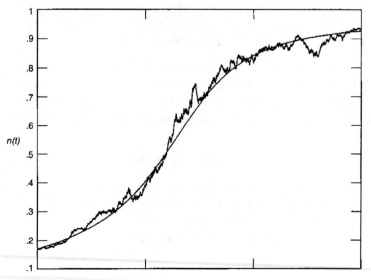

Figure 4.1.   The market share of the new firm ($\alpha_0 = .1$, $\mu_L = 2$, $\mu_H = 6$, $s = 4$, $h = 1$).

product is high and state the results first. Again, symmetric results hold for the case when the true value of the new product is low. By analogy we denote $\hat{p}_i(t) \equiv p_i(\hat{\alpha}(t))$.

Proposition 6 (Prices over time):

(i) *The price process $\hat{p}_1(t)$ is strictly decreasing in t, first at a decreasing then at an increasing rate.*

(ii) *The price process $\hat{p}_2(t)$ is strictly increasing in t, first at an increasing then at a decreasing rate.*

(iii) *The concavity of $\hat{p}_1(t)$ prevails strictly longer than the convexity of $\hat{p}_2(t)$.*

(iv) *The convexity of $\hat{p}_2(t)$ prevails strictly longer than the convexity of $\hat{n}(t)$.*

Proof: See the Appendix.   □

The movements of the price processes can again be decomposed along the state and time space dimension, and together they reveal a rather rich dynamic process. The changes in the prices $\hat{p}_i(t)$ as described in (i) and (ii) follow a pattern parallel to the market shares of the respective sellers. But if we compare the price processes across the sellers and similarly if we compare the price and share processes for a given seller, then the different rates of adjustment elucidate the strategic incentives of the sellers over time.

The drift in the price process $\hat{p}_1(\cdot)$ of the established seller continues to decrease longer than the drift in the price $\hat{p}_2(\cdot)$ of the new seller continues to increase. In other words, as $\hat{\alpha}(t)$ increases, the downward movement in the price $\hat{p}_1(\cdot)$ accelerates, even when the upward movement of $\hat{p}_2(\cdot)$ decelerates. This result translates the increasing responsiveness of the established seller to the success of his competitor, documented in Proposition 3, into the time profile of the pricing policy of the established seller.

The dynamic movement of the prices is displayed in Figure 4.2 in the behavior of the mean prices, along with an actual sample path of the prices. The underlying sample path of $\alpha(t)$ is the same as the one in Figure 4.1.

Finally, the trade-off for the new seller between higher market share and higher price is changing over time. As $\alpha(t)$ increases over time, the expected gains in market share $n(\cdot)$ start to decrease earlier than the gains in the prices $p_2(\cdot)$. So for a larger market share, the new seller is using his competitive edge to increase prices rather than to increase market shares. Again, we display the dynamics for the mean processes as well as for an actual sample path in Figure 4.3.

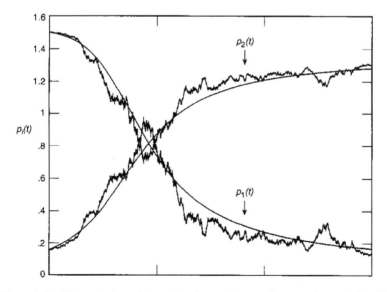

Figure 4.2.   The evolution of the prices ($\alpha_0 = .1$, $\mu_L = 2$, $\mu_H = 6$, $s = 4$, $h = 1$).

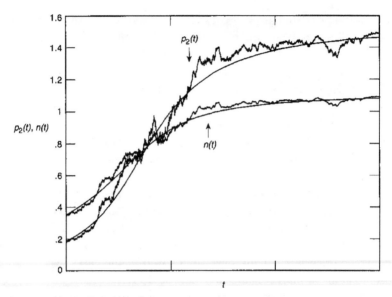

Figure 4.3.   The evolution of price and market share ($\alpha_0 = .1$, $\mu_L = 2$, $\mu_H = 6$, $s = 4$, $h = 1$).

6. Conclusion

In this article we analyzed the dynamic aspects of pricing policies and market shares following the introduction of a new product with uncertain value. The strategic aspect of the learning process led to an increase in prices in the dynamic competition relative to the static competition with given beliefs. The (uncertain) value of the new product was not a choice variable of its seller and, in particular, it was assumed to be constant over time. While this restriction should clearly be removed in future research on dynamic innovation, the comparison between entry with a certain or uncertain value is nonetheless suggestive. The premium to uncertain innovation for the new firm, as well as the positive externality the innovation has on the competing seller, implies that stochastic innovation reduces competition and increases profits in oligopolistic markets. This suggests excessive innovations in differentiated markets, as the search for innovations is the dynamic equivalent of the static vertical differentiation in Shaked and Sutton (1982).

We recall in this context that the established seller valued information about the true quality of the new product higher than the new seller did and this in spite of the purely strategic role of information. If the established firm could use the information beyond its strategic aspect, say for product improvements of its own, then the value of information would *a fortiori* even be higher, and might relax competition even further.

A final remark concerns the information structure in our model. Buyers and sellers make their choices over time under imperfect but symmetric information. As the new product is an experience good, it may be natural to ask what would happen if the information-aggregation device. While sufficient care is required in specifying the signal structure with a continuum of agents, the surprising answer is that a model with private signals would have the same structure and dynamics as the current model. As a first approximation we observe that the private signals create heterogeneity among the buyers similar to the heterogeneity due to the horizontal differentiation. As the strategic weight of each buyer is small relative to the market, the bias in their purchase decision due to the effort to manipulate the sellers' beliefs is small, and zero with a continuum of buyers. The sales in the subsequent period then reveal to the sellers the average experience of the buyers and enable the sellers to update their beliefs about the true quality of the new product.[12] Thus many of the basic properties of the model, in particular the structure of the inefficiency and the qualitative behavior of the diffusion rates, would be present in the private-signal model. The basic difference is the endogenous change in the heterogeneity over time. The influence of a single private signal on the posterior decreases over time, since the posteriors become more concentrated around the true value of the alternative, and so the heterogeneity among buyers decreases over time as well.

Appendix

Proofs to Propositions 2–6 follow.

Lemma A1 (long-run averages): The long-run average in the efficient program is given by

$$v(\alpha) = \frac{\left(s + \mu(\alpha) + \frac{3}{2}h\right)}{2} + (1 - \alpha)\frac{(\mu_L - s)^2}{4h} + \alpha\frac{(\mu_H - s)^2}{4h}. \qquad (A1)$$

---

[12] Judd and Riordan (1994) essentially analyze this private-information model with a monopolist under the additional complication that the aggregate sales are private information to the monopolist.

The long-run averages of the two sellers in the strategic program are given by

$$v_1(\alpha) = (1 - \alpha)\frac{\left(\frac{1}{3}(s - \mu_L) + h\right)^2}{2h} + (\alpha)\frac{\left(\frac{1}{3}(s - \mu_H) + h\right)^2}{2h} \qquad (A2)$$

and

$$v_2(\alpha) = (1 - \alpha)\frac{\left(\frac{1}{3}(\mu_L - s) + h\right)^2}{2h} + (\alpha)\frac{\left(\frac{1}{3}(\mu_H - s) + h\right)^2}{2h}. \qquad (A3)$$

Proof: The long-run average values $v(\alpha)$ and $v_i(\alpha)$ are equal to the expected full-information payoffs:

$$v_i(\alpha) = (1 - \alpha)v_i(0) + \alpha v_i(1),$$

as $n(\alpha)$ and $n^*(\alpha)$ are bounded away from zero under the condition (2.4). But $v_i(0)$ and $v_i(1)$ are the values to the static allocation and equilibrium problems when the value of the new product is known to be either μ_L or μ_H. The composite values are then computed immediately. In particular, the equilibrium prices and allocations are

$$p_1 = \frac{1}{3}(s - \mu_i) + h \quad \text{and} \quad p_2 = \frac{1}{3}(\mu_i - s) + h,$$

and the market share of the new firm is

$$n = \frac{\frac{1}{3}(\mu_i - s) + h}{2h}. \qquad \square$$

Proof of Proposition 2: The first-order condition for the efficient program,

$$\max_{n(\alpha)} \frac{s + \frac{h}{2} - v(\alpha)}{n(\alpha)} - hn(\alpha),$$

is given by

$$hn(\alpha)^2 + s + \frac{1}{2}h - v(\alpha) = 0,$$

and the positive root of the quadratic from is the solution with

$$n^*(\alpha) = \sqrt{\frac{v(\alpha) - s - \frac{h}{2}}{h}}. \qquad \square$$

Proof of Proposition 3: The first-order conditions for the sellers are given by

$$\frac{2h}{p_1(\alpha) - p_2(\alpha) + \mu(\alpha) - s + h}$$
$$- \frac{2h(p_1(\alpha) - v_1(\alpha))}{(p_1(\alpha) - p_2(\alpha) + \mu(\alpha) - s + h)^2} - 1 = 0 \qquad (A4)$$

and

$$- \frac{2hv_2(\alpha)}{(p_1(\alpha) - p_2(\alpha) + \mu(\alpha) - s + h)^2} + 1 = 0. \qquad (A5)$$

Rearranging the first-order condition (A5),

$$v_2(\alpha)2h = (p_1(\alpha) - p_2(\alpha) + \mu(\alpha) - s + h)^2,$$

and by the equilibrium share condition (4.3) we have

$$n(\alpha) = \sqrt{\frac{v_2(\alpha)}{2h}}. \qquad (A6)$$

By rewriting (A4) and using again the equilibrium share condition (4.3) we obtain

$$2hn(\alpha) + v_1(\alpha) - p_1(\alpha) = 2hn^2(\alpha),$$

which after using (A6) gives us

$$p_1(\alpha) = \sqrt{2hv_2(\alpha)} + v_1(\alpha) - v_2(\alpha). \qquad (A7)$$

Finally, the equilibrium share condition (4.3) gives us an explicit expression for the price of the new seller by using (A6) and (A7):

$$p_2(\alpha) = \mu(\alpha) - s + v_1(\alpha) - v_2(\alpha) + h. \qquad (A8)$$

As $v_1(\alpha) - v_2(\alpha) = 2/3(s - \mu(\alpha))$ by Lemma A1, we obtain the unique equilibrium prices as in (4.8) and (4.9). The martingale and supermartingale characterizations follow directly from the linearity of $v_i(\alpha)$ in α. $\qquad \square$

Proof of Proposition 4: We first prove the strict monotonicity and then the single-crossing property. By

$$n^*(\alpha) - n(\alpha) = \sqrt{\frac{v(\alpha) - s - \frac{h}{2}}{h}} - \sqrt{\frac{v_2(\alpha)}{2h}}. \tag{A9}$$

A sufficient condition for $n^*(\alpha) - n(\alpha)$ to be strictly increasing is that (i) $n^*(0) < n(0)$ and (ii) $2v'(\alpha) > v_2'(\alpha)$. By Lemma A1, (ii) is equivalent to

$$\mu_H - \mu_L + \frac{(\mu_H - s)^2}{2h} - \frac{(\mu_L - s)^2}{2h}$$
$$> \frac{\left(\frac{1}{3}(\mu_H - s) + h\right)^2}{2h} - \frac{\left(\frac{1}{3}(\mu_L - s) + h\right)^2}{2h},$$

which in turn is equivalent to

$$\frac{2}{9}(\mu_H - \mu_L)\frac{2\mu_L + 3h - 4s + 2\mu_H}{h} > 0. \tag{A10}$$

Since $\mu_H - \mu_L$ is strictly positive,

$$(3h + 2(\mu_H + \mu_L) - 4s) > 0, \tag{A11}$$

the inequality (A10) is implied by condition (2.4). The single-crossing property is satisfied after verifying that

$$2v(0) - 2s - h < v_2(0) \tag{A12}$$

and

$$2v(1) - 2s - h > v_2(1), \tag{A13}$$

which can be computed directly from the long-run averages given in Lemma A1. □

Proof of Proposition 5: (i) The evolution of the mean posterior $\hat{\alpha}(t)$ as a function of time is given by

$$d\hat{\alpha}(t) = (\mu_H - \mu_L)^2 \hat{\alpha}(t)(1 - \hat{\alpha}(t))^2 n(\hat{\alpha}(t))dt.$$

By Proposition 3 we have

$$n(\hat{\alpha}(t)) = \sqrt{\frac{v_2(\hat{\alpha}(t))}{2h}}.$$

By Lemma A1, $v_2(\hat{\alpha}(t))$ is linear in $\hat{\alpha}(t)$ and we can express $n(\hat{\alpha}(t))$ for the moment simply as the root of a linear function

$$n(\hat{\alpha}(t)) = \sqrt{b_1 + b_2\hat{\alpha}(t)}, \tag{A14}$$

with b_1, $b_2 > 0$ as positive constants. The second time derivative is then given by

$$\frac{d^2\hat{\alpha}(t)}{dt^2} = (\mu_H - \mu_L)^2 \left((1 - \hat{\alpha}(t))^2 \sqrt{b_1 + b_2\hat{\alpha}(t)} \right.$$

$$\left. - 2\hat{\alpha}(t)(1 - \hat{\alpha}(t))\sqrt{b_1 + b_2\hat{\alpha}(t)} + \frac{1}{2}\hat{\alpha}(t)\frac{(1 - \hat{\alpha}(t))^2}{\sqrt{b_1 + b_2\hat{\alpha}(t)}}b_2 \right).$$

It can then be verified that there exists $\bar{\alpha} \in (1/3, 2/3)$ such that

$$\frac{d^2\hat{\alpha}(t)}{dt^2} > 0 \Leftrightarrow \hat{\alpha}(t) < \hat{\alpha} \quad \text{and} \quad \frac{d^2\hat{\alpha}(t)}{dt^2} < 0 \Leftrightarrow \hat{\alpha}(t) > \bar{\alpha}.$$

(ii) The evolution of the mean market share of the new seller as a function of time is given by

$$dn(\hat{\alpha}(t)) = n'(\hat{\alpha}(t))n(\hat{\alpha}(t))(\mu_H - \mu_L)^2\hat{\alpha}(t)(1 - \hat{\alpha}(t))^2 dt. \tag{A15}$$

presents the curvature of the market share of the new firm as a function of time. We may replace $n(\hat{\alpha}(t))$ again by (A14) to obtain

$$\frac{dn(\hat{\alpha}(t))}{dt} = \frac{1}{2}b_2(\mu_H - \mu_L)^2\hat{\alpha}(t)(1 - \hat{\alpha}(t))^2.$$

The second time derivative is given by

$$\frac{d^2n(\hat{\alpha}(t))}{dt^2} = \frac{1}{2}b_2(\mu_H - \mu_L)^2 \left((1 - \hat{\alpha}(t))^2 - 2\hat{\alpha}(t)(1 - \hat{\alpha}(t)) \right) \frac{d\hat{\alpha}(t)}{dt}.$$

As we have $[d\hat{\alpha}(t)/dt] > 0$ for all $\hat{\alpha}(t) \in (0, 1)$, the sign of the curvature is determined by the first term, from which we infer directly that

$$\frac{d^2n(\hat{\alpha}(t))}{dt^2} > 0 \Leftrightarrow \hat{\alpha}(t) \in \left(0, \frac{1}{3}\right). \qquad \square$$

Proof of Proposition 6: The proofs of the four claims rely on the same technique as the ones in Proposition 5 and are therefore omitted. □

References

AGHION, P., ESPINOSA, M.P, and JULLIEN, B. "Dynamic Duopoly with Learning Through Market Experimentation." *Economic Theory*, Vol. 3 (1993), pp. 517–539.

BERGEMANN, D. and VÄLIMÄKI, J. "Market Experimentation and Pricing." Cowles Foundation Discussion Paper, Yale University, 1996.

BOLTON, P. and HARRIS, C. "Strategic Experimentation." STICERD Working Paper no. TE/93/261, London School of Economics, 1993.

CAMINAL, R. and VIVES, X. "Why Market Shares Matter: An Information-Based Theory." *RAND Journal of Economics*, Vol, 27 (1996), pp. 221–239.

DIXIT, A.K. and PINDYCK, R.S. *Investment Under Uncertainty.* Princeton: Princeton University Press, 1994.

DUTTA, P. "What Do Discounted Optima Converge To?" *Journal of Economic Theory*, Vol. 55 (1991), pp. 64–94.

GORT, M. and KLEPPER, S. "Time Paths in the Diffusion of Product Innovation." *Economic Journal*, Vol. 92 (1982), pp. 630–653.

HARRINGTON, J.E. "Experimentation and Learning in a Differentiated-Products Duopoly." *Journal of Economic Theory*, Vol. 66 (1995), pp. 275–288.

HARRISON, J.M. *Brownian Motion and Stochastic Flow Systems.* New York: John Wiley & Sons, 1985.

JUDD, K.L. and RIORDAN, M.H. "Price and Quality in a New Product Monopoly." *Review of Economic Studies*, Vol. 61 (1994), pp. 773–789.

LIPTSER, R.S. and SHIRYAEV, A.N. *Statistics of Random Processes I.* New York: Springer Verlag, 1977.

MANSFIELD, E. *Industrial Research and Technological Innovation.* New York: W.W. Norton, 1968.

MASKIN, E. and TIROLE, J. "Markov Perfect Equilibrium." Mimeo, Harvard University, 1995.

ROB, R. "Learning and Capacity Expansion Under Demand Uncertainty." *Review of Economic Studies*, Vol. 58 (1991), pp. 655–675.

SCHLEE, E. "Learning About Product Quality and Introductory Pricing." Mimeo, Arizona State University, 1996.

SHAKED, A. and SUTTON, J. "Relaxing Price Competition Through Product Differentiation." *Review of Economic Studies*, Vol. 49 (1982), pp. 3–13.

VETTAS, N. "Demand and Supply in New Markets: Diffusion with Bilateral Learning." *RAND Journal of Economics*, forthcoming.

Chapter 5

Entry and Vertical Differentiation[*]

Dirk Bergemann[†] and Juuso Välimäki[‡]

[†]*Department of Economics, and Cowles Foundation of Research in Economics, Yale University, New Haven, CT 06520-8268, USA*
dirk.bergemann@yale.edu
[‡]*Department of Economics, Aalto University School of Business, 02150 Espoo, Finland*
juuso.valimaki@aalto.fi

This paper analyzes the entry of new products into vertically differentiated markets where an entrant and an incumbent compete in quantities. The value of the new product is initially uncertain and new information is generated through purchases in the market. We derive the (unique) Markov perfect equilibrium of the infinite horizon game under the strong long run average payoff criterion. The qualitative features of the optimal entry strategy are shown to depend exclusively on the relative ranking of established and new products based on current beliefs. Superior products are launched relatively slowly and at high initial prices whereas substitutes for existing products are launched aggressively at low initial prices. The robustness of these results with respect to different model specifications is discussed. *Journal of Economic Literature* Classification Numbers: C72, C73, D43, D83. © 2002 Elsevier Science (USA)

Keywords: Entry; duopoly; quantity competition; vertical differentiation; Bayesian learning; Markov perfect equilibrium; experimentation; experience goods.

1. Introduction

1.1. *Motivation*

In this paper, we analyze the optimal entry strategies for different types of experience goods in a dynamic Cournot duopoly with vertically

---

[*]This chapter is reproduced from *Journal of Economic Theory*, **106**, 91–125, 2002.

differentiated buyers. Our main goal is to obtain a characterization of the features of the new product that lead to qualitatively different entry strategies. We show that a new product that represents a certain improvement to an existing product is launched in the market at prices above the static equilibrium level and sales quantities below the static level. A new product that has a positive probability of being the leading brand in the market, but also a positive probability of being revealed to be inferior to the current product, is launched with a more aggressive strategy where the initial prices are low and initial sales exceed the static equilibrium quantities.

The firms compete in a continuous time model with an infinite horizon. The uncertainty about the new product is common to all buyers in the market. Additional information about the quality of the new product is generated only through experiments, i.e. through purchases in the market. The information generated is assumed to be public, and while the exact mechanism of information transmission is left unmodeled it is motivated by considerations such as word of mouth communication between the buyers and consumer report services. As a consequence all buyers have identical beliefs about the new product, and we can represent the stage game as a vertically differentiated quantity game parametrized by the common belief about the new product. Examples of markets where the assumptions of common value (aside from the aspect of vertical differentiation) and common information may be valid include markets for transportation or communications services. To take a precise example, suppose buyers evaluate the services of an airline carrier on the basis of the probability of on-time departure and arrival and/or the probability of lost or misplaced baggage. The uncertainty about the quality of the service is then common to all customers of the airline, provided that the uncertainties are not route specific. The (expected) performance or reliability of the new service is then best predicted by aggregate and publicly available statistics such as the percentage of on-time performances by an airline. In particular, all idiosyncratic experiences are of equal value in providing information and can therefore be replaced by sufficient aggregate statistics. Our model only requires that all consumers rank reliability of the airlines or congestion in the provision of internet services according to the same scale, yet they can differ in their willingness to pay for different service qualities. Hence the model displays vertical but not horizontal differentiation.

We have chosen a model of quantity competition as the stage game. With this choice, we extend the scope of viable new products. In particular, quantity competition allows for the possibility of launching an innovation which brings the two competitors closer to each other without a change in the leadership. In a model of price competition, such innovations would never be profitable, and as a result improved substitutes would never be observed. In those models, the profits of both firms vanish as the substitutability of the two products increases, and as a result the static profit functions of the two firms are nonmonotonic in the level of differentiation. We believe that a model where each firm's profit is increasing in its own quality is better suited for a dynamic investigation of a market with vertical differentiation.

In order to simplify the analysis, we assume that there is no discounting. As we want to stay close to the model with small discounting, we use the strong long-run average criterion as defined in Dutta [10] as the intertemporal evaluation criterion. This criterion can be justified as the limit of models where the discount rate is tending to zero, and it retains the recursive formulation of standard discounted dynamic programming. Under the assumptions of no discounting and quantity competition, it is surprisingly simple to examine the Markov perfect equilibria of the model. In Section 5, we show that for quite general demand structures the comparisons between static and dynamic equilibrium policies can be based exclusively on information about static payoff functions. It is hoped that the simplicity of the technique of undiscounted dynamic programming as used here will prove useful in other applications beyond the scope of this paper.

In Section 4, we assume that the underlying stage game is the standard linear model used in the literature on vertical differentiation. This allows us to interpret the dynamic equilibria in an economically intuitive manner. Using the curvature properties of the stage game profit functions, we show that aggressive entry corresponds to relatively low (current) expected quality of the entrant's product while cautious entry corresponds to higher expected quality. In the linear model, we can also solve the dynamic equilibrium policies explicitly and, as a result, we get a set of empirically testable predictions for the model.

The paper proceeds as follows. Section 2 introduces the basic model and the learning environment. Section 3 derives the benchmark results of the static duopoly game. The main results are then presented for the standard linear demand specification in Section 4 where we derive the Markov perfect equilibrium of the intertemporal game. In Section 5, we extend the model beyond the linear specification and show that the qualitative conclusions extend to much more general demand structures. All the proofs are relegated to an appendix.

Initially all market participants have a common prior belief α_0 that the new product has a high valuation, or

$$\alpha_0 = \Pr(\mu = \mu_H).$$

The expected value given a belief $\alpha(t)$ in period t is denoted by $\mu(\alpha(t))$, where

$$\mu(\alpha(t)) \triangleq \alpha(t)\mu_H + (1 - \alpha(t))\mu_L.$$

Since the buyers are nonatomic they have no individual effect on prices and quantities, and as a result they choose according to their myopic preferences at each stage.[1] To complete the description of the stage game payoffs we need to specify the profit functions for the two firms. The flow profits resulting from a vector of quantities $(q_I(t), q_E(t))$ are given by $p_i(t)q_i(t)$ for $i = I, E$, where the $p_i(t)$ are obtained from static market clearing conditions.

The uncertainty about the new product can only be resolved over time by experience with the new product. We assume that the evolution of the belief about the quality of the new product is governed by the diffusion process

$$d\alpha(t) = \sqrt{\frac{q_E(t)\alpha(t)(1 - \alpha(t))(\mu_H - \mu_L)}{\sigma^2}} \, dB(t), \qquad (1.1)$$

where $B(t)$ is the standard Wiener process. In the appendix, we provide a microfoundation for this particular form of the evolution of the beliefs. There we derive the diffusion process $\alpha(t)$ from a discrete time model with a finite number of buyers, where each buyer is sampling from a normal distribution with known variance σ^2 and unknown mean μ, which is either μ_L or μ_H.

Observe that, being a posterior belief, $\alpha(t)$ follows a martingale; i.e. has a zero drift. The variance of the process is at its largest when $\alpha(t)$ is away from its boundaries as the marginal impact of new information is at its largest when the posterior is relatively imprecise. The economic assumption behind the form of this particular process is that the variance in the posterior belief is linear in the quantity of sales by the entrant and thus the informativeness of the market experiment grows linearly in the sales of the entrant. The remaining term in the expression, $(\mu_H - \mu_L)/\sigma^2$, is

---

[1]We are implicitly assuming that the firms' information sets consist of all past market observations, i.e. all past prices and quantities.

sometimes referred to as the signal-to-noise ratio as it measures the strength of the signal $\mu_H - \mu_L$ compared to the inherent noise in the observation structure, σ^2. Define

$$\Sigma(\alpha(t)) \equiv \sqrt{\frac{\alpha(t)(1-\alpha(t))(\mu_H - \mu_L)}{\sigma^2}}.$$

From (1.1), we see that as long as $q_E(t)$ is bounded away from 0 for all t, $\alpha(t)$ converges to $\alpha^* \in \{0,1\}$ almost surely. In fact, the convergence is fast enough to make the limit

$$\lim_{T \to \infty} \mathbb{E}_{\alpha_0}\left[\int_0^T [\phi(\alpha(t)) - \phi(\alpha^*)]dt\right],$$

where $\phi(\cdot)$ is an arbitrary continuous and piecewise smooth function of α, finite almost everywhere. This result allows us to use the strong long-run average as the intertemporal evaluation criterion in our model.

3. Static Equilibrium

In this section, we derive some of the basic equilibrium properties in the static model for the case where $f(\theta)$ is the uniform density. The safe product is worth s and the new product is worth $\mu(\alpha)$ for a given α. In the description of the equilibrium conditions we shall assume that $\mu(\alpha) \leqslant s$. The corresponding results for $\mu(\alpha) > s$ are symmetric and stated in the relevant proposition as well. Define α_m as the belief at which the expected value of the new product is equal to the established one:

$$\mu(\alpha_m) = s \Leftrightarrow \alpha_m = \frac{s - \mu_L}{\mu_H - \mu_L}.$$

The static prices and quantities are denoted by P_E, P_I, Q_E, and Q_I for the entrant and the incumbent respectively. The *equilibrium* prices and quantities are denoted by $P_i(\alpha)$ and $Q_i(\alpha)$ as we are interested in the comparative static behavior of the equilibrium variables as a function of the belief α. The equilibrium conditions are given by the profit maximization conditions of the firms and the indifference conditions of the marginal buyers. The latter can be stated as

$$(1 - Q_I)s - P_I = (1 - Q_I)\mu(\alpha) - P_E$$

and

$$(1 - Q_I - Q_E)\mu(\alpha) - P_E = 0.$$

The first indifference condition implies that at the equilibrium prices buyers with valuations $\theta \in [1-Q_I, 1]$ prefer the incumbent. The second indifference condition implies that buyers with valuations $\theta \in [1 - Q_I - Q_E, 1 - Q_I]$ prefer the entrant. It also follows that all buyers get a nonnegative expected utility from their purchases, but the segment with the lowest valuations may not buy at all. The market clearing prices for given quantities $\{Q_E, Q_I\}$ are

$$P_E = \mu(\alpha)(1 - Q_I - Q_E)$$

and

$$P_I = s(1 - Q_I) - \mu(\alpha)Q_E.$$

Since the derivation of the static equilibrium is completely standard, the derivation of the results is relegated to the appendix.

Proposition 1 (Static Policies):

1. $P_E(\alpha)$, $Q_E(\alpha)$, and $P_E(\alpha)Q_E(\alpha)$ are increasing in α.
2. $P_I(\alpha)$, $Q_I(\alpha)$, and $P_I(\alpha)Q_I(\alpha)$ are decreasing in α.

Proof: See the appendix.   □

As expected, the quantity and the price of the entrant are increasing in α. The entrant can increase his sales as well as his margins as the quality is improved. The incumbent responds to an increase in the value of the competing product by lowering his sales as well as his margins. It is worthwhile to point out that the monotonicity result extends over the entire range of posterior beliefs and holds also around the point α_m where the leadership between the two firms is changing. This is one instance where the model with quantity competition behaves in a more regular manner than the one with price competition, which displays nonmonotonicities in the prices and quantities around the switching point α_m.

Proposition 2 (Curvatures):

1. *For* $\mu(\alpha) < s$,

    (a) $P_E(\alpha)$, $Q_E(\alpha)$, and $P_E(\alpha)Q_E(\alpha)$ are convex in α;
    (b) $P_I(\alpha)$, $Q_I(\alpha)$, and $P_I(\alpha)Q_I(\alpha)$ are concave in α.

2. *For $\mu(\alpha) > s$,*

   (a) *$P_E(\alpha)$, $Q_E(\alpha)$, and $P_E(\alpha)Q_E(\alpha)$ are concave in α;*
   (b) *$P_I(\alpha)$, $Q_I(\alpha)$, and $P_I(\alpha)Q_I(\alpha)$ are convex in α.*

Proof: See the appendix. □

An intuition for the curvature as well as for the change in the curvature of the policies and revenues can be given as follows. For low posterior beliefs where $\mu(\alpha) < s$, a marginal increase in α increases the profit of the entrant through two channels. First, it increases the price for fixed quantities. This direct effect is the same at all levels of α as long as the quantities supplied are unchanged. There is also the indirect effect from a stronger competitive position of the entrant and the corresponding reduction in the quantity of the incumbent. This effect is strongest when α is close to α_m, and vanishes for very low values of α. The combination of these two effects leads to a convex profit function as long as $\mu(\alpha) < s$. As α increases beyond α_m, the position of the entrant resembles increasingly that of a monopolist. The indirect effect then becomes weaker and it is only the ability of the new firm to increase its prices which increases its profits.

4. Dynamic Equilibrium

In Subsection 4.1 we consider the dynamic optimization problems of the firms, and we also introduce the model without discounting using the strong long-run average criterion for evaluating payoffs. In Subsection 4.2 we then characterize the unique equilibrium and the associated equilibrium policies.

4.1. *Dynamic optimization*

In a discounted model the entrant's value function is the solution to the Hamilton–Jacobi–Bellman equations[2]

$$rV_E(\alpha) = \max_{q_E(\alpha)} \left\{ p_E(\alpha)q_E(\alpha) + \frac{1}{2}q_E(\alpha)\Sigma^2(\alpha)V_E''(\alpha) \right\}. \qquad (4.1)$$

If we tried to solve this equation jointly with the corresponding one for the incumbent, we would obtain a nonlinear system of second-order differential

---

[2]See Dixit and Pindyck [9] or Harrison [17] for a complete derivation of the dynamic programming equation in continuous time when uncertainty is represented by a Brownian motion.

equations. An analytical solution of such systems is, in general, impossible. For this reason we consider the limiting case as the discount rate vanishes or $r \to 0$ and then derive the equilibrium policies under the strong long-run average criterion.[3]

The *strong* long-run average criterion has the important property that the optimal policies under this criterion are the unique limits to the associated policies under discounting. The equilibrium policies to be derived therefore maintain all the qualitative properties of the equilibrium with small, but positive, discount rates $r > 0$. In particular, they preserve the intertemporal trade-off of the experimentation policies under discounting.

We start by fixing the policies of all other players to a set of arbitrary (Markovian) policies and consider the decision problem of the entrant. In the next subsection, we return to the full equilibrium problem. The reformulation for the incumbent only requires the obvious substitutions. The long run average payoff for the entrant under an initial belief α_0 is given by

$$v_E(\alpha_0) = \sup_{q_E(\alpha(t))} \lim_{T \to \infty} \frac{1}{T} \mathbb{E}_{\alpha_0} \left[\int_0^T q_E(\alpha(t)) p_E(\alpha(t)) dt \right].$$

Since $\alpha(t)$ converges almost surely to zero or one, the long-run average starting at any arbitrary belief α_0 is given by[4]

$$v_E(\alpha_0) = (1 + \alpha_0) v_E(0) + \alpha_0 v_E(1).$$

As $v_E(0)$ and $v_E(1)$ are simply the full information payoffs associated with the static payoffs at $\alpha = 0$ or $\alpha = 1$, the long-run average can be computed exclusively on the basis of the static problem. In contrast, the *strong* long-run average is defined through the optimization problem:

$$V_E(\alpha_0) = \sup_{q_E(\alpha(t))} \lim_{T \to \infty} \mathbb{E}_{\alpha_0} \left[\int_0^T (q_E(\alpha(t)) p_E(\alpha(t)) - v_E(\alpha(t))) dt \right]. \quad (4.2)$$

Thus the strong long-run average criterion maximizes the expected return net of the long-run average. The limit as $T \to \infty$ in (4.2) is well-defined and finite. The strong long-run average hence discriminates between policies based on finite time intervals as well. The infinite horizon problem (4.2) can

---

[3]See Dutta [10] for a detailed analysis of the link between optimality criteria under discounting and no discounting.

[4]It is easy to see that the sales of both firms are bounded away from zero at all points in time.

be represented by a dynamic programming equation,[5]

$$v_E(\alpha) = \max_{q_E(\alpha)} \left\{ q_E(\alpha)p_E(\alpha) + \frac{1}{2}q_E(\alpha)\Sigma^2(\alpha)V_E''(\alpha) \right\}. \qquad (4.3)$$

The difference between the dynamic programming equation under discounting (4.1) and under no discounting (4.3) is simply that the flow payoff, $rV_E(\alpha)$, is replaced by the long-run average payoff, $v_E(\alpha)$, whereas the right-hand side of the equation remains identical. However, as the long-run average is independent of the current policy $q_E(\alpha)$, we can rewrite (4.3) to read

$$0 = \max_{q_E(\alpha)} \left\{ q_E(\alpha)p_E(\alpha) - v_E(\alpha) + \frac{1}{2}q_E(\alpha)\Sigma^2(\alpha)V_E''(\alpha) \right\}.$$

After dividing the entire expression through $q_E(\alpha)$ (assuming that $q_E(\alpha) > 0$ can be guaranteed), the optimality equation can be rewritten as

$$0 = \max_{q_E(\alpha)} \left\{ p_E(\alpha) - \frac{v_E(\alpha)}{q_E(\alpha)} \right\} + \frac{1}{2}\Sigma^2(\alpha)V_E''(\alpha). \qquad (4.4)$$

This last expression demonstrates the advantage of analyzing the undiscounted program rather than the discounted one. The first-order conditions do not involve the second derivative of the value function any more. The only modification relative to the static program is the introduction of the long-run average, but as we saw above it can be computed on the basis of the static equilibrium as well.

4.2. Equilibrium analysis

Consider now the entire set of equilibrium conditions under no discounting. The dynamic programming equation for the entrant is

$$0 = \max_{q_E(\alpha)} \left\{ p_E(\alpha)q_E(\alpha) - v_E(\alpha) + \frac{1}{2}q_E(\alpha)\Sigma^2(\alpha)V_E''(\alpha) \right\}, \qquad (4.5)$$

and for the incumbent the equation is by extension

$$0 = \max_{q_I(\alpha)} \left\{ p_I(\alpha)q_I(\alpha) - v_I(\alpha) + \frac{1}{2}q_E(\alpha)\Sigma^2(\alpha)V_I''(\alpha) \right\}, \qquad (4.6)$$

where $v_I(\alpha)$ and $v_E(\alpha)$ are the long-run average payoffs of the sellers.

---

[5]For the full details on the strong long-run average in continuous-time dynamic programming models see Krylov [21], and for a derivation of Bellman's equation for the problem in a related application see Bolton and Harris [6].

Since each buyer is of negligible size, her decision doesn't influence the market experiment and hence her value of information is independent of her decision. The purchase decision of each buyer is therefore exclusively determined by the current payoff offered by the various alternatives. In consequence, the sorting of buyers in the intertemporal equilibrium will display the same structure as in the static equilibrium: for $\mu(\alpha) \leqslant s$,

$$
\begin{aligned}
p_E(\alpha) &= \mu(\alpha)(1 - q_I(\alpha) - q_E(\alpha)), \\
p_I(\alpha) &= s(1 - q_I(\alpha)) - \mu(\alpha)q_E(\alpha);
\end{aligned}
\tag{4.7}
$$

and symmetrically for $\mu(\alpha) > s$,

$$
\begin{aligned}
p_E(\alpha) &= \mu(\alpha)(1 - q_E(\alpha)) - sq_I(\alpha), \\
p_I(\alpha) &= s(1 - q_I(\alpha) - q_E(\alpha)).
\end{aligned}
\tag{4.8}
$$

Definition 1 (Markov Perfect Equilibrium): A Markov perfect equilibrium is a pair of functions $\{q_E(\alpha), q_I(\alpha)\}$ such that the equations (4.5)–(4.8) are satisfied for all $\alpha \in [0, 1]$.

By substituting the prices into the value functions, we obtain the value functions of the firms as a function of the quantities $\{q_E(\alpha), q_I(\alpha)\}$,

$$
\begin{aligned}
0 = \max_{q_E(\alpha)} \Big\{ &\mu(\alpha)(1 - q_E(\alpha)) - m(\alpha)q_I(\alpha))q_E(\alpha) - v_E(\alpha) \\
&+ \frac{1}{2}q_E(\alpha)\Sigma^2(\alpha)V_E''(\alpha) \Big\},
\end{aligned}
\tag{4.9}
$$

and

$$
\begin{aligned}
0 = \max_{q_I(\alpha)} \Big\{ &(s(1 - q_I(\alpha)) - m(\alpha)q_E(\alpha))q_I(\alpha) - v_I(\alpha) \\
&+ \frac{1}{2}q_E(\alpha)\Sigma^2(\alpha)V_I''(\alpha) \Big\}.
\end{aligned}
\tag{4.10}
$$

We can then solve for the unique equilibrium quantities by the methods from the previous subsection to get

$$
q_E(\alpha) = \sqrt{\frac{v_E(\alpha)}{\mu(\alpha)}}
\tag{4.11}
$$

and

$$q_I(\alpha) = \frac{1}{2} - \frac{1}{2}\frac{m(\alpha)}{s}\sqrt{\frac{v_E(\alpha)}{\mu(\alpha)}}, \tag{4.12}$$

where $m(\alpha) \triangleq \min\{s, \mu(\alpha)\}$. It can be verified that the quantity $q_I(\alpha)$ is continuous at $\alpha = \alpha_m$, but not differentiable. The equilibrium prices $p_E(\alpha)$ and $p_I(\alpha)$ follow from the indifference conditions (4.7) and (4.8) of the marginal buyers. The monotonicity properties of quantities and prices as a function of the posterior belief α, which we observed in the static equilibrium (as a comparative static result), are preserved in the dynamic model.

Proposition 3 (Prices and Quantities):

1. $p_E(\alpha)$, $q_E(\alpha)$, and $p_E(\alpha)q_E(\alpha)$ *are increasing in* α.
2. $p_I(\alpha)$, $q_I(\alpha)$, and $p_I(\alpha)q_I(\alpha)$ *are decreasing in* α.

Proof: See the appendix. □

Next we want to contrast the dynamic entry policies with the static policies. To this end, suppose that the value of information to the entrant is zero at some critical posterior belief α_c, or $V_E''(\alpha_c) = 0$. From (4.9), his dynamic best-response function at α_c is identical to the static one. As intertemporal considerations in terms of $V_E''(\alpha)$ or $V_I''(\alpha)$ enter the best response function of the incumbent only indirectly through the choices of the entrant (see (4.10)), it follows that if $V_E''(\alpha_c) = 0$, then necessarily $q_i(\alpha_c) = Q_i(\alpha_c)$ and $p_i(\alpha_c) = P_i(\alpha_c)$ for all $i \in \{E, I\}$. Moreover, since the dynamic programming equation (4.9) has to hold it follows that at α_c the *flow revenues* (static or intertemporal) of the entrant have to be equal to his *long-run average* $v_E(\alpha_c)$.

Recall that the long-run average $v_E(\alpha)$ at α_c is the expected value of the static equilibrium revenues at $\alpha = 0$ and $\alpha = 1$ weighted with $1 - \alpha_c$ and α_c respectively. Thus, even if we don't know $V_E(\alpha)$ or $V_I(\alpha)$ we can find the points where static and intertemporal values coincide through a comparison of the static values with the long-run average. Conversely, at all points where static revenues and the long-run average diverge we can expect to see discrepancies between static and intertemporal policies.

Proposition 4 (Single Crossing):

1. *The difference* $p_E(\alpha)q_E(\alpha) - v_E(\alpha)$ *crosses zero at most once and only from below.*

2. *The critical point α_c satisfies $\alpha_c > \alpha_m$.*
3. *A necessary condition for crossing to occur is $\mu_L < s < \mu_H$.*
4. *A necessary and sufficient condition for crossing to occur is*

$$[P_E(0)Q_E(0)]' - v_E'(0) < 0, \quad \text{and} \quad [P_E(1)Q_E(1)]' - v_E'(1) < 0.$$

Proof: See the appendix.   □

The proof proceeds by establishing the above properties first for the static revenues $P_E(\alpha)Q_E(\alpha)$ and then extending them to the intertemporal flow revenues $P_E(\alpha)q_E(\alpha)$. Thus there is at most one critical point where the value of information for the entrant is zero. As the equilibrium policies we derived earlier as well as the long-run average are continuous, it follows that the preference of the entrant toward information represented by $V_E''(\alpha)$ changes signs at most once. As the sign of the term $V_E''(\alpha)$ determines the bias in the intertemporal policy relative to the static policy, the proposition shows that this bias changes sign at most once, and in fact a necessary condition for the change is that there is uncertainty about the ranking of the alternatives, or $\mu_L < s < \mu_H$.

Observe that the necessary and sufficient condition for a change of sign is given entirely in terms of the static profit functions. The long-run average $v_E(\alpha)$ is linear in α and satisfies $v_E(\alpha) = P_E(\alpha)Q_E(\alpha)$ for $\alpha \in \{0,1\}$. We showed earlier (in Proposition 2), that the static profit function of the entrant is convex whenever $\mu(\alpha) < s$ and concave whenever $\mu(\alpha) > s$. Hence it is sufficient to compare the local behavior of the static profit to the long-run average around the endpoints. The equilibrium revenue and long-run average for the entrant are displayed in Fig. 5.1 for the case that the necessary and sufficient condition is satisfied.

The value of information for the entrant is represented by the second derivative of the value function $V_E''(\alpha)$. The dynamic programming equation (4.9) immediately shows that the value of information has the opposite sign of $p_E(\alpha)q_E(\alpha) - v_E(\alpha)$. This allows us to establish directly how the presence of market learning affects the equilibrium policies of the firms on either side of the critical value α_c.

Proposition 5 (Static vs. Dynamic Strategies):

1. *For $\alpha < \alpha_c$,*

   (a) $q_E(\alpha) > Q_E(\alpha)$ and $p_E(\alpha) < P_E(\alpha)$;
   (b) $q_I(\alpha) < Q_I(\alpha)$ and $p_I(\alpha) < P_I(\alpha)$.

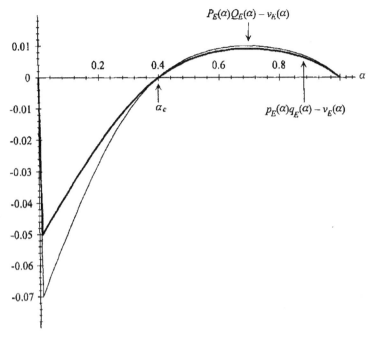

Figure 5.1.   Equilibrium revenue minus long-run average for entrant for $\mu_L = 99/100$, $s = 1$, $\mu_H = 2$.

2. For $\alpha > \alpha_c$,

(a) $q_E(\alpha) < Q_E(\alpha)$ and $p_E(\alpha) > P_E(\alpha)$;

(b) $q_I(\alpha) > Q_I(\alpha)$ and $p_I(\alpha) > P_I(\alpha)$.

Proof: See the appendix.   □

The behavior of the equilibrium prices and the quantities is displayed (in their differences) in Fig. 5.2 for the same environment as that in Fig. 5.1.

The curvature properties of the equilibrium policies provide us with valuable additional information about the intertemporal properties as the curvature properties can be directly translated into a time series profile by exploiting the fact that α is a martingale.

Proposition 6:

1. $q_E(\alpha)$ is concave in α,
2. $p_E(\alpha)$ is convex if $\mu(\alpha) \leqslant s$ and concave if $\mu(\alpha) > s$,
3. $q_I(\alpha)$ is convex in α, and
4. $p_I(\alpha)$ is convex in α.

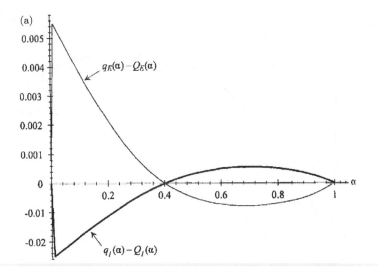

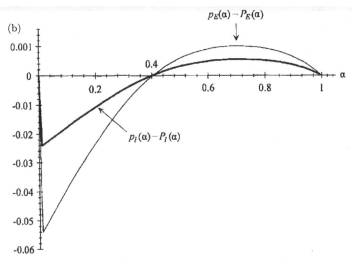

Figure 5.2. Static and dynamic equilibrium policies for $\mu_L = 99/100$, $s = 1$, and $\mu_H = 2$; (a) $q_E(\alpha) - Q_E(\alpha)$, $q_I(\alpha) - Q_I(\alpha)$, (b) $p_E(\alpha) - p_E(\alpha)$, $p_I(\alpha) - P_I(\alpha)$.

Proof: See the appendix. □

It is an immediate consequence of the previous proposition that $q_E(\alpha)$ is a submartingale, whereas $q_I(\alpha)$ and $p_I(\alpha)$ are supermartingales. Hence the expected sales and the expected prices of the incumbent rise over time whereas the expected sales of the entrant fall over time.

When we combine the time series behavior of the equilibrium with the properties of the equilibrium policies relative to their static counterparts, a rather complete picture regarding the entrance and deterrence behavior emerges. As the policies depend essentially on the current position of the firms in the quality spectrum, it is useful to consider the two polar cases relative to the intermediate case where $\mu_L < s < \mu_H$. If $\mu_L < \mu_H \leqslant s$ we refer to the new product as a *substitute*, and if $s \leqslant \mu_L < \mu_H$ then we refer to it as an *improvement*. A substitute is at best equal to the established product, whereas an improvement is at least as good as the established product. The first scenario may represent the introduction of a generic pharmaceutical or a no-name product, whereas the second may represent a new version of a current product with additional features whose (positive) contribution is yet uncertain.

With a substitute entry is aggressive and the equilibrium price of the entrant is below the static price. Over time, the expected equilibrium price of the entrant is increasing and the expected supply is decreasing as the entrant becomes more established and less aggressive. The effect of entry with uncertain valuations on the incumbent is that sales as well as prices are uniformly lower for the incumbent. But the submartingale property of both equilibrium variables then shows that sales and prices are expected to increase over time.

The entry strategy with an improved product is substantially different. The supply is at all times lower than with a static equilibrium, as the new firm will lose more through a (gradual) decrease in the posterior than through a (gradual) increase. In consequence, the new firm will start with lower than myopic quantities and will be essentially cream-skimming. Over time, its expected price is decreasing and the expected sales and revenues of the incumbent are increasing. Thus the aggressiveness of the strategy is almost entirely predicated by the position of the new firm relative to the established firm.

Finally, we may ask why the value of the information of the entrant has different signs for a substitute and for an improvement. The intuition behind this result can be obtained by considering the strategic incentives in the static game. With uncertainty, sales by the entrant lead to the release of more information to the market participants. This release of information can be thought of as inducing a zero mean lottery over posterior beliefs. From Proposition 2, the entrant's static revenue is convex in α with a substitute and concave in α with an improvement. Thus, if there were a single possibility to acquire additional information, the entrant would prefer more information as represented by more variance in the posterior belief with a convex static equilibrium profit function and less information with a concave one. The results above show that this preference for additional information in a single experiment also holds in the general dynamical model where information is acquired at all instants.

5. Robustness

In this section we discuss in some detail how robust our equilibrium results are to different modeling assumptions. In Subsection 5.1 we remove the assumption of a uniform distribution on θ and extend the analysis to more general inverse demand functions. In Subsection 5.2 we discuss how our qualitative results would be changed by considering price competition.

5.1. *Quantity competition and general distributions*

Consider a general distribution $F(\theta)$ over the unit interval. Associated with any $F(\theta)$ and an initial belief α is a static profit function $\pi_i(Q_E, Q_I|\alpha)$ for firm i. In addition, denote by $\pi_E(Q_E|\alpha)$ the profit function of the entrant when he faces a competitive fringe with quality s rather than a single competitor. We make the following three assumptions about the behavior of the static profit functions for the remainder of this section:

1. $\pi_i(Q_E, Q_I|\alpha)$ is concave in Q_i for all i and all α.
2. $\pi_E(Q_E|\alpha)$ is concave in Q_E for all α.
3. The static best response functions satisfy the stability condition: $-1 < Q_i'(Q_j) < 0, \forall i$.

As our main interest is in the dynamic aspects of the model, we do not attempt to present the most general conditions on $F(\theta)$ which would guarantee that the above fairly standard assumptions on the static profit functions are met. Yet it can be verified that a sufficient condition for all three assumptions jointly is that the distribution function $F(\theta)$ is convex, which includes the uniform density model analyzed so far.

We proceed to show that the qualitative properties of the entry and deterrence behavior can be derived in this general setting based exclusively on the interaction between static profit functions and long-run average values.

As before, dynamic programming equations characterize the Markov perfect equilibria; i.e.

$$0 = \max_{q_E} \left\{ \pi_E(q_E, q_I|\alpha) - v_E(\alpha) + \frac{1}{2}q_E \Sigma^2(\alpha) V_E''(\alpha) \right\},$$

and

$$0 = \max_{q_I} \left\{ \pi_I(q_E, q_I|\alpha) - v_I(\alpha) + \frac{1}{2}q_E \Sigma^2(\alpha) V_I''(\alpha) \right\},$$

where $v_E(\alpha)$ and $v_I(\alpha)$ are the long-run average revenues under the static profit functions $\pi_E(Q_E, Q_I|\alpha)$ and $\pi_I(Q_E, Q_I|\alpha)$. For $q_E > 0$, we may divide the above equations by q_E to obtain:

$$0 = \max_{q_E} \left\{ \frac{\pi_E(q_E, q_I|\alpha) - v_E(\alpha)}{q_E} \right\} + \frac{1}{2}\Sigma^2(\alpha) V_E''(\alpha), \qquad (5.1)$$

$$0 = \max_{q_I} \left\{ \frac{\pi_I(q_E, q_I|\alpha) - v_I(\alpha)}{q_E} \right\} + \frac{1}{2}\Sigma^2(\alpha) V_I''(\alpha). \qquad (5.2)$$

In order to facilitate the comparison with the static equilibrium which is a solution to

$$\max_{q_i} \{ \pi_i(Q_E, Q_I|\alpha) \}, \quad \forall i,$$

we consider the first-order conditions to the dynamic programming equations (5.1) and (5.2):

$$q_E \frac{\partial}{\partial q_E} \pi_E(q_E, q_I|\alpha) = \pi_E(q_E, q_I|\alpha) - v_E(\alpha) \qquad (5.3)$$

and

$$\frac{\partial}{\partial q_I} \pi_I(q_E, q_I|\alpha) = 0. \qquad (5.4)$$

We observe that the first-order condition of the incumbent leads to the same best response function as his static one. Moreover, if the right-hand side in (5.3) vanishes, then the equations (5.3) and (5.4) reduce to the static equilibrium conditions. Hence we know that the dynamic equilibrium conditions are satisfied at the static equilibrium values of $\{Q_E(\alpha), Q_I(\alpha)\}$ if and only if $\pi_E(Q_E(\alpha), Q_I(\alpha)|\alpha) = v_E(\alpha)$. Thus the coincidence of static and dynamic equilibrium policies is in general linked to the equality of the static equilibrium profit and the long-run average for the *entrant*.

Denote by $Q_i(q_f)$ the myopic best response of firm i to firm j's quantity, where we omit the dependence of Q_i on α for notational simplicity. As indicated by Eq. (5.4), the static and the dynamic best response are identical for the incumbent, or $q_I(q_E) = Q_I(q_E)$. All dynamic equilibria must therefore lie on the reaction curve of the incumbent, $\{q_E, q_I(q_E)\}$. Assumptions 1 and 3 guarantee that there is a single stable static equilibrium, and thus we know that for all $q_E > Q_E(\alpha)$, $q_E > Q_E(Q_I(q_E))$ and hence, by the strict concavity of $\pi_E(q_E, q_I)$ in q_E,

$$\frac{\partial \pi_E(q_E, q_I(q_E)|\alpha)}{\partial q_E} < 0, \tag{5.5}$$

for all $q_E > Q_E(\alpha)$. A similar argument can be made for $q_E < Q_E(\alpha)$ to show that

$$\frac{\partial \pi_E(q_E, q_I(q_E)|\alpha)}{\partial q_E} > 0. \tag{5.6}$$

As the first-order condition of the entrant in the dynamic equilibrium requires that

$$\mathrm{sgn}\left(\frac{\partial \pi_E(q_E, q_I(q_E)|\alpha)}{\partial q_E}\right) = \mathrm{sgn}(\pi_E(q_E, q_I(q_E)|\alpha) - v_E(\alpha)), \tag{5.7}$$

a local argument around the static equilibrium quantities $\{Q_E(\alpha), Q_I(\alpha)\}$ based on (5.5) and (5.6) suggests the direction in which dynamic quantities deviate from static ones. In fact, the argument is easy for the case that $\pi_E(Q_E(\alpha), Q_I(\alpha)|\alpha) < v_E(\alpha)$. More care is required in the case where $\pi_E(Q_E(\alpha), Q_I(\alpha)|\alpha) > v_E(\alpha)$ if we want to guarantee that *all* equilibria have the desired property. Therefore assume initially the following relation between the static equilibrium profits and the long-run average:

$$\pi_E(Q_E(\alpha), Q_I(\alpha)|\alpha) < v_E(\alpha).$$

To determine the location of the dynamic equilibrium we must determine $\mathrm{sgn}(\pi_E(q_E, q_I(q_E)|\alpha) - v_E(\alpha))$ on the locus $\{q_E, q_I(q_E)\}$. The claim is that every quantity q_E which satisfies the dynamic equilibrium conditions must imply that $q_E > Q_E(\alpha)$. To see this we observe that we have either

$$\pi_E(q_E, q_I(q_E)|\alpha) < v_E(\alpha) \quad \text{for all } q_E \tag{5.8}$$

or

$$\pi_E(q_E, q_I(q_E)|\alpha) \geqslant v_E(\alpha) \Rightarrow q_E > Q_E(\alpha). \qquad (5.9)$$

In the first case, the static profit function remains below the long-run average for all pairs $\{q_E, q_I(q_E)\}$, and the first-order condition (5.7) together with condition (5.5) implies that $q_E(\alpha) > Q_E(\alpha)$. Consider next the case of (5.9). If there exist values $\{q_E, q_I(q_E)\}$ such that the static profit exceeds the long-run average, then condition (5.5) shows that (5.7) cannot possibly hold at $q_E > Q_E(\alpha)$. Hence we can conclude that whenever $\pi_E(Q_E, Q_I|\alpha) < v_E(\alpha)$, the dynamic equilibrium quantity sold by the new firm exceeds the static equilibrium quantity, or $q_E(\alpha) > Q_E(\alpha)$. To establish this argument we only used the stability condition of the static best response function. The complementary results for

$$\pi_E(Q_E(\alpha), Q_I(\alpha)|\alpha) > v_E(\alpha),$$

are proved in the appendix under the additional concavity assumptions.

Proposition 7: *Suppose that assumptions 1–3 hold. Then,*

1. $\pi_E(Q_E(\alpha), Q_I(\alpha)|\alpha) < v_E(\alpha) \Rightarrow (q_E(\alpha) > Q_E(\alpha),$
2. $\pi_E(Q_E(\alpha), Q_I(\alpha)|\alpha) > v_E(\alpha) \Rightarrow (q_E(\alpha) < Q_E(\alpha).$

Proof: See the appendix. $\qquad\qquad\qquad\qquad\qquad\qquad\qquad\qquad\square$

Under Assumptions 1–3, the predictions for the dynamic model are then straightforward. To determine whether the equilibrium quantities of the new firm exceed or fall short of the myopic quantities, all we need to do is to compare the myopic equilibrium profits to the long-run average profits. If the static equilibrium profits are below the long-run average profits, then the new firm will adopt an aggressive sales policy and by the property of the best response function the incumbent will adopt a more defensive stance. In contrast, if the static equilibrium profits are above the long-run average revenues, the entrant will proceed cautiously with the introduction of the new product and the incumbent will increase his supply to the market. The dynamic programming equations also inform us that the entry strategies are always associated with $V_E''(\alpha) > 0$ and $V_E''(\alpha) < 0$, respectively. The change in the entry strategy can therefore generally be located as in the uniform model analyzed earlier at the intersection $\pi_E(Q_E(\alpha), Q_I(\alpha)|\alpha)v_E(\alpha)$, where all the necessary data can be computed on the basis of the static profit function alone.

5.2. *Price competition*

Finally, we sketch how the qualitative results would be affected by a model of price competition within the linear specification. We show that despite some fundamental differences in the static equilibrium profit functions, the dynamic equilibria of the two models share very similar properties.

The most important change in terms of the static equilibria of the two models is that the equilibrium profits are no longer monotone in the quality of the new product. As emphasized in the literature on vertical differentiation, the competitor with a lower quality product doesn't want to increase the quality of his product if this brings the inferior product too close to the superior product In consequence, the equilibrium prices and revenues are not monotone in α either, rather they display a global minimum at $\alpha = \alpha_m$. At the point α_m, price competition with identical products leads to the Bertrand outcome with marginal cost pricing. The static equilibrium profit functions display a kink at α_m, but on file intervals $[0, \alpha_m)$ and (α_m, ∞) they are concave for the entrant as well as for the incumbent.

In the dynamic model, the strategic interaction is more complex with price competition. With quantity competition, the only variable which affects the evolution of future states, i.e. the level of sales by the entrant, is directly a decision variable of the entrant. In obtaining the dynamic best response of the incumbent, we can therefore ignore the impact of his current decision on future states. But this implies that the best response of the incumbent to any output decision by the entrant is the same in the static and the dynamic models. As a result, all comparisons can be carried out by analyzing the shifts in the best response function of the entrant. In a model with price competition, the price decisions by the firms *jointly* determine the sales level of the entrant. In consequence, we have to analyze the joint effects of changes in the two best response functions on the dynamic equilibrium. To see how this interaction is resolved in equilibrium, we check how the static policies are modified by intertemporal considerations. Figure 5.3 illustrates the static equilibrium revenues as well as the long run average revenues for the case that the value of the new product can either be lower or higher than the established product. Due to the local minimum at $\alpha = \alpha_m$, the long-run average is always above the static revenues. Thus if the static policies were in fact the dynamic equilibrium policies, then the respective Bellman equations would indicate that $V_E''(\alpha) > 0$ as well as $V_I''(\alpha) > 0$. But this would imply that both firms

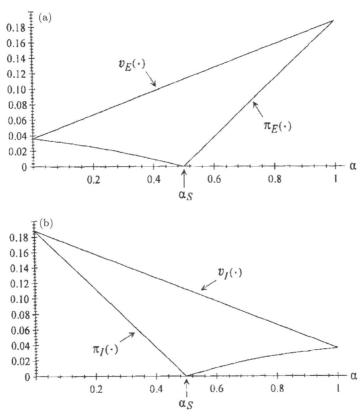

Figure 5.3.   Static equilibrium profits and long-run average (I), with $\mu_L = 3/2$, $s = 2$, and $\mu_H = 5/2$: (a) $\pi_E(Q_E(\alpha), Q_I(\alpha))$, and $v_E(\alpha)$; (b) $\pi_I(Q_E(\alpha), Q_I(\alpha))$ and $v_I(\alpha)$.

would like to see more sales by the entrant relative to the static equilibrium. We can therefore conjecture that the entrant will lower and the incumbent will raise its price relative to the static equilibrium price. In consequence, sales by the entrant must be larger (and the incumbent's sales must be lower) than in the static equilibrium. If, on the other hand, the product is an improvement and $s < \mu_L < \mu_H$, then the long-run average revenue is lower than the static equilibrium revenue, as Fig. 5.4 illustrates.

By the same intuition as that above we can then infer from the value functions that if the static policies were indeed equilibrium policies in the dynamic model, then it would have to be that $V_E''(\alpha) < 0$ as well as $V_I''(\alpha) < 0$. But this implies that both firms perceive sales by the entrant as

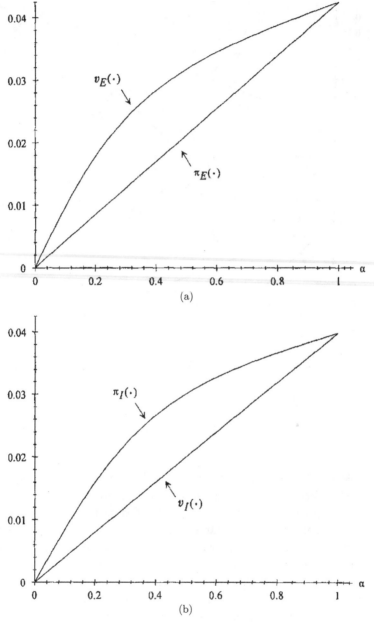

Figure 5.4. Static equilibrium profits and long-run average (II), with $s = \mu_L$ and $\mu_H = 2$: (a) $\pi_E(Q_E(\alpha), Q_I(\alpha))$ and $v_E(\alpha)$; (b) $\pi_I(Q_E(\alpha), Q_I(\alpha))$ and $v_I(\alpha)$.

carrying a negative value of information. The strategic response relative to the static solution for the new firm is to raise its price, and for the incumbent it is to decrease its price. This leads to lower quantities for the entrant and higher quantities for the incumbent. Thus the qualitative behavior of entrant and incumbent are similar in a model for quantity competition.

The only difference between the two models arises when $\mu_L < \mu_H < s$. By the concavity of the static profit function, the static revenues of the new firm are always below the long-run average. Observe, however, that a marginal improvement actually brings the new firm closer to its competitor in the quality spectrum and this leads to lower profits. If we interpret the random product quality as reflecting the uncertain value of some new features in the product, it would be unlikely that these features would be included in the product if $\mu_L < \mu_H < s$.

Our earlier paper, [4], analyzed a model of horizontal differentiation with price competition. In that model, an increase in the entrant's expected quality leads to a lower equilibrium price for the incumbent. This clearly has an adverse effect on the entrant's profit, but as the expected quality increases the entrant's profits depend to a lesser extent on the incumbent's decisions. As a result, the entrant's static profit increases in α at an increasing rate or, in other words, the static profit function is convex in α. A similar analysis applies to the incumbent, and as a result both the incumbent and the entrant have a positive value for information. In the current model with purely vertical differentiation, the incumbent's price increases with the entrant's expected quality in the static equilibrium in the region where $\mu(\alpha) > s$. As a consequence, the entrant's static profit function is concave in that region. Hence the differences in the two papers depend crucially on the economic distinction between the two models (i.e. the type of differentiation) rather than on the more arbitrary decision of price vs quantity competition.

6. Conclusion

This paper analyzed the entry game in a model with vertical differentiation. The precise location of the new product relative to the existing product was initially uncertain and was learned over time through experience. We derived the optimal entry and deterrence strategies for the competitors. It was shown that their qualitative properties depend on the current position of the new product in the quality spectrum. This allowed particularly sharp characterization results for the polar cases of a substitute or an improvement, respectively. By focusing on the Markov perfect equilibrium of the game, we derived a set of time series implications which may be amenable to empirical tests.

The current analysis faced some restrictions by the very nature of the model. First, we assumed that the value *and* the uncertainty about the new product were common to all buyers, after controlling for the element of vertical differentiation. It may be interesting to pursue how the equilibrium strategies would be affected if the experience by the buyers would contain an idiosyncratic element (see Milgrom and Roberts [25] for a simple monopoly model). The second limitation is the "once and for all" nature of the innovation presented by the new product. This was reflected in the model by the fact the posterior beliefs converged to either of the absorbing states $\alpha \in \{0, 1\}$ almost surely.

The techniques employed in this paper, however, generalize beyond the present model. The use of the undiscounted optimization criterion, and in particular the notion of the long-run average, allowed us to make a series of predictions based almost exclusively on the static equilibrium behavior. While the long-run average here was computed on the basis of the absorbing and mutually exclusive posterior beliefs, the technique extends naturally to ergodic distributions of the state variables. This should make the methodology used in this paper an attractive candidate for a much richer class of strategic models such as investment games and models of industry evolution, for which there are very few explicit solutions currently known (e.g. Ericson and Pakes [12]). In particular, it would seem feasible to combine dynamic competition models such as the one analyzed here with an ongoing process of innovation.

Appendix

We first present a derivation of the Bayesian filtering equation (1.1) based on a discrete time model with a finite number of buyers. The limiting behavior of the discrete learning model will lead to the Brownian motion depicted in (1.1) as the number of buyers becomes large and the time elapsed between any two periods converges to zero. Suppose therefore in an economy with N buyers, that each individual experiment with the new product by buyer i is an independent and identically distributed random variable $\tilde{x}_i$, with a normal distribution of unknown mean $(\mu_N = \mu/N)$ and known variance $(\sigma_N^2 = \sigma^2/N)$. The parameter μ can take on the value

μ_L or μ_H. Note that the mean as well as the variance of the individual experiment is scaled with respect to the total number of buyers N. The utility for buyer i is then given by $\theta_i x_i$, where x_i is a sample realization of $\tilde{x}_i$. Based on the individual experiences of all buyers, we can describe the aggregate or market experience in every period. As the informational content in every realization x_i is independent of the willingness to pay θ_i of individual i, we take the market experience to be the sum of the individual random variables while omitting the weights θ_i; i.e.

$$\tilde{x}(N) = \sum_{i=1}^{N} \tilde{x}_i.$$

As the mean and variance of the random variable $\tilde{x}_i$ are normalized by the number of buyers in the market, the aggregate mean and aggregate variance of the market experiment $\tilde{x}(N)$ are independent of the number N of buyers and are given by (μ, σ^2). If only a number k of buyers experiment with the new product, where $k \leqslant N$, then the aggregate experiment is given by the random variable

$$\tilde{x}(k) = \sum_{i=1}^{k} \tilde{x}_i,$$

which is again normally distributed with mean $\frac{k}{N}\mu$ and variance $\frac{K}{N}\sigma^2$. If we take the limit as N goes to infinity, the distribution of an aggregate experiment with a fraction n of the buyers, where

$$n = \frac{k}{N},$$

is given by

$$\tilde{x}(n) \sim N(n\mu, n\sigma^2).$$

Next we take the limit as the time between any two periods converges to zero. In the continuous time limit the market experiment then becomes a Brownian motion which can be described by the stochastic differential equation

$$dx(n(t)) = n(t)\mu dt + \sigma\sqrt{n(t)}dB(t), \quad t \in [0, \infty).$$

The flow realization in period t is given by the true mean μ weighted by the fraction of buyers participating in the experiment and the random term of the standard Brownian motion $dB(t)$ weighted by the standard deviation $\sigma\sqrt{n(t)}$.

Based on the evolution of the market experiment the market can update the prior belief α_0 to the posterior belief $\alpha(t)$. Based on the standard result for Bayesian updating in continuous time, it can be shown that the posterior belief $\alpha(t)$ also evolves as a Brownian motion.[6] It can be represented by

$$d\alpha(t) = \sqrt{\frac{n(t)\alpha(t)(1 - \alpha(t))(\mu_H - \mu_L)}{\sigma^2}}\,dB(t),$$

and as in equilibrium $n(t) = q_E(t)$, Eq. (1.1) follows.

Proof of Proposition 1. The static Nash equilibrium of the duopoly is obtained by solving simultaneously for profit maximizing $\{Q_E(\alpha), Q_1(\alpha)\}$.[7] By solving the best response functions simultaneously, we get

$$Q_E(\alpha) = \frac{\mu(\alpha) + M(\alpha) - m(\alpha)}{4M(\alpha) - m(\alpha)} \quad \text{and} \quad Q_I(\alpha) = \frac{s + M(\alpha) - m(\alpha)}{4M(\alpha) - m(\alpha)},$$

$$\text{(A.1)}$$

where $m(\alpha)$ and $M(\alpha)$ are defined as follows:

$$m(\alpha) \triangleq \min\{s, \mu(\alpha)\}, \quad M(\alpha) \triangleq \max\{s, \mu(\alpha)\}.$$

The equilibrium prices follow from the market clearing conditions and the monotonicity properties follow directly from the relevant derivatives. ☐

Proof of Proposition 2. The curvature properties follow directly from the second derivatives of the equilibrium (A.1) and prices. ☐

The next lemma records the construction of the long-run averages for the firms.

---

[6]See Lipster and Shiryayev [22, Chap. 9] for the derivation of the filtering equation for the continuous time, Brownian motion model.

[7]With the linear demand specification, the profit function of each firm is concave in its own quantity, and therefore first-order conditions are also sufficient for optimality.

Lemma 1 (Long-Run Averages): *The long-run averages are given by*

$$v_E(\alpha) = (1-\alpha)\frac{\mu_L(\mu_L + M(0) - m(0))^2}{(4M(0) - m(0))^2} + \alpha\frac{\mu_H(\mu_H + M(1) - m(1))^2}{(4M(1) - m(1))^2}$$

$$(A.2)$$

and

$$v_I(\alpha) = (1-\alpha)\frac{s(s + M(0) - m(0))^2}{(4M(0) - m(0))^2} + \alpha\frac{s(s + M(1) - m(1))^2}{(4M(1) - m(1))^2}.$$

Proof: The long-run average values $v_i(\alpha)$ are equal to the expected full-information payoffs

$$v_i(\alpha) = (1-\alpha)v_i(0) + \alpha v_i(1),$$

if $q_E(\alpha)$ is bounded away from zero for all α. It can be verified from (4.11) that this indeed guaranteed in equilibrium. As $v_i(0)$ and $v_i(1)$ are simply the values of the full information static equilibrium problems with $\alpha \in \{0, 1\}$, the composite values follow. □

Next we record without proof some properties of ratios and products of $\mu(\alpha)$ and $v_E(\alpha)$.

Lemma 2:

1. *The ratios $v_E(\alpha)/\mu(\alpha)$ and $\sqrt{v_E(\alpha)/\mu(\alpha)}$ are increasing and concave in α.*
2. *The product $v_E(\alpha)\mu(\alpha)$ is increasing and convex in α.*
3. *The product $\sqrt{v_E(\alpha)\mu(\alpha)}$ is increasing and concave in α.*

Proof of Proposition 3. The first-order conditions associated with (4.9) and (4.10) deliver the solutions for $q_E(\alpha)$ and $q_I(\alpha)$ given in (4.11) and (4.12). The market clearing conditions (4.7) and (4.8) lead to the equilibrium prices:

$$p_E(\alpha) = \mu(\alpha)\left(\frac{1 - \sqrt{v_E(\alpha)}}{\mu(\alpha)}\right) - m(\alpha)\left(\frac{1}{2} - \frac{1}{2}\frac{m(\alpha)}{s}\sqrt{\frac{v_E(\alpha)}{\mu(\alpha)}}\right)$$

and

$$p_I(\alpha) = \frac{s}{2} - \frac{m(\alpha)}{2}\sqrt{\frac{v_E(\alpha)}{\mu(\alpha)}}.$$

Next we prove the monotonicity properties. Consider first $\mu(\alpha) \geqslant s$ or $m(\alpha) = s$. A necessary and sufficient condition for $q_E(\alpha)$ to be increasing is that $v_E(1)\mu(0) \geqslant v_E(0)\mu(1)$, which is equivalent to

$$\frac{\mu_L + M(0) - m(0)}{4M(0) - m(0)} \leqslant \frac{\mu_H + M(1) - m(1)}{4M(1) - m(1)},$$

which holds for all values of μ_L, μ_H, and s. It follows directly that $q_I(\alpha)$ and $p_I(\alpha)$ are decreasing in α. It remains to show that $p_E(\alpha)$ is increasing. Suppose initially that μ_L, $\mu_H \geqslant s$. It is sufficient to show that $\mu(\alpha) - \sqrt{\mu(\alpha)v_E(\alpha)}$ is increasing in α. As $\mu(\alpha) > v_E(\alpha)$ for all α, it suffices to show that

$$\frac{\mu'(\alpha)}{v'_E(\alpha)} \geqslant \sqrt{\frac{\mu(\alpha)}{v_E(\alpha)}}.$$

By Lemma 2, the RHS is convex and decreasing, and evaluating the inequality at $\alpha = 0$ is sufficient as $\mu'(\alpha)$ and $v'_E(\alpha)$ are constant. We then obtain

$$\frac{\mu_H - \mu_L}{\frac{\mu_H(2\mu_H - s)^2}{(4\mu_H - s)^2} - \frac{\mu_L(2\mu_L - s)^2}{(4\mu_L - s)^2}} \geqslant \frac{4\mu_L - s}{2\mu_L - s}. \tag{A.3}$$

As the LHS is increasing in μ_H, it is sufficient to evaluate it as $\mu_H \downarrow \mu_L$, and (A.3) reads as

$$(4\mu_L - s)^2 \geqslant 8\mu_L^2 - 2\mu_L s + s^2,$$

which is satisfied by the hypothesis of $\mu_L \geqslant s$. Suppose next that $\mu_L < s < \mu_H$. Then with (A.3) the argument changes only slightly as $v_E(\alpha)$ has a different form, or

$$\frac{\mu_H - \mu_L}{\frac{\mu_H(2\mu_H - s)^2}{(4\mu_H - s)^2} - \frac{\mu_L(s)^2}{(4s - \mu_L)^2}} \geqslant \frac{4s - \mu_L}{s}.$$

As the LHS is now decreasing in μ_H, it is sufficient to evaluate it in the limit as $\mu_H \to \infty$, where the inequality is satisfied as it reads

$$4 \geqslant \frac{4s - \mu_L}{s}.$$

Consider next $\mu(\alpha) \leqslant s$. The price is then given by

$$p_E(\alpha) = \frac{1}{2}\mu(\alpha) - \sqrt{\mu(\alpha)v_E(\alpha)} + \frac{\mu(\alpha)}{2s}\sqrt{\mu(\alpha)v_E(\alpha)}.$$

Suppose initially that μ_L, $\mu_H < s$. It is now sufficient to show that

$$\frac{1}{2}\mu(\alpha) - \sqrt{\mu(\alpha)v_E(\alpha)} \tag{A.4}$$

is increasing in α. By the multiplication rule this is equivalent to showing that

$$\mu'(\alpha)\sqrt{\mu(\alpha)v_E(\alpha)} \geqslant \mu'(\alpha)v_E(\alpha) + \mu(\alpha)v'_E(\alpha).$$

As the term in (A.4) is concave in α, it remains to show that the inequality holds at $\alpha = 0$ or that

$$\frac{\mu_H - \mu_L}{4s - \mu_L} \geqslant \frac{(\mu_H - \mu_L)s}{(4s - \mu_L)^2} + \frac{\mu_H s}{(4s - \mu_H)^2} - \frac{\mu_L s}{(4s - \mu_L)^2}.$$

As the RHS term is increasing faster in μ_H than the LHS, it is sufficient to evaluate it at $\mu_H = s$, or

$$\frac{s - \mu_L}{4s - \mu_L} \geqslant \frac{(s - \mu_L)s}{(4s - \mu_L)^2} + \frac{1}{9} - \frac{\mu_L s}{(4s - \mu_L)^2},$$

which is satisfied for all $\mu_L \leqslant s$. Suppose now that $\mu_L < s < \mu_H$; it is then sufficient to show that $p'_E(\alpha) > 0$ at $\alpha = 0$ by Lemma 2, which is equivalent to showing that at $\alpha = 0$

$$\mu'(\alpha)\sqrt{\mu(\alpha)v_E(\alpha)} + \frac{1}{2s}(3\mu(\alpha)\mu'(\alpha)v_E(\alpha) + (\mu(\alpha))^2 v'_E(\alpha))$$

$$\geqslant \mu'(\alpha)v_E(\alpha) + \mu(\alpha)v'_E(\alpha).$$

Since the LHS is increasing faster in μ_H than the RHS it is sufficient to evaluate the inequality at $\mu_H = s$, and again it can be verified that the inequality holds for all $\mu_L \leqslant s$. $\qquad\square$

The proof of Proposition 4 relies on the following two lemmas. The first states that the difference $P_E(\alpha)Q_E(\alpha) - v_E(\alpha)$ satisfies the same single-crossing properties as $p_E(\alpha)q_E(\alpha) - v_E(\alpha)$ does and the second shows that the crossing points of the two differences coincide.

Denote by A_c the crossing point for the static revenue function.

Lemma 3:

1. *The difference $P_E(\alpha)Q_E(\alpha) - v_E(\alpha)$ crosses zero at most once and only from below.*
2. *The critical point A_c satisfies $A_c > \alpha_m$.*
3. *A necessary condition for crossing is $\mu_L < s < \mu_H$.*
4. *A necessary and sufficient condition for crossing to occur is:*

$$[P_E(0)Q_E(0)]' - v'_E(0) < 0 \quad and \quad [P_E(1)Q_E(1)]' - v'_E(1) < 0.$$

Proof: (1) Observe initially that $P_E(0)Q_E(0) - v_E(0) = 0$ and $P_E(1)Q_E(1) - v_E(1) = 0$. We first show that if $\mu_L < \mu_H \leqslant s$, or $s \leqslant \mu_L < \mu_H$, then $P_E(\alpha)Q_E(\alpha) - v_E(\alpha)$ never crosses at any $\alpha \in (0,1)$. By Lemma 1, $v_E(\alpha)$ is linear in α, and by Proposition 2, $P_E(\alpha) Q_E(\alpha)$ is either convex or concave, respectively. This together with the behavior at the end points excludes an interior crossing point. Consider next $\mu_L < s < \mu_H$, then the revenue function $P_E(\alpha) Q_E(\alpha)$ changes curvature behavior exactly once at $\alpha = \alpha_m$. As the curvature changes from convex to concave, the boundary behavior then implies that $P_E(\alpha)Q_E(\alpha) - v_E(\alpha)$ has to cross from below and can cross zero at most once.

(2) It is easily verified that at $\alpha = \alpha_m$, $P_E(\alpha_m)Q_E(\alpha_m) - v_E(\alpha_m) < 0$.

(3) The necessary condition follows from the arguments given for (1).

(4) The necessary and sufficient conditions follow from the curvature and boundary behavior of the static and long-run revenue functions. $\qquad\square$

Lemma 4: $\alpha_c = A_c$.

Proof: As $p_E(\alpha)q_E(\alpha)$ and $v_E(\alpha)$ are continuous, a change in sign for $p_E(\alpha)q_E(\alpha) - v_E(\alpha)$ requires a point $\alpha = \alpha_c$ at which

$$p_E(\alpha_c)q_E(\alpha_c) - v_E(\alpha_c) = 0. \qquad (A.5)$$

At such a point α_c, either $q_E(\alpha_c) = Q_E(\alpha_c)$ or $q_E(\alpha_c) \neq Q_E(\alpha_c)$. Suppose first that $q_E(\alpha_c) = Q_E(\alpha_c)$ were to hold, then it follows by the equilibrium conditions (4.9) and (4.7)–(4.8) that $p_E(\alpha_c) = P_E(\alpha_c)$ as well. But then it is has to be the case that $\alpha_c = A_c$. Suppose to the contrary that $q_E(\alpha_c) \neq Q_E(\alpha_c)$ would hold, then we show that (A.5) can't hold. Since $q_E(\alpha_c) \neq Q_E(\alpha_c)$, it has to be the case that $V_E''(\alpha_c) \neq 0$, by the first-order conditions from the Bellman equation (4.9). But then the hypothetical policies at α_c don't satisfy the Bellman equation and hence cannot be equilibrium conditions. Thus if $\alpha_c \in (0,1)$ it has to be that $\alpha_c = A_c$. It remains to show that if $P_E(\alpha)Q_E(\alpha) - v_E(\alpha)$ changes sign, then $p_E(\alpha)q_E(\alpha) - v_E(\alpha)$ necessarily changes signs as well. This is established easily as at α_c, $q_E(\alpha_c) = Q_E(\alpha_c)$ is a solution to the first-order condition (4.9), and as the solution is unique the claim follows. $\qquad\square$

Proof of Proposition 4. (1–3) By Lemma 3, the difference $p_E(\alpha)q_E(\alpha) - v_E(\alpha)$ shares the single-crossing behavior with the difference $P_E(\alpha)Q_E(\alpha) - v_E(\alpha)$. By Lemma 4, they also share the crossing point.

(4) As the myopic and intertemporal policies are identical at the endpoints, or $q_E(\alpha) = Q_E(\alpha)$ and $p_E(\alpha) = P_E(\alpha)$ for $\alpha \in \{0,1\}$, it follows

that the gradient of the flow revenues at the endpoints are necessary and sufficient conditions as well. □

Proof of Proposition 5. The asymmetry in the relationship between myopic and intertemporal quantities for the sellers follows directly from the best response function based on (4.9). It is therefore sufficient to consider the relationship between $q_E(\alpha)$ and $Q_E(\alpha)$. It follows from the first-order condition of the entrant that $q_E(\alpha) > Q_E(\alpha)$ if and only if $V_E''(\alpha) > 0$. Likewise $q_E(\alpha) < Q_E(\alpha)$ if and only if $V_E''(\alpha) < 0$. The results concerning the equilibrium quantities follow then directly from Proposition 4.

For the equilibrium prices consider first the interval $\alpha \in [0, \alpha_m]$. As the inequality $q_E(\alpha) > Q_E(\alpha)$ leads to $q_I(\alpha) < Q_I(\alpha)$, the best response function based on (4.9) implies together with the market clearing condition (4.7) that $p_E(\alpha) < P_E(\alpha)$, which in turn leads to $p_I(\alpha) < P_I(\alpha)$. Consider next the interval $\alpha \in [\alpha_m, \alpha_c]$. The inequality $q_E(\alpha) > Q_E(\alpha)$ leads to $q_I(\alpha) < Q_I(\alpha)$. The best response function based on (4.9) together with the market clearing condition (4.8) implies that $p_I(\alpha) < P_I(\alpha)$, which in turn leads to $p_E(\alpha) < P_E(\alpha)$. In the remaining interval $\alpha \in [\alpha_c, 1]$, the inequality $q_E(\alpha) < Q_E(\alpha)$ leads to $q_I(\alpha) > Q_I(\alpha)$. The best response function (4.9) together with the market clearing condition (4.8) implies that $p_I(\alpha) > P_I(\alpha)$, which in turn leads to $p_E(\alpha) > P_E(\alpha)$.

Proof of Proposition 6. (1) By Lemma 2.

(2) It follows directly from Lemma 2 that $p_E(\alpha)$ is convex for $\mu(\alpha) < s$ and concave for $\mu(\alpha) > s$.

(3) By Lemma 2.

(4) By Lemma 2. □

Proof of Proposition 7. The case of $\pi_E(Q_E(\alpha), Q_I(\alpha)|\alpha) < v_E(\alpha)$ was argued in the text. Suppose now that $\pi_E(Q_E(\alpha), Q_I(\alpha)|\alpha) > v_E(\alpha)$. Suppose first that $q_E < Q_E(\alpha)$, then we want to show that at q_E, $\pi_E(q_E, q_I(q_E)|\alpha) > v_E(\alpha)$. The argument is by contradiction. Suppose not; then it would follow from the Bellman equation that $V_E''(\alpha) > 0$, but then $q_E < Q_E(\alpha)$ cannot be an equilibrium as die entrant would have an incentive to deviate and increase the quantity. By a similar argument, we can exclude the possibility of $q_E > Q_E(\alpha)$ where $\pi_E(q_E, q_I(q_E)|\alpha) > v_E(\alpha)$ holds simultaneously.

Finally, we present sufficient conditions to rule out possible equilibria in the region where $q_E > Q_E$ and $\pi_E(q_E, q_I(q_E)|\alpha) < v_E(\alpha)$. Observe that for all q_E sufficiently close to $Q_E(\alpha)$:

$$q_E \frac{\partial \pi_E(q_E, q_I(q_E)|\alpha)}{\partial q_E} < \pi_E(q_E, q_I(q_E)|\alpha) - v_E(\alpha). \tag{A.6}$$

A sufficient condition to rule out equilibria with $q_E > Q_E(\alpha)$ is therefore that the derivative of the LHS is always below the derivative of the RHS for $q_E > Q_E(\alpha)$. Using the fact that

$$\pi_E(q_E, q_I(q_E)|\alpha) = p_E(q_E, q_I(q_E)q_E)$$

we may rewrite the inequality (A.6) as

$$q_E^2 \frac{\partial p_E(q_E, q_I(q_E))}{\partial q_E} < -v_E(\alpha).$$

The sufficient condition can then be written as

$$2\frac{\partial p_E(q_E, q_I(q_E))}{\partial q_E} + q_E \left(\frac{\partial^2 p_E(q_E, q_I(q_E))}{\partial q_E^2} + \frac{\partial^2 p_E(q_E, q_I(q_E))}{\partial q_I \partial q_E} q_I'(q_E) \right) < 0. \tag{A.7}$$

As the first term is strictly negative independent of $F(\theta)$, it is sufficient to show that

$$\frac{\partial^2 p_E(q_E, q_I(q_E))}{\partial q_E^2} + \frac{\partial^2 p_E(q_E, q_I(q_E))}{\partial q_I \partial q_E} q_I'(q_E) \leqslant 0.$$

Consider first $\mu(\alpha) \leqslant s$, then the equilibrium price of the entrant can be written as

$$p_E = \mu(\alpha) F^{-1}(1 - q_E - q_I),$$

and hence

$$\frac{\partial^2 p_E(q_E, q_I(q_E))}{\partial q_E^2} = \frac{\partial^2 p_E(q_E, q_I(q_E))}{\partial q_I \partial q_E}.$$

The sufficient condition (A.7) can then be written as

$$2\frac{\partial p_E(q_E, q_I(q_E))}{\partial q_E} + q_E \frac{\partial^2 p_E(q_E, q_I(q_E))}{\partial q_E^2}(1 + q_I'(q_E)) < 0. \tag{A.8}$$

By the assumption of concavity of the profit function of the duopolist.

$$2\frac{\partial p_E(q_E, q_I(q_E))}{\partial q_E} + q_E \frac{\partial^2 p_E(q_E, q_I(q_E))}{\partial q_E^2} < 0.$$

Now if

$$\frac{\partial^2 p_E(q_E, q_I(q_E))}{\partial q_E^2} > 0,$$

then (A.8) holds since $q_I'(q_E) < 0$ by the stability of the best response. On the other hand, if

$$\frac{\partial^2 p_E(q_E, q_I(q_E))}{\partial q_E^2} < 0,$$

then (A.8) holds since

$$\frac{\partial p_E(q_E, q_I(q_E))}{\partial q_E} < 0$$

and $1 + q_I'(q_E) > 0$.

Next suppose that $\mu(\alpha) > s$. Then the price of the entrant is given by

$$p_E = \mu(\alpha) F^{-1}(1 - q_E) + s[F^{-1}(1 - q_E - q_I) - F^{-1}(1 - q_E)]$$

Let $H(\cdot)$ be the inverse function of F, or $H = F^{-1}$, and let h be the first derivative of H. The condition (A.7) can be written as:

$$-2[h(1 - q_E)(\mu(\alpha) - s) + h(1 - q_E - q_I)s]$$
$$+ q_E[h'(1 - q_E)(\mu(\alpha) - s) + (1 + q_I'(q_E))h'(1 - q_E - q_I)s] < 0.$$
$$(A.9)$$

By the concavity of the profit function of the monopolist, we know that

$$-2h(1 - q_E - q_I)(\mu(\alpha) - s) + q_E h'(1 - q_E - q_I)(\mu(\alpha) - s) < 0 \quad (A.10)$$

and also that

$$-2h(1 - q_E)(\mu(\alpha) - s) + q_E h'(1 - q_E)(\mu(\alpha) - s) < 0. \quad (A.11)$$

But since $0 < 1 + q_I'(q_E) < 1$,

$$(-2h(1 - q_E - q_I)(\mu(\alpha) - s) + q_E h'(1 - q_E - q_I)(\mu(\alpha) - s))(1 + q_I'(q_E)) < 0,$$

and therefore

$$-2h(1 - q_E - q_I)s + (1 + q_I'(q_E))q_E h'(1 - q_E - q_I)s < 0. \quad (A.12)$$

Finally, adding (A.11) and (A.12) yields (A.9). $\qquad\square$

Acknowledgments

The authors thank Phillipe Aghion, Glenn Ellison, Ezra Friedman, Ariel Pakes, and Robin Mason, in particular, as well as two anonymous referees for many helpful comments. Financial support from NSF Grants SBR 9709887 and 9709340, respectively, is gratefully acknowledged. The first author thanks the Department of Economics at the University of Mannheim for its hospitality and the Sloan Foundation for financial support through a Faculty Research Fellowship. The second author thanks ESRC for financial support through Grant R000223448.

References

1. P. Aghion, M. P. Espinosa, and B. Jullien, Dynamic duopoly with learning through market experimentation, *Econ. Theory* **3** (1993), 517–539.
2. P. Aghion, P. Bolton, C. Harris, and B. Jullien, Optimal learning by experimentation, *Rev. Econ. Stud.* **58** (1991), 621–654.
3. K. Bagwell and M. Riordan, High and declining prices signal product quality, *Amer. Econ. Rev.* **81** (1991), 224–239.
4. D. Bergemann and J. Välimäki, Market diffusion with two-sided learning, *Rand J. Econ.* **28** (1997), 773–795.
5. P. Bolton and C. Harris, Strategic experimentation, *Econometrica* **67** (1999), 349–374.
6. P. Bolton and C. Harris, Strategic experimentation: The undiscounted case, *in* "Incentives and Organization: Essays in Honour of Sir James Mirrlees" (P. J. Hammond and G. D. Myles, Eds.), forthcoming.
7. G. Bonnano, Vertical differentiation with Cournot competition, *Econ. Notes* **15** (1986), 68–91.
8. A. Ching, Dynamic equilibrium in the U.S. prescription drug market after patent expiration, mimeo, University of Minnesota.
9. A. Dixit and R. Pindyck, "Investment under Uncertainty," Princeton Univ. Press, Princeton, NJ, 1994.
10. P. Dutta, What do discounted optima converge to?, *J. Econ. Theory* **55** (1991), 64–94.
11. D. Easley and N. M. Kiefer, Controlling a stochastic process with unknown parameters, *Econometrica* **56** (1988), 1045–1064.
12. R. Ericson and A. Pakes, Markov-perfect industry dynamics: A framework for empiricial work, *Rev. Econ. Stud.* **62** (1995), 53–82.
13. J. Gabszewicz and J.-F. Thisse, Price competition, quality, and income disparities, *J. Econ. Theory* **20** (1979), 340–359.
14. J. Gabszewicz and J. F. Thisse, Entry and (exit) in a differentiated industry, *J. Econ. Theory* **22** (1980), 327–338.
15. E. Gal-Or, Quality and quantity competition, *Bell J. Econ.* **14** (1983), 590–600.

16. J. E. Harrington, Experimentation and learning in a differentiated-products duopoly, *J. Econ. Theory* **66** (1995), 275–288.
17. J. Harrison, "Brownian Motion and Stochastic Flow Systems," Krieger, Malabar, 1985.
18. K. Judd and M. Riordan, price and quality in a new product monopoly, *Rev. Econ. Stud.* **61** (1994), 773–789.
19. G. Keller and S. Rady, Market experimentation in a dynamic differentiated-goods duopoly, mimeo, Stanford University and LSE, 1998.
20. R. Kihlstrom, L. Mirman, and A. Postlewaite, Experimental consumption and the Rothschild Effect, *in* "Bayesian Models of Economic Theory" (M. Boyer and R. Kihlstrom, Eds.), North-Holland/Elsevier, Amsterdam, 1984.
21. N. V. Krylov, "Controlled Diffusion Processes," Springer-Verlag, New York, 1980.
22. R. S. Liptser and A. N. Shiryayev, "Statistics of Random Process, I," Springer-Verlag, New York, 1980.
23. D. McFadden and K. Train, Consumers' evaluation of new products: Learning from self and others, *J. Polit. Econ.* **104** (1996), 683–703.
24. A. McLennan, Price dispersion and incomplete learning in the long-run, *J. Econ. Dynam. Control* **7** (1984), 331–347.
25. P. Milgrom and J. Roberts, Prices and advertising signals of product quality, *J. Polit. Econ.* **94** (1986), 796–821.
26. L. Mirman, L. Samuelson, and A. Urbano, Monopoly experimentation, *Int. Econ. Rev.* **34** (1993), 549–564.
27. E. C. Prescott, The multiperiod control problem under uncertainty, *Econometrica* **40** (1972), 1043–1058.
28. M. Rothschild, A two-armed bandit theory of market pricing, *J. Econ. Theory* **9** (1974), 185–202.
29. A. Shaked and J. Sutton, Relaxing price competition through product differentiation, *Rev. Econ. Stud.* **49** (1982), 3–13.
30. A. Shaked and J. Sutton, Natural oligopolies, *Econometrica* **51** (1983), 1469–1484.
31. D. Treffler, The ignorant monopolist: Optimal learning with endogenous information, *Int. Econ. Rev.* **34** (1993), 565–581.

Chapter 6

Dynamic Pricing of New Experience Goods*

Dirk Bergemann[†] and Juuso Välimäki[‡]

[†]*Department of Economics, and Cowles Foundation of Research in Economics, Yale University, New Haven, CT 06520-8268, USA*
dirk.bergemann@yale.edu
[‡]*Department of Economics, Aalto University School of Business, 02150 Espoo, Finland*
juuso.valimaki@aalto.fi

We develop a dynamic model of experience goods pricing with independent private valuations. We show that the optimal paths of sales and prices can be described in terms of a simple dichotomy. In a mass market, prices are declining over time. In a niche market, the optimal prices are initially low followed by higher prices that extract surplus from the buyers with a high willingness to pay. We consider extensions of the model to integrate elements of social rather than private learning and turnover among buyers.

1. Introduction

In this paper, we develop a simple and tractable model of optimal pricing for a monopolist that sells a new experience good over time to a population of heterogeneous buyers. Our main result shows that it is possible to classify all markets to be either *mass markets* or *niche markets* with qualitatively different equilibrium patterns for prices and quantities. These two different markets lead to different pricing strategies that have been called skimming and penetration pricing, respectively, in the earlier literature.

To classify the market as a mass or a niche market, it is sufficient to consider the intertemporal incentives of a new buyer who is uncertain about her tastes for the product. For this calculation, we *assume* that the

---

*This chapter is reproduced from *Journal of Political Economy*, **114**, 713–743, 2006.

market price is at its static monopoly level. The buyer compares (potential) current losses to future informational benefits from the purchase. In a mass market, she is willing to purchase at the static monopoly price, whereas in a niche market she refuses such trades. We show that in a mass market, the dynamic equilibrium prices decrease over time and uninformed buyers purchase in all periods. In a niche market, uninformed buyers do not buy at the static monopoly price. As a result, the monopolist must offer low initial prices to capture a larger share of the uninformed at the expense of targeting the more attractive informed segment of the market. This is in contrast to the mass market, where the monopolist skims in the early stages the more attractive part of the market (i.e. the uninformed buyers). In both cases, prices converge in the long run to static monopoly price levels. Yet in a mass market, the sales also converge to the static monopoly level, but in a niche market, quantities remain below that level as some uninformed buyers drop out of the market.

The model in this paper is an infinite-horizon, continuous-time model of monopoly pricing. There is a continuum of ex-ante identical consumers who have a unit demand per period for the purely perishable good. At each instant of time, the monopolist offers a spot price and the buyers decide whether to purchase or not. In the beginning, as the new product is introduced, all the buyers are uncertain about their valuation. By consuming the product, they learn their true valuation for the product in a stochastic manner. For analytical convenience, we assume that a perfectly revealing signal arrives according to a Poisson process to the active buyers in the market. We also assume that the aggregate distribution of preferences in the population is common knowledge.

In an experience goods market, the seller is facing two different submarkets simultaneously. The demand curve in the part of the population that has already learned its preferences is similar to the standard textbook case. Those buyers who are uncertain about the true quality of the product behave in a more sophisticated manner. Each purchase incorporates an element of information acquisition that is relevant for future decisions. The value of this information is endogenously determined in the market. If future prices are high, purchases are unlikely to yield information that results in future consumer surpluses. If future prices are low, it may be in the buyers' best interest to forgo purchases in the current period since future prices are attractive regardless of the true value of the product. As a result, current and future prices determine simultaneously the sales in the informed segment of the market and the value of information in the uninformed segment.

The market for new pharmaceuticals illustrates the key features of our model. In the pharmaceutical industry, each new drug undergoes an extensive period of prelaunch testing to determine its performance in the overall population. The aggregate uncertainty relating to the product is therefore negligible at the moment of introduction. Nevertheless, many drugs differ in their effectiveness and the incidence of side effects across patients. This idiosyncratic uncertainty provides a motive for the individual agent to experiment. A recent empirical study by Crawford and Shum (2005) regarding the dynamic demand behavior in pharmaceutical markets documents the important role of idiosyncratic uncertainty and learning in explaining demand.[1] For a data set of anti-ulcer descriptions, they observe substantial uncertainty about the idiosyncratic effectiveness of the individual drugs and high precision in the signals received through consumption experience. We model the effectiveness of the new treatment to an individual new patient as a random event. The time of response to the drug is random, and the response may be either positive or negative (successful recovery from the illness or severe side effects).

To our knowledge, the current paper is the first to address the issue of experimental consumption in a fully dynamic model with a population of heterogeneous buyers. We provide a tractable analytical framework and demonstrate the flexibility of our framework by outlining extensions of the basic model in Sections 5 and 6. We should mention that our model allows an alternative interpretation, which we spell out in Section 6. We can rephrase the random arrival of information as a random arrival of a consumption opportunity. With this specification, we assume that the buyers learn their true preferences upon their first purchase. In addition to being commonly used in other papers on experience goods, this assumption allows us to apply the model to a wider range of markets with experience goods such as the market for many professional services (e.g., hospitality, law, and Internet services).

Besides the earlier cited work on learning in markets for new pharmaceutical products, there is a growing literature on Bayesian learning

---

[1] In closely related work on learning in pharmaceutical markets, Coscelli and Shum (2004) estimate the impact of uncertainty and learning for the introduction of a new drug, and Ching (2002) provides structural, dynamic demand estimates when there is learning among patients about a new (generic) pharmaceutical with common values.

in consumer markets with experience goods.[2] Of particular interest for the current model is the recent work by Israel (2005), who uses the random arrival of information as an identification strategy in the context of the automobile insurance market. In his empirical analysis, automobile insurance is an experience good that is provided continuously to the buyer, punctuated by distinct "learning events," namely, the occurrence of an accident. If the insuree files a claim following the accident, he effectively has a random opportunity to learn about the quality of the service, the automobile insurance. The pattern of consumer departures from the insurer before and after such events therefore reveals the impact of learning.

The estimation of Bayesian models of experience goods is inherently difficult since it requires a dynamic model *with* substantial heterogeneity to estimate the Bayesian learning process by idiosyncratic agents. A common limitation in the above empirical contributions is their focus on the optimal behavior of the buyers paired with an exogenous pricing policy. A contribution of our paper is to present a parsimonious model of joint equilibrium behavior with forward-looking buyers and seller. The model has a number of features that will facilitate the empirical analysis of experience good markets. In our model, the equilibrium converges to the static optimal price. The true underlying demand can therefore be estimated as in standard static discrete-choice models (see Berry 1994). This in turn plausibly allows the identification and estimation of the learning rate λ, given the discount rate r, or the identification of the ratio λ/r from the dynamics of the price path. The result here may provide an important step to extend the estimation from models of Bayesian learning to models of Bayesian equilibrium behavior. The simple characterization in terms of niche and mass markets makes an empirical verification of the fit of the model feasible.

---

[2]A partial list of recent papers includes Erdem and Keane (1996), Ackerberg (2003), Erdem, Imai, and Keane (2003), Israel (2005), Osborne (2005), and Goettler and Clay (2006). The empirical evidence that emerges from these papers is that the initial information and the rate of learning differ substantially across markets. Ackerberg (2003) and Crawford and Shum (2005) find a fast learning rate for yogurt and pharmaceuticals, respectively. Erdem and Keane (1996) find little evidence of experimental consumption for laundry detergents. In the same market, Osborne (2005) finds evidence of initially similar expectations but different individual experiences in a model that allows for heterogeneous buyers. In Goettler and Clay (2006) and Israel (2005), learning is slow for an online grocery and automobile insurance, respectively.

The empirical work on dynamic pricing is surprisingly sparse, even in the absence of the experience good aspect.[3] An important exception is the paper by Lu and Comanor (1998), which investigates the determinants and the intertemporal evolution of prices for a large panel of new pharmaceutical products. The authors find that pharmaceuticals that represent a large therapeutic improvement (as measured by Food and Drug Administration ranking and surveys) over previous products tend to follow a skimming strategy. They start with a high price and then lower it over time. On the other hand, pharmaceuticals with a small improvement over previous products tend to follow a penetration strategy by first offering low prices and eventually switching to higher prices. These two different price patterns correspond to our distinction between mass and niche markets. A product with a substantial improvement provides value to the average consumer, and hence we expect the mass market condition to hold. On the other hand, a product with a small improvement naturally constitutes a niche market and hence follows a penetration strategy. The panel in Lu and Comanor's study is split between products designed for acute and chronic ailments. They show that products for chronic ailments are more likely to follow a skimming strategy. This empirical finding is consistent with our theoretical prediction since more frequent purchases, all else equal, lead to a faster arrival of information and hence a higher λ. Alternatively, more frequent consumption acts imply that the time elapsed between uses is smaller and hence the effective discount rate r is smaller, again a factor that contributes positively to a mass market.

Related literature — We now briefly discuss the relationship between prior theoretical contributions and the current model. An important early paper on the dynamic pricing of experience goods is Shapiro (1983). The author investigates the optimal price policy of a monopolist in a two-period model when each consumer learns the true value of the product through experience. The analysis and the results differ substantially from the current contribution since Shapiro imposes two specific assumptions: (i) each consumer acts myopically, and (ii) the expectation of the quality is biased with respect to the true expected quality. The myopic consumer fails to evaluate the option value of the experiment, and with the natural

---

[3] Recently Gowrisankaran and Rysman (2005) and Nair (2005) made substantial progress toward estimating dynamic price policies. They suggest structural models to estimate the dynamic pricing policies of firms in the context of durable good markets. They provide estimates for console video games and digital video disc players and digital cameras.

benchmark of expectationally unbiased buyers, his model predicts constant prices.

Cremer (1984) considers a model with initially identical buyers and idiosyncratic experience to explain the use of coupons for repeat buyers in a two-period setting. In contrast to our analysis, he considers the optimal commitment price path and does not analyze the time-consistent pricing policy.

In subsequent work by Farrell (1986), Milgrom and Roberts (1986), and Tirole (1988), the dynamic selling policy for experience goods has been investigated without biased expectations. A common feature of these three models is that the buyers differ ex-ante in their willingness to pay for quality and the true quality has either a zero or a positive value. When one of the possible quality values is zero, the identity of the marginal buyer is constant across the two periods. In consequence, the marginal buyer does not receive a positive surplus in either period and never associates an option value to early purchases. As a result, the marginal buyer acts myopically, and his decision does not reflect any intertemporal trade-off. Further, the equilibrium price path is always increasing, independent of the shape of the demand or the discount factor. In contrast, our model allows for the possibility that the monopolist discriminates intertemporally in the market in a more flexible manner, and as a result, our conclusions are quite different from those in the earlier literature.

Villas-Boas (2004) considers the equilibrium pricing of experience goods in a *duopoly model* with differentiation along a location and a taste dimension. The location is known at the outset, whereas tastes are learned through experience. The analysis is concerned mostly with brand loyalty, that is, whether buyers return to the seller they bought from in the past. It presents a sufficient condition on the skewness of the distribution under which brand loyalty exists in equilibrium.

In a recent and complementary contribution, Johnson and Myatt (2006) use the distinction between niche and mass markets to investigate under which conditions a (mean-preserving) spread of the valuation increases the revenue in the *static* optimal pricing problem. They interpret "advertising" and "marketing" as providing information that increases the variance of the valuations by the customers. In contrast, in our model the shape of the distribution is influenced by the pricing policy itself rather than being determined through separate instruments.

The paper is organized as follows. Section 2 sets up the basic model and discusses the appropriate solution concepts. Section 3 presents the problem of optimal demand management for the seller. Section 4 analyzes the properties of the optimal price path. Section 5 discusses the social efficient allocation and the role of idiosyncratic versus social learning. Section 6 provides extensions of the model, including inflow of new buyers, commitment to price paths, and random purchasing opportunities for the buyer. Section 7 presents conclusions. The proofs of all the results are collected in the Appendix.

2. Model

We consider a continuous-time model with $t \in [0, \infty)$ and a positive discount rate $r > 0$. A monopolist with a zero marginal cost of production offers a single product for sale in a market consisting of a continuum of consumers. For analytical simplicity, we assume that the buyers have unit demand for the product within periods and that the product is not storable. We also abstract from the possibility of price differentiation within periods. At each instant, the monopolist offers a spot price. Upon seeing the price, each consumer decides whether to purchase or not.

Every consumer is characterized by his idiosyncratic willingness to pay for the product, denoted by θ. The good is an experience good, and the true value of θ is initially unknown to the buyer and the seller. The ex-ante distribution of valuations is given by a continuously differentiable distribution function $F(\theta)$ with support $[\theta_l, \theta_h] \subset \mathbb{R}$. This distribution is assumed to be common knowledge and reflects our assumption that there is no aggregate uncertainty in the model. As the focus in this paper is on private individual experiences, we abstract from possible common sources of uncertainty. To simplify the analysis, we also require that $\theta[1 - F(\theta)]$ be strictly quasi-concave in θ. This assumption guarantees that the full-information profit maximization problem is well behaved.

All buyers are ex-ante identical, and their expected utility from consuming the product prior to learning their type is given by v:

$$v \triangleq \int_{\theta_l}^{\theta_h} \theta dF(\theta).$$

Throughout the paper, we assume that a perfectly informative signal (e.g., the emergence of side effects in a drug therapy) arrives at a constant Poisson rate λ for all buyers who purchase the product in a time interval of length dt.[4] In this case, the posterior distribution on θ remains constant at the prior until the signal is observed.[5] The most important analytical consequence of this assumption is that if the buyers have not observed a signal, they remain identical. After observing the signal, the buyers are heterogeneous, and the monopolist's key objective is to manage the endogenous composition of these two market segments.

As we analyze the dynamic behavior of the market, it is natural to use dynamic programming tools to derive the equilibrium conditions for the model. We assume that the only publicly observable variables are the prices and aggregate quantities. This is in line with the assumption that each individual buyer is small and has no strategic impact on the aggregate outcomes. The state variable of the model at time t is the fraction of informed buyers at t, denoted by $\alpha(t) \in [0, 1]$. Even though $\alpha(t)$ is not directly observable to the players, it can be calculated from the equilibrium purchasing strategies. If the uninformed buy in period t, the state variable $\alpha(t)$ evolves according to

$$\frac{d\alpha(t)}{dt} = \lambda[1 - \alpha(t)],$$

since in period t there are $1 - \alpha(t)$ currently uninformed buyers and a fraction λdt of them become informed in a time interval of length dt.

A Markovian pricing strategy for the seller is denoted by $p(\alpha)$. The uninformed buyer has a Markovian purchasing strategy $d^u(\alpha, p)$ that depends on the state variable α as well as the current price p. Similarly, the informed buyer with valuation θ adopts a Markovian purchasing strategy $d^\theta(\alpha, p)$.

The monopolist maximizes her expected discounted profit over the horizon of the game. The buyers maximize the expected discounted value of their utilities from consumption net of price. As there is no aggregate uncertainty, the price and aggregate sales processes are deterministic. The individual buyer, however, faces uncertainty regarding his true valuation and the random time at which he will receive the information.

---

[4]It might be natural to allow for cases in which λ depends on t. In the pharmaceutical example, such a time-varying arrival rate might reflect, e.g., the decline in the probability of a treatment being eventually successful given a number of unsuccessful trials. We have analyzed this possibility, but given that the qualitative features of the model remain the same, we report only the constant case.

[5]This assumption is made for ease of exposition only. We have computed the model for posteriors with positive and negative drift. The qualitative features of the equilibrium remain as in the case of a constant prior.

3. Demand Management

The basic issue in the introduction of a new product is the dynamic demand management. In the early stages, the majority of the buyers are inexperienced and uninformed. Over time, the segment of informed buyers grows. As the relative sizes of these two market segments change, the seller adapts her policy and shifts her attention to the more important segment. More precisely, the type of the marginal buyer whose willingness to pay determines the equilibrium price changes over time. With a new product, the marginal buyer is inevitably uninformed in the early stages. As the informed segment grows, the marginal buyer is more likely to come from that segment. Optimal demand management then determines the switch between these two market segments. The dynamic pricing policy of the seller therefore contains at its core an optimal stopping problem. The stopping point identifies the time at which the marginal buyer ceases to be uninformed.

After stopping, the marginal buyer is informed. What needs to be determined, however, is whether the uninformed buyers keep on purchasing the product. Either the uninformed buyers are priced out of the market or they stay in the market as inframarginal buyers. Whether the uninformed buyer will eventually stay in or drop out of the market is essentially a question of the size of the market in equilibrium.

The demand management problem of the seller is more subtle than a pure optimal stopping problem. In the canonical optimal stopping problem, the alternative payoffs do not depend on future policy choices. In the optimal pricing problem here, buyers are forward looking, and their willingness to pay today depends on future prices. The current revenue of the seller therefore depends on his future prices.

3.1. *Market size*

As the number of informed buyers increases in the market, the optimal price is determined by the distribution of the valuations, $F(\theta)$. When the seller ignores the uninformed buyers, the optimal monopoly price $\hat{p}$ maximizes the flow revenues from the informed buyers:

$$\hat{p} = \arg \max_{p \in \mathbb{R}_+} \{p[1 - F(p)]\}.$$

Of course, the price $\hat{p}$ is also the optimal price in the static monopoly problem in which each buyer knows his valuation for the object. The corresponding equilibrium quantity is denoted by $\hat{q} = 1 - F(\hat{p})$.

The key comparison for the analysis is between the willingness to pay of the uninformed buyers and the static equilibrium price $\hat{p}$. In a static setting the expected value v of the product is the *average* willingness to pay in the market. It follows that if the optimal price $\hat{p}$ is below the average willingness to pay, then the uninformed buyers stay in the market and eventually become informed.

In the intertemporal setting, the willingness to pay of the uninformed actually exceeds v in most cases. In addition to the expected flow value from consumption, the uninformed buyer also has a chance to learn more about his true valuation for the product. If the future price stays constant and equal to $\hat{p}$, then his willingness to pay is the value of a purchase today, or

$$\hat{w} = v + \frac{\lambda}{r}\mathbb{E}_\theta \max\{\theta - \hat{p}, 0\}. \qquad (3.1)$$

The uninformed buyer becomes informed at rate λ. When informed, she purchases the product if and only if $\theta - \hat{p} \geq 0$. Finally, the future benefits are discounted at rate r. The value of a purchase today is then simply the sum of the expected value of the flow consumption, v, and the expected value of information, $(\lambda/r)\mathbb{E}_\theta \max\{\theta - \hat{p}, 0\}$.

The monopoly price in the informed segment $\hat{p}$ and the expected value of information both depend on the distribution $F(\theta)$. We now distinguish between a *niche market* and a *mass market* by comparing the willingness to pay of the uninformed buyers, $\hat{w}$, with the optimal static price $\hat{p}$.

Definition 1 (Niche market and mass market):

1. The market is said to be a niche market if $\hat{w} < \hat{p}$.
2. The market is said to be a mass market if $\hat{w} \geq \hat{p}$.

In a mass market, the price $\hat{p}$ is so low that new buyers are willing to enter the market. The monopoly price $\hat{p}$ is independent of λ and r, and hence the mass market condition is more likely to occur if the rate of information arrival λ is large or the discount rate r is small.

We can gain further insight into the notions of niche and mass markets by a comparative static analysis. For a given r and λ, let us consider a family of distributions $F(\theta; \sigma^2)$, parameterized by variance σ^2 with a fixed mean. In this environment, the market is more likely to be a mass market if the variance is small and more likely to be a niche market if the variance is large. For a small variance, the seller can increase the sales by lowering his price just below the mean. In contrast, if the variance is large, the seller prefers to sell at a price above the mean to the upper tail of the market. The size of this particular segment is sufficiently large whenever the variance is large. This intuition is exact within the classes of binary, uniform, and normal distribution (with constant mean). In other words, for any such family of distributions, there exists a critical value $\bar{\sigma}^2$ such that for all $\sigma^2 > \bar{\sigma}^2$ the market is a mass market and for all $\sigma^2 > \bar{\sigma}^2$ the market is a niche market.[6]

3.2. *Optimal switching*

We first describe the intertemporal decision problems in terms of the familiar dynamic programming equations in Markov strategies. The solution of the optimal stopping problem will then follow from the optimality conditions. The size of the segment of informed buyers, $\alpha(t) \in [0,1]$, in period t is the state variable of the model. We omit the indexation with respect to time and simply write all value functions as a function of α rather than $\alpha(t)$. We start with the simple decision problem of the informed buyers. These buyers have complete information about their true valuation θ of the object. For a given price policy $p(\alpha)$ by the monopolist, we can determine the value function V^θ of the informed buyer from the Bellman equation

$$rV^\theta(\alpha) = \max\{\theta - p(\alpha), 0\} + \frac{dV^\theta}{d\alpha}\frac{d\alpha}{dt}. \tag{3.2}$$

The decision whether to buy or not to buy is solved by the myopic decision rule: buy whenever θ exceeds the current price $p(\alpha)$. The only intertemporal component in this equation (the second term) reflects the effect of a change in the composition of the market segments, represented by α, on the future utilities. Future utilities are affected by changes in α as future prices respond to changes in aggregate demand. These changes are beyond the control of any single (informed) buyer, and hence the myopic rule characterizes optimal behavior.

---

[6]We would like to thank the editor, Robert Shimer, who suggested the variance analysis. For arbitrary distributions, the comparison between the willingness to pay $\hat{w}$ and the optimal price $\hat{p}$ is more difficult since $\hat{w}$ and $\hat{p}$ are determined by different properties of the distribution function: $\hat{p}$ is determined by the first-order condition of the static revenue function and hence by the hazard rate at $\hat{p}$, whereas $\hat{w}$ is (partially) determined by the expectation over all valuations exceeding $\hat{p}$. This conditional expectation of the upper tail of the distribution can change in either direction with an increase in variance or related measures, such as dispersion or second-order stochastic dominance. Within the above parameterized classes of distributions, the behavior of the tail expectation is in balance with the change in the local hazard rate condition.

For the uninformed buyers, a purchase of the new product represents a bundle, consisting of the flow of consumption *and* information. Their value function $V^u(\alpha)$ is given by

$$rV^u(\alpha) = \max\{v - p(\alpha) + \lambda[\mathbb{E}_\theta V^\theta(\alpha) - V^u(\alpha)], 0\} + \frac{dV^u}{d\alpha}\frac{d\alpha}{dt}. \qquad (3.3)$$

The main difference between these two value functions reflects the value of information to the uninformed buyers. A purchase in the current period generates an inflow of information at rate λ. Conditional on receiving the signal, the uninformed becomes informed. In consequence the new value function becomes $V^\theta(\alpha)$ for some θ. From the point of a currently uninformed buyer, there is uncertainty about his true valuation θ. He estimates the expected gain from the information by taking the expectation with respect to θ. The informational gain attached to a current purchase is given by

$$\lambda[\mathbb{E}_\theta V^\theta(\alpha) - V^u(\alpha)].$$

The value function of the seller is denoted by $V(\alpha)$. We describe the seller's dynamic programming equation in two parts to separate the intertemporal considerations as cleanly as possible. The basic trade-off facing the firm is that sales are made at a single price in two separate market segments. If the firm decides to sell to the uninformed buyers as well as some informed ones, the relevant equation is given by

$$rV(\alpha) = \max_{p(\alpha)\in\mathbb{R}_+} \{p(\alpha)\{1 - \alpha + \alpha[1 - F(p(\alpha))]\}\} + \frac{dV}{d\alpha}\frac{d\alpha}{dt} \qquad (3.4)$$

subject to

$$p(\alpha) \le v + \lambda\mathbb{E}_\theta[V^\theta(\alpha) - V^u(\alpha)].$$

Here $1 - \alpha$ is the share of uninformed buyers in the population and $\alpha[1 - F(p(\alpha))]$ is the fraction of informed buyers who are willing to buy at prices $p(\alpha)$. The constraint on the price $p(\alpha)$ guarantees that the uninformed buyers are indeed willing to purchase at prices $p(\alpha)$.

If the monopolist sells to the informed segment only, then her value function satisfies

$$rV(\alpha) = \max_{p(\alpha)\in\mathbb{R}_+} \{p(\alpha)\alpha[1 - F(p(\alpha))]\}. \qquad (3.5)$$

In this latter case, the size of the informed segment, α, remains constant and $d\alpha/dt = 0$, since the flow of information to the uninformed buyers has stopped. The Markovian prices in this regime must hence remain constant in all future periods. With these preliminaries, we can state the following definition.

Definition 2 (Markov-perfect equilibrium): A Markov-perfect equilibrium of the dynamic game is a triple (d^u, d^θ, p) such that the problems (3.2)–(3.5) are simultaneously solved for all α and θ.

We now employ the dichotomy between a niche market and a mass market to find the optimal launch strategy as the solution to a specific stopping problem. We denote the size of the informed market segment at the stopping point by $\hat{\alpha}$.

3.3. *Niche market*

In the niche market the willingness to pay of the uninformed buyers is below the static optimal price: $\hat{w} < \hat{p}$. It follows that if the seller sets prices optimally in the informed segment, then the uninformed stop buying. In consequence, the seller has to decide how long she wishes to serve the uninformed market segment.

We now describe the marginal conditions that characterize the stopping point $\hat{\alpha}$. After $\hat{\alpha}$ is reached, the optimal dynamic price equals $\hat{p}$. At the stopping point, the uninformed buyers purchase the new product for the last time. Their willingness to pay at the stopping point is therefore exactly equal to $\hat{w}$. At the stopping point, the seller must be indifferent between charging $\hat{p}$ and $\hat{w}$:

$$\hat{\alpha}[1 - F(\hat{p})]\hat{p} = \{(1 - \hat{\alpha}) + \hat{\alpha}[1 - F(\hat{w})]\}\hat{w} + \frac{\lambda(1 - \hat{\alpha})}{r}[1 - F(\hat{p})]\hat{p}. \quad (3.6)$$

The indifference condition compares the revenue from $\hat{p}$ relative to revenue from $\hat{w}$. If the seller were to offer $\hat{p}$, then only those informed buyers who have a true valuation $\theta \geq \hat{p}$ purchase the product, leading to a sales volume of $\hat{\alpha}[1 - F(\hat{p})]$. On the other hand, if the seller were to offer $\hat{w}$, then all uninformed buyers would stay in the market and all informed buyers with $\theta \geq \hat{w}$ would also buy the object, leading to a larger sales volume of $(1 - \hat{\alpha}) + \hat{\alpha}[1 - F(\hat{w})]$. At price $\hat{w}$, $\lambda(1 - \hat{\alpha})$ currently uninformed customers become informed, and hence they will add to the revenue from the informed customers for all future periods. If we denote by $\pi(p, \alpha)$ the flow profit to the monopolist from price p when α is the fraction of informed buyers, then the above equation can be written as

$$\pi(\hat{p}, \hat{\alpha}) - \pi(\hat{w}, \hat{\alpha}) = \frac{\lambda(1 - \hat{\alpha})[1 - F(\hat{p})]\hat{p}}{r}. \tag{3.7}$$

The left-hand side represents the differential gains from extracting surplus from the informed agents, and the right-hand side represents the benefits from building up future demand. The latter is the long-term gain from an additional inflow of $\lambda(1 - \alpha)$ informed buyers of whom $1 - F(\hat{p})$ are willing to purchase at price $\hat{p}$. As the right-hand side is positive, we conclude that with niche markets, the monopolist sacrifices current profits to build up future demands.

Proposition 1 (Equilibrium stopping in the niche market): *If* $\hat{w} < \hat{p}$, *then*

1. $\hat{\alpha} < 1$ *and*
2. $\hat{\alpha}$ *is increasing in* λ *and decreasing in* r.

In the calculation of $\hat{\alpha}$, the buyers' optimality conditions are reflected only through $\hat{w}$. As a result, it is quite straightforward to extend the model to allow for different discount factors for the buyers and the seller. The discount rate of the buyer determines $\hat{w}$ through (3.1) and the seller's discount rate determines the long-run gains from additional goodwill customers. In Sections 5 and 6, we formulate models of social learning and random purchasing opportunities in which the separability of the problems with respect to different discount rates is useful.

3.4. *Mass market*

Initially, the informed segment does not yet exist. The monopolist then offers high prices, which leave the uninformed agents just indifferent between buying and not buying. The monopolist can thus extract initially all the surplus from the current purchases of the uninformed agents. As the informed segment grows, any price that leaves the uninformed indifferent results in revenue losses in the informed segment relative to pricing at $\hat{p}$. The monopolist's problem is therefore to determine the stopping point at which he starts to leave surplus to the uninformed buyers.

In contrast to the niche market, the uninformed buyers continue to purchase in the mass market. After the stopping point, they become inframarginal rather than marginal buyers. As a result, the optimal stopping condition can (almost) exclusively be described in terms of the flow revenue for the seller. Until the stopping point $\hat{\alpha}$, uninformed buyers are marginal. In consequence, the equilibrium price makes the uninformed buyer just indifferent between buying and not buying. We can express this in terms of the equilibrium value function of the buyer, using (3.3):

$$p(\alpha) = v + \lambda[\mathbb{E}_\theta V^\theta(\alpha) - V^u(\alpha)]. \tag{3.8}$$

The flow revenue of the seller at a given price p and a fraction α of informed buyers is

$$\pi(p, \alpha) = (1 - \alpha)p + \alpha[1 - F(p)]p \quad \text{for } p \le p(\alpha). \tag{3.9}$$

The seller sets prices to make the uninformed customers marginal as long as the marginal revenue from increasing the price at $p(\alpha)$ is nonnegative, or

$$\frac{\partial \pi(p(\alpha), \alpha)}{\partial p} \ge 0. \tag{3.10}$$

As long as the uninformed buyer is the marginal buyer, the marginal flow revenue $\partial \pi / \partial p$ at $p = p(\alpha)$ can well be strictly positive. The reason is that the true payoff function has a discontinuity at $p = p(\alpha)$ reflecting the positive mass of uninformed buyers that drop out of the market at prices above $p(\alpha)$. The optimal stopping point $\hat{\alpha}$ is hence derived from

$$\frac{\partial \pi(p(\hat{\alpha}), \hat{\alpha})}{\partial p} = 0. \tag{3.11}$$

In a mass market, the stopping condition can therefore be expressed in terms of the flow revenues. Still, the problem contains an intertemporal element since the equilibrium price $p(\alpha)$ before and at the stopping point is based on the equilibrium continuation values; see equation (3.8). After the stopping point $\hat{\alpha}$, the equilibrium price is computed as the price that maximizes the flow revenue, or

$$\frac{\partial \pi(p, \alpha)}{\partial p} = 0. \tag{3.12}$$

Even though this equation has the flavor of a static optimization condition, the dynamics of the model still enter into the determination of prices through the evolution of α.

Proposition 2 (Equilibrium stopping in the mass market): *If $\hat{w} \ge \hat{p}$, then*

1. $\hat{\alpha} \le 1$ *and*
2. $\hat{\alpha}$ *is decreasing in λ and increasing in r.*

If we consider the comparative statics results in propositions 1 and 2, then it is worth observing that the respective stopping points for niche and mass markets move in opposite directions as a function of λ and r. Suppose that we consider a distribution $F(\theta)$ with the property that it may be either a niche or a mass market depending on the values of λ and r and consider a comparative statics exercise in λ. For very low values of λ, the willingness to pay by the uninformed is low, the market is a niche market, learning stops early, and few buyers become informed. As λ increases, the uninformed buyers are willing to pay more. In turn the seller will offer introductory prices for a longer period. Eventually λ will reach a point at which the willingness to pay of the uninformed exactly equals the static optimal price, or $\hat{w} = \hat{p}$. At this knife-edge case the optimal dynamic price will in fact be constant for all t and $\hat{\alpha} = 1$. At this point, the market turns from a niche market into a mass market. The willingness to pay of the uninformed is now high enough for them to stay in the market until they become informed. In consequence, the uninformed buyer becomes an inframarginal buyer. The stopping point $\hat{\alpha}$ now decreases in λ as the uninformed customers become inframarginal earlier. The comparative statics are simply reversed for r. In Figure 6.1, the increasing part of the graph corresponds to the niche market case and the decreasing part belongs to the mass market.

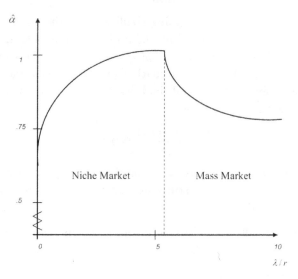

Figure 6.1.   Stopping point for varying λ/r.

4. Equilibrium Pricing

We are now in a position to characterize the complete equilibrium pricing policy on the basis of the equilibrium stopping point. Before the stopping point $\hat{\alpha}$, the marginal buyer is the uninformed buyer. We showed earlier that the price before stopping is

$$p(\alpha) = v + \lambda \mathbb{E}[V^\theta(\alpha) - V^u(\alpha)].$$

We first discuss the pricing policy in the early market and then describe the equilibrium conditions for the mature market.

Proposition 3 (Early market):

1. *The price $p(\alpha)$ satisfies for all $\alpha \leq \hat{\alpha}$*

$$\frac{dp}{d\alpha}\frac{d\alpha}{dt} = r[p(\alpha) - v] - \lambda \mathbb{E}_\theta \max\{\theta - p(\alpha), 0\}. \qquad (4.1)$$

2. *The price $p(\alpha)$ is decreasing in α for all $\alpha < \hat{\alpha}$.*
3. *The equilibrium sales $q(\alpha)$ are decreasing for small α and convex in α.*

The differential equation (4.1) describes the evolution of the price in the early market. When we rearrange the equation, it becomes apparent that it represents the trade-off that the uninformed buyer is facing in his purchase decision:

$$\lambda \mathbb{E}_\theta \max\{\theta - p(\alpha), 0\} = r[p(\alpha) - v] - \frac{dp}{d\alpha}\frac{d\alpha}{dt}.$$

The left-hand side represents the net benefit of buying the new product today rather than tomorrow. In particular, a purchase today generates an informative signal at rate λ and allows the buyer to make an informed decision. The right-hand side represents the net benefits of buying tomorrow rather than today. The net benefit has two components. First, buying tomorrow allows the buyer to postpone the net cost of a purchase, $p(\alpha) - v$, by an instant; and second, it may change the price the buyer has to pay for the information.

In the early market, the price contains a premium for the option value generated by the information. As time goes by, this option value decreases, and in consequence the price will have to decline to offset the lower option value. The rate at which the price decreases depends directly on the variation of valuations in the market. In particular, the price decreases at a faster rate if there is more riskiness in the distribution in the sense of second-order stochastic dominance. For a given p and v, the rate at which the price decreases is determined by the second term, namely $-\lambda \mathbb{E}_\theta \max\{\theta - p(\alpha), 0\}$, which is a concave function in θ. Consequently, its expected value decreases if the riskiness in the sense of second-order stochastic dominance increases. Thus we expect to see a more rapid price decline in niche markets relative to mass markets. In niche markets the seller initially pursues a more aggressive pricing strategy to build up a clientele before he eventually switches to extract the surplus from his target market, namely the niche of high-valuation buyers.[7]

Proposition 4 (Mature market):

1. *In the niche market, the price $p(\alpha)$ jumps up and stays at $p(\hat{\alpha}) = \hat{p}$ at the stopping point $\hat{\alpha}$.*
2. *In the mass market, $p(\alpha)$ is decreasing for all $\alpha \geq \hat{\alpha}$ and $\lim_{\alpha \to 1} p(\alpha) = \hat{p}$.*

The value of information before the stopping point, as expressed by $p(\alpha) - v$, is decreasing over time. While the evolution of the price is governed by the same differential equation for the niche and the mass markets, the source of the decrease in the value of information is different in the niche and the mass markets. In the niche market, the stopping point is the end of a phase of introductory prices. After this, the seller increases her price to $\hat{p}$ and a large fraction of surplus is extracted from the informed buyers. Hence the initially high value of information results from the relatively low initial prices.

In the mass market, the seller stops extracting all the surplus from the uninformed buyers and lowers her price to attract more informed buyers with lower valuations for the object. The value of information is now decreasing because with lower future prices, the option value that arises from the possibility of rejecting the product when θ is low is smaller.

Our interpretation of the two qualitatively different price paths goes as follows. In the niche market, the monopolist makes introductory offers to increase the number of goodwill customers once the price is raised. In the mass market, the monopolist skims the high-valuation buyers in the market (the uninformed buyers) with a high and declining price. It should be noted

---

[7]We would like to thank the editor and a referee for inquiring about the relationship between the rate of price change and the variance in the valuations.

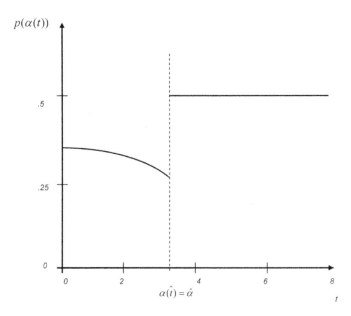

Figure 6.2.   Equilibrium price for the niche market.

that in the niche as well as in the mass market, the prices do not change by large amounts before $\hat{\alpha}$, and as a result, adjustment costs to changing prices might well force the monopolist to adopt a two-price regime with low initial prices followed by higher prices in the niche market and high prices followed by low prices in the mass market.

The intertemporal pricing policies are graphically depicted in Figures 6.2 and 6.3 for the niche and mass markets, respectively. With the niche market, the introductory price slowly decreases until it reaches a value equal to the willingness to pay, and at that point, the seller ceases to pursue new customers and sells only to informed customers with sufficiently high valuations. In the mass market, the discount factor r is small, and hence the option value for the uninformed buyer is almost constant. In consequence, the price declines very slowly until the seller begins to seek sales more aggressively from the informed customers. At this point, the price begins to decrease more rapidly and eventually converges to the static monopoly price.

Notice also that our model provides a theoretical prediction for the joint movements of prices and equilibrium quantities. These effects should be taken into account when estimating the demand for new products. If one

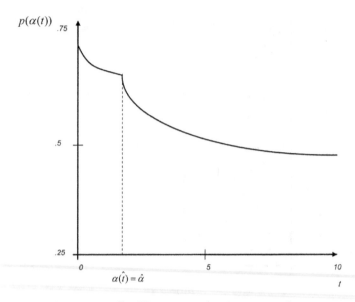

Figure 6.3.   Equilibrium price for the mass market.

were to estimate a static demand function for a product using data that include observations of prices for $\alpha < \hat{\alpha}$, then it is immediate that the resulting estimators would be biased.

The equilibrium analysis in this paper focused on the notion of a Markov-perfect equilibrium and derived the unique equilibrium in this class. The uniqueness result extends to a much larger class of equilibria. In Bergemann and Välimäki (2004), we show that the Markov-perfect equilibrium remains the only sequential equilibrium outcome as long as the information sets of the players include only their own past actions, observations, and the aggregate market data. With this much weaker restriction, the continuation paths of play are still independent of the choices of any individual buyer, but they may depend on past prices in an arbitrary manner.

5. Idiosyncratic Versus Social Learning

We now contrast the equilibrium allocation with the socially efficient allocation. In the presence of idiosyncratic learning, the socially efficient outcome can be implemented in a competitive equilibrium with marginal cost pricing. We then augment the analysis of idiosyncratic learning with an element of social learning. The social learning naturally introduces informational externalities among the buyers, and we show that the earlier welfare ranking between competitive and monopolistic market structure may be reversed in the presence of social learning.

5.1. *Social efficiency*

The socially efficient policy maximizes the sum of the expected discounted value across all agents. As learning and the resolution of uncertainty are purely idiosyncratic, the socially optimal policy can be determined as the solution for a representative consumer. The buyers and the seller have quasi-linear preferences, and hence the socially efficient policy simply maximizes expected social surplus. The optimal consumption policy can be determined for informed and uninformed consumers separately. For a given constant marginal cost c of producing the object, the (social) value function W^θ of the informed customer is

$$rW^\theta = \max\{\theta - c, 0\},$$

and for the uninformed consumer it is given by W^u, or

$$rW^u = \max\{v + \lambda \mathbb{E}_\theta[W^\theta - W^u] - c, 0\}. \tag{5.1}$$

For the informed consumer, the socially efficient decision is simply to consume if and only if his net value $\theta - c$ is positive. For the uninformed consumer, he should buy if the *current* social net benefit, $v - c$, and the *future* social net benefit, $\lambda \mathbb{E}[W^\theta - W^u]$, exceed zero. The expected value of the informed consumer, on the other hand, is

$$\mathbb{E}[W^\theta] = \frac{1}{r} \mathbb{E}_\theta \max\{\theta - c, 0\}. \tag{5.2}$$

By inserting the value function for the informed consumer, or (5.2), into (5.1), we obtain the critical value v so that the social value of consuming an additional unit of the new good is larger than or equal to zero:

$$v + \frac{\lambda}{r} \mathbb{E}_\theta \max\{\theta - c, 0\} \geq c.$$

The socially optimal policy is therefore to consume the new good as long as the expected net value today, $v - c$, and the value of information from current consumption are positive. It is worth emphasizing that it may be efficient to try the new product even if the expected net value of the object, $v - c$, or even the expected gross value v, is negative.

The monopoly position of the seller introduces static as well as intertemporal distortions away from the socially optimal level of consumption. The static distortionary element comes from the standard revenue considerations of the seller. Her objective is to maximize the revenue rather than the social welfare. In consequence, she will typically set the price above marginal cost and hence fail to offer the product to some buyers who would have a positive contribution to the social surplus. The dynamic element of the distortion arises in the niche market as well as in the mass market, but with different consequences. In the niche market, the willingness to pay of the uninformed is low, and hence the seller will eventually stop selling to the uninformed in pursuit of higher per unit revenue from the informed buyers. In the mass market, efficient learning takes place. But as the uninformed agents are willing to pay a premium over and above their expected value for the object, the seller maintains a high price even relative to static monopoly price $\hat{p}$. This implies that during the launch phase, many informed buyers will not purchase the object even though their purchase would generate a positive social surplus.

In contrast to this, a competitive market would support the efficient allocation in this model of idiosyncratic learning. In a competitive market, the object is offered at the marginal cost c. In consequence, the objective function of the buyer in a competitive market coincides with the social objective for both the informed and the uninformed consumers. The important ingredient of the model that leads to an agreement of the competitive and the efficient outcomes is the idiosyncratic nature of the learning experience. We next discuss a minimal extension of our model to introduce an element of social learning. We shall see that the qualitative insights of the equilibrium analysis will essentially carry over but that the welfare comparison between competitive and monopoly markets will now be different.

5.2. *Social learning*

We introduce social learning into our model by extending every single buyer to a group of k buyers who consume individually but share the same true valuation for the object and share the information about the valuation among the members of the group. We may think of the group as a family, a department, or a neighborhood with similar preferences and the ability to communicate among each other. At each point in time, we randomly select one buyer out of the group of k members to consider a purchase. If that member decides to purchase the object, then he receives the consumption value of the object privately but shares all resulting information from his purchase with the other members of his group.

The equilibrium conditions of the idiosyncratic learning model are easily adapted to accommodate social learning. The value function $V^\theta(\alpha)$ of an informed agent is given by

$$rV^\theta(\alpha) = \frac{1}{k}\max\{\theta - p(\alpha), 0\} + \frac{dV^\theta}{d\alpha}\frac{d\alpha}{dt}. \tag{5.3}$$

The only change from the earlier formulation is that the frequency of purchases is reduced by factor $1/k$. Similarly, the value function of an uninformed buyer is given by

$$rV^u(\alpha) = \frac{1}{k}\max\{v - p(\alpha) + \lambda[\mathbb{E}_\theta V^\theta(\alpha) - V^u(\alpha)], 0\}$$

$$+ \frac{k-1}{k}\lambda[\mathbb{E}_\theta V^\theta(\alpha) - V^u(\alpha)] + \frac{dV^u}{d\alpha}\frac{d\alpha}{dt}. \tag{5.4}$$

The element of social learning enters the dynamic programming equation of the individual buyer since he now has two sources of information: either he can directly purchase the product as an occasion arises (at the rate $1/k$) or he can benefit from the information generated by the other members of the group (at the rate $k - [1/k]$).

The social learning has two effects on the pricing policy. First, the uninformed are less willing to pay a premium for the information, since there is a chance to learn from the experiences of others. On the other hand, the seller is willing to sponsor the experiment of each agent since she knows that the information will spread among all members of the group. If they value the object sufficiently highly, the seller can extract their surplus in the future. For an individual buyer, the frequency $1/k$ by which an opportunity for a purchase arrives acts as a (stochastic) increase in the discount rate. From the point of view of the seller, the actual market size in each period remains unchanged. In consequence, the stopping conditions for $\hat{\alpha}$ are determined by using a modified discount rate, namely $r \cdot k$, for the buyer. In consequence, the basic distinctions in terms of launch strategies remain unchanged, and the equilibrium policies of the seller are exactly as if he would face buyers with a larger discount rate.

With this informational externality the comparison between the competitive market outcome and the monopoly outcome becomes quite different. The monopolist partially internalizes the informational externality since he is aware that the release of information today leads to a higher revenue in future periods. In a competitive setting, the product is offered at marginal cost in each period. As each buyer takes into account only the private benefits from learning, the social optimum may not be reached in the competitive market. In particular, if the expected value of the current consumption v falls below marginal cost c, then the only reason to purchase the product is to acquire more information. If the informational externality between the buyers is increased by increasing k, then the uninformed will not purchase at the marginal cost pricing. The socially optimal policy is independent of k, and the size of the market for the monopolist is independent of k as well. It follows that the monopolist may now sustain the market and induce socially beneficial learning even though the competitive market cannot do so.

Proposition 5 (Monopoly versus competitive market): *If $v < c$ and it is socially efficient to adopt the new product, then*

1. *there exists $\bar{k} > 0$ such that, for all $k \geq \bar{k}$, the new product is never sold in the competitive market;*
2. *if $v - c + (\lambda/r)\hat{p}[1 - F(\hat{p})] > 0$, then it is optimal for the monopolist to launch the product for all k.*

6. Extensions

Random purchases — We first describe an alternative interpretation for the informational structure of our basic model. We assume as before that there is a continuum of buyers who are initially uncertain about their tastes for the new product. In contrast to the basic model, we now assume that the consumption opportunities arrive at random time intervals. For analytical convenience, we assume that these arrivals follow a Poisson process with parameter λ.

In this reformulation of the model, it makes sense to assume that the buyers learn their true tastes upon consuming the first unit of the good. Even though this assumption is less realistic in the context of the pharmaceuticals market, it may fit better consumer goods such as cereals, cosmetics, and so forth. In the initial periods, the monopolist is facing demand mostly from uninformed buyers. The fraction of repeat buyers increases as the good stays in the market, and eventually most of the buyers are repeat buyers. The only change in the dynamic programming formulations is the random purchase rate, which acts like an increase in the discount rate. The problem of the informed buyers is now

$$(r + \lambda)V^\theta(\alpha) = \lambda \max\{\theta - p(\alpha), 0\} + \frac{dV^\theta(\alpha)}{d\alpha}\frac{d\alpha}{dt}.$$

Correspondingly, the value function of the uninformed buyers is given by

$$(r + \lambda)V^u(\alpha) = \lambda \max\{v - p(\alpha) + \mathbb{E}_\theta V^\theta(\alpha) - V^u(\alpha), 0\} + \frac{dV^u(\alpha)}{d\alpha}\frac{d\alpha}{dt}.$$

These two value functions correspond to the original model in which the buyers' discount rate is set to $(r + \lambda)/\lambda$ and the arrival rate of information is set to one. As was explained in Section 3, the analysis of the original model can be carried out with different interest rates for the buyers and the seller, and hence this alternative interpretation is included as a special case of the original model.

Commitment — New distortionary effects arise from the intertemporal trade-offs faced by a time-consistent monopolist. In order to understand these dynamic effects, it is interesting to ask whether they could be overcome if the seller had the ability to commit himself to an entire future price path at the beginning of the game.

The analytical techniques of the solution with commitment are quite different from the equilibrium solution, since we must abandon the requirement of sequential rationality or time consistency in characterizing the optimal commitment paths. For this reason, we simply report the basic results, which are formally stated in an earlier working paper version (Bergemann and Välimäki 2004).

The optimal commitment solution is qualitatively different from the time-consistent equilibrium solution in the niche market. With the ability to commit, the seller can promise future prices below $\hat{p}$ in exchange for higher prices today. This allows her to maintain future prices that are low enough so that *all* uninformed customers will stay in the market until they are informed. The optimal price path is then a constant price $\hat{w}$, which can be obtained as the solution from the willingness to pay of the uninformed (see eq. [3.1]):

$$\bar{w} = v + \frac{\lambda}{r}\mathbb{E}_\theta \max\{\theta - \bar{w}, 0\}.$$

In consequence, the uninformed customers receive zero expected surplus in the commitment solution, but their purchasing policy is now socially efficient. It follows that commitment increases social welfare in the niche market. The consumers do not benefit since the ability to commit enables the seller simply to extract more surplus over time.

In the mass market, the ability to commit has negative consequences on social welfare. In the time-consistent policy, the seller eventually cuts her price to sell more to the informed buyers. With the ability to commit, the seller can commit to higher prices in the future. As such higher prices increase the value of information for the uninformed, prices can be higher in the initial periods as well. Thus in a mass market, commitment allows the seller to set higher prices everywhere and hence further depress socially efficient purchases.

Stationary model — The current model describes the optimal pricing for the introduction of a new product. The buyers started with ex-ante identical information regarding the new product, and over time their personal experiences lead them to have heterogeneous and idiosyncratic valuations. It is then natural to expand the scope of our analysis to a market with a constant inflow and outflow of consumers. In fact, many consumer products face a constant renewal in their customer base, because of either the aging of the customers or other systematic changes to the agents' preferences. In Bergemann and Valimaki (2004), we analyze the steady-state equilibrium in a market with idiosyncratic learning. We model the change in the population by constant entry rate γ of new customers and equal exit rate γ of old customers. The new customers are all initially uninformed and become informed according to the same information technology as in the current paper.

The steady-state equilibrium with a constant renewal of customers displays the same basic features as the current model of the launch of a new good to an entirely new market. The willingness to pay of the new customers is modified in the obvious way by the birth and death rate γ, or

$$\hat{w} = v + \frac{\lambda}{r + \gamma} \mathbb{E}_\theta \max\{\theta - \hat{p}, 0\}.$$

With this modification, the optimal policy can again be described in terms of mass market versus niche market. If, as before, the complete information price $\hat{p}$ is below $\hat{w}$, then the market is a mass market, and the seller will offer a price in between the statically optimal price $\hat{p}$ and the willingness to pay of the new customers, $\hat{w}$, so as to balance his revenue management. Importantly, all new customers will enter the market and learn more about the new product.

If, on the other hand, the willingness to pay is below the statically optimal price, or $\hat{p} > \hat{w}$, then we are again in the situation of a niche market. Now, the seller does not want to sell to the new customers all the time, but rather extract surplus from the informed agents. Yet, with a constant inflow of new customers, the seller cannot abandon the new customers altogether since this would imply a diminishing customer base.

The resolution of the trade-off is that the seller offers probabilistically both low and high prices. The high price is given by $\hat{p}$ and optimally extracts surplus from the informed agents, and the low price allows the new customers to learn and become acquainted with the new product. In this way, the seller balances revenue objectives yet maintains a constant informed clientele for his product. The dispersed prices in the niche market are common in models of durable goods with entry of new buyers (see, e.g., Sobel 1991). In our setting, the product is effectively a bundle, consisting of a perishable element in the form of the immediate consumption benefit, but also a durable element in the form of the information obtained with the purchase. From this perspective, the relationship to the durable good pricing problem appears inherently.

7. Conclusion

In this paper, we have shown that the optimal sales policy of a monopolist in a model of experience goods is qualitatively different depending on whether the market is a mass market or a niche market. In a niche market, it is in the monopolist's best interest to build up a sufficient base of goodwill clientele. To achieve this, the monopolist sacrifices current profit by pricing low in order to find new future buyers. The durability of information about the product quality thus plays a key role in this situation. In a mass market, managing information is less important since the uninformed buyers are willing to buy at the static optimal prices. In this case the monopolist's optimal price path can be seen as an attempt to skim the uninformed segment of the market until the informed segment becomes large.

We have kept the model as simple as possible in order to highlight the dynamics of price setting. The modeling strategy of the current paper could be used to investigate models with more general specifications for either the buyers' valuations of the product or the strategic environment. An interesting instance of a more general demand structure would be one in which the buyers differ in their willingness to pay for quality, but the perceived quality is idiosyncratic and must be learned over time. Regarding the competitive structure of the model, the natural next step would be to consider the role of idiosyncratic learning in a strategic environment against either a known or a similarly unknown product. An interesting variation of the current model and of special importance for the pharmaceutical market would be to consider optimal pricing in which a competitor will appear only in T periods hence, induced by the expiration of the patent (see Berndt, Kyle, and Ling [2003] for an empirical analysis of this situation).

Appendix

This appendix collects the proofs of all the results in the main body of the paper. For notational convenience, we shall adopt a standard notation from probability theory by writing

$$(\theta - p)^+ \triangleq \max\{\theta - p, 0\}.$$

Proof of Proposition 1

Part 1. The optimal stopping point is the solution to (3.6) and is given by

$$\hat{\alpha} = \frac{\hat{w} + (\lambda/r)\hat{p}[1 - F(\hat{p})]}{\hat{w}F(\hat{w}) + \hat{p}[1 - F(\hat{p})][1 - (\lambda/r)]}. \tag{A.1}$$

For simplicity we define $\eta \triangleq \lambda/r$, and with this we rewrite equation (A1) as

$$\hat{\alpha} = \frac{\hat{w} + \eta\hat{p}[1 - F(\hat{p})]}{\hat{w} + (1 + \eta)\hat{p}[1 - F(\hat{p})] - \hat{w}[1 - F(\hat{w})]}, \tag{A.2}$$

which shows that $\hat{\alpha} < 1$ since $\hat{p}[1 - F(\hat{p})] - \hat{w}[1 - F(\hat{w})] > 0$.

Part 2. After replacing $\hat{w}$ by its explicit expression given in (3.1), we rearrange (A2) to get

$$\hat{\alpha}\{[v + \eta\mathbb{E}(\theta - \hat{p})^+]F(v + \eta\mathbb{E}(\theta - \hat{p})^+) + \hat{p}[1 - F(\hat{p})](1 - \eta)\}$$
$$= v + \eta\mathbb{E}(\theta - \hat{p})^+ + \eta\hat{p}[1 - F(\hat{p})]. \tag{A.3}$$

Differentiating the equality (A3) implicitly with respect to η yields

$$\frac{d\hat{\alpha}}{d\eta}\{\hat{w}F(\hat{w}) + \hat{p}[1 - F(\hat{p})](1 + \eta)\}$$
$$+ \hat{\alpha}\{[\hat{w}f(\hat{w}) + F(\hat{w})]\mathbb{E}(\theta - \hat{p})^+ + \hat{p}[1 - F(\hat{p})]\}$$
$$= \mathbb{E}(\theta - \hat{p})^+ + \hat{p}[1 - F(\hat{p})],$$

and hence

$$\frac{d\hat{\alpha}}{d\eta} = \frac{\mathbb{E}(\theta - \hat{p})^+[1 - \hat{\alpha}F(\hat{w}) - \hat{\alpha}\hat{w}f(\hat{w})] + (1 - \hat{\alpha})\hat{p}[1 - F(\hat{p})]}{\hat{w}F(\hat{w}) + \hat{p}[1 - F(\hat{p})](1 + \eta)}.$$

The denominator is clearly positive. For the numerator, we observe that $1 - \hat{\alpha}F(\hat{w}) - \hat{\alpha}\hat{w}f(\hat{w})$ is the derivative of the profit function $p\{1 - \hat{\alpha} + \hat{\alpha}[1 - F(p)]\}$ evaluated at $\hat{w}$, which is positive by the assumed quasi concavity of $p[1 - F(p)]$ together with the fact that $\hat{w} < \hat{p}$. Therefore, the numerator is also positive, as needed. $\square$

Proof of Proposition 2

Part 1. Suppose that, for all $\alpha \geq \hat{\alpha}$, the marginal buyer is an informed buyer. The equilibrium price $p(\alpha)$ is then given as the solution of the static revenue maximization problem (3.9). The value function of an informed buyer at $\alpha(t) = \alpha$ is

$$V^\theta(\alpha) = \int_t^\infty e^{-r(\tau-t)}[\theta - p(\alpha(\tau))]^+ d\tau.$$

The expected value of the informed buyer at $\alpha(t) = \alpha$ is

$$\mathbb{E}_\theta[V^\theta(\alpha)] = \int_{\theta_l}^{\theta_h} \left\{ \int_t^\infty e^{-r(\tau-t)}[\theta - p(\alpha(\tau))]^+ d\tau \right\} dF(\theta).$$

In contrast, the value function of the uninformed buyer for a particular realization $T \geq t$ of the signal arrival time is given by

$$\int_t^T e^{-r(\tau-t)}[v - p(\alpha(\tau))]d\tau + \int_{\theta_l}^{\theta_h} \left\{ \int_T^\infty e^{-r(\tau-t)}[\theta - p(\alpha(\tau))]^+ d\tau \right\} dF(\theta).$$

(A.4)

The value function of the uninformed buyer is obtained from (A4) by taking the expectation over all signal arrival times $T \geq t$, or $V^u(a)$ is given by

$$\int_t^\infty \int_{\theta_l}^{\theta_h} \left\{ \int_t^T e^{-r(\tau-t)}[v - p(\alpha(\tau))]d\tau \right.$$
$$\left. + \int_T^\infty e^{-r(\tau-t)}[\theta - p(\alpha(\tau))]^+ d\tau \right\} dF(\theta)\lambda e^{-\lambda T} dT.$$

The willingness to pay for all $\alpha \geq \hat{\alpha}$ is given by

$$w(\alpha) = v + \lambda[\mathbb{E}_\theta V^\theta(\alpha) - V^u(\alpha)].$$

From the above expressions, the difference in the value functions, $\mathbb{E}_\theta V^\theta(\alpha) - V^u(\alpha)$, can therefore be written as

$$\mathbb{E}_\theta V^\theta(\alpha) - V^u(\alpha)$$
$$= \int_t^\infty \int_{\theta_l}^{\theta_h} \int_t^T e^{-r(\tau-t)}[p(\alpha(\tau)) - \theta]^+ d\tau dF(\theta)\lambda e^{-\lambda T} dT. \quad \text{(A.5)}$$

The gain of the informed vis-à-vis the uninformed buyer arises in all those instances in which the uninformed buyer accepts the offer by the seller even though his true valuation is below the equilibrium price. It follows that $w(\alpha)$ is decreasing in α (and t) since $p(\alpha)$ is decreasing in α (and t). We can then run $w(\alpha)$ backward as long as $w(\alpha) \geq p(\alpha)$.

The stopping point $\hat{\alpha}$ is the smallest α at which

$$w(\alpha) = p(\alpha). \tag{A.6}$$

We next argue that there is a unique stopping point $\hat{\alpha}$ by showing that $w(\alpha)$ and $p(\alpha)$ are single crossing. By the hypothesis of $\hat{w} > \hat{p}$, we have

$$w(1) > p(1). \tag{A.7}$$

Observe further that

$$\lim_{\alpha \downarrow 0} p(\alpha) = +\infty.$$

The maximal willingness to pay, $w(\alpha)$, is a constant v and a discounted average over future prices $p(\alpha)$, represented by (A5). It therefore follows that, provided that $p(\alpha)$ is monotone,

$$|p'(\alpha)| > w'(\alpha),$$

which together with (A7) is sufficient to guarantee existence and uniqueness of the stopping point.

Part 2. We observe first that $p(\alpha)$ is independent of r. It follows that r affects the expression $\mathbb{E}_\theta V^\theta(\alpha) - V^u(\alpha)$, only through discounting. As an increase in r decreases $w(\alpha)$, it follows that the intersection (A6) is reached later and thus at a higher value of $\hat{\alpha}$. The argument for λ is similar except for the obvious reverse in the sign. □

Proof of Proposition 3

Part 1. The differential equation (4.1) for the full extraction prices has a unique rest point, $\dot{p}(t) = 0$, at $p(t) = w$ since w uniquely solves

$$0 = r(w - v) - \lambda \mathbb{E}_\theta \max\{\theta - w, 0\}.$$

Part 2. We show the monotonicity of $p(\alpha)$ separately for $\hat{p} < \hat{w}$ and $\hat{p} \geq \hat{w}$. We start with the latter case and argue by contradiction. Thus suppose that $p(0) > w$. Then $p(\alpha) > w > \hat{w}$ for all α. It follows that at $\alpha = \hat{\alpha}$, we have $p(\hat{\alpha}) > w > \hat{w}$; but at $a(t) = \hat{\alpha}$, we have to have $p(\hat{\alpha}) = \hat{w}$ for the uninformed buyer to be willing to buy, and this leads to the desired contradiction.

Consider then $\hat{p} < \hat{w}$ and consequently $\hat{p} < w < \hat{w}$. Suppose that $p(0) > w$ and hence by the differential equation $p(\alpha) > w$ for all t with $\alpha < \hat{\alpha}$. We also recall that as the equilibrium price path is continuous, it follows that at α such that $p(\alpha) = w > \hat{p}$, we have $w(\alpha) > p(\alpha)$. From proposition 2 we recall that the equilibrium during the full extraction phase satisfies

$$p(\alpha) = v + \lambda[\mathbb{E}_\theta V^\theta(\alpha) - V^u(\alpha)],$$

or, more explicitly,

$$p(\alpha) = v + \lambda \int_t^\infty \int_{\theta_l}^{\theta_h} \int_t^T e^{-r(\tau - t)}[p(\alpha(\tau)) - \theta]^+ d\tau dF(\theta)\lambda e^{-\lambda T} dT$$

For notational ease we shall denote by $R(t)$ the forgone utility benefit from being uninformed in period t (i.e., the regret) or

$$R(t) \triangleq \int_{\theta_l}^{\theta_h} [p(\alpha(t)) - \theta]^+ dF(\theta),$$

and the equilibrium price is then given by

$$p(t) = v + \lambda \int_t^\infty \int_t^T e^{-r(\tau - t)} R(\tau)d\tau \lambda e^{-\lambda T} dT.$$

By hypothesis, $p(t)$ is strictly increasing until $t = \hat{t}$, where $\alpha(\hat{t}) = \hat{\alpha}$ and is decreasing thereafter. It is immediate that $R(t)$ shares the monotonicity properties with $p(t)$. We next show that $p(t)$ cannot be monotone increasing for all $t < \hat{t}$. After integrating with respect to T, we get

$$p(t) = v + \lambda \int_t^\infty e^{-(r+\lambda)(\tau - t)} R(\tau)d\tau.$$

Differentiating with respect to t, we get

$$p'(t) = \lambda \left[-R(t) + (r + \lambda) \int_t^\infty e^{-(r+\lambda)(\tau - t)} R(\tau)d\tau \right],$$

which has to turn negative as $t \uparrow \hat{t}$ by the hypothesis of an increasing price for all $t < \hat{t}$ and the continuity of the price path. This delivers the desired contradiction. The concavity in t follows immediately from $p(0) < w$ and the differential equation (4.1).

Part 3. The equilibrium sales are given by

$$q(t) = (1 - \alpha) + \alpha[1 - F(p(t))],$$

as long as the uninformed buyers are participating. The equilibrium sales are governed by the following differential equation:

$$q'(t) = -(1 - \alpha)F(p(t))\lambda - \alpha f(p(t))p'(t).$$

It follows that, even though $p'(t) < 0$, for all a sufficiently small, $q'(t) < 0$. The second derivative is given by

$$q''(t) = (1 - \alpha)F(p(t))\lambda^2 - 2(1 - \alpha)\lambda f(p(t))p'(t)$$
$$- \alpha\{f'(p(t))[p'(t)]^2 + f(p(t))p''(t)\},$$

and again for all a sufficiently close to zero, the convexity of the sales follows directly from the decreasing price. □

Acknowledgments

We would like to thank Kyle Bagwell, Ernst Berndt, Steve Berry, Patrick Bolton, Sven Rady, Mike Riordan, Klaus Schmidt, K. Sudhir, and Ferdinand von Siemens for very helpful comments and Colin Stewart for excellent research assistance. The comments of the editor, Robert Shimer, and two referees also improved the paper. We benefited from discussions at the Industrial Organization Day at New York University and seminars at the University of Darmstadt, European University Institute, London School of Economics, University of München, University of Nürnberg, the Research Institute of Industrial Economics (Sweden), University of Oulu, and University College London. Bergemann gratefully acknowledges financial support from National Science Foundation grant SES 0095321 and the DFG Mercator Research Professorship at the Center of Economic Studies at the University of Munich. Välimäki gratefully acknowledges support from the Economic and Social Research Council through grant R000223448.

References

Ackerberg, Daniel A. 2003. "Advertising, Learning, and Consumer Choice in Experience Good Markets: An Empirical Examination." *Internat. Econ. Rev.* 44 (August): 1007–40.

Bergemann, Dirk, and Juuso Välimäki. 2004. "Monopoly Pricing of Experience Goods." Tech. Report no. 1463, Yale Univ., Cowles Found. Res. Econ.

Berndt, Ernst R., Margaret K. Kyle, and Davina C. Ling. 2003. "The Long Shadow of Patent Expiration: Generic Entry and Rx-to-OTC Switches." In *Scanner Data and Price Indexes*, edited by Robert C. Feenstra and Matthew D. Shapiro. Chicago: Univ. Chicago Press.

Berry, Steven T. 1994. "Estimating Discrete-Choice Models of Product Differentiation." *Rand J. Econ.* 25 (Summer): 242–62.

Ching, Andrew T. 2002. "Consumer Learning and Heterogeneity: Dynamics of Demand for Prescription Drugs after Patent Expiration." Manuscript, Univ. Toronto.

Coscelli, Andrea, and Matthew Shum. 2004. "An Empirical Model of Learning and Patient Spillovers in New Drug Entry." *J. Econometrics* 122 (October): 213–46.

Crawford, Gregory S., and Matthew Shum. 2005. "Uncertainty and Learning in Pharmaceutical Demand." *Econometrica* 73 (July): 1137–73.

Cremer, Jacques. 1984. "On the Economics of Repeat Buying." *Rand J. Econ.* 15 (Autumn): 396–403.

Erdem, Tulin, Susumu Imai, and Michael P. Keane. 2003. "Brand and Quantity Choice Dynamics under Price Uncertainty." *Quantitative Marketing and Econ.* 1 (1): 5–64.

Erdem, Tulin, and Michael P. Keane. 1996. "Decision-Making under Uncertainty: Capturing Dynamic Brand Choice Processes in Turbulent Consumer Goods Markets." *Marketing Sci.* 15 (1): 1–20.

Farrell, Joseph. 1986. "Moral Hazard as an Entry Barrier." *Rand J. Econ.* 17 (Autumn): 440–49.

Goettler, Ronald L., and Karen Clay. 2006. "Price Discrimination with Experience Goods: A Structural Econometric Analysis." Manuscript, Carnegie Mellon Univ.

Gowrisankaran, Gautam, and Marc Rysman. 2005. "Determinants of Price Declines for New Durable Consumer Goods." Manuscript, Washington Univ. and Boston Univ.

Israel, Mark. 2005. "Service as Experience Goods: An Empirical Examination of Consumer Learning in Automobile Insurance." *A.E.R.* 95 (December): 1444–63.

Johnson, Justin P., and David P. Myatt. 2006. "On the Simple Economics of Advertising, Marketing, and Product Design." *A.E.R.*, forthcoming.

Lu, Z. John, and William S. Comanor. 1998. "Strategic Pricing of New Pharmaceuticals." *Rev. Econ. and Statis.* 80 (February): 108–18.

Milgrom, Paul, and John Roberts. 1986. "Price and Advertising Signals of Product Quality." *J.P.E.* 94 (August): 796–821.

Nair, Harikesh. 2005. "Intertemporal Price Discrimination with Forward Looking Consumers: Application to the US Market for Console Video Games." Manuscript, Stanford Univ.

Osborne, Martin. 2005. "Consumer Learning, Habit Formation and Heterogeneity: A Structural Examination." Manuscript, Univ. Chicago.

Shapiro, Carl. 1983. "Optimal Pricing of Experience Goods." *Bell J. Econ.* 14 (Autumn): 497–507.

Sobel, Joel. 1991. "Durable Goods Monopoly with Entry of New Consumers." *Econometrica* 59 (September): 1455–85.

Tirole, Jean. 1988. *The Theory of Industrial Organization.* Cambridge, MA: MIT Press.

Villas-Boas, J. Miguel. 2004. "Consumer Learning, Brand Loyalty, and Competition." *Marketing Sci.* 23 (Winter): 134–45.

Chapter 7

Dynamic Common Agency*

Dirk Bergemann[†] and Juuso Välimäki[‡]

[†] *Department of Economics, and Cowles Foundation of Research in Economics,*
Yale University, New Haven, CT 06520-8268, USA
dirk.bergemann@yale.edu
[‡] *Department of Economics, Aalto University School of Business,*
02150 Espoo, Finland
juuso.valimaki@aalto.fi

A general model of dynamic common agency with symmetric information is considered. The set of truthful Markov perfect equilibrium payoffs is characterized and the efficiency properties of the equilibria are established. A condition for the uniqueness of equilibrium payoffs is derived for the static and the dynamic game. The payoff is unique if and only if the payoff of each principal coincides with his marginal contribution to the social value of the game. The dynamic model is applied to a game of agenda setting.

Keywords: Common agency, dynamic bidding, marginal contribution, Markov perfect equilibrium.

1. Introduction

Common agency refers to a broad class of problems in which a single individual, the agent, controls a decision that has consequences for many individuals with distinct preferences. The other affected parties, the principals, influence the agent's decisions by promising payments contingent on the action chosen. The static model of common agency under perfect information was introduced by Bernheim and Whinston [4] as a model of an auction where bidders are submitting a menu of offers to the

---

*This chapter is reproduced from *Journal of Economic Theory*, **111**, 23–48, 2003.

auctioneer. Since then it has gained prominence in many applications, such as procurement contracting, models of political economy [6, 7], as well as strategic international trade [9].

In this paper, we examine the structure of dynamic common agency problems. The extension of the model beyond the static version is of particular interest for the applications above. Political choices are rarely made only once, and the future implications of a current policy are often more important than its immediate repercussions. If the politician and the lobbyists cannot commit to future actions and transfers, a dynamic perspective is needed. Similarly, many procurement situations involve staged development with bidding occurring at each stage of the process.

The dynamic perspective also broadens the reach of common agency models. Consider for example a dynamic matching problem where the employee works in each period for at most a single employer, but may change employers over time. In the language of common agency, the employee has only one principal in each period. If human capital is acquired over time within jobs, then future employers may have preferences over the current career choices of the employee. Thus the intertemporal element introduces a more subtle aspect of common agency to various allocations problems such as career choice and job matching models.

We start our analysis with the static common agency model of Bernheim and Whinston [4], who concentrated on a refinement of Nash equilibrium, called *truthful equilibrium*. A strategy is said to be truthful relative to a given action if it reflects accurately the principals' willingness to pay for any other action relative to the given action. For the static game, we show that the truthful equilibrium payoff is unique if and only if the marginal contributions of the principals to the value of the grand coalition are weakly superadditive. We show that in such equilibria, all principals receive their marginal contributions as payoffs. We call equilibria satisfying this property *marginal contribution equilibria*.

The second part of the paper derives conditions for the existence of a marginal contribution equilibrium in the dynamic framework. Since we assume that the players lack commitment power over periods, the interplay between the payoffs received at different stages of the game becomes important. The first model we analyze is perhaps the simplest of all dynamic common agency models and illustrates the importance of changes in the stage game. This is a two-period game where in the first period, the agent chooses the available actions for the second stage. In the second stage, the common agency game is played with the set of actions as determined in the first period. We refer to this game as the agenda game as the agent initially determines the set of actions (agenda) from which the choice will be eventually made. Depending on the context this may represent the choice of the relevant policy alternatives by a political decision maker or the choice of the auction format by an auctioneer, such as how different licenses or rights should be bundled in a multi-unit auction.

The equilibrium payoff to the agent depends on the degree of competition between the principals in the second stage. Since the competitiveness in turn depends on the set of available actions, the agent is sometimes able to increase her second period rent by excluding the efficient action. This leads to an inefficiency in the overall game unless we allow the principals to lobby the first period choices of the agent as well. With lobbying in both periods, the overall efficiency is restored, but the payoff to the agent is higher in the two-stage game than in the static game where the first period choice of actions is ignored. In the following sections, we show that the essential features of the agenda game, in particular the aspect of intertemporal rent extraction, carry over to a general model of dynamic common agency.

In the rest of the paper, we can be quite general in formulating the states of nature governing the transitions between stage games. In particular, we can accommodate deterministic as well as stochastic transitions between states. We require that the current state depends only on the previous state and the previous action by the agent. The former of these assumptions is not crucial to the argument and is made to simplify the exposition. The second assumption is more substantial and it is made to preserve the flavor of the static model where the agent is the only player that affects the utilities directly. We concentrate on Markov strategies since we want to study the effects of changes in the stage game in isolation from the effects created by conditioning on payoff irrelevant histories. In the spirit of Bernheim and Whinston [4], we are particularly interested in truthful Markovian policies.

We start by proving the existence of truthful Markov equilibria in the dynamic model. As in the static case, all truthful Markov perfect equilibria of the dynamic game are efficient. The truthful equilibrium payoffs are also unique if and only if the game possesses a marginal contribution equilibrium. We characterize the necessary and sufficient conditions for the dynamic game to have a marginal contribution equilibrium in terms of the trade-off between efficiency and rent extraction for the agent.

The papers that are the most closely related to the current paper are those by Bernheim and Whinston [4], Dixit *et al.* [7] and Laussel and Le Breton [11], (as well as [10]). Our paper extends the static model of common agency in Bernheim and Whinston [4] to a dynamic setting. The second major point of departure is our focus on marginal contribution equilibria as an interesting solution concept in this class of games. Another recent extension of the basic model of common agency may be found in [7], where the assumption of quasi-linear preferences is dropped. Whereas the motivation for that extension is based on concerns relating to the distribution of the payoffs within the single period of analysis, our motivation is based on the distribution of payoffs between the players over time when commitment is precluded. The work by Laussel and Le Breton analyzes the payoffs received by the agent in a class of static common agency games. A recent paper by Prat and Rustichini [14] extends the common agency game to many common agents.

The paper is organized as follows. In Section 2 we introduce the common agency model in its dynamic version. The notion of marginal contribution is introduced here as well. Section 3 introduces the basic results for the static model of common agency. Section 4 analyzes the agenda setting game. Section 5 presents the main results for the dynamic common agency. The characterization of the truthful Markov perfect equilibrium is given here and necessary and sufficient conditions for its uniqueness are stated as well.

2. Model

2.1. *Payoffs*

We extend the common agency model of Bernheim and Whinston [4] to a dynamic setting. The set of players is the same in all periods, but actions available to them as well as payoffs resulting from the actions may change from period to period.

The principals are indexed by $i \in \mathscr{I} = \{1, \ldots I\}$. Time is discrete and indexed by $t = 0, 1, \ldots, T$, where T can be finite or infinite. In period 0, the agent selects an action a_0 from a finite set of available actions $\mathscr{A}_0$, and principal i offers a reward schedule $r_i(a_0) \in \mathbb{R}_+$. The stage game may change from period to period and the payoff relevant state of the world (in the sense of Maskin and Tirole [13]) in period t is denoted by $\theta_t \in \Theta$. For simplicity, we assume that $\Theta = \{\theta_1, \ldots, \theta_K\}$ for some $K < \infty$.[1] In the spirit

---

[1]The only place that uses this finiteness assumption is the existence proof in Section 5. Under appropriate continuity properties for state transitions, more general state spaces

of Bernheim and Whinston [4], only the agent makes directly payoff relevant choices in any of the periods. To that effect, we assume that the transition function, $q(\theta_{t+1}|a_t, \theta_t)$ is Markovian in the sense that the distribution of the payoff relevant state in period $t+1$ depends only on the current action, a_t, and the current state θ_t. By setting $h_1 = (\theta_0, a_0, r_0, \theta_1)$, the histories for $t > 1$ in the game are given by

$$h_t = (h_{t-1}, a_{t-1}, \mathbf{r}_{t-1}, \theta_t),$$

where $\mathbf{r}_t$ is the profile of reward schedules in period t and $a_t \in \mathscr{A}(\theta_t)$, the set of available actions in state θ_t. For simplicity, we assume that $\bigcup_{\theta \in \Theta} \mathscr{A}(\theta)$ is finite. We maintain the assumption made in the common agency literature that the principals can commit in every period to the reward schedules.

The cost of action a_t in period t to the agent is given by $c(a_t, \theta_t)$. The benefit to principal i is $v_i(a_t, \theta_t)$, which may again depend on θ_t. After a history h_t the aggregate reward paid by a subset of principals $S \subset \mathscr{I}$ for an action a_t is

$$r_S(a_t, h_t) \triangleq \sum_{i \in S} r_i(a_t, h_t),$$

and the aggregate benefits for the principals are

$$v_S(a_t, \theta_t) \triangleq \sum_{i \in S} v_i(a_t, \theta_t).$$

For $S = \mathscr{I}$, the aggregate rewards and benefits are denoted by $r(a_t, h_t) \triangleq r_{\mathscr{I}}(a_t, h_t)$ and $v(a_t, \theta_t) \triangleq v_{\mathscr{I}}(a_t, \theta_t)$, respectively. Without loss of generality we shall assume that $v_i(a_t, \theta_t) \geqslant 0$ and $c(a_t, \theta_t) \geqslant 0$ for all a_t and θ_t. We also assume the existence of a (default) action $a_t \in \mathscr{A}_t(\theta_t)$ such that $c(a_t, \theta_t) = 0$ for all θ_t.

All players maximize expected discounted value and their common discount factor for future periods is $\delta \in (0, 1)$.

2.2. *Social values*

With transferable utility between the agent and the principals, Pareto efficiency coincides with surplus maximization. The value of the socially

---

can be incorporated as well. An example with a continuous state space is the job matching model given in an earlier version of this paper.

efficient program is denoted by

$$W(\theta_t) \triangleq W_{\mathscr{I}}(\theta_t),$$

and the value of the efficient program with a subset S of principals and the agent is denoted by $W_S(\theta_t)$. These values are obtained from a familiar dynamic programming equation:

$$W_S(\theta_t) = \max_{a_t \in \mathscr{A}(\theta_t)} \mathbb{E}\{v_S(a_t, \theta_t) - c(a_t, \theta_t) + \delta W_S(\theta_{t+1})\}.$$

The efficient action for the set S in state θ_t is denoted by $a_S^* \triangleq a_S^*(\theta_t)$ and for the entire set $\mathscr{I}$, it is $a^* \triangleq a_{\mathscr{I}}^*(\theta_t)$. The value of a set of firms $\mathscr{I} \setminus S$ is similarly denoted by $W_{-S}(\theta_t)$.

The marginal contribution of principal i is defined by

$$M_i(\theta_t) \triangleq W(\theta_t) - W_{-i}(\theta_t). \tag{2.1}$$

The marginal contribution of a subset of principals $S \subset \mathscr{I}$ is by extension

$$M_S(\theta_t) \triangleq W(\theta_t) - W_{-S}(\theta_t). \tag{2.2}$$

The marginal contribution of a set of principals S measures the increase in the total value to the grand coalition which results from adding the set S of principals. We emphasize that for all social values $W_s(\theta_t)$ the agent is always implicitly included in the set S of principals, whereas for all marginal contributions $M_s(\theta_t)$, the agent is always excluded from set S.

3. Static Common Agency

3.1. *Equilibrium characterization*

This section presents the equilibrium concept and new characterization results for the static common agency game. As we discuss the static model here, the state variable θ_t is omitted in this section. The basic model and the equilibrium notions were first introduced by Bernheim and Whinston [4].

A strategy for principal i is a reward function $r_i : \mathscr{A} \to \mathbb{R}_+$ by which the principal offers a reward to the agent contingent on the action chosen by her. The net benefit from action a to principal i is $n_i(a) \triangleq v_i(a) - r_i(a)$. The vector of net benefits is $n(a) = (n_1(a), \ldots, n_I(a))$ and the aggregate benefits for a subset S is $n_s(a) \triangleq \sum_{i \in S} n_i(a)$. The net benefit to the agent is given by $r(a) - c(a)$.

Definition 1 (Best response):

1. An action a is a best response to the rewards $r(\cdot)$ if

$$a \in \arg\max_{a' \in \mathscr{A}}(a') - c(a').$$

2. A reward function $r_i(\cdot)$ is best response to the rewards $r_{-i}(\cdot)$, if there does not exist another reward function $r_i'(\cdot)$ and action a' such that

$$v_i(a') - r_i'(a') > v_i(a) - r_i(a),$$

where a and a' are best responses to $(r_i(\cdot), r_{-i}(\cdot))$ and $(r_i'(\cdot), r_{-i}(\cdot))$, respectively.

Definition 2 (Nash equilibrium): A *Nash equilibrium* of the common agency game is an I-tuple of reward functions $\{r_i(\cdot)\}_{i=1}^I$ and an action a such that $r_i(\cdot)$ and a are best responses.

Bernheim and Whinston suggest that the focus be put on a subset of the Nash equilibria where all strategies satisfy an additional restriction, called truthfulness.

Definition 3 (Truthful strategy):

(1) A reward function $r_i(\cdot)$ is said to be truthful relative to a if for all $a' \in \mathscr{A}$, either

    (a) $n_i(a') = n_i(a)$, or,
    (b) $n_i(a') < n_i(a)$, and $r_i(a') = 0$.

(2) The strategies $\{\{r_i(\cdot)\}_{i=1}^I, a\}$ are said to be a truthful Nash equilibrium if they form a Nash equilibrium and $\{r_i(\cdot)\}_{i=1}^I$ are truthful strategies relative to a.

The main result of Bernheim and Whinston [4] describes the set of truthful equilibrium payoffs as follows.

Theorem 1 (Bernheim and Whinston): *A vector* $\mathbf{n} \in \mathbb{R}^I$ *is a vector of net payoffs for some truthful equilibrium if and only if*

1. *for all* $S \subseteq \mathscr{I}$,

$$\sum_{i \in S} n_i \leqslant M_S, \text{ and}$$

2. *for all $i \in \mathscr{I}$, there exists an $S \subseteq \mathscr{I}$ such that $i \in S$ and*

$$\sum_{j \in S} n_j = M_S.$$

Using these (in-)equalities, Bernheim and Whinston [4] prove that all truthful equilibria are efficient and coalition-proof. Next we provide an additional characterization of the set of truthful Nash equilibrium payoffs for the static game.

Definition 4 (Marginal contribution equilibrium): A *marginal contribution equilibrium* of the common agency game is a truthful Nash equilibrium with $n_i = M_i$ for all i.

In other words, all principals receive their marginal contribution to the social welfare as their equilibrium net payoff in a marginal contribution equilibrium. The following theorem characterizes the games that have marginal contribution equilibria and their truthful equilibrium payoffs.

Theorem 2 (Existence and uniqueness):

1. *A marginal contribution equilibrium exists if and only if*

$$\forall S \subseteq \mathscr{I}, \quad \sum_{i \in S} M_i \leqslant M_S. \tag{3.1}$$

2. *The truthful Nash equilibrium payoff set is a singleton if and only if the game has a marginal contribution equilibrium.*[2]
3. *If M_S is superadditive:*

$$\forall S, T, S \cap T = \emptyset, \quad M_S + M_T \leqslant M_{S \cup T}, \tag{3.2}$$

then the truthful equilibrium is unique.

Proof: See appendix.   □

Condition (3.1) requires that the sum of the marginal contributions of each principal $i \in S$ to $\mathscr{I}$ is less than the marginal contribution of the entire set S to $\mathscr{I}$. Condition (3.1) is referred to as *weak superadditivity* of the marginal contributions. Superadditivity condition (3.2) is a sufficient

---

[2]The truthful Nash equilibrium is also unique in a generic set of games. To see that this cannot be always the case, consider an example where $\mathscr{I} = \{1, 2\}$, $A = \{1, 2\}$, $c(a) = 0$ for all a, $u_i(a) = v$ if $a = i$ and $u_i(a) = 0$ if $a \neq i$ for $i \in \{1, 2\}$. The set of truthful equilibria for this game is $r_1 = r_2 = v$, $a \in \{1, 2\}$.

condition for (3.1) and it agrees with (3.1) if $|\mathscr{I}| = 2$. Laussel and LeBreton [11] give the following equivalent form of the sufficient condition (3.2). If for $\forall S, T \subset \mathscr{I}$, such that $S \cap T = \emptyset$,

$$ W \leqslant W_{-S} + W_{-T} - W_{-(S \cup T)}, $$

then the truthful equilibrium is unique. This is, however, not a necessary condition for uniqueness.

4. Agenda Setting

The static common agency model is almost exclusively a game among the principals. The outcome of the bidding game decides which of the feasible actions the agent should pursue. As the agent herself is made a take-it-or-leave-it offer, her strategic role in the game is minimal, and under truthful bidding strategies, she will indeed select the socially efficient action.

In this section, we change the static common agency minimally by allowing the set of actions available to the agent to be endogenous. More precisely, we allow the agent to set the agenda in the initial period by selecting a subset of the exogenously given set of feasible actions. In the subsequent period, the principals bid on the actions in the selected subset. In order to preserve the spirit of the common agency game, we allow the principals to influence the selection in the initial stage. The resulting game, referred to as agenda setting game, extends over two-periods and is arguably the minimal extension from a static to a dynamic problem of common agency. The initial choice has no direct payoff consequences for any player and with no discounting between the periods, the set of feasible payoffs in the agenda game is the same as in the static common agency game.

The game proceeds as follows. In period 0, each principal bids on the subset A chosen by the agent from the set of feasible actions, $\mathscr{A}$. The agent receives a reward $r_i(A)$ from principal i if she selects the subset A for the second stage. The choice of A is costless to the agent and has no immediate payoff consequences for the principals. The eventual choice of the agent in period 1 is, however, restricted to the subset A. The payoffs in period 1 game are as in the previous section.

This simple model is sufficient to introduce the conceptual difference between static and dynamic common agency. By selecting an action today, the agent can change the nature of competition among the principals tomorrow. The agent prefers early stage actions that increase competition between the principals and therefore give rise to higher equilibrium payoffs to the agent in the later stage. In general, the socially optimal action does not have to be included in the subset of actions that maximize the equilibrium payoff to the agent in the final stage. Since the principals' future payoffs are reflected in their bid schedules in the initial period, the agent faces a trade-off between efficiency and rent extraction in period 0.

The set of available actions in the initial period is the set of all subsets of $\mathscr{A}$, or $2^{\mathscr{A}}$. As we want to focus on the interaction between period 0 and period 1 outcomes, we assume that a marginal contribution equilibrium in period 1 exists for all subsets A. The payoffs in period 1 then given directly by their marginal contribution relative to the set A and by the residual rent for the agent. For this purpose, let

$$W_S(A) \triangleq \max_{a \in A}\{v_S(a) - c(a)\},$$

and

$$M_S(A) \triangleq W(A) - W_{-S}(A).$$

By assumption, every set A induces a marginal contribution equilibrium in the second stage with payoffs given by $M_i(A)$ for principal i and by $W(A) - \sum_{i \in \mathscr{I}} M_i(A)$ for the agent. By backward induction, we may then take the equilibrium payoffs tomorrow as the gross payoffs associated with the selection of set A today and analyze the agenda game as a single-period common agency game with the payoff structure as just defined. This intertemporal structure suggests a recursive notion of marginal contribution. The social value the set S of principals can achieve jointly with the agent is denoted by

$$\widehat{W}_S \triangleq \max_{A \in 2^{\mathscr{A}}} \left\{ W(A) - \sum_{i \notin S} M_i(A) \right\}.$$

It is a recursive notion as it incorporates the fact that every principal outside of S will be able to claim his marginal contribution in the future. The *recursive contribution* of a set S of principals is then defined as

$$\widehat{M}_S \triangleq \widehat{W} - \widehat{W}_{-S}.$$

When we compare $\widehat{W}_{-S}$ with W_{-S}, where the former is social value of set $\mathscr{I} \backslash S$ in the static game, we observe that they differ as the contribution of the set S in the static game is computed directly from the payoffs, whereas in the agenda game, it is computed from the (future) marginal contributions. It is easy to verify that $W = \widehat{W}$, $W_{-i} = \widehat{W}_{-i}$, and $M_i = \widehat{M}_i$, but the equalities do not hold in general for other $S \subseteq \mathscr{I}$. The discrepancy arises as the contributions of the coalitions are now computed on the basis of their equilibrium continuation payoffs rather than their gross payoffs. In particular, if $\sum_{i \in S} M_i < M_S$, then $\widehat{M}_S < M_S$.

The agenda game can now be analyzed as a static game of common agency. The first result is that all truthful equilibria of the agenda game are efficient. As all subsets A which include the efficient action a^* permit the realization of the efficient surplus tomorrow, the equilibrium choice of A is not unique, but rather includes all subset which include a^*. Denote by $\mathscr{A}^*$ the set of all such subsets:

$$\mathscr{A}^* = \{A \in 2^{\mathscr{A}} | a^* \in A\}.$$

Whether each principal gets his marginal contribution, which is his equilibrium payoff in the static common agency game, depends on the ability of the agent to structure the agenda to her advantage.

Theorem 3 (Agenda game):

1. *Every truthful equilibrium of the agenda game is efficient:* $A \in \mathscr{A}^*$, $a = a^*$.
2. *The following three statements are equivalent:*

   (a) *the agenda game has a marginal contribution equilibrium;*
   (b) *for all $S \subseteq \mathscr{I}$:*

$$\sum_{i \in S} \widehat{M}_i \leqslant \widehat{M}_S \qquad (4.1)$$

   (c) *for all $A \subseteq \mathscr{A}$ and all $S \subseteq \mathscr{I}$:*

$$W(\mathscr{A}) - \sum_{i \in S} M_i(\mathscr{A}) \geqslant W(A) - \sum_{i \in S} M_i(A). \qquad (4.2)$$

Proof: See appendix.   ☐

Condition (2b) in Theorem 3 is the familiar condition of superadditivity, but now stated in terms of the recursive contributions for the agenda game. Of more interest is the equivalent condition (2c). The inequality fails to hold if there exist subsets A and S such that

$$W(\mathscr{A}) - W(A) < \sum_{i \in S} M_i(\mathscr{A}) - \sum_{i \in S} M_i(A). \tag{4.3}$$

Observe first that a subset A which satisfies this inequality cannot be an element of $\mathscr{A}^*$. If it were the case that $A \in \mathscr{A}^*$, then $W(\mathscr{A}) = W(A)$, and moreover $M_i(\mathscr{A}) \leqslant M_i(A)$. The last inequality holds as a strict subset $A \subset \mathscr{A}$ reduces the choice set and therefore the social values display in general $W_S(A) \leqslant W_S(\mathscr{A})$, and in particular, $W_{-i}(A) \leqslant W_{-i}(\mathscr{A})$. In consequence, the marginal contribution of i satisfies $M_i(A) \geqslant M_i(\mathscr{A})$ as long as $A \in \mathscr{A}^*$.

Consider therefore a set $A \notin \mathscr{A}^*$ and relative to this set A, partition the set of principals into two groups, with

$$S = \{i \in \mathscr{I} \,|\, M_i(\mathscr{A}) > M_i(A)\},$$

and its complement, S^c:

$$S^c = \{i \in \mathscr{I} \,|\, M_i(\mathscr{A}) \leqslant M_i(A)\}.$$

If inequality (4.3) is to hold for the set A and some set S', then it certainly holds for the set S as just defined. By the truthfulness of the equilibrium strategies, all principals $i \in S^c$, bid for agenda A in such a way that they are indifferent between agenda A and $\mathscr{A}$. Inconsequence, the agent acts as if she were the residual claimant after conceding surplus to all principals in the set S. In other words, she acts as if she were to maximize the joint objective of all principals outside of S and her private objective.[3] Hence if (4.3) holds, there exist sets $A \in 2^{\mathscr{A}}$ and $S \subset \mathscr{I}$, such that the efficiency losses caused by the restriction to A are smaller than the increase in the rent extraction from the subset S of principals.

Yet, in the truthful equilibrium of the agenda game, agenda and action choice will be efficient. But the option to increase rent extraction tomorrow allows the agent to extract more surplus from the principals than she could in the static game. The next corollary identifies the principals who will see a decrease in their equilibrium payoff due to the possibility of agenda setting by the agent. It suffices to describe the payoffs associated with $A = \mathscr{A}$, as truthfulness implies that for all $A, A' \in \mathscr{A}^*$, the net return for each

---

[3] A different line of argument, based on the notion of coalition-proof equilibrium, would lead to the same conclusion.

principal is constant, or

$$M_i(A) - r_i(A) = M_i(A') - r_i(A'),$$

and we recall that $M_i(\mathscr{A}) = M_i$. The characterization of the equilibrium transfers follows directly from the properties of static equilibrium established in Theorems 1 and 2, after using the recursive notion of marginal contribution, $\widehat{M}_i$ and $\widehat{M}_S$.

Corollary 1 (Equilibrium payoffs in agenda game): *The equilibrium transfers* $\{r_i(\mathscr{A})\}_{i \in \mathscr{I}}$ *have the properties:*

1. $\forall S, \sum_{i \in S}(M_i - r_i(\mathscr{A})) \leqslant \widehat{M}_S$;
2. $\forall i, \exists S \text{ s.th. } i \in S \text{ and } \sum_{j \in S}(M_j - r_j(\mathscr{A})) = \widehat{M}_S.$

Notice the close analogy of properties 1 and 2 to the characterization of static payoffs in Theorem 1. Note, however that since $\widehat{M}_S < M_S$ in some games, property (2.2) of Theorem 1 does not hold in general.

The equilibrium transfers in the initial period are such that the superadditivity of the net payoffs is maintained in the agenda game. If principal i does not belong to any subset S of principals for which the agent's gain in rent extraction exceeds the efficiency losses, then i receives his marginal contribution and offers no rewards to the agent for keeping the choice set unrestricted, or

$$\forall i, \text{ such that } \forall S \text{ with } i \in S, \quad \sum_{j \in S} M_j \leqslant \widehat{M}_S \Rightarrow r_i(\mathscr{A}) = 0.$$

On the other hand if a subset S exists where rent extraction would be sufficiently increased by an inefficient restriction of the agenda, then in equilibrium the principals belonging to the set S, have to jointly provide transfers so as to make the agent indifferent between $\mathscr{A}$ and an inefficient agenda $A' \notin \mathscr{A}^*$, or

$$\forall S, \text{ such that } \sum_{j \in S} M_j > \widehat{M}_S, \quad \exists i \in S, r_i(\mathscr{A}) > 0.$$

The preceding results are illustrated by a simple example. There are two principals, 1 and 2. The agent can choose among three actions $\mathscr{A} = \{a, b, c\}$ at zero cost. The actions a and b are most favored by principals 1 and 2, respectively, whereas the action c represents a compromise. The gross payoffs are (with slight abuse of notation):

$$v_1(a) = a, \quad v_1(b) = 0, \quad v_1(c) = c,$$

and

$$v_2(a) = 0, \quad v_2(b) = b, \quad v_2(c) = c.$$

The ranking of the payoffs is $a \geqslant b > c > 0$, and the social payoffs are ranked $2c > a \geqslant b$. The static common agency has a unique equilibrium in which the efficient action c is selected and supported by the equilibrium transfers $r_1(c) = b - c$ and $r_2(c) = a - c$. However, by excluding the compromise from the set of available actions, the agent could extract a larger rent from the principals. Indeed, inequality (4.2) fails to hold for $A = \{a, b\}$ and $S = \mathscr{I}$. In the agenda game, the principals therefore have to lobby the agent to keep the compromise c on the agenda. In consequence, their transfers in period 0 have to be large enough to make the agent forego the sharper conflict between the principals. This requires that

$$r_1(\mathscr{A}) + r_2(\mathscr{A}) = 2c - a.$$

The sharing of the burden to maintain c as a feasible solution is subject to the participation constraints:

$$M_1(\mathscr{A}) - r_1(\mathscr{A}) \geqslant M_1(\{a, b\}) = a - b,$$

and

$$M_2(\mathscr{A}) - r_2(\mathscr{A}) \geqslant M_2(\{a, b\}) = 0.$$

The ability to set the agenda then allows the agent to extract higher rents, b rather than $a + b - 2c$. Remarkably, the differential value of setting the agenda increases as a decreases, or as the competition between 1 and 2 intensifies.

5. Dynamic Common Agency

The agenda game introduced some of the new aspects of the common agency game which arise in a dynamic setting. In this section we present the main results for a general dynamic model of common agency. The equilibrium of the dynamic game is defined in Section 5.1, where we also prove its existence. The general characterization is given in Section 5.2, and necessary and sufficient conditions for the uniqueness of the truthful equilibrium are given in Section 5.3.

5.1. *Truthful equilibrium*

In the dynamic game, a reward strategy for principal i is a sequence of reward mappings

$$r_i : A_t \times H_t \to \mathbb{R}_+$$

assigning to every action $a_t \in \mathscr{A}_t$ a nonnegative reward, possibly contingent on the entire past history of the game. A strategy by the agent is a sequence of actions over time

$$a_t : \overset{I}{\underset{i=1}{\times}} \mathbb{R}_+^{|\mathscr{A}_t|} \times H_t \to \mathscr{A}_t,$$

depending on the profile of reward schedules in period t and history until period t. Strategies that depend on h_t only through θ_t are called Markov strategies. Since our main objective in this paper is to analyze how changes in the stage game influence the strategic positions of the players, we restrict our attention to Markov equilibria. Under this modeling choice, behavior cannot depend on payoff irrelevant features of the past path of play. Notice also that as θ_t depends on a_t, but not directly on r_t, the principals influence the future state of the world only through inducing different choices by the agent.

The expected discounted payoff with a history h_t for a given sequence of reward policies $\mathbf{r}$ and action profiles $\mathbf{a}$ is denoted by $V_0(h_t)$ for the agent and $V_i(h_t)$ for principal i. When $\mathbf{a}$ and $\mathbf{r}$ are Markov policies, then the values are given by $V_0(\theta_t)$ and $V_i(\theta_t)$ if the state is θ_t in period t. In this context, $\mathbb{E}V_i(a_t, \theta_t)$ represents the expectation of the continuation value in period $t+1$ if in period t the action was a_t and the state was θ_t. While the transition from θ_t to θ_{t+1} may be stochastic, we shall omit the expectations operator $\mathbb{E}[\cdot]$ for simplicity and all values are henceforth understood to represent expected values.

Definition 5 (Markov perfect equilibrium): The strategies $\{r_i(a_t, \theta_t)\}_{i \in \mathscr{I}}$ and $a(r(\cdot), \theta_t)$ form a Markov perfect equilibrium (MPE) if

1. $\forall \theta_t, \forall r'(\cdot), a(r'(\cdot), \theta_t)$ is a solution to

$$\max_{a_t \in \mathscr{A}_t} \{r'(a_t, \theta_t) - c(a_t, \theta_t) + \delta V_0(a_t, \theta_t)\},$$

2. $\forall i$, $\forall \theta_t$, there is no other reward function $r_i'(a_t, \theta_t)$ such that

$$v_i(a', \theta_t) - r_i'(a', \theta_t) + \delta V_i(a', \theta_t) > v_i(a, \theta_t) - r_i(a, \theta_t) + \delta V_i(a, \theta_t),$$

where a and a' are best responses to $(r_i(\cdot), r_{-i}(\cdot))$ and $(r_i'(\cdot), r_{-i}(\cdot))$, respectively.

Truthful strategies are defined as in the static game by the property that they reflect accurately each principal's net willingness to pay. The major difference to the static definition is that the allocation relative to which truthfulness is defined is now an action a_t *and* a state θ_t. The intertemporal net benefit $n_i(a_t, \theta_t)$ of an allocation a_t in the state θ_t is the flow benefit $v_i(a_t, \theta_t) - r_i(a_t, \theta_t)$ and the continuation benefit $\delta V_i(a_t, \theta_t)$:

$$n_i(a_t, \theta_t) \triangleq v_i(a_t, \theta_t) - r_i(a_t, \theta_t) + \delta V_i(a_t, \theta_t). \tag{5.1}$$

With this extension to the dynamic framework, the definition of a truthful (Markov) strategy and an associated MPE in truthful strategies is immediate.

Definition 6 (Truthful (Markov) strategy):

1. A reward function $r_i(a_t, \theta_t)$ is truthful relative to (a, θ_t) if for all $a_t \in \mathscr{A}(\theta_t)$, either:

   (a) $n_i(a_t, \theta_t) = n_i(a, \theta_t)$, or,
   (b) $n_i(a_t, \theta_t) < n_i(a, \theta_t)$, and $r_i(a_t, \theta_t) = 0$.

2. The strategies $\{r_i(\cdot)\}_{i=1}^{I}$ and $a(r(\cdot), \theta_t)$ are an MPE in truthful strategies if they are an MPE and $\{r_i(\cdot)\}_{i=1}^{I}$ are truthful strategies relative to $a(r(\cdot), \theta_t)$.

With this definition at hand, we can prove the existence result for our solution concept.

Theorem 4: *A Markov perfect equilibrium in truthful strategies exists.*

Proof: See appendix. $\square$

The proof runs along familiar lines in games with discounted payoffs. In the first step, the existence of an MPE in truthful strategies is proved for finite games. In the second step, it is shown that because of continuity at infinity, the limit of finite equilibria forms an equilibrium in the infinite game. As the limit of finite equilibria may still depend on calendar time, we also provide an independent proof, suggested by Michel LeBreton, to prove the existence of a stationary MPE in the infinite horizon game.

5.2. *Characterization*

The characterization of the set of truthful equilibria relies as in the static model on the marginal contribution of each principal. The marginal contribution of principal i is, as defined earlier,

$$M_i(\theta_t) \triangleq W(\theta_t) - W_{-i}(\theta_t). \tag{5.2}$$

We are now in a position to develop the recursive argument sketched in the agenda game to its full extent. The recursion developed in the agenda game can be extended from terminal payoffs to arbitrary continuation payoffs. By the principle of optimality, this allows us to show in the next theorem that all truthful equilibria have to be efficient. To this effect, we define $W(\theta_t|a_t)$ to be the social value of the program which starts with an arbitrary and not necessarily efficient action a_t, but thereafter chooses an intertemporally optimal action profile. Similarly, let $M_i(\theta_t|a_t) \triangleq W(\theta_t|a_t) - W_{-i}(\theta_t|a_t)$.

   If the equilibrium continuation play is indeed efficient, the distribution of the surplus along the equilibrium path can also be determined recursively by the sequence of residual claims the agent can establish. The maximal value the agent and a subset $\mathscr{I}\backslash S$ of principals can achieve along the equilibrium path is obtained by selecting a_t so as to solve

$$\max_{a_t} \left\{ W(\theta_t|a_t) - v_S(a_t, \theta_t) - \sum_{i \in S} \delta V_i(a_t, \theta_t) \right\}. \tag{5.3}$$

The net value $n_s(\theta_t)$ of the set S of principals in truthful equilibrium must then satisfy the following inequality in every period:

$$n_S(\theta_t) \leqslant W(\theta_t) - \max_{a_t \in \mathscr{A}_t} \left\{ W(\theta_t|a_t) - v_S(a_t, \theta_t) - \sum_{i \in S} \delta V_i(a_t, \theta_t) \right\}.$$

By relating the equilibrium continuation values $V_i(\theta_t)$ recursively to the marginal contributions $M_i(\theta_t)$, we obtain the following:

Theorem 5 (Efficiency):

1. *All MPE in truthful strategies are efficient.*
2. *For all $S \subseteq \mathscr{I}$,*

$$\sum_{i \in S} V_i(\theta_t) \leqslant M_S(\theta_t). \tag{5.4}$$

Proof: See appendix. $\qquad\qquad\qquad\qquad\qquad\qquad\qquad\qquad\qquad$ □

An important qualification for the efficiency result is the participation issue which did not arise in the static game. For the equilibrium to be efficient, it has to be the case that every principal i who might realize some nontrivial payoff $v_i(a_\tau, \theta_\tau)$ at some future time τ, participates in the game in all periods t prior to τ. For if he were absent in some period, the agent and the remaining principals would only seek to maximize their current and future payoff, and fail to internalize the impact of their decision on principal i. For the same reason, the theorem cannot accommodate a change in the identity of the agent, unless of course the sequence of agents would have perfectly dynastic preferences. The efficiency failure with varying participation is related to the observation made in Aghion and Bolton [1], where the collusion between an incumbent and a buyer against a potential future entrant may result in welfare losses.

In the static game, there is an additional result relating the equilibrium payoffs to the marginal contributions. For every i, there is a set S, with $i \in S$ such that the joint equilibrium payoff of the set S of principals equals their marginal contribution, or

$$\sum_{j \in S} V_j = M_S. \tag{5.5}$$

The analysis of the agenda game in the previous section provided an example of a dynamic environment where the above equality failed to hold for every i. As the principals lobbied to keep the option of a compromise open, neither a single principal nor the principals jointly could realize their marginal contribution.

While the focus of our analysis is on the efficiency and the existence of a unique truthful equilibrium, Bernheim and Whinston [4] also gave a complete description of the set of equilibrium payoffs in the absence of a unique marginal contribution equilibrium. The characterization relied on a set of inequalities relating the social payoffs of various subsets of principals (see their Theorem 2). For all finite horizon games, such as the agenda game discussed in the previous section, a straightforward recursive extension of their inequalities would give a similar characterization for dynamic games.

5.3. *Marginal contribution equilibrium*

The intertemporal aspects of the game weakens the position of the principals as neither an individual principal nor any group of principals can receive their marginal contribution in general. In this section, we give necessary and sufficient conditions for a marginal contribution equilibrium to exist. A *marginal contribution equilibrium* is a Markov perfect equilibrium in truthful strategies where the payoff of each principal coincides with his marginal contribution, or for all i and all θ_t,

$$V_i(\theta_t) = M_i(\theta_t).$$

The previous efficiency theorem already indicated that the weak superadditivity of the marginal contributions $M_i(\theta_t)$ remains a necessary condition in the dynamic game as the inequality

$$\sum_{i \in S} V_i(\theta_t) \leqslant M_S(\theta_t),$$

can only hold with $V_i(\theta_t) = M_i(\theta_t)$ if indeed

$$\sum_{i \in S} M_i(\theta_t) \leqslant M_S(\theta_t).$$

However the agenda game already illustrated that it cannot be a sufficient condition anymore. The analysis of the agenda game also suggested that if a marginal contribution equilibrium is to exist, then the current decision by the agent should not be biased too strongly by her interest to depress the future shares of the principals relative to the shares along the efficient path. Since the agent is the residual claimant after the principals receive their marginal contribution, a formal statement of this requirement is that the social loss from a deviation from the efficient policy exceeds the loss in the marginal contributions of the principals, or $\forall a_t \in \mathscr{A}(\theta_t)$, $\forall S \subseteq \mathscr{I}$:

$$W(\theta_t|a_t) - W(\theta_t) \leqslant \sum_{i \in S}(M_i(\theta_t|a_t) - M_i(\theta_t)).$$

Recall that $W(\theta_t|a_t)$ is the social value of the program which starts with an arbitrary action a_t, but thereafter chooses an intertemporally optimal action profile, and it follows that $W(\theta_t|a_t) - W(\theta_t) < 0$ for all $a_t \neq a^*$.

Theorem 6 (Marginal contribution equilibrium): *The marginal contribution equilibrium exists if and only if*

$$\sum_{i \in S}(M_i(\theta_t) - M_i(\theta_t|a_t)) \leqslant W(\theta_t) - W(\theta_t|a_t), \quad \forall a_t, \forall \theta_t, \forall S. \qquad (5.6)$$

Proof: See appendix. $\qquad\qquad\qquad\qquad\qquad\qquad\qquad\qquad\qquad\qquad\qquad\square$

Inequality (5.6) can be directly interpreted as the trade-off between rent extraction and efficiency gains. The LHS of the inequality describes the opportunities of the agent to extract additional rents from the principals by deviating from the efficient path, whereas the RHS describes the social losses associated with such a deviation. The rent of the agent here is not due to informational asymmetries, but rather to the changing nature of the competition between the principals. In the case of a repeated common agency game, condition (5.6) reduces to the condition of weak superadditivity of the marginal contributions in the static game as the transition from period t to $t + 1$ is of course independent of the action chosen in period t.

The equilibrium characterization by the inequality is particularly useful in applications. Since all values entering the inequality can be obtained from appropriate efficient (continuation) programs, the inequality can be established independently of any equilibrium considerations. As efficient programs are in general easier to analyze than dynamic equilibrium conditions, the technique suggested here may be usefully applied to a wide class of dynamic bidding models. In an earlier version of this paper [3a] we analyzed a job-matching model with n firms. The value of the match between the worker and any of the firms is initially uncertain and the issue is whether equilibrium wages can induce intertemporally efficient matching. The technique developed here, and in particular the theorem above allows us to prove efficiency and characterize the equilibrium payoffs. Earlier work on this class of models by Bergemann and Välimäki [2] and Felli and Harris [8] could prove efficiency without the use of the common agency framework only for two firms.

In a recent contribution Bergemann and Välimäki [3b] show how the equilibrium argument for spot prices rather than menus can be extended to many firms by using the marginal contributions as value functions in the dynamic programming equations.

A reformulation of the rent extraction inequality (5.6) provides a link between the static and the dynamic conditions for the existence of a marginal contribution equilibrium. For any state θ_t define:

$$\widehat{M}_S(\theta_t) \triangleq W(\theta_t) - \max_{a_t \in \mathscr{A}_t} \left\{ W(\theta_t | a_t) - v_S(a_t, \theta_t) - \sum_{i \in S} \delta M_i(a_t, \theta_t) \right\}$$

as the *recursive contribution* of a subset S of principals. For all singleton subsets, we have $M_i(\theta_t) = \widehat{M}_i(\theta_t)$, but in general the equality fails to holds for sets S with a cardinality exceeding one. The main difference between the marginal contribution, $M_S(\theta_t)$, and the recursive contribution, $\widehat{M}_S(\theta_t)$, is the different treatment of the continuation value associated with a subset S. While $M_S(\theta_t)$ attributes the entire future marginal contribution of coalition S to its members, $\widehat{M}_S(\theta_t)$ attributes only the sum of individual marginal contributions. These two notions are equivalent if and only if the marginal contributions are additive. Likewise, if the marginal contributions are (strictly) superadditive, or $\sum_{i \in S} M_i(\theta_t) < M_S(\theta_t)$, then using the definition given above, one can show that $\widehat{M}_S(\theta_t) < M_S(\theta_t)$. We obtain necessary and sufficient conditions for a marginal contribution equilibrium more in the spirit of the static condition as follows:

Corollary 2: *A marginal contribution equilibrium exists if and only if* $\forall \theta_t, \forall S$:

$$\sum_{i \in S} \widehat{M}_i(\theta_t) \leqslant \widehat{M}_S(\theta_t). \tag{5.7}$$

Proof: See appendix. □

The disadvantage of condition (5.7), when compared to the rent extraction inequality, is that it is based on two nested optimization problems rather than a single one based entirely on the social value of the program. An immediate implication of the reformulation in terms of the recursive contribution is that the relation between the uniqueness of the truthful equilibrium and the marginal contribution equilibrium is still valid in the dynamic game.

Corollary 3 (Uniqueness): *A MPE in truthful strategies is unique if and only if it is a marginal contribution equilibrium.*

As the marginal contribution equilibrium is by definition a truthful equilibrium, the 'if' part of the corollary states that if there is a marginal contribution then it is also the unique MPE in truthful strategies.

6. Conclusion

This paper considered common agency in a general class of dynamic games with symmetric information. By focusing on Markovian equilibria, a detailed characterization of the equilibrium strategies and payoffs was possible for this class of games. As in the static analysis by Bernheim and Whinston [4], the link between truthful strategies and the social value of various coalitions was central in obtaining the results. In the dynamic context the link is even more valuable. The continuation payoffs which determine the current bidding strategies, are themselves endogenous to the equilibrium and hence of little help in determining the equilibrium strategies. In contrast, the marginal contributions are defined independently of equilibrium considerations.

As in the static game, a connection can be made between the truthful equilibria and coalition proof equilibria. It is relatively straightforward to extend the notion of coalition-proof equilibrium period by period in a finite horizon game. The notion becomes a bit more problematic in an infinite horizon model. We refer the reader to the previous version of this paper [3a] for a minimal notion of coalition proofness in an infinite horizon model.

We restricted our analysis to symmetric information environments. Bernheim and Whinston [4], however, observed that in the static context with two bidders for a single good, the principals net payoffs are equivalent to the equilibrium net payoffs of the Vickrey–Clarke–Groves mechanism with incomplete information. In this paper, we showed that in fact whenever the principals receive their marginal contribution, their equilibrium net payoff is equal to the Vickrey–Clarke–Groves payoff. This suggests that the techniques presented here could possible be extended to sequential allocation problems with asymmetric information, or more generally, dynamic efficient mechanism design problems. In this context, it should be noted that the case of private information for the principals is distinct from the analysis of Bernheim and Whinston [5], Martimort [12] or Stole [15], where moral hazard or adverse selection is due to a better informed agent.

Acknowledgments

We thank Michel Le Breton, John Geanakoplos, David Pearce, Grazia Rapizarda, Steve Tadelis, Tim van Zandt, Eyal Winter, an anonymous reader and seminar participants at Carnegie Mellon University, Columbia University, Georgetown University, New York University, MIT, University of Pennsylvania and Warwick University for helpful comments. The authors acknowledge support from NSF Grant SBR 9709887 and 9709340, respectively. Dirk Bergemann thanks the Department of Economics at MIT, where the first version of the paper was written, for its hospitality.

Appendix

Proof of Theorem 2. (1) (if) Suppose that $\sum_{i \in S} M_i \leqslant M_S$. Set $n_i = M_i$ for all i, and by hypothesis $n_s \leqslant M_s$. Thus the vector $\mathbf{n} = (n_1, \ldots, n_I)$ satisfies the condition of Theorem 1 and hence a marginal contribution equilibrium exists.

(only if) Suppose $n_i = M_i$ is the net payoff for every principal i in a truthful Nash equilibrium. Then by Theorem 1.1, the sum of the net payoffs have to satisfy for all S, $\sum_{i \in S} n_i \leqslant M_S$. But as $n_i = M_i$, this implies that $\sum_{i \in S} M_i \leqslant M_S$.

(2) (if) By part 1 of this theorem, the existence of a marginal contribution equilibrium is equivalent to: $\forall S$, $\sum_{i \in S} M_i \leqslant M_S$. In a marginal contribution equilibrium we set $n_i = M_i$ for all i, and by hypothesis $n_S \leqslant M_S$. Thus if there were to be a different equilibrium payoff vector $\mathbf{n} = (n_1, \ldots, n_I)$, there would have to be a subset S such that $\forall i, i \in S$, $n_i < M_i$, as $n_i \leqslant M_i$ has to hold by Theorem 1.1. But notice that the decrease in the equilibrium net payoff for all i in some subset S does not permit the increase of any other n_j, $j \notin S$, as $n_j = M_j$ is a binding constraint and hence the uniqueness of the payoffs follows. By the definition of a truthful strategy, this also determines uniquely the equilibrium strategies.

(only if) We prove the contrapositive. Suppose for some $S \subseteq \mathscr{I}$, $\sum_{i \in S} M_i > M_S$, then we show that the equilibrium net payoff vector cannot be unique. The proof is by construction using the greedy algorithm. Define $n_1 \triangleq M_1$, and in general,

$$n_k \triangleq \min_{\{S | k \in S \wedge S \subseteq \{1, 2, \ldots, k\}\}} M_S^k,$$

where

$$M_S^k \triangleq M_S - \sum_{j \in S \setminus k} n_j.$$

It can be verified that the induced allocation $\{n_1, n_2, \ldots, n_I\}$ is an equilibrium allocation with, by hypothesis, $n_{i'} < M_{i'}$, for some $i' > 1$. Consider next a permutation $\sigma : \mathscr{I} \to \mathscr{I}$ such that $i' \mapsto 1$. By applying the greedy algorithm to the new ordering, we again obtain an equilibrium allocation, but clearly $n_{\sigma(i')} = M_{\sigma(i')}$, which is distinct from the previous allocation.

(3) It is enough to prove that (3.2) is a sufficient condition for (3.1). Consider sets $S, T \subset \mathscr{I}$, and suppose that (3.2) holds, then we have for

any S_1, $S_2 \subset S$, $S_1 \cap S_2 = \emptyset$, $S \cap T = \emptyset$,

$$M_{S_1} + M_{S_2} \leqslant M_S,$$

and thus

$$M_{S_1} + M_{S_2} + M_T \leqslant M_{S \cup T}.$$

As we continue to split up S and T until we have singleton sets consisting of single principals on the left-hand side, we obtain

$$\sum_{i \in S \cup T} M_i \leqslant M_{S \cup T},$$

which completes the claim.   □

Proof of Theorem 3. (1) The efficiency of the action choice $a = a^*$ in period 1 is immediate if $A \in \mathscr{A}^*$. By assumption every $A \in 2^{\mathscr{A}}$ induces a unique truthful equilibrium in period 1. By backward induction, we take the continuation payoffs following an agenda choice A to be the static payoffs in period 1 associated with agenda A. In consequence we can analyze the game in period 0 as a static common agency game with the action set being the set of all possible agendas, $2^{\mathscr{A}}$. By Theorem 2 of Bernheim and Whinston [4], it then follows that $A \in \mathscr{A}^*$.

(2) By backward induction and the uniqueness of the continuation payoffs in period 1, the equilibrium net payoffs are decided by the choice of an agenda A in period 0. The gross payoffs in period 0 can then taken to be $M_i(A)$ and $W(A) - \sum_{i \in \mathscr{A}} M_i(A)$. By the definition of $\widehat{M}_S$, the necessary and sufficient condition for a marginal contribution equilibrium of Theorem 2, can then be written as $\forall S \subseteq \mathscr{I}$, $\sum_{i \in S} \widehat{M}_i \leqslant \widehat{M}_S$.

As we have $\widehat{M}_i = M_i(\mathscr{A})$, and $\widehat{M}_S = W(\mathscr{A}) - \max_{A \in 2^{\mathscr{A}}} \{W(A) - \sum_{i \in S} M_i(A)\}$, we can in turn write the inequality (4.1) as

$$W(\mathscr{A}) - \sum_{i \in S} M_i(\mathscr{A}) \geqslant \max_{A \in 2^{\mathscr{A}}} \left\{ W(A) - \sum_{i \in S} M_i(A) \right\},$$

and since the RHS has to hold for all $A \in 2^{\mathscr{A}}$, the equivalence between (4.1) and (4.2) follows directly.   □

Proof of Theorem 4. The existence of a truthful MPE for finite horizon games follows from the existence result in [4] by backwards induction.

Consider next the case of $T = \infty$. The existence of a truthful MPE follows by a limiting argument from the case of $T < \infty$ along the lines of

Maskin and Tirole [13]. Unfortunately, the equilibrium thus obtained may fail to be in stationary strategies and it could be time dependent. As a result, we prove the existence of a stationary equilibrium directly without using the previous result. For this, we need some preliminary definitions.

Let $\mathscr{A}$ be an arbitrary finite action set and denote an arbitrary static common agency game on the action set $\mathscr{A}$ by Γ, with

$$\Gamma = \{\mathscr{I}, \{v_i(a)\}_{i \in \mathscr{I}, a \in \mathscr{A}}, \{c(a)\}_{a \in \mathscr{A}}\}.$$

For the static common agency game Γ, define $W_\Gamma(S)$ as follows for all $S \subseteq \mathscr{I}$:

$$W_\Gamma(S) = \max_{a \in \mathscr{A}} \left\{ \sum_{i \in S} v_i(a) - c(a) \right\}.$$

Let $W_\Gamma = (W_\Gamma(S))_{S \subseteq \mathscr{I}}$ and thus W_Γ is a vector in $\mathbb{R}^{2^I - 1}$. Let $V(W_\Gamma) = (V_0(W_\Gamma), \dots, V_I(W_\Gamma))$ be the vector of net payoffs for the agent and the principals, where the payoffs for the principals satisfy the Bernheim–Whinston inequalities and the payoffs are obtained by the greedy algorithm defined in the proof of Theorem 2 for the order $1, 2, \dots, I$. The payoff for the agent is then defined by $V_0(W_\Gamma) = W_\Gamma(\mathscr{I}) - \sum_{i \in \mathscr{I}} V_i(W_\Gamma)$.

Lemma 1: *The mapping $V : \mathbb{R}^{2^I - 1} \to \mathbb{R}^{I+1}$ is continuous.*

Proof: The greedy algorithm assigns payoff

$$V_1(W_\Gamma) = W_\Gamma(\mathscr{I}) - W_\Gamma(\mathscr{I} \backslash \{1\})$$

to $i = 1$. This is clearly continuous in W_Γ. Since

$$V_i(W_\Gamma) = W_\Gamma(\mathscr{I}) - W_\Gamma(\mathscr{I} \backslash \{1, \dots, i\}) - \sum_{j=1}^{i-1} V_j(W_\Gamma),$$

an inductive argument establishes continuity for all i. The continuity of the agent's payoff follows similarly. $\square$

Next suppose that we augment the fixed static payoff functions $c(a)$ and $v_i(a)$, for agent and principals respectively, with a nonnegative vector

$\lambda = (\lambda_0, \lambda_1, \ldots, \lambda_I)$ and define

$$c(a; \lambda) = c(a) - \lambda_0,$$

and

$$v_i(a; \lambda) = v_i(a) + \lambda_i.$$

Let $\Gamma(\lambda)$ denote the resulting common agency game parametrized by λ.

Lemma 2: *The mapping $\lambda \to W_{\Gamma(\lambda)}$ is continuous in λ.*

Proof: Since

$$W_{\Gamma(\lambda)}(S) = \max_{a \in \mathcal{A}} \left\{ \sum_{i \in S} v_i(a; \lambda) - c(a; \lambda) \right\}$$

for all $S \in \mathcal{I}$, the continuity property follows from the theorem of the maximum. $\qquad \square$

The role of the additive term λ in the next step is to represent the continuation values in a dynamic common agency game. To see this consider next K different common agency games, denoted by Γ_k for every state θ_k, with $k \in \{1, \ldots, K\}$. We assume that the action set $\mathcal{A}$ does not depend on θ,[4] and that the payoffs are parametrized by a matrix

$$v = [v_i(\theta_k)]_{0 \leqslant i \leqslant I, 1 \leqslant k \leqslant K}.$$

The gross payoffs in the common agency game Γ_k are defined by the following payoff functions:

$$v_0(a, \theta_k; v) = -c(a, \theta_k) + \delta \sum_{k'=1}^{K} q(\theta_{k'} | a, \theta_k) v_0(\theta_{k'}),$$

$$v_i(a, \theta_k; v) = v_i(a, \theta_k) + \delta \sum_{k'=1}^{K} q(\theta_{k'} | a, \theta_k) v_i(\theta_{k'}), \quad \forall i = 1, \ldots, I, \quad (A.1)$$

where $q(\theta_{k'} | a, \theta_k)$ are the state transition probabilities defined earlier. The payoff functions in all of these games are clearly continuous in v. For every given matrix v we can then look at the net payoffs of the common agency game Γ_k as described for Lemma 1 by $V(W_{\Gamma_k})$.

---

[4]This is without loss of generality since the action set for each state could always be taken to be the union of all possible actions where the payoffs to the actions not available at that state are assigned arbitrarily large negative payoffs.

Lemma 3: *The mapping* $\Phi : \mathbb{R}^{(I+1)K} \to \mathbb{R}^{(I+1)K}$ *defined by*

$$\Phi(v) = (V(W_{\Gamma_1}), \ldots, V(W_{\Gamma_K}))$$

is continuous in v.

Proof: This follows from Lemmas 1 and 2. $\square$

Finally observe that the image under the mapping Φ of the hypercube, where the later is defined by

$$\left[0, \max_{\{i,k,a\}} \frac{v_i(a, \theta_k)}{1 - \delta} \right]^{(I+1)K}$$

is contained in the hypercube itself, or

$$\Phi\left(\left[0, \max_{\{i,k,a\}} \frac{v_i(a, \theta_k)}{1 - \delta} \right]^{(I+1)K} \right) \subset \left[0, \max_{\{i,k,a\}} \frac{v_i(a, \theta_k)}{1 - \delta} \right]^{(I+1)K}.$$

Hence $\Phi(v)$ satisfies the conditions for Brouwer's fixed point theorem and we have:

Lemma 4: *There is a* v *such that* $v = \Phi(v)$.

Thus the proof of the theorem is complete if we can show that v is a truthful and stationary Markov perfect equilibrium payoff of the dynamic game. Observe that by the construction of Φ and the payoff functions $v_i(a, \theta_k)$ for $1 \leqslant i \leqslant I$ and $v_0(a, \theta_k)$, all strategies obtained in the construction of the truthful equilibria of the Γ_k game are stationary Markovian as well as truthful for all $\theta \in \Theta$ and furthermore there are no profitable one-shot deviations in the dynamic game. As the dynamic game satisfies continuity at infinity, this is sufficient for establishing that v is indeed a truthful stationary Markov perfect equilibrium payoff of the dynamic game. $\square$

Proof of Theorem 5. (1) By the assumption of Markovian strategies, the continuation values for the agent and the principals depend only on the action a_t inducing the transition from θ_t to θ_{t+1}. This implies by Theorems 2 and 3 of Bernheim and Whinston [4] efficiency.

(2) The equilibrium value function $V_i(\theta_t)$ of principal i are required to satisfy the following set of equalities, $\forall i$,

$$V_i(\theta_t) \leqslant \max_{a_t} \left\{ v(a_t, \theta_t) - c(a_t, \theta_t) + \sum_{k=0}^{I} \delta V_k(a_t, \theta_t) \right\}$$

$$- \max_{a_t} \left\{ v_{-i}(a_t, \theta_t) - c(a_t, \theta_t) + \sum_{k \neq i} \delta V_k(a_t, \theta_t) \right\}, \quad \text{(A.2)}$$

and inequalities $\forall S \subseteq \mathscr{I}$,

$$\sum_{i \in S} V_i(\theta_t) \leqslant \max_{a_t} \left\{ v(a_t, \theta_t) - c(a_t, \theta_t) + \sum_{k=0}^{I} \delta V_k(a_t, \theta_t) \right\}$$

$$- \max_{a_t} \left\{ v_{-S}(a_t, \theta_t) - c(a_t, \theta_t) + \sum_{k \notin S} \delta V_k(a_t, \theta_t) \right\}. \quad \text{(A.3)}$$

Since all truthful equilibria are efficient by part 1 of this theorem, we have the identity:

$$W(\theta_t) = \max_{a_t} \left\{ v(a_t, \theta_t) - c(a_t, \theta_t) + \sum_{k=0}^{I} \delta V_k(a_t, \theta_t) \right\}.$$

Next we argue by contradiction. Suppose inequality (5.4) does not hold for some S, but inequalities (A.2) and (A.3) are still satisfied. Then there $\exists \varepsilon > 0$ such that

$$\sum_{i \in S} V_i(\theta_t) - M_S(\theta_t) > \varepsilon, \quad \text{(A.4)}$$

and a fortiori

$$W(\theta_t) - \max_{a_t} \left\{ v_{-S}(a_t, \theta_t) - c(a_t, \theta_t) + \sum_{k \notin S} \delta V_k(a_t, \theta_t) \right\} - M_S(\theta_t) > \varepsilon.$$

$$\text{(A.5)}$$

Since the inequality in (A.5) holds for the maximizing a_t in (A.5), it has to hold for a^*_{-S} as well, so that (A.5) may be rewritten in this instance as

$$\delta W(a^*, \theta_t) - \sum_{k \notin S} \delta V_k(a^*_{-S}, \theta_t) - \delta W_{-S}(a^*, \theta_t)$$

$$+ \delta W_{-S}(a^*_{-S}, \theta_t) - \delta M_S(a^*, \theta_t) > \varepsilon, \quad \text{(A.6)}$$

and since

$$\delta W(a^*, \theta_t) = \delta W_{-S}(a^*, \theta_t) + \delta M_S(a^*, \theta_t),$$

it follows from (20) that

$$W_{-S}(a_{-S}, \theta_t) - \sum_{k \notin S} V_k(a_{-S}, \theta_t) > \frac{\varepsilon}{\delta},$$

which is equivalent to

$$\sum_{i \in S} V_i(a^*_{-S}, \theta_t) - M_S(a^*_{-S}, \theta_t) > \frac{\varepsilon}{\delta}.$$

But by repeating the argument, which we started at (A.4), it then follows that the equilibrium value for the set S of principals increases without bound along some path $(\theta_t, \theta_{t+1}, \ldots)$ which delivers the contradiction as the value of the game is finite.

Proof of Theorem 6. It suffices to show that $\{M_i(\theta_t)\}_{i \in \mathscr{I}}$ satisfy the following set of equalities, $\forall i$,

$$M_i(\theta_t) = \max_{a_t} \left\{ v(a_t, \theta_t) - c(a_t, \theta_t) + \sum_{k=0}^{I} \delta V_k(a_t, \theta_t) \right\}$$

$$- \max_{a_t} \left\{ v_{-i}(a_t, \theta_t) - c(a_t, \theta_t) + \sum_{k \neq i} \delta V_k(a_t, \theta_t) \right\}, \quad \text{(A.7)}$$

and inequalities $\forall S \subseteq \mathscr{I}$,

$$\sum_{i \in S} M_i(\theta_t) \leqslant \max_{a_t} \left\{ v(a_t, \theta_t) - c(a_t, \theta_t) + \sum_{k=0}^{I} \delta V_k(a_t, \theta_t) \right\}$$

$$- \max_{a_t} \left\{ v_{-S}(a_t, \theta_t) - c(a_t, \theta_t) + \sum_{k \notin S} \delta V_k(a_t, \theta_t) \right\}, \quad \text{(A.8)}$$

if and only if the inequalities represented by (5.6) hold. By hypothesis $V_k(a_t, \theta_t) = M_k(a_t, \theta_t)$ for all $k > 0$. For notational ease, we omit that a_t is restricted to $a_t \in \mathscr{A}(\theta_t)$. We start with the set of equalities (A.7). Since all truthful equilibria are efficient by Theorem 5 we have the identity:

$$W(\theta_t) = \max_{a_t} \left\{ v(a_t, \theta_t) - c(a_t, \theta_t) + \sum_{k=0}^{I} \delta V_k(a_t, \theta_t) \right\}.$$

Consider next the term

$$\max_{a_t} \left\{ v_{-i}(a_t, \theta_t) - c(a_t, \theta_t) + \sum_{k \notin i} \delta V_k(a_t, \theta_t) \right\},$$

which can be written as

$$\max_{a_t} \{ v_{-i}(a_t, \theta_t) - c(a_t, \theta_t) + \delta V(a_t, \theta_t) - \delta M_i(a_t, \theta_t) \} = W_{-i}(\theta_t),$$

where the equality follows from the definition of the marginal contribution in (2.1) and hence the equality in (A.7) is satisfied. Consider next the set of inequalities (A.8)

$$\sum_{i \in S} M_i(\theta_t) \leqslant W(\theta_t) - \max_{a_t} \left\{ v_{-S}(a_t, \theta_t) - c(a_t, \theta_t) + \sum_{k \notin S} \delta V_k(a_t, \theta_t) \right\}.$$

$$(A.9)$$

If for any set S, $a^*_{-S} = a^*$, it follows that

$$\sum_{i \in S} M_i(\theta_t) - \delta \sum_{i \in S} M_i(a_t, \theta_t) = v_S(a^*, \theta_t),$$

and hence the set as an aggregate is not making any net contributions to $\mathscr{I} \backslash S$, and (A.9) is satisfied. Suppose next that $a^*_{-S} \neq a^*$, then (A.9) is equivalent to

$$\sum_{i \in S} M_i(\theta_t) \leqslant W(\theta_t) - \max_{a_t} \left\{ W(\theta_t | a_t) - v_S(a_t, \theta_t) - \sum_{i \in S} \delta M_i(a_t, \theta_t) \right\}.$$

$$(A.10)$$

Since the inequality has to hold for the action a^*_{-S} which maximizes the payoff inside the braces, it follows a fortiori that the inequality has to hold for an arbitrary action a_t. Then we may write (A.10) as

$$\sum_{i \in S} M_i(\theta_t) - v_S(a_t, \theta_t) - \sum_{i \in S} \delta M_i(a_t, \theta_t) \leqslant W(\theta_t) - W(\theta_t | a_t),$$

or equivalently

$$\sum_{i \in S} (M_i(\theta_t) - M_i(\theta_t | a_t)) \leqslant W(\theta_t) - W(\theta_t | a_t),$$

which completes the proof. □

Proof of Corollary 2. It is sufficient to show the equivalence between (5.6) and (5.7). Starting with (5.7), we can write the inequality as

$$W(\theta_t) - \sum_{i \in S} M_i(\theta_t) \geqslant \max_{a_t} \left\{ W(\theta_t | a_t) - v_S(a_t, \theta_t) - \sum_{i \in S} \delta M_i(a_t, \theta_t) \right\}.$$

As

$$M_i(\theta_t | a_t) = v_i(a_t, \theta_t) + \delta M_i(a_t, \theta_t),$$

it follows that

$$W(\theta_t) - \sum_{i \in S} M_i(\theta_t) \geqslant \max_{a_t} \left\{ W(\theta_t | a_t) - \sum_{i \in S} M_i(\theta_t | a_t) \right\}. \tag{A.11}$$

As inequality (5.6) has to hold for all S and all a_t, it has to hold in particular for the maximand of the RHS of (A.11), which establishes the result. □

References

[1] P. Aghion, P. Bolton, Contracts as a barrier to entry, Amer. Econ. Rev. 77 (1987) 388–401.

[2] D. Bergemann, J. Välimäki, Learning and strategic pricing, Econometrica 64 (1996) 1125–1149.

[3a] D. Bergemann, J. Välimäki, Dynamic common agency, Northwestern University Center for Mathematical Studies in Economics and Management Science, Discussion Paper 1259, 1999.

[3b] D. Bergemann, J. Välimäki, Dynamic Price Competition, Cowles Foundation Discussion Paper, Yale University, 2003.

[4] B.D. Bernheim, M.D. Whinston, Menu auctions, resource allocation, and economic influence, Quart. J. Econ. 101 (1986a) 1–31.

[5] B.D. Bernheim, M.D. Whinston, Common agency, Econometrica 54 (1986b) 923–942.

[6] A.K. Dixit, Special-interest lobbying and endogeneous commodity taxation, Eastern Econ. J. 22 (1996) 375–388.

[7] A. Dixit, G. Grossman, E. Helpman, Common agency and coordination: General theory and application to government policy making, J. Polit. Econ. 105 (1997) 752–769.

[8] L. Felli, C. Harris, Job matching, learning and firm-specific human capital, J. Polit. Econ. 104 (1996) 838–868.

[9] G.M. Grossman, E. Helpman, Trade wars and trade talks, J. Polit. Econ. 103 (1995) 675–708.

[10] D. Laussel, M. LeBreton, Complements and substitutes in common agency, Rich. Econ. 50 (1996) 325–345.

[11] D. Laussel, M. Le Breton, Conflict and cooperation: The structure of equilibrium payoffs in common agency, J. Econ. Theory 100 (2001) 93–128.

[12] D. Martimort, Exclusive dealing, common agency and multi-principals incentive theory, Rand J. Econ. 27 (1996) 1–31.

[13] E. Maskin, J. Tirole, Markov perfect equilibrium I: Observable actions, J. Econ. Theory 100 (2001) 191–219.

[14] A. Prat, A. Rustichini, Games played through agents, Tilburg Center for Economic Research Discussion Paper No. 9968, August 1999.

[15] L. Stole, Mechanism design under common agency, Mimeo, GSB, University of Chicago, 1997.

Chapter 8

Dynamic Price Competition*

Dirk Bergemann[†] and Juuso Välimäki[‡]
[†]*Department of Economics, and Cowles Foundation of Research in Economics,*
Yale University, New Haven, CT 06520-8268, USA
dirk.bergemann@yale.edu
[‡]*Department of Economics, Aalto University School of Business,*
02150 Espoo, Finland
juuso.valimaki@aalto.fi

We consider the model of price competition for a single buyer among many sellers in a dynamic environment. The surplus from each trade is allowed to depend on the path of previous purchases, and as a result, the model captures phenomena such as learning by doing and habit formation in consumption.

We characterize Markovian equilibria for finite and infinite horizon versions of the model and show that the stationary infinite horizon version of the model possesses an efficient equilibrium where all the sellers receive an equilibrium payoff equal to their marginal contribution to the social welfare.

JEL classification: D81; D83

Keywords: Dynamic competition, marginal contribution, Markov perfect equilibrium, common agency.

1. Introduction

In this paper, we consider a model of Bertrand price competition for a single buyer among many sellers in a dynamic environment. We analyze the existence and efficiency of equilibria in models where the stage game payoffs to the buyer as well as the sellers may depend on the history of past purchases. Examples of dependence of this type include learning by doing for the sellers and habit formation for the buyer.

---

*This chapter is reproduced from *Journal of Economic Theory*, **127**, 232–263, 2006.

Bertrand price competition provides an attractive modeling approach for markets with differentiated commodities as it places the bargaining power in the hands of the players on the long side of the market. This results in a nontrivial sharing of the surplus arising from trades between the buyer and the sellers. If the buyer has a unit demand, the equilibrium in the static game is efficient, and the sharing of economic surplus can be studied independently of any economic distortions. We extend the static model to a general dynamic environment. Issues such as surplus sharing and efficiency then require more careful analysis. If current choices have an impact on future surpluses, the intertemporal aspects of surplus sharing gain in importance. Consider for example an industry where an entrant has a technology that will achieve lower costs of production than the incumbent's technology, but whose initial costs are quite high. It may well be that the seller must sell at prices below costs in the initial periods in the expectation of future profits. The ultimate success of the entrant depends on the degree to which the costs and benefits of the initial periods can be shared between the participants in the market.

In the model of this paper, a finite number of sellers offer differentiated products to a single buyer with unit demand over a discrete time horizon of either finite or infinite length. At the beginning of each period, the sellers choose simultaneously prices for their products and the buyer chooses the seller to supply the product (or possibly she chooses not to buy in that period). Because of assuming unit demands, we have also the alternative interpretation of the model as one where a number of firms compete in spot wage contracts for a given worker over time.[1] All the players discount future with the same discount factor δ. In order to allow for dynamic elements in the model, we allow the surplus from each trade to depend on the sequence of trades made in the previous periods. In the job matching model, such dynamic features arise naturally from learning on the job and participation in training programs. Hence the scope of the model is much larger than the simple repetition of the static price competition game across many periods.

The finite horizon model is analyzed first. By a simple example, we show that the existence of a pure strategy equilibrium cannot be taken for granted. If the surplus resulting from a purchase from seller i depends on the history of sales by sellers other than i, the model has a direct intertemporal externality. Hence there is really no reason to expect that a model

---

[1]Models of this type include [10, 12] in a competitive market and [5] in a duopsonistic labor market.

with (spot) prices as the only feasible transfers would be well behaved with respect to the efficiency of the equilibrium allocation. The restrictive element in price competition is that seller i can offer (positive or negative) transfers to the buyer only in conjunction with a purchase of the product of seller i. Yet with externalities, it is conceivable that seller i would sometimes like to induce the buyer to purchase from j and would be willing to support the purchase of product j with a subsidy. To rule out this class of problems, we assume that the surplus generated by the purchase of a given seller's product depends only on the number of past purchases from that seller. This is consistent with the examples of habit formation and learning by doing and it also accommodates job-specific learning in the job matching model. Surprisingly, the equilibria in this case may be inefficient as well and there may be a multiplicity of them.

In contrast, the results that we obtain in the stationary infinite horizon version of the model are much more in line with the static model. In particular, the model always has an efficient equilibrium and the payoffs are uniquely determined in a large class of games. We show that in the efficient equilibrium, the payoff to each seller coincides with her marginal contribution to the social surplus in the model.

The equilibrium in the infinite horizon model is derived by using the well-known "guess and verify" method from dynamic programming. But in stark contrast to the received use of this method, our guess does not rely on any functional or parametric specification of the value function. The novelty in our argument is that the value function of each seller is guessed to be the difference of two general value functions generated by different socially optimal programs. The equilibrium is then established by demonstrating that the constituting social programs have certain structural properties. This allows us to analyze the value functions of the sellers as if they were the result of single agent optimization problems rather than multi-agent strategic interactions.

In order to keep the arguments as simple as possible, we consider only deterministic models. However, we conjecture that the techniques we use extend to the stochastic case. This suggests that models of Bayesian learning about the match quality might also be analyzed with the tools developed in this paper. In the conclusion, we briefly discuss how the argument ought to extend to the stochastic case.

This paper is related to two of our earlier papers. In [2], we analyzed an infinite horizon model of dynamic price competition with *two* sellers and uncertainty. Felli and Harris [5] considered a similar model, yet set in a labor market environment as a job matching model, in which two buyers (the employers) compete for a single seller (the employee) in an infinite horizon continuous time model with Brownian motion. In the current paper, we identify directions in which the results of these two papers extend (the number of sellers), and also dimensions along which the results cannot be pushed any further (finite horizon models). Felli and Harris [6] consider an extension of their job matching model in which the worker can be assigned to one of two different divisions inside each firm, either a training or a sales division. Inside each firm, the performance in one division is correlated with the performance in the other division. The model thus presents an instance where the independent rewards assumption fails and the resulting equilibrium is inefficient. In a recent paper, Bar-Isaac [1] presents a dynamic learning model where the current value of each alternative also depends on past transactions, but learning occurs through private rather than symmetric public information.

In [3], we analyze a general stochastic dynamic environment but there the sellers compete in a menu auction rather than in prices. Formally, that model is a dynamic and stochastic version of the common agency model first analyzed in [4]. In a menu auction the sellers are allowed to offer transfers contingent on the purchase decision of the buyer. Thus seller i is allowed to offer a transfer to the buyer in a particular period even in the event that the buyer purchases from seller j in that period. Each seller can therefore cross subsidize the buyer for her purchases with other sellers. This rich set of transfers allows us to establish the efficiency of the dynamic allocation in the presence of intertemporal externalities. As such transfers and subsidies are not very common in actuality, we investigate in the current paper when simple spot prices are sufficient for guaranteeing the efficient allocation in the dynamic model. Conversely, if there are (intertemporal) externalities across the sellers, then we already know from the static analysis that simple price competition does not lead to efficient allocation in general and hence we should expect more complex contractual arrangements to arise.

The paper is organized as follows. Section 2 describes the model, defines the Markov Perfect equilibrium and introduces the notion of marginal contribution. Section 3 analyzes the finite horizon model. Section 4 considers the infinite horizon case. It provides sufficient conditions for the existence of an efficient equilibrium in terms of properties of the marginal contributions. Section 5 considers an example and establishes that the marginal contributions satisfy the desired properties. Section 6 shows that in a wide class of environments, the marginal contribution equilibrium is the unique Markov Perfect equilibrium. Section 7 concludes and suggests further questions for research.

2. The Model

We consider the following stage game model of price competition for a buyer with a unit demand. There are I sellers in the market, and we denote the set of sellers by $\mathcal{I} = \{1, \ldots, I\}$. Denote the surplus generated in the purchase of seller $i's$ product by x_i. The sellers set simultaneously prices p_i for their products. At the end of the stage game, the buyer chooses either one of the sellers or does not purchase at all. We denote the buyers choice by $a \in \{0, 1, \ldots, I\}$, where $a = 0$ is interpreted as no purchase and $a = i$ denotes a purchase from seller i. The stage game payoffs for the sellers are given by

$$\pi_i(p_1, \ldots, p_I, a) = \begin{cases} p_i & \text{if } a = i, \\ 0, & a \neq i. \end{cases}$$

The buyer's stage game payoff is given by

$$u_i(p_1, \ldots, p_I, a) = x_i - p_i \quad \text{if} \quad a = i.$$

By setting $x_0 = p_0 = 0$, we incorporate the payoffs from the case of no purchase as well.

This (extensive form) stage game is repeated in a discrete time model with $t = 0, 1, \ldots, T$. We analyze separately the cases where $T < \infty$ and $T = \infty$. All of the players discount future at discount factor $\delta \leqslant 1$. In the infinite horizon model, we assume that $\delta < 1$ and that the payoff criterion is the discounted sum of payoffs. At stage t, the actions in all previous periods are observable to all players. A history in the game is a sequence of prices and decisions. More formally, we define histories inductively by letting $h^0 = \emptyset$ and $h^t = h^{t-1} \cup \{p_0^{t-1}, p_1^{t-1}, \ldots, p_I^{t-1}, a^{t-1}\}$, where $\{p_0^{t-1}, p_1^{t-1}, \ldots, p_I^{t-1}, a^{t-1}\}$ are the actions chosen in period $t-1$. Let H^t denote the set of all possible histories in period t and $H = \cup_{t=0}^{\infty} H^t$.

We are interested in a dynamic version of the stage game. Hence we allow the stage game surpluses x_i to depend on h^t. At the same time, we do not want to make the surpluses dependent on past pricing decisions by the sellers. As we also want to avoid calendar time having a direct effect on the surpluses, we are led to consider the vector

$$\theta(h) \triangleq (t_0(h), \ldots, t_I(h)),$$

where $t_i(h)$ counts the number of times that alternative i was chosen by the buyer along history h, as the relevant state variable summarizing the history. Notice that with this choice for a state variable, we ignore the importance of the order in which the sellers made their sales. Thus we assume that

$$x_i(h) = x_i(\theta(h)) \text{ for all } i \in \{0, \ldots, I\} \text{ and } h \in H.$$

In fact, in most of this paper we make the stronger assumption that the payoffs from a given seller depend only on the past purchases with that seller.

Definition 1 (Independent rewards): The payoffs display independent rewards if for all i and $\theta : x_i(\theta(h)) = x_i(t_i(h))$.

From now on, we define a dynamic price competition game Γ to be a collection of functions $\{x_0(\theta), x_1(\theta), \ldots, x_I(\theta); T\}$ for all θ such that $\sum_i t_i(\theta) \leqslant T$.

A pure behavior strategy for seller i is a sequence of functions $\mathbf{p}_i = \{p_i^t\}_{t=0}^\infty$, where

$$p_i^t : H^t \to \mathbf{R}.$$

The buyer's pure behavior strategy is similarly a sequence $\mathbf{a} = \{a^t\}_{t=0}^\infty$, where

$$a^t : \mathbf{R}^I \times H^t \to \{0, 1, \ldots, I\}.$$

We are interested in the impact of the payoff relevant history on future play. In other words, we want to consider only Markov perfect equilibria of the game, as defined by [11]. In order to get a precise definition for the payoff relevant state variable, we have to define a set of equivalence classes on the set of all possible states Θ. Fix the dynamic price competition game Γ. Each possible θ induces a continuation price competition game $\Gamma(\theta)$ in the standard fashion. We partition Θ into a (possibly infinite) family of subsets $\{\Theta_k\}_{k=1}^H$ by the requirement that

$$\Gamma(\theta) = \Gamma(\theta') \Leftrightarrow \theta, \theta' \in \Theta_k \text{ for some } k.$$

In the generic case, a purchase from seller i leads to a payoff relevant state that is different from that following a purchase from seller $j \neq i$.

The collection $\{\Theta_k\}_{k=1}^{H}$ forms then the payoff relevant set of states for the dynamic price competition game. We say that a strategy for seller i is Markovian if for all h and h' such that $\theta(h)$, $\theta(h') \in \Theta_k$ for some k, we have $p_i(h) = p_i(h')$. The buyer's strategy is Markovian if for all h and h' such that $\theta(h)$, $\theta(h') \in \Theta_k$ for some k, and for all p, we have $a(h, p) = a(h', p)$.

Definition 2 (Markov perfect equilibrium): A collection $(\mathbf{p}_1, \ldots, \mathbf{p}_I, \mathbf{a})$ is a Markov perfect equilibrium if

(1) for all i, $\mathbf{p}_i$, is a best response to $(\mathbf{p}_{-i}, \mathbf{a})$ after all histories and $\mathbf{a}$ is a best response to $(\mathbf{p}_1, \ldots, \mathbf{p}_I)$ after all histories;
(2) all players use Markovian strategies.

In much of what follows, we concentrate on a refinement of the Markov perfect equilibrium called a cautious equilibrium. For an arbitrary history h we write the continuation payoffs to the buyer and seller i respectively as $V_B(h)$ and $V_i(h)$.

Definition 3 (Cautious equilibrium): A Markov perfect equilibrium is a cautious equilibrium if for $a(\theta, p_1, \ldots, p_I) \in \{0, 1, \ldots, I\}$, and all $i \neq a(\theta, p_1, \ldots, p_I)$,

$$\delta V_i(\theta, a(\theta, p_1, \ldots, p_I)) = p_i(\theta) + \delta V_i(\theta, i),$$

where $a(\theta, p_1, \ldots, p_I)$ denotes the equilibrium choice rule of the buyer and (θ, j) denotes the state vector after state τ followed by the choice of alternative $j \in \{0, 1, \ldots, I\}$.

The basic idea behind this definition is that no seller should be willing to offer prices that make the seller worse off relative to the equilibrium if accepted. In the static version of this price competition model, equilibria in cautious strategies are payoff equivalent to equilibria in weakly undominated strategies.[2]

In the equilibrium analysis, we shall make repeated use of arguments based on the social efficiency of paths along the game. We therefore conclude this section by defining socially optimal payoffs for different versions of the game. These concepts will be used repeatedly in the sections that follow. We introduce the indicator function $\mathbf{1}_{\{a^t = i\}}$ to describe the realized payoffs as a function of the choice behavior in period t. More precisely, let $\mathbf{1}_{\{a^t = i\}} = 1$

---

[2]The same equilibrium notion has been used in the equilibrium analysis of [2, 5].

if $a^t = i$ and $\mathbf{1}_{\{a^t=i\}} = 0$ otherwise. The total surplus along path $\{a^t\}$ is given by

$$\sum_{t=0}^{T} \delta^t \left(\sum_{i=1}^{I} x_i(\theta(h)) \mathbf{1}_{\{a^t=i\}} \right).$$

This social surplus is split between the buyer and the sellers. The intertemporal payoff to the buyer from a sequence of choices $\{a^t\}$ is simply

$$\sum_{t=0}^{T} \delta^t \left(\sum_{i=1}^{I} (x_i(\theta(h)) - p_i(\theta(h))) \mathbf{1}_{\{a^t=i\}} \right),$$

and the intertemporal payoff to seller i is given by

$$\sum_{t=0}^{T} \delta^t (p_i(\theta(h)) \mathbf{1}_{\{a^t=i\}}).$$

Because of the quasi-linear payoff specification, Pareto efficiency coincides with total surplus maximization in the game. Therefore, we let

$$W(\theta_0) \triangleq \max_{\{a^t\}\in I^T} \sum_{t=0}^{T} \delta^t \left(\sum_{i=1}^{I} x_i(\theta(h)) \mathbf{1}_{\{a^t=i\}} \right),$$

denote the *social value* of the game in period 0 at state θ_0. We can similarly define the continuation values $W(\theta)$ from an arbitrary state vector θ onwards. We will also make use of social values to the game where some sellers have been excluded. We let

$$W^{-S} \triangleq \max_{\{a^t\}\in(\mathcal{I}\backslash S)^T} \sum_{t=0}^{T} \delta^t \left(\sum_{i\in(\mathcal{I}\backslash S)} x_i(\theta(h)) \mathbf{1}_{\{a^t=i\}} \right)$$

denote the social value in the game where all sellers in the set $S \subset \mathcal{I}$ have been removed.

The next definition is of key importance for the rest of this paper. The *marginal contribution of seller i* to the social welfare at state θ is defined as

$$M_i(\theta) \triangleq W(\theta) - W^{-i}(\theta).$$

We may also define the marginal contribution of seller i in the game where seller j has been removed as

$$M_i^{-j}(\theta) \triangleq W^{-j}(\theta) - W^{-i \cup j}(\theta).$$

By $M_S(\theta)$, we denote the marginal contribution of a coalition S of sellers:

$$M_S(\theta) \triangleq W(\theta) - W^{-S}(\theta).$$

Finally, we introduce notation

$$W(\theta|k), \quad M_i(\theta|k), \quad \text{etc.}$$

to describe the social values and marginal contributions along paths that start with an arbitrary alternative k in the initial period, but follow the (conditionally) socially optimal path in all subsequent periods.

3. Finite Horizon Equilibrium

In this section, we present three examples that illustrate how changes in the competitive positions of the sellers can have problematic consequences for the efficiency of the equilibrium. The first example addresses direct intertemporal externalities between the sellers. The second and third examples consider independent reward payoffs and illustrate the role of fixed finite horizons for the competition between the sellers.

Consider first an example where three sellers, $i \in \{1, 2, 3\}$ are selling a good to a buyer with unit demand per period in a two-period economy. To fix ideas, assume that the first firm sells champagne, the second red wine and the third a dessert wine. Assume also that the first glass results in a utility of u for the buyer regardless of the specific choice, or:

$$x_i(0) = u > 0 \text{ for all } i.$$

The twist in this example comes from the fact that after consuming the wine of seller i, the consumer only wants to consume the wine of seller $i+1$ whose product yields utility v, and we assume that $\delta v > u$. The second glass of wine thus has a higher discounted utility than the first glass. (All summations and subtractions on the set of sellers are to be interpreted as modulo 3 in this section.) No other choice gives any utility to the buyer in the second period.

Observe that in this model there are no externalities in the stage game between the different sellers, but obviously the sales of one producer affect the values of other sellers in future periods. We solve for equilibria in the model by backwards induction. Denote by $V_B(j)$, $V_i(j)$ the continuation payoffs to the buyer and seller i in the second period, respectively, if j was the seller in period 0. The cautious equilibrium payoffs in period 1 are given by $V_i(j) = v$ if $j = \{i - 1\}$ and $V_i(j) = 0$ otherwise. In other words, only the seller who has the desirable product in period 1 makes a positive profit equal to the value of the product. In the cautious equilibrium, we also have that $V_B(j) = u$ if $j = 0$, and $V_B(j) = 0$ otherwise. Notice that if the buyer chose not to make a selection in period 0, then the sellers remain symmetric in period 1 and competition leaves the buyer with all the surplus.

In order to show that this game does not have any cautious equilibria, it is sufficient (by symmetry) to show that there is no cautious equilibrium where seller 1 makes the sale in the first period. To this end, assume to the contrary that seller 1 makes a sale in the first period in a cautious equilibrium. Then by Bertrand pricing, the buyer must be indifferent between buying from seller 1 and another action. There are three possibilities: the indifference can hold vis a vis the second seller, the third seller or the option of not buying. The following argument shows that none of these alternatives is consistent with a cautious equilibrium.

First assume that the buyer is indifferent between seller 1 and 2. The equilibrium indifference condition for the buyer is given by

$$u - p_1 + \delta 0 = u - p_2 + \delta 0,$$

and hence requires that

$$p_1 = p_2.$$

On the other hand, cautious pricing implies for seller 2 that he is indifferent between making a sale today and making a sale tomorrow, or

$$p_2 = \delta v.$$

But by assumption $\delta v > u$, and hence we are lead to

$$u - p_1 = u - p_2 < 0,$$

contradicting optimality for the buyer.

Next assume that the indifference is between sellers 1 and 3. Then again $p_1 = p_3$ by buyer indifference and $p_3 = 0$ by cautious pricing, as seller 3 will not be able to make a sale after seller 1 was chosen in period 0. But then seller 1 can gain by deviating to a higher price, leaving the buyer to seller 3 in period 0 and allowing seller 1 to extract a payoff of $\delta v > u$ in the subsequent period.

The remaining possibility is that the buyer is indifferent between seller 1 and not buying. The buyer's indifference in this case requires that

$$u - p_1 = \delta u, \quad \text{or} \quad p_1 = (1 - \delta)u.$$

But then seller 3 can undercut profitably by setting

$$p_3 = (1 - \delta)u - \varepsilon.$$

As a result, this two-period game does not have cautious equilibria. It should be noted that this game does have a subgame perfect equilibrium in pure strategies. For example, first period prices $p_1 = p_2 = p_3 = 0$ are consistent with equilibrium if the buyer is to choose seller 1 on equilibrium path and following any deviations by either seller 2 or 3. If seller 1 deviates, then the buyer should buy from seller 2. It is easily verified that this configuration is sequentially rational, but seller 2 is not using a cautious strategy.

The problem arises in this example because sales by seller i in period t have a direct impact on the rewards from the sales by seller j in period $t+1$. Such dependence is a clear manifestation of an intertemporal externality between the sellers, and there is no a priori reason to believe that such externalities can be dealt with in a model where transaction prices are the only transfer instruments. For this reason we ruled out externalities of this type with the assumption of independent rewards. It removes direct intertemporal externalities between the sellers.

The next two examples are meant to show that within finite horizon models, there is a subtle, but important indirect externality even under the assumption of independent rewards. Consider first a model with three firms where the payoffs are given by the following matrix:

	$x_1(\cdot)$	$x_2(\cdot)$	$x_3(\cdot)$
$t_i = 0$	2	2	0
$t_i = 1$	0	0	3

The table reads as follows for alternative 1. The payoff from alternative 1 is 2 when it has never been used before, or $t_1 = 0$ and 0 when it has been used once before, or $t_1 = 1$. The fixed time is assumed to be given by $T = 1$ and the discount factor δ is close to one. The efficient allocation would prescribe seller 1 and 2 to each realize their high valuation 2, however in the unique cautious equilibrium seller 3 is successful in period 0 and 1. To see why there cannot be an efficient equilibrium note that if alternative 1 or 2 is chosen in the first period, then the outside option value from alternative 3 vanishes altogether. This implies that the buyer cannot guarantee herself any surplus in period 1. By setting price $-2(1 - \delta) - \varepsilon$ in $t = 0$, seller 3 can guarantee the buyer a higher total payoff than what is individually rational for of the sellers 1 and 2. Two points are worth observing here. First, because of the finite time horizon, the outside option value offered by a given seller changes over time even when that seller is not chosen. Second, the equilibria would be quite different if the sellers were allowed to write binding contracts with the buyer.

The next example shows that even when the game has an efficient equilibrium, it need not be unique. Consider the following two alternatives with a payoff stream as described below:

$$x_1(\cdot) \ x_2(\cdot)$$
$$t_i = 0 \ \ 1+\varepsilon \ \ \ 1$$
$$t_i = 1 \ \ \ \ 0 \ \ \ \ \ \ 0$$

where $\varepsilon > 0$. The fixed time horizon is again given by $T = 1$ and the discount factor δ is again close to one. In this game, it is possible to support a period 0 choice of 1 as well as a choice of 2 in cautious equilibrium. To see how the inefficient equilibrium arises, we observe first that the buyer can always guarantee herself a payoff of δ by refusing all offers in period 0. By refusing all offers in period 0, the buyer puts the sellers in a very competitive position in period 0 and will receive a (discounted) payoff of $\delta \cdot 1$ in the unique cautious equilibrium of the continuation game. We can now establish a cautious equilibrium in period 0 where seller 2 offers a price $p_2 = 1 - \delta$ and seller 1 prices cautiously at $p_1 = \delta(1 + \varepsilon)$. It is easy to verify that for $\delta \approx 1$, these prices induce the buyer to select seller 2 in period 0. Obviously a similar equilibrium can be constructed where period 0 sales are made by seller 1. In this example, multiplicity arises since sales by firm i increase future profitability of firm j and as a result, there is little incentive to compete for the buyer in the first period. Notice that in both of the examples above, the buyer has a period 0 choice available that reduces the rents of the efficient sellers. In the first example, this choice (i.e. seller 3) is exercised in equilibrium, in the second example, this choice (seller 0) serves only as an outside option. But in both cases, the competitiveness of the situation was raised by reducing the number of periods in which the sellers could offer their product. In contrast, in an infinite horizon environment, the buyer can only postpone, but never completely eliminate periods of sale.

Finally, it is interesting to note that if the firms were allowed to bid in menu contracts as in [3], then all of the equilibria in both games would satisfy allocative efficiency. Hence we conclude that restricting our attention to equilibria in dynamic (spot) prices may induce efficiency losses in the model, and this should motivate the use of more sophisticated contracts in such environments.

4. Infinite Horizon Equilibrium

In demonstrating the existence of an efficient equilibrium, we make use of the guess and verify method of dynamic programming. In this approach, we assume that the equilibrium path is socially efficient after all possible histories. Furthermore, we assume that each seller is paid her marginal contribution in equilibrium. In other words, each seller j gets as her payoff the difference of the social surplus in the model and the social surplus in the model where seller j has been removed. These assumptions pin down the price of the successful seller and the buyer's purchasing decision at all histories. The remaining prices can be recovered from the requirement of cautious pricing. The main task is then to verify that no individual in the model has an incentive to deviate from this guess.

In this section we derive sufficient conditions for the existence of an efficient cautious equilibrium where the sellers' equilibrium payoffs coincide with their marginal contributions. These conditions relate to the properties of the social value function and the marginal contributions. We shall then verify these properties in Section 5 which studies the social programs rather than the equilibrium programs. We assume from now on that the model satisfies independent rewards.

The equilibrium conditions can be written as follows. The (weak) indifference conditions for the buyer with i as the successful seller are:

$$x_i(\theta) - p_i(\theta) + \delta V_B(\theta, i) \geqslant x_j(\theta) - p_j(\theta) + \delta V_B(\theta, j) \quad \forall j \neq i, \qquad (4.1)$$

where the inequality holds for at least one seller, say $k \neq i$, as an equality,

$$x_i(\theta) - p_i(\theta) + \delta V_B(\theta, i) = x_k(\theta) - p_k(\theta) + \delta V_B(\theta, k). \qquad (4.2)$$

The conditions imposed by cautiousness on the pricing policies are

$$p_j(\theta) = \delta V_j(\theta, i) - \delta V_j(\theta, j), \tag{4.3}$$

and

$$p_i(\theta) \geqslant \delta V_i(\theta, k) - \delta V_i(\theta, i). \tag{4.4}$$

In this section, we derive conditions for an efficient equilibrium and i shall always identify the socially efficient seller.

We now guess that the equilibrium value function for each seller j is her marginal contribution, or

$$V_j(\theta_t) = M_j(\theta_t). \tag{4.5}$$

We then verify that the guess, expressed by (4.5), actually satisfies the efficient equilibrium conditions (4.1)–(4.4). With the guess we can rewrite the conditions as

$$x_i(\theta) - p_i(\theta) + \delta \left(W(\theta, i) - \sum_{l=1}^{I} M_l(\theta, i) \right)$$

$$\geqslant x_j(\theta) - p_j(\theta) + \delta \left(W(\theta, j) - \sum_{l=1}^{I} M_l(\theta, j) \right) \quad \forall j \neq i, \tag{4.6}$$

with at least one inequality holding as equality, and

$$p_j(\theta) = \delta M_j(\theta, i) - \delta M_j(\theta, j), \tag{4.7}$$

as well as

$$p_i(\theta) \geqslant \delta M_i(\theta, k) - \delta M_i(\theta, i).$$

We first derive sufficient conditions for the indifference condition of the buyer. We do this by guessing that inequality (4.1) is satisfied as an equality for the seller who would be the socially optimal seller in the absence of i. We denote this seller by j^{-i}. As the marginal contribution property is conjectured to hold in every continuation game, the price of the winning seller i in period t has to be

$$p_i(\theta) = x_i(\theta) - x_{j-1}(\theta) + \delta W^{-i}(\theta, i) - \delta W^{-i}(\theta, j^{-i}). \tag{4.8}$$

When we insert the equilibrium prices (4.7) and (4.8) into the indifference condition (4.6) of the buyer, we get an expression involving only the social program and the marginal contributions

$$x_{j^{-i}}(\theta) - \delta W^{-i}(\theta, i) + \delta W^{-i}(\theta, j^{-i}) + \delta \left(W(\theta, i) - \sum_{l=1}^{I} M_l(\theta, i) \right)$$

$$= x_{j^{-i}}(\theta) - \delta M_j(\theta, i) + \delta M_j(\theta, j^{-i}) + \delta \left(W(\theta, j^{-i}) - \sum_{l=1}^{I} M_l(\theta, j^{-i}) \right).$$

Using the definition of $W^{-i}(\cdot)$ and after cancellations on both sides, we are left with the equality

$$\sum_{l \neq i, j^{-i}} \delta M_l(\theta, i) = \sum_{l \neq i, j^{-i}} \delta M_l(\theta, j^{-i}). \tag{4.9}$$

This equality involves the marginal contributions of all sellers with the exception of the two most efficient sellers, i and j^{-i}. It states that the sum of the marginal contributions of these sellers is the same whether the current selection is i or j^{-i}. As all these sellers are less efficient than either i or j^{-i}, each one of them will be chosen along the efficient path only after i and j^{-i} have been selected initially. From the point of view of these less-efficient sellers then, both i and j^{-i} will precede them and hence their contribution will only arise after the selection of i and j^{-i}. Hence from their point of view, it should not matter whether i or j^{-i} is chosen first. We therefore conjecture that the marginal contribution of each seller l is unaffected by the order in which i and j^{-i} are chosen and thus

$$M_l(\theta, i) = M_l(\theta, j^{-i}),$$

which would clearly be sufficient to support equality (4.9).

Let us next consider the remaining indifference conditions for the buyer, namely his choice between i and all sellers with the exception of j^{-i}. Naturally it is now sufficient to establish that the indifference conditions hold as inequalities rather than equalities. To achieve this, we do not insert the candidate equilibrium price of the winning seller i but a hypothetical price which allows for an easier comparison. Use $p_{i,j}(\theta)$ to denote the following

$$p_{i,j}(\theta) \triangleq x_i(\theta) - x_j(\theta) + \delta W^{-i}(\theta, i) - \delta W^{-i}(\theta, j). \tag{4.10}$$

The price $p_{i,j}(\theta)$ represents the differential social value of i compared to j if i were to be removed from the set of alternatives beginning tomorrow.

We get the following relation between $p_{i,j}(\theta)$ and $p_i(\theta)$:

$$p_{i,j}(\theta) - p_i(\theta) = x_{j^{-i}}(\theta) + \delta W^{-i}(\theta, j^{-i}) - x_j(\theta) - \delta W^{-i}(\theta, j) \geqslant 0.$$

The last inequality holds since alternative j^{-i} is the efficient alternative in the absence of i and hence the social value generated by the choice of j^{-i} is larger than the choice of any other alternative j in the absence of i. As we insert prices (4.7) and (4.10) into the indifference condition (4.6), we arrive again at an expression which involves only the social values and the marginal contributions, namely

$$x_{j^{-i}}(\theta) + \delta W^{-i}(\theta, j^{-i}) - x_j(\theta) - \delta W^{-i}(\theta, j)$$
$$\geqslant \delta \sum_{l \neq i,j} [M_l(\theta, i) - M_l(\theta, j)].$$

By adding $M_i(\theta) - M_i(\theta|j)$ on both sides, we return to the social value with all sellers, including i and the above inequality becomes

$$W(\theta) - W(\theta|j) \geqslant \sum_{l \neq j} [M_l(\theta) - M_l(\theta|j)]. \qquad (4.11)$$

Inequality (4.11) is our second sufficient condition and it has an intuitive interpretation. It states that the social gains of moving from an inefficient allocation j to the efficient allocation i is larger than the gains arising for the same change in the marginal contributions. Since we want to interpret the marginal contributions as the payoffs to the sellers, the condition simply says that the efficiency loss in choosing an inefficient seller exceeds the reduction in future payoffs to other sellers, or in other words, efficiency losses outweigh rent extraction gains.

The verification of the equilibrium condition for the sellers is straightforward. In a cautious equilibrium, the losing sellers (weakly) prefer sales by i to making sales on their own by construction. It remains to verify that the winning seller i prefers to make a seller rather than to concede the market to another seller j. By the equilibrium hypothesis, seller i receives his marginal contribution in every subgame, and thus a sufficient condition for optimality can be stated as

$$M_i(\theta) \geqslant \delta M_i(\theta, l) \quad \forall l \neq i.$$

This is the final sufficient condition that we need for our construction of an equilibrium. It simply says that the marginal contribution of agent i is maximized along the efficient path. We thus have established the following result:

Theorem 1 (Existence): *An MPE in cautious strategies which is* (i) *efficient and* (ii) *displays marginal contribution payoffs exists, provided that the marginal contributions satisfy:*

(1) $M_j(\theta|k) = M_j(\theta)$ *if k is chosen prior to j on the efficient path;*
(2) $M_j(\theta|k) \leqslant M_j(\theta)$ *if j is chosen prior to k on the efficient path;*
(3) $W(\theta) - W(\theta|k) \geqslant \sum_{j \in \mathcal{I} \backslash k} (M_j(\theta) - M_j(\theta|k))$.

In the next section we verify that in the model with independent rewards, the marginal contributions satisfy all three properties listed in Theorem 1.

5. Marginal Contributions

The verification of the sufficient conditions derived in Theorem 1 involves the comparisons of four different, yet related, social programs. The determination of the marginal contribution of each alternative occurs through the value of the social program when all alternatives are available and the social program when all alternatives but alternative j are available. The verification of the equilibrium choice of the buyer requires a comparison of the social welfare loss with the private welfare losses of the sellers. The potential social loss is induced by a single suboptimal deviation towards the alternative k, which requires the computation of the conditional optimal value following the choice of k. Finally, as the private welfare losses are measured by the marginal contributions and as those are obtained as the difference of social values, we also have to establish the value of the program without j and an initial suboptimal choice of k. The (conditionally) optimal assignments which arise in these various problems are denoted respectively by $a : \Theta \to \mathcal{I}$, $a_j : \Theta \to \mathcal{I} \backslash \{j\}$, $a^k : \Theta \to \mathcal{I}$ and $a_j^k : \Theta \to \mathcal{I} \backslash \{j\}$. The relationship between them are represented in the following diagram:

$$
\begin{array}{cc}
\text{marginal} \\
\text{contribution} \\
\textit{of } j
\end{array}
\left\{
\begin{array}{cc}
\overbrace{}^{\text{social loss}} \\
a(\cdot) \leftrightarrow a^k(\cdot) \\
\updownarrow \qquad \updownarrow \\
a_j(\cdot) \quad a_j^k(\cdot)
\end{array}
\right\}
\begin{array}{c}
\text{marginal} \\
\text{contribution} \\
\textit{of } j \\
\text{(with suboptimal } k)
\end{array}
$$

The main difference between these policies arises with respect to the *calendar time* at which the t_ith realization of alternative i will be employed. The argument that we present needs careful tracking of the calendar time t as well as the usage time t_i of seller i. We recall that we denote by $t_i(\theta)$ the number of times that alternative i has been used at state θ. We describe by $t(i, t_i)$ the *calendar time* in which the t_ith realization of alternative i is used in the optimal program. Similarly, we denote by $t(i, t_i| - j)$, $t(i, t_i|k)$ and $t(i, t_i|k, -j)$ the calendar time in which the t_ith realization of alternative i is employed in the program without j, in the suboptimal program starting with k, and in the suboptimal program starting with k and without alternative j, respectively. The comparison between the different programs can then be usefully reduced to a comparison of the calendar times in which a given realization of alternative i is employed across the different programs.

5.1. *Example*

Before we set out to verify the validity of the sufficient conditions within the general model, we consider a very simple environment. The simplicity of the example facilitates the tracking of the calendar times at which the alternatives are employed across the different programs. The evident structure of the allocation problem then provides transparency to the proof of the sufficient conditions. Consider the following specification:

$$x_i(t_i) = \begin{cases} x_i & \text{if } t_i = 0, \\ 0 & \text{if } t_i > 0, \end{cases}$$

and suppose further that

$$x_0 > x_1 > \cdots > x_I > 0.$$

In other words, each alternative generates a positive value at its first use and thereafter generates zero value. The socially optimal *continuation* policy is therefore to employ in every period the alternative j with the highest remaining valuation. The socially optimal policy starting at $t = 0$ is to select each alternative exactly in the order of their valuations. The path of the optimal policy is thus described by $a(t) = t$. Here, the descending order of the alternatives allows us to identify each alternative i with the time period in which it is employed along the efficient path.

The social value of the efficient program in period 0 can then be written as

$$W(\theta_0) = \sum_{t=0}^{\infty} \delta^t x_t.$$

As there is only a finite number of sellers and hence strictly positive realizations, for all $t \geqslant I$, we have $x_t = 0$.

The marginal contribution of seller j is given by

$$M_j(\theta_0) = W(\theta_0) - W^{-j}(\theta_0) = \sum_{t=0}^{\infty} \delta^t x_t - \sum_{t=0}^{j-1} \delta^t x_t - \sum_{t=j+1}^{\infty} \delta^{t-1} x_t$$

$$= \sum_{t=j}^{\infty} \delta^t (x_t - x_{t+1}). \tag{5.1}$$

As one might have expected, the removal of alternative j does not change the value in the programs before the arrival of j in the socially efficient program. However, if we remove alternative j, the immediate consequence is that we have to use the next best alternative, which is alternative $j + 1$. By extension, we will be forced in all future periods to move one of the less-efficient alternatives up by exactly one period, and this accounts for the sum of discounted differences starting in period j. In other words, the social benefit of seller j propagates into all future periods as the existence of alternative j permits all subsequent and less-valuable alternatives to make their appearance exactly one period later.

The social value of the suboptimal program which starts with alternative k can also be represented as a variation of the efficient social program by simply forwarding the appropriate time indices as

$$W(\theta_0|k) = x_k \sum_{t=0}^{k-1} \delta^{t+1} x_t + \sum_{t=k+1}^{\infty} \delta^t x_t.$$

The suboptimal anticipation of seller k changes the social values in two ways: (i) seller k appears k periods too early and (ii) all sellers ranked before k appear one period too late relative to the social optimum. After $k + 1$ periods, the optimal social policy catches up with the suboptimal policy and the values thereafter are identical. The rearrangement of the order in which the alternatives are selected due to the initial suboptimal deviation is depicted in the following Table 8.1 for $k = 3$:

Table 8.1. Allocation times along optimal and sub-optimal paths

$t = 0\ \ 1\ 2\ 3\ 4\ 5 \ldots$	
$a(t) = 0\ \ 1\ 2\ 3\ 4\ 5 \ldots$	
$a(t\mid3) = 3\ \ 0\ 1\ 2\ 4\ 5 \ldots$	

Finally, the marginal contribution of seller j in a suboptimal program can be represented by a combination of the forward and backward operation. Consider first the case in which alternative j would be employed after alternative k in the efficient program, or $j > k$. The marginal contribution of j is given by

$$M_j(\theta_0|k) = W(\theta_0|k) - W^{-j}(\theta_0|k)$$

$$= \sum_{t=j}^{\infty} \delta^t(x_t - x_{t+1}). \tag{5.2}$$

As alternative k is used in the optimal as well as in the suboptimal program before alternative j, the marginal contribution of alternative j should remain identical across these two program and this is easily verified by comparing (5.1) and (5.2).

If alternative j arrives before alternative k in the efficient program, then the marginal contribution in the suboptimal program is

$$M_j(\theta_0|k) = W(\theta_0|k) - W^{-j}(\theta_0|k)$$

$$= \sum_{t=j}^{k-2} \delta^{t+1}(x_t - x_{t+1}) + \delta^k(x_{k-1} - x_{k+1}) + \sum_{t=k+1}^{\infty} \delta^t(x_t - x_{t+1}).$$

The effect of a suboptimal allocation on the marginal contribution of seller j is now similar to the one imposed on the social value. The distinction arises due to the fact that the marginal contribution is expressed in differences rather than absolute values: (i) the sub-optimal anticipation of k delays the arrival of benefits due to seller j by one period, (ii) as k has been anticipated the marginal benefit in period k is then between x_{k-1} and the next available alternative x_{k+1}, (iii) at $t = k + 1$ the socially optimal program catches up with the suboptimal program and the third term is identical in both expressions. The difference in the marginal contributions for $j < k$ can then be expressed as

$$M_j(\theta_0) - M_j(\theta_0|k) = (1 - \delta) \sum_{t=j}^{k-1} \delta^t (x_t - x_{t+1}). \tag{5.3}$$

An initial and inefficient assignment of seller k then depresses the contribution of seller j through a sequence of one period delays in the accrual of the marginal values of seller j.

Based on these simple computations we can now verify that all three sufficient conditions in Theorem 1 hold. The first sufficient condition, namely that for $j > k$, $M_j(\theta_0) = M_j(\theta_0|k)$, is verified by simply comparing (5.1) and (5.2). The second sufficient condition, $M_j(\theta_0) - M_j(\theta_0|k) \geqslant 0$ for $j < k$, was established by (5.3) coupled with the observation that $x_t - x_{t+1} > 0$ for all t. The third and final sufficient condition

$$W(\theta_0) - W(\theta_0|k) \geqslant \sum_{j \in \mathcal{I}} (M_j(\theta_t) - M_j(\theta_t|k)) \tag{5.4}$$

is readily established as well. The difference in the value between the optimal and the suboptimal program is given by

$$W(\theta_0) - W(\theta_0|k) = (1 - \delta) \sum_{t=0}^{k-1} \delta^t x_t - (1 - \delta^k) x_k. \tag{5.5}$$

Inequality (5.4) can be expressed, using (5.3) and (5.5), and dividing both sides by $(1 - \delta)$ as

$$\sum_{j=0}^{k-1} (\delta^j x_j - \delta^k x_k) \geqslant \sum_{j=0}^{k-1} \sum_{t=j}^{k-1} \delta^t (x_t - x_{t+1}).$$

It is easy to verify that the inequality holds when we consider every element indexed with j separately, or

$$\delta^j x_j - \delta^k x_k \geqslant \sum_{t=j}^{k-1} \delta^t (x_t - x_{t+1}). \tag{5.6}$$

We notice that without discounting, i.e. for $\delta = 1$, both sides equalize as the RHS is simply a telescopic expansion of the difference of the LHS. But for $\delta < 1$, the inequality becomes in fact strict. The LHS of (5.6) expresses for every x_j with $j < k$, the value difference between x_j and x_k *weighted* with the appropriate discount factors of the optimal program

$$\delta^j x_j - \delta^k x_k > 0.$$

The RHS also presents for every x_j a differential expression between x_j and x_k, but it proceeds in steps $\delta^t(x_t - x_{t+1})$ which are increasingly discounted. This reflects the value difference between x_j and x_k but now in terms of the marginal contribution. Since the marginal contribution only picks up the *inframarginal* differences in every period, it follows directly that inequality (5.6) holds.

5.2. *Optimal index policies*

We now proceed to prove the three properties of the marginal contributions within the general payoff environment. The basic argument follows exactly the logic suggested by the example. But to pursue this argument, we have to use the structure of the optimal policies to bring the general model closer to the example. We do this in three steps. The first step, carried out in Lemmas 1 and 2, will show that the decreasing sequence of payoffs in the example is in essence without loss of generality. The second step, carried out in Lemmas 3 and 4, shows that among all possible continuation paths following an initial and suboptimal use of alternative k, the one which returns immediately to the initial and unconditionally optimal path is the critical path. The third step, carried out in Lemma 5, shows that the marginal contributions *and* the difference in the marginal contribution from optimal and suboptimal path are superadditive. This final argument allows us to reduce the rent extraction inequality to a particularly simple case. Finally it should be emphasized that all arguments in this section are based on the properties of optimal policies in single agent allocation problems and no equilibrium arguments are needed.

We recall that the payoff of each alternative i is a function of its own past use only. We defined the payoff stream $x_i(t_i)$ to depend only on the number of times, t_i, alternative i has been used in the past. For each alternative i and each state t_i we can define an index of its future value through an optimal stopping problem. The optimal stopping time for alternative i in state t_i, denoted by $\tau_i(t_i)$, is defined as the solution to the following problem:

$$\tau_i(t_i) \in \arg\max_{\tau_i \geqslant (t_i)} \left\{ \frac{\sum_{s_i=t_i}^{\tau_i} \delta^{s_i} x_i(s_i)}{\sum_{s_i=t_i}^{\tau_i} \delta^{s_i}} \right\}. \tag{5.7}$$

If the maximization problem (5.7) allows for multiple solutions for τ_i, then we identify $\tau_i(t_i)$ to be the largest time among the maximizers. We define the index of alternative i, $X_i(t_i)$, as the discounted average under the optimal stopping time, or

$$X_i(t_i) \triangleq \max_{\tau_i \geq t_i} \left\{ \frac{\sum_{s_i=t_i}^{\tau_i} \delta^{s_i} x_i(s_i)}{\sum_{s_i=t_i}^{\tau_i} \delta^{s_i}} \right\}. \tag{5.8}$$

With these preliminaries in place, it is straightforward to characterize the (conditionally) optimal programs.

Lemma 1 (Optimal policies): (1) *The (conditionally) optimal assignment in state θ is determined by*

$$\arg\max_i \{X_i(t_i(\theta))\}.$$

(2) *For all i j, k, l (all distinct) and all t_i, t_j:*

$$t(i, t_i) < t(j, t_j) \Leftrightarrow t(i, t_i|k) < t(j, t_j|k)$$
$$\Leftrightarrow t(i, t_i| - l) < t(j, t_j| - l)$$
$$\Leftrightarrow t(i, t_i|k, -l) < t(j, t_j|k, -l).$$

Proof: See appendix. □

The first part of Lemma 1 simply restates the celebrated Gittins index theorem for deterministic payoff streams. The index characterization depends only on the properties of the payoff streams of the individual alternatives. As a consequence of the index property, the subsequent parts state that the order in which the alternatives are optimally employed is invariant to the removal of some alternatives or the suboptimal initial use of alternative k.

A further consequence of the optimality of the index policy is stated next. For each alternative i, we define inductively a sequence of stopping times $\{\tau_i^n\}_{n=0}^{\infty}$ as follows. Define $\tau_i^0 \triangleq -1$, and let

$$\tau_i^{n+1} \in \arg\max_{s > \tau_i^n} \left\{ \frac{\sum_{t_i=\tau_i^n+1}^{s} \delta^{t_i} x_i(t_i)}{\sum_{t_i=\tau_i^n+1}^{s} \delta^{t_i}} \right\}. \tag{5.9}$$

As before, if the maximization problem (5.9) allows for multiple solutions for τ_i^{n+1}, then we identify τ_i^{n+1} to be the largest time among the maximizers.

With this inductively defined sequence of stopping times, we can associate average rewards between stopping times as follows

$$x_i^{n+1} \triangleq \frac{\sum_{t_i=\tau_i^{n+1}}^{\tau_i^{n+1}} \delta^{t_i} x_i(t_i)}{\sum_{t_i=\tau_i^{n+1}}^{\tau_i^{n+1}} \delta^{t_i}}. \tag{5.10}$$

In contrast to the earlier stopping times and average discounted rewards in (5.7) and (5.8), the stopping times defined by (5.9) are not recalculated at every clock time t_i, but only from stopping time τ_i^n to stopping time τ_i^{n+1}. Similarly, the average discounted rewards defined by (5.10) are only calculated between stopping times.

Lemma 2 (Decreasing valuations): (1) *The average returns x_i^n are decreasing in n.*

(2) *The tail realizations, with $\tau_i^n < s \leqslant \tau_i^{n+1}$, satisfy*

$$\frac{\sum_{t_i=s}^{\tau_i^{n+1}} \delta^{t_i} x_i(t_i)}{\sum_{t_i=s}^{\tau_i^{n+1}} \delta^{t_i}} \geqslant x_i^{n+1}.$$

Proof: See appendix.   □

The relevance of Lemma 2 is easiest to understand if we start with the second result. It says that between any two stopping times, τ_i^n and τ_i^{n+1}, the discounted average reward until τ_i^{n+1} must be at least as large as it was when evaluated starting from time $t_i = \tau_i^n$. The optimal index policy, stated in Lemma 1, then implies that if it was optimal to use i starting at $t_i = \tau_i^n$, then it will remain optimal to use alternative i uninterrupted at least until $t_i = \tau_i^{n+1}$ is reached. The first result of Lemma 2 then states that when the discounted averages are taken along the stopping times, then the average returns from alternative i are decreasing over time. By the index policy, it follows that the sequence of averages across agents will also decrease over time.

Next we consider a particular alternative k which is used in the suboptimal program $W(\theta_0|k)$. We know from Lemma 1 that the initial use of the suboptimal k will not change the order in which the remaining alternatives are selected in the optimal continuation program, but it may change repeatedly the time at which they are selected. This is due to the fact that the initial, even if suboptimal use of k, may make the future use of k more desirable and thus further delay the employment of the other alternatives. The source of this complication is the fact that after the initial use of $x_k = x_k(0)$, the average value of the tail realization of $x_k(0)$ may increase and be larger than $x_k(0)$ so that for $s > 0$:

$$\frac{\sum_{t_k=s}^{\tau_k^1} \delta^{t_k} x_k(t_k)}{\sum_{t_k=s}^{\tau_k^1} \delta^{t_k}} > x_k(0).$$

In this case, the index policy may recommend the use of alternative k earlier than it would have if $x_k(0)$ had not been removed through its suboptimal early use in period 0. (The payoff $x_k(0)$ now constitutes a sunk cost in order to reach the higher payoffs $x_k(t_k)$.) As the index of alternative k for all tail realizations stays above the original index, we cannot rule out that the initial suboptimal use of k leads to many early (relative to the optimal program) usage times of the alternative k. The occurrence of such repeated changes makes the comparison between $W(\theta_0)$ and $W(\theta_0|k)$ notationally cumbersome. It is therefore desirable to find instances for the payoff stream of $x_k(t_k)$ in which the changes between $W(\theta_0)$ and $W(\theta_0|k)$ are minimal. The only payoff stream for alternative k which can guarantee this and still obtain an average return of x_k^1 for an arbitrary stopping time τ_k^1 is the constant sequence

$$x_k(0) = x_k(1) = \cdots = x_k(\tau_k^1) = x_k^1, \tag{5.11}$$

where the last equality is the result of the averaging over a constant sequence. With such a constant sequence, the average value and hence index of all tail realization is constant and equal to the initial value. In consequence, the optimal continuation path after the initial suboptimal choice of k is easy to describe. It will simply continue where the optimal path would have started. Moreover, beginning with the time period where the optimal continuation path uses the alternative k again, the path of optimal and suboptimal path will be identical again.

In this respect, if we compare the program $W(\theta_0)$ and $W(\theta_0|k)$, the only realizations of alternative j which matter are those which are realized optimally before the alternative k is used for the first time. It is then useful to introduce an auxiliary allocation problem, where all realizations of alternative j which would occur after the first use of alternative k are set equal to zero, or

$$x_j^k(t_j) = \begin{cases} x_j(t_j) & \text{if } t(t_j) < t(t_k = 0), \\ 0 & \text{if } t(t_j) > t(t_k = 0) \end{cases} \tag{5.12}$$

and all realizations of alternative k are set equal to x_k^1:

$$x^k(t_k) = x_k^1 \quad \text{for all } t_k. \tag{5.13}$$

We refer to the allocation problem with the payoffs in (5.12) and (5.13) as the k-truncated allocation problem. We add the superscript k to the payoff realizations and the social values, $W^k(\theta_0)$, to indicate the modification in the payoffs.

We now establish a relationship between our original allocation model and this specifically modified allocation model.

Lemma 3 (Truncated allocation problem): *If $x_k(t_k)$ displays constant payoffs for all t_k with $0 \leqslant t_k \leqslant \tau_k^1$, then:*

(1) *for all k,*

$$\frac{1}{1-\delta}(W(\theta_0) - W(\theta_0|k)) = W^k(\theta_0) - \frac{1}{1-\delta}x_k^1;$$

(2) *for all k and all i,*

$$\frac{1}{1-\delta}(M_i(\theta_0) - M_i(\theta_0|k)) = M_i^k(\theta_0).$$

Proof: See appendix.   □

We can now show that the constant sequence of realizations for alternative k is the critical sequence to analyze for the purpose of establishing the sufficient conditions of the marginal contribution equilibrium.

Lemma 4 (Minimal loss payoff stream): (1) *For all k, $W(\theta_0|k)$ is maximized with $x_k(0) = \cdots = x_k(\tau_k^1)$.*
     (2) *For all i and k, $M_i(\theta_0|k)$ is maximized with $x_k(0) = \cdots = x_k(\tau_k^1)$.*

Proof: See appendix.   □

A final useful fact about the marginal contributions is stated next.

Lemma 5 (Superadditive marginal contributions): (1) *For all i and j,*

$$M_i(\theta_0) + M_j(\theta_0) \leqslant M_{i \cup j}(\theta_0).$$

(2) *For all i, j and k, all distinct,*

$$M_i(\theta_0) - M_i(\theta_0|k) + M_j(\theta_0) - M_j(\theta_0|k) \leqslant M_{i \cup j}(\theta_0) - M_{i \cup j}(\theta_0|k).$$

Proof: See appendix. □

The first results says that the marginal contribution of i and j jointly exceed the sum of the marginal contributions of i and j individually. This is rather intuitive as the removal of alternative i leaves the social program with the possibility to use j, but once i and j are removed jointly, the social program will have to immediately use the possible inferior alternative k. It can be shown that the superadditivity property of the marginal contributions is equivalent to

$$M_i(\theta_0) \leqslant M_i^{-j}(\theta) \quad \text{for all } i \text{ and } j.$$

This inequality is perhaps even more intuitive as it says that the contribution of alternative i to the social program becomes more valuable after the removal of alternative j from the choice set. The superadditivity property extends to the difference of the marginal contributions arising from optimal and suboptimal program, and it is this inequality which we will use for the next theorem. With these preliminaries in place, we are ready to prove the main result of this section.

Theorem 2 (Marginal contributions): *The marginal contributions of the infinite horizon game with independent rewards satisfy the sufficient conditions of Theorem 1:*

(1) $M_i(\theta_0|k) = M_i(\theta_0)$ *if* $t(k,1) < t(i,1)$;
(2) $M_i(\theta_0|k) \leqslant M_i(\theta_0)$ *if* $t(i,1) < t(k,1)$;
(3) $W(\theta) - W(\theta_0|k) \geqslant \sum_{j \in \mathcal{I} \backslash k}(M_j(\theta) - M_j(\theta_0|k))$.

Proof: See appendix. □

This theorem establishes that the infinite horizon model with independent rewards always possesses an equilibrium with strong welfare predictions. The equilibrium path is socially efficient and all the sellers receive their marginal contribution. A further welfare consequence of the marginal contribution property is that (i) the sellers have the socially optimal ex-ante incentives for making surplus enhancing investments and (ii) the buyer has the socially correct incentives for investments that increase surplus in all purchases. In the job market context this implies that efficiency results if the firms pay for job-specific training and the workers pay for general training.

In the next section, we turn to the issue of uniqueness of these equilibria.

6. Uniqueness

In this section, we address the issue of uniqueness in the infinite horizon model. We assume first that for all i, there exists a $T_i < \infty$ and an $\bar{x}_i$ such that for all $t_i \geq T_i$, $x_i(t_i) = \bar{x}_i$. This assumption implies that along any Markov perfect equilibrium path, the payoffs become static after some finite number of periods. Formally, let

$$T_i = \min\{t_i | x_i(s_i) = \bar{x}_i \text{ for some } \bar{x}_i \text{ and all } s_i \geq t_i\}. \qquad (6.1)$$

With this assumption, we analyze the game by backwards induction in the state space.

Theorem 3 (Uniqueness): *The marginal contribution equilibrium is the unique cautious Markov perfect equilibrium of the dynamic price competition game if for all i, there exists $T_i < \infty$ and $\bar{x}_i$ such that for all $t_i \geq T_i$, $x_i(t_i) = \bar{x}_i$.*

Proof: See appendix.   □

It should be pointed out that we do not know of any counterexamples to the uniqueness of equilibria in games which do not satisfy the above assumption of eventual constancy. While we have not been able to prove the uniqueness without the use of backward induction arguments based on continuation payoffs, we conjecture that the uniqueness holds even in games where the constancy assumption (6.1) fails, provided, of course, we maintain the independent rewards condition. The basic difficulty in proving uniqueness for the general model stems from the fact that when the continuation paths are inefficient, very little can be said about the exact form of individual continuation payoffs. The best way to see this is to recall the examples in Section 3. For an arbitrary continuation path, there is no reason to believe that the cautious equilibrium choice in period t would be uniquely defined. It might seem that by letting $T_i \to \infty$ for all i, we could make use of arguments based on continuity at infinity.[3] Yet, while this approach would supply us with an alternative way for proving the *existence* of a marginal contribution equilibrium in the general model, it does not provide us with an argument for uniqueness.

The final point to note is that the scope of the uniqueness result is narrower than in the main result of [2]. The uniqueness here pertains only to Markov perfect equilibria in cautious strategies whereas the previous result

---

[3]See e.g. [7, 9] for discussions of such arguments.

held for *all* MPE of the game with two sellers. The reason for this is that with three or more sellers, continuation payoffs resulting from strategies that do not satisfy cautiousness may violate the marginal contribution property. To see this, consider the following simple example and assume for simplicity that $\delta \approx 1$:

$$
\begin{array}{cccc}
x_1(\cdot) & x_2(\cdot) & x_3(\cdot) & \\
t_i = 0 \quad 1 + \varepsilon & 1 & \varepsilon & . \\
t_i = 1 \quad 0 & 0 & 0 &
\end{array}
$$

Efficiency requires that 1 is chosen first, 2 next and 3 in the final period. The game has, however, inefficient Markov perfect equilibria in strategies that are not cautious. To see this, observe that seller 3 can transfer all the surplus to the buyer by suitable non-cautious pricing. If he uses such a strategy in the continuation game following the choice of 3 in the first period, and if all other continuation games are played according to the marginal contribution equilibrium, then for ε small, there is an equilibrium where seller 3 is chosen in the first period.

7. Conclusion

This paper shows that much of the intuition obtained in static price competition games extends to the stationary infinite horizon case as long as externalities between sellers are ruled out. In particular, the stationary infinite horizon model possesses an efficient equilibrium where all of the sellers receive their marginal contribution as their equilibrium payoff. The finite horizon case is, however, quite different. Efficient equilibria do not exist in general, and the games may have multiple cautious equilibria.

The arguments in the stationary infinite horizon case are based on a version of the Gittins index theorem for the optimal scheduling of tasks. The most important consequence of this is that the optimal paths in the model satisfy the following invariance property: if along the optimal path, a purchase is made from seller i prior to seller j, then it cannot be the case that j is used prior to i in a game where the stage rewards of players other than i and j have been modified. Since the Gittins index theorem is also valid in the stochastic case (as long as the stage rewards are statistically independent across sellers), the arguments given in the paper should remain valid in that case as well. The major modification needed is that in the calculation of the optimal number of times that a seller is used, we would have to consider random rather than deterministic stopping times.

There are a number of directions for extending our analysis. A possible formulation that would be consistent with our insistence on no static or dynamic externalities between the sellers would be to allow $x_i(\cdot)$ to depend on calendar time in addition to t_i. For example, one could have a time to build before a new sale can be made (this would be relevant for many industries such as ship building where capacity utilization and securing a constant stream of purchases are of primary importance). Such models could also capture many of the dynamic issues arising in the sales of renewable resources. Pursuing these extensions is left for future work.

Acknowledgments

We would like to thank seminar participants at University of Helsinki for their comments. The authors gratefully acknowledge financial support from NSF Grant SBR 9709887 and A.P. Sloan Foundation Fellowship and NSF Grant SBR 9709340 respectively. The first author would like to thank the Center of Economic Studies at the University of Munich for its hospitality during the completion of this manuscript.

Appendix

The proofs for all results are collected in the appendix.

Proof of Lemma 2. (1) This is simply a statement of the optimality of the Gittins index policy in the special case of deterministic payoff streams, see [13] or [8] for a statement in the general stochastic case.

(2) The relationships follow immediately from the nature of the index policy. As the index of alternative i and j are computed on the basis of their respective payoff streams exclusively, their index is invariant to the addition or removal of additional alternative. As the indices are invariant, the relative order in which alternatives i and j are being employed stays invariant too. The same argument applies naturally for the optimal continuation path of i and j, when the initial suboptimal choice was k. □

Proof of Lemma 3. (1) The argument is by contradiction. Suppose not and hence there exists n such that $x_i^n < x_i^{n+1}$. This directly contradicts the optimality of τ_i^n as a solution to (5.9) since by hypothesis of $x_i^n < x_i^{n+1}$:

$$\frac{\sum_{t_i=\tau_i^{n-1}+1}^{\tau_i^{n+1}} \delta^{t_i} x_i(t_i)}{\sum_{t_i=\tau_i^{n-1}+1}^{\tau_i^{n+1}} \delta^{t_i}} > x_i^n,$$

but this shows that τ_i^n cannot be a solution to (5.9) as setting its value to be equal to τ_i^{n+1} is feasible and would obtain a larger average value.

(2) The argument is again by contradiction. Suppose therefore that there exists an s with $\tau_i^{n-1} < s < \tau_i^n$ such that

$$\frac{\sum_{t_i=s}^{\tau_i^n} \delta^{t_i} x_i(t_i)}{\sum_{t_i=s}^{\tau_i^n} \delta^{t_i}} < x_i^n.$$

It follows from averaging that the average during the complement time has to satisfy

$$\frac{\sum_{t_i=\tau_i^{n-1}+1}^{s-1} \delta^{t_i} x_i(t_i)}{\sum_{t_i=\tau_i^{n-1}+1}^{s-1} \delta^{t_i}} > x_i^n,$$

for the joint interval to achieve the average x_i^n. But it now follows immediately that τ_i^n cannot be the solution to (5.9) as $s-1$ achieves a higher average, contradicting the fact that τ_i^n is a solution to the maximization problem. $\qquad\square$

Proof of Lemma 4. (1) For a constant payoff stream of $x_k(t_k)$ for all t_k with $0 \leqslant t_k \leqslant \tau_k^1$, the difference $W(\theta_0) - W(\theta_0|k)$ can be written as

$$W(\theta_0) - W(\theta_0|k) = (1 - \delta)W^{t_k}(\theta_0) - x_k^1(1 - \delta^{t_k}),$$

and after dividing by $(1 - \delta)$ as

$$\frac{1}{1-\delta}W(\theta_0) - W(\theta_0|k)) = W^k(\theta_0) - \frac{x_k^1}{1-\delta}.$$

(2) We can write the difference $M_j(\theta_0) - M_j(\theta_0|k)$ generally as

$$M_j(\theta_0) - M_j(\theta_0|k)$$
$$= (W(\theta_0) - W^{-j}(\theta_0)) - (W(\theta_0|k) - W^{-j}(\theta_0|k))$$
$$= (W(\theta_0) - W(\theta_0|k)) - (W^{-j}(\theta_0) - W^{-j}(\theta_0|k)). \qquad (7.1)$$

For the case of a constant payoff stream of $x_k(t_k)$ for all t_k with $0 \leqslant t_k \leqslant \tau_k^1$, the last line in (7.1) can be written by the first part of this lemma as

$$\left(W^k(\theta_0) - \frac{x_k^1}{1-\delta} \right) - \left(W^{-i,k}(\theta_0) - \frac{x_k^1}{1-\delta} \right) = W^k(\theta_0) - W^{-i,k}(\theta_0),$$

which proves the second part of this lemma. $\qquad\square$

Proof of Lemma 5. (1) Suppose not and thus the maximizing solution is obtained with a different payoff stream by alternative k. By the optimality of the stopping times $\{\tau_k^n\}_{n=0}^{\infty}$, this different payoff stream must have some realization $x_k(t_k) > x_k(0)$ with $t_k \leqslant \tau_k^1$. Consider then a modified version of this payoff stream, denoted by $\hat{x}_k(0)$ and $\hat{x}_k(t_k)$, satisfying:

$$\hat{x}_k(0) \triangleq x_k(0) + \varepsilon \delta^{t_k}, \tag{7.2}$$

and

$$\hat{x}_k(t_k) \triangleq x_k(t_k) - \varepsilon, \tag{7.3}$$

with $\varepsilon > 0$. Clearly, the modified payoff stream has the same average return as the original stream, and for sufficiently small ε, it maintains the same stopping times $\{\tau_k^n\}_{n=0}^{\infty}$ as the original stream.

Consider then the original term $W(\theta_0|k)$ and the same term under the modification defined by (7.2) and (7.3), and denoted, by extension, as $\widehat{W}(\theta_0|k)$. As the stopping times and the average rewards of all alternatives remain unchanged, the timing in the employment remains unchanged as well. The difference between the two terms is therefore simply

$$\widehat{W}(\theta_0|k) - W(\theta_0|k) = \varepsilon \delta^{t_k} \cdot 1 - \varepsilon \cdot \delta^{t(k,t_k|k)}. \tag{7.4}$$

As $t_k \leqslant t(k, t_k|k)$, it follows that the constant stream for alternative k can never decrease the payoff of the program $W(\theta_0|k)$ and increases it if the t_kth realization of alternative k comes only after some other alternatives, say j, have been realized as the inequality $t_k < t(k, t_k|k)$ then becomes strict.

(2) The marginal contribution of i under the suboptimal choice of k is given by $M_i(\theta_0|k) = W(\theta_0|k) - W^{-i}(\theta_0|k)$. Again, suppose that the maximizing solution is obtained with a different payoff stream of alternative k. We then look at a modified payoff stream of alternative k

as described by (7.2) and (7.3). Consider then the difference

$$\widehat{M}_i(\theta_0|k) - M_i(\theta_0|k)$$
$$= [\widehat{W}(\theta_0|k) - \widehat{W}^{-i}(\theta_0|k)] - [W(\theta_0|k) - W^{-i}(\theta_0|k)]$$
$$= [\widehat{W}(\theta_0|k) - W(\theta_0|k)] - [\widehat{W}^{-i}(\theta_0|k) - W^{-i}(\theta_0|k)].$$

By the same argument as in (7.4), this leads to

$$[\varepsilon \delta^{t_k} \cdot 1 - \varepsilon \cdot \delta^{t(k,t_k|k)}] - [\varepsilon \delta^{t_k} \cdot 1 - \varepsilon \cdot \delta^{t(k,t_k|k,-i)}]$$
$$= \varepsilon \cdot (\delta^{t(k,t_k|k,-i)} - \delta^{t(k,t_k|k)}) \geqslant 0,$$

where the last inequality follows from $t(k, t_k|k, -i) \leqslant t(k, t_k|k)$. $\qquad \square$

Proof of Lemma 6. (1) Since

$$M_{i \cup j}(\theta_0) = W(\theta_0) - W^{-i \cup j}(\theta_0)$$
$$= W(\theta_0) - W^{-j}(\theta_0) + W^{-j}(\theta_0) - W^{-i \cup j}(\theta_0)$$
$$= M_j(\theta_0) + M_i^{-j}(\theta_0),$$

the claim follows if we can show:

$$M_i^{-j}(\theta_0) \geqslant M_i(\theta_0). \tag{7.5}$$

To establish the above inequality, we replace the payoffs of alternative j by the average payoffs $\{x_j^n\}_{n=1}^{\infty}$ computed along the stopping times $\{\tau_j^n\}_{n=0}^{\infty}$, as defined earlier in (5.9) and (5.10):

$$x_j(t_j) = x_j^n \text{ for } \tau_j^{n-1} < t \leqslant \tau_j^n. \tag{7.6}$$

By the property of the tail realizations, established in Lemma 3, and the optimal index policy, as stated in Lemma 2, we know that the optimal values in the allocation problem and hence the marginal contribution of alternative i are unaffected if we modify the payoffs of alternative j's as done in (7.6).

Next we split alternative j into $n = 1, \ldots, \infty$ (sub-)alternatives, denoted by j_n, along the above-mentioned stopping times $\{\tau_j^n\}_{n=0}^{\infty}$.

The payoffs by alternative j_n are defined as follows

$$x_{j_n}(t_{j_n}) = \begin{cases} x_j^n & \text{if } 1 \leqslant t_{j_n} \leqslant \tau_j^n - \tau_j^{n-1}, \\ 0 & \text{otherwise.} \end{cases}$$

The marginal contribution of alternative i is unaffected by this split of alternative j.

We now prove inequality (7.5) in the form of $M_i(\theta_0) \leqslant M_i^{-j_n}(\theta_0)$ for all j_n. Now,

$$M_i(\theta_0) - M_i^{-j_n}(\theta_0) = W(\theta_0) - W^{-i}(\theta_0) - (W^{-j_n}(\theta_0) - W^{-i \cup j_n}(\theta_0))$$

$$= W(\theta_0) - W^{-j_n}(\theta_0) - [W^{-i}(\theta_0) - W^{-i \cup j_n}(\theta_0)).$$

$$(7.7)$$

By the Gittins index theorem as stated in Lemma 2, the order of optimal choices for alternatives different from i and j_n is unaffected by the removal of j_n. Denote by W_{j_n} and $W_{j_n}^{-i}$ the continuation value to the social program, with and without i, after using alternative j_n. Let $\sigma_j^n \triangleq \tau_j^n - \tau_j^{n-1}$ be the number of periods alternative j_n has a positive contribution x_j^n. By elementary calculations, it follows that

$$W(\theta_0) - W^{-j_n}(\theta_0) = \delta^{t(j_n,1)} \left(\frac{1 - \delta^{\sigma_j^n}}{1 - \delta} x_j^n - (1 - \delta^{\sigma_j^n}) W_{j_n} \right), \qquad (7.8)$$

and

$$W^{-i}(\theta_0) - W^{-i \cup j_n}(\theta_0) = \delta^{t(j_n,1|-i)} \left(\frac{1 - \delta^{\sigma_j^k}}{1 - \delta} x_j^n - (1 - \delta^{\sigma_j^n}) W_{j_n}^{-i} \right). \quad (7.9)$$

After inserting (7.8) and (7.9) into (7.7) and observing that $t(j_k, 1| - i) \leqslant t(j_k, 1)$, by Lemma 3, and that $W_{j_n}^{-i} \leqslant W_{j_n}$, by virtue of being value functions, the claim is established.

(2) It is sufficient to show that if we combine any two alternatives j and j' into a single alternative $j \cup j'$, then

$$M_j(\theta_0) - M_j(\theta_0|k) + M_{j'}(\theta_0) - M_{j'}(\theta_0|k)$$

$$\leqslant M_{j \cup j'}(\theta_0) - M_{j \cup j'}(\theta_0|k). \qquad (7.10)$$

A repeated application of the same inequality would then eventually lead to merge all alternatives but k under a single identity and lead to the desired result. As

$$M_{j\cup j'}(\theta_0) - M_{j'}(\theta_0) = M_j^{-j'}(\theta_0), \qquad (7.11)$$

we can write inequality (7.10) equivalently as

$$M_j(\theta_0) - M_j(\theta_0|k) \leqslant M_j^{-j}(\theta_0) - M_j^{-j}(\theta_0|k). \qquad (7.12)$$

The identity in (7.11) follows directly from the definition of marginal contribution

$$
\begin{aligned}
M_{j\cup j'}(\theta_0) - M_{j'}(\theta_0) &= (W(\theta_0) - W^{-j\cup j'}(\theta_0)) - (W(\theta_0) - W^{-j'}(\theta_0)) \\
&= W^{-j'}(\theta_0) - W^{-j\cup j'}(\theta_0) \\
&= M_j^{-j'}(\theta_0).
\end{aligned}
$$

Rearranging inequality (7.12)

$$(M_j(\theta_0) - M_j(\theta_0|k)) - (M_j^{-j}(\theta_0) - M_j^{-j}(\theta_0|k)) \leqslant 0, \qquad (7.13)$$

we show next that the left-hand side is maximized for every k when $x_k(0) = \cdots = x_k(\tau_k^1)$. The argument is by contradiction and uses again the construction of (7.2) and (7.3). We thus look at the difference between the payoff generated by the modified and the original sequence of payoffs, or

$$
\begin{aligned}
&((\widehat{M}_j(\theta_0) - \widehat{M}_j(\theta_0|k)) - (\widehat{M}_j^{-j'}(\theta_0) - \widehat{M}_j^{-j'}(\theta_0|k))) \\
&-((M_j(\theta_0) - M_j(\theta_0|k)) - (M_j^{-j}(\theta_0) - M_j^{-j}(\theta_0|k))).
\end{aligned}
$$

The payoffs from the marginal contribution of the efficient program remain unchanged, and it suffices to evaluate

$$(\widehat{M}_j^{-j'}(\theta_0|k) - M_j^{-j'}(\theta_0|k)) - (\widehat{M}_j(\theta_0|k) - M_j(\theta_0|k)),$$

or identically expressing it in terms of the social values

$$
\begin{aligned}
&((\widehat{W}^{-j'}(\theta_0|k) - W^{-j'}(\theta_0|k)) - (\widehat{W}^{-j\cup j'}(\theta_0|k) - W^{-j\cup j'}(\theta_0|k)) \\
&-((\widehat{W}(\theta_0|k) - W(\theta_0|k)) - (\widehat{W}^{-j}(\theta_0|k) - W^{-j}(\theta_0|k)). \qquad (7.14)
\end{aligned}
$$

With the construction of (7.2) and (7.3), the payoffs resulting from each term inside the respective parenthesis are identical with the exception of the first and the t_kth realization of alternative k. As the timing of the realizations remains unchanged, it follows that difference (7.14) can be written as

$$((\varepsilon\delta^{t_k} - \varepsilon\delta^{t(k,t_k|k,-j')}) - (\varepsilon\delta^{t_k} - \varepsilon\delta^{t(k,t_k|k,-j\cup j')}))$$
$$-((\varepsilon\delta^{t_k} - \varepsilon\delta^{t(k,t_k|k)}) - (\varepsilon\delta^{t_k} - \varepsilon\delta^{t(k,t_k|k,-j)})),$$

and after the obvious cancellations

$$\varepsilon((\delta^{t(k,t_k|k,-j\cup j')} - \delta^{t(k,t_k|k,-j')}) - (\delta^{t(k,t_k|k,-j)} - \delta^{t(k,t_k|k)})).$$

By Lemma 2, we know that

$$t(k,t_k|k,-j\cup j') - t(k,t_k|k,-j') = t(k,t_k|k,-j) - t(k,t_k|k),$$

as well as

$$t(k,t_k|k,-j') \leqslant t(k,t_k|k).$$

It then follows from discounting with $\delta \in (0,1)$, that

$$(\delta^{t(k,t_k|k,-j\cup j')} - \delta^{t(k,t_k|k,-j')}) \geqslant (\delta^{t(k,t_k|k,-j)} - \delta^{t(k,t_k|k)}),$$

which leads to the desired contradiction. Hence it is sufficient to evaluate inequality (7.13) for constant payoff streams of alternative k. We can write the difference $M_j(\theta_0) - M_j(\theta_0|k)$ by Lemma 4 as

$$M_j(\theta_0) - M_j(\theta_0|k) = M_j^k(\theta_0).$$

It follows that we can rewrite inequality (7.13) as

$$M_j^k(\theta_0) \leqslant M_j^{-j',k}(\theta_0),$$

which holds by the first part of this lemma. $\square$

Proof of Theorem 7. (1) By hypothesis, $t(k,1) < t(i,1)$. We can therefore write the marginal contributions of i, $M_i(\theta_0)$ and $M_i(\theta_0|k)$, respectively as

$$M_i(\theta_0) = \delta^{t(i,1)} M_i(\theta_{t(i,1)}),$$

and

$$M_i(\theta_0|k) = \delta^{t(i,1|k)} M_i(\theta_{t(i,1|k)}).$$

By Lemma 2, it further follows that $t(i,1) = t(i,1|k)$ and that $\theta_{t(i,1)} = \theta_{t(i,1|k)}$ which establishes the claim.

(2) By Lemma 5.2, it is sufficient to evaluate the inequality $M_i(\theta_0) \geqslant M_i(\theta_0|k)$ when the payoff stream of alternative k is constant for all t_k with $0 \leqslant t_k \leqslant \tau_k^1$. By Lemma 4.2, the difference then satisfies the relation

$$\frac{1}{1-\delta}(M_i(\theta_0) - M_i(\theta_0|k)) = M_i^k(\theta_0),$$

and as the marginal contribution of any alternative to any program is weakly positive, the weak inequality follows.

(3) By Lemma 6, it follows that it is sufficient to evaluate the inequality in Theorem 1.3 for the case that there are only two sellers, namely k, and $\mathcal{I}/\{k\}$. In this case, the inequality can be written after expressing the marginal contributions through the social values as

$$W(\theta_0) - W(\theta_0|k)$$
$$\geqslant (W(\theta_0) - W_{-\mathcal{I}/\{k\}}(\theta_0)) - (W(\theta_0|k) - W_{-\mathcal{I}/\{k\}}(\theta_0|k)),$$

and as all terms cancel in this case, it follows that the inequality is always satisfied and in the extreme case of two sellers in fact as an equality. $\quad\square$

Proof of Theorem 8. The existence of the marginal contribution equilibrium was proved in the previous section. If $t_i(\theta) \geqslant T_i$ for all i, then the continuation game is in fact a repeated static game of price competition. Define next

$$Z(\theta) \triangleq \sum_{i=1}^{I} \min(t_i(\theta), T_i).$$

It is clear that for all states θ' such that $Z(\theta') \sum_{i=1}^{I} T_i$, the marginal contribution equilibrium is the unique cautious equilibrium. The induction hypothesis that we use is that the claim is also true for all states θ such that $Z(\theta) \geqslant Z'$, where $Z' \leqslant \sum_{i=1}^{I} T_i$. The claim is then proved by induction if we can show that the claim is also true for all states such that $Z(\theta) = Z'-1$.

Consider any state θ such that $Z(\theta) = Z' - 1$. Denote the equilibrium choice of the buyer at that state by $a(\theta)$. If $a(\theta) = i$ for some i such that $t_i(\theta) < T_i$, then the new state θ' is such that $Z(\theta') = Z'$ and by the induction hypothesis, the continuation game $\Gamma(\theta')$ has a unique cautious Markov perfect equilibrium payoff vector that coincides with the vector of marginal contributions of the players. Let $M_i(\theta, j)$ denote the marginal contribution and hence by the induction hypotheses the equilibrium continuation payoff to seller i if $a(\theta) = j$. If $a(\theta) = i$ for some i

such that $t_i(\theta) \geqslant T_i$, then θ and the new state θ' have the same continuation games, and by the Markov restriction, i will be chosen in all future periods.

We claim first that the cautious Markov perfect equilibrium payoff is unique for each fixed first period choice by the buyer. To see this, it is enough to show that the first period prices are uniquely pinned down by the continuation payoffs. If $a(\theta) = i$, then $p_j(\theta) = \delta V_j(\theta, i) - \delta V_j(\theta, j)$ for all $j \neq i$ by cautiousness. If $t_i(\theta) < T_i$, then $V_j(\theta, i) = M_j(\theta, i)$ by the induction hypothesis. If $t_i(\theta) \geqslant T_i$, then $V_j(\theta, i) = 0$ since the strategies are Markovian. Similarly, if $t_j(\theta) < T_j$, $V_j(\theta, i) = M_j(\theta, i)$ by the induction hypothesis and if $t_j(\theta) \geqslant T_j$, then $V_j(\theta, j) = \delta V(\theta, i)$ if $t_i(\theta) < T_i$, and $V_j(\theta, j) = 0$ if $t_i(\theta) \geqslant T_i$. Hence in all cases, $p_j(\theta)$ is uniquely determined. Finally, $p_i(\theta)$ is determined by $p_j(\theta)$, $x_i(\theta)$, $x_j(\theta)$ and the buyer's indifference condition.

Hence the remaining task is to show that for all θ such that $Z(\theta) = Z' - 1$, in all equilibria, $a(\theta) = i(\theta)$, where $i(\theta)$ is the socially efficient choice at state θ. We argue by contradiction. To do this, we suppose that $a(\theta) = k \notin i(\theta)$ in some equilibrium of the game starting at θ. Since the marginal contribution equilibrium also exists, there must be two separate sets of equilibrium prices at θ. Denote the marginal contribution equilibrium prices by $p_i(\theta)$ and the prices in the other equilibrium by $\hat{p}_i(\theta)$. We know by construction that in the marginal contribution equilibrium, the buyer is indifferent between $i(\theta)$ and $j^{-i}(\theta)$. Hence we have for the buyer the equilibrium conditions:

$$V_B(\theta) = x_{i(\theta)}(\theta) - p_{i(\theta)}(\theta) + \delta \left[W(\theta, i(\theta)) - \sum_{l \in \mathcal{I}} M_l(\theta, i(\theta)) \right]$$

$$= x_{j^{-i}(\theta)}(\theta) - p_{j^{-i}(\theta)}(\theta) + \delta \left[W(\theta, j^{-i}(\theta)) - \sum_{l \in \mathcal{I}} M_l(\theta, j^{-i}(\theta)) \right],$$

$$\geqslant x_k(\theta) - p_k(\theta) + \delta \left[W(\theta, k) - \sum_{l \in \mathcal{I}} M_l(\theta, k) \right]$$

for all $k \neq i(\theta), j^{-i}(\theta)$,

and for the efficient seller:

$$V_{i(\theta)}(\theta) = p_{i(\theta)}(\theta) + \delta M_{i(\theta)}(\theta, i(\theta)) \geqslant \delta M_{i(\theta)}(\theta, j^{-i}(\theta)),$$

and for all other sellers:

$$p_k(\theta) = \delta M_k(\theta, i) - \delta M_k(\theta, k) \text{ for all } k \neq i(\theta).$$

In the other equilibrium, let $j(k)$ be the seller that the buyer considers as good as k. Then the buyer's indifference condition is

$$\widehat{V}_B(\theta) = x_k(\theta) - \widehat{p}_k(\theta) + \delta \left[W(\theta, k) - \sum_{l \in \mathcal{I}} M_l(\theta, k) \right]$$

$$= x_{j(\kappa)}(\theta) - \widehat{p}_{j(\kappa)}(\theta) + \delta \left[W(\theta, j(\kappa)) - \sum_{l \in \mathcal{I}} M_l(\theta, j(\kappa)) \right],$$

and for the winning seller k

$$\widehat{p}_k(\theta) = \delta M_k(\theta, k) \geqslant \delta M_k(\theta, j(k)),$$

and all other sellers $l \neq k$:

$$\widehat{p}_l(\theta) = \delta M_l(\theta, k) - \delta M_l(\theta, l).$$

We derive the contradiction for the case of

$$t(i(\theta), t_{i(\theta)}) < t(j^{-i}(\theta), t_{j^{-i}(\theta)}) < t(k, t_k) < t(j(k), t_{j(k)}).$$

The remaining cases, i.e. those with

$$t(i(\theta), t_{i(\theta)}) < t(j^{-i}(\theta), t_{j^{-i}(\theta)}) < t(j(k), t_{j(k)}) < t(j, t_j)$$

and the ones where

$$k = j^{-i}(\theta) \quad \text{or} \quad j(k) \in \{i(\theta), j^{-i}(\theta)\}$$

are handled similarly. By Theorem 7.1 and the equilibrium conditions

$$\widehat{p}_{j(k)}(\theta) = p_{j(k)}(\theta).$$

As a result, we have

$$V_B(\theta) \geqslant \widehat{V}_B(\theta). \tag{7.15}$$

By Theorem 7.2,

$$V_{i(\theta)}(\theta) > \widehat{V}_{i(\theta)}(\theta). \tag{7.16}$$

But (7.15) and (7.16) imply jointly that $i(\theta)$ can capture the buyer and increase the profit by offering $p_{i(\theta)}(\theta) - \varepsilon$ instead of $\widehat{p}_{i(\theta)}(\theta)$, contradicting the equilibrium requirements. $\qquad \square$

References

[1] H. Bar-Isaac, Reputation and survival: learning in a dynamic signalling model, Rev. Econ. Stud. 70 (2003) 231–251.

[2] D. Bergemann, J. Välimäki, Learning and strategic pricing, Econometrica 64 (1996) 1125–1149.

[3] D. Bergemann, J.Välimäki, Dynamic common agency, J. Econ. Theory 111 (2003) 23–48.

[4] B. Bernheim, M. Whinston, Menu auctions, resource allocation, and economic influence, Quart. J. Econ. 101 (1986) 1–31.

[5] L. Felli, C. Harris, Job matching, learning and firm-specific human capital, J. Polit. Economy 104 (1996) 838–868.

[6] L. Felli, C. Harris, Firm specific training, Technical Report Working Paper No. 38, Institute of Advanced Studies, Princeton, 2004.

[7] D. Fudenberg, D. Levine, Subgame-perfect equilibrium of finite- and infinite-horizon games, J. Econ. Theory 31 (1983) 251–268.

[8] J. Gittins, Allocation Indices for Multi-Armed Bandits, London, Wiley, 1989.

[9] C. Harris, A characterization of the perfect equilibria of infinite horizon games, J. Econ. Theory 37 (1985) 99–127.

[10] B. Jovanovic, Job search and the theory of turnover, J. Polit. Economy 87 (1979) 972–990.

[11] E. Maskin, J. Tirole, Markov perfect equilibrium, J. Econ. Theory 100 (2001) 191–219.

[12] R. Miller, Job matching and occupational choice, J. Polit. Economy 92 (1984) 1086–1120.

[13] P. Whittle, Optimization Over Time, vol. 1, Wiley, Chichester, 1982.

Chapter 9

Venture Capital Financing, Moral Hazard, and Learning*

Dirk Bergemann[†] and Ulrich Hege[‡]

[†]*Department of Economics, and Cowles Foundation of Research in Economics, Yale University, New Haven, CT 06520-8268, USA*
dirk.bergemann@yale.edu
[‡]*Toulouse School of Economics, 31015 Toulouse, France*
ulrich.hege@tse-fr.eu

We consider the provision of venture capital in a dynamic agency model. The value of the venture project is initially uncertain and more information arrives by developing the project. The allocation of the funds and the learning process are subject to moral hazard. The optimal contract is a time-varying share contract which provides intertemporal risk-sharing between venture capitalist and entrepreneur. The share of the entrepreneur reflects the value of a *real option*. The option itself is based on the control of the funds. The dynamic agency costs may be high and lead to an inefficient early stopping of the project. A positive liquidation value explains the adoption of strip financing or convertible securities. Finally, relationship financing, including monitoring and the occasional replacement of the management improves the efficiency of the financial contracting.

JEL classification: D83; D92; G24; G31

Keywords: Venture capital, optimal stopping, learning, dynamic financial contracts, share contracts, security design, relationship finance.

---

*This chapter is reproduced from *Journal of Banking and Finance*, **22**, 703–735, 1998: joint with Ulrich Hege.

1. Introduction

1.1. *Motivation*

Venture capital has become a major vehicle for the funding of start-up firms. In many countries, most notably the United States, venture capital is now the financing mode of choice for projects where "learning" and "innovation" are important. Because of their innovative nature, venture projects carry a substantial risk of failure. Only a minority of start-ups are high-return investments: 20% or less are frequent estimates for the fraction of projects where investors can successfully "cash out", mostly through IPO's. Of the remaining, a majority is sold off privately or merged which can mean anything between a modest success and scantly disguised failure with substantial losses. A minority is liquidated implying a complete write-off of the investment.[1]

One of the most challenging problems in venture financing is to determine when to release funds for continued development and when to abandon a project. Many aspects of the venture capital industry suggest that practitioners are well aware that they face a sequence of starting and stopping problems in the financing of a venture.[2] An essential feature of any venture project is the necessity to fund the project in order to learn more about the uncertain return of the venture. This process often starts with the provision of seed financing to set up a business plan. The simultaneity of the financing decision and the acquisition of information about the investment project is characteristic for ventures and more generally for the financing of innovation. Surprisingly, the dynamic interaction of both aspects has received little attention in the literature.

This paper proposes a simple model to analyze the optimal financing of venture projects when learning and moral hazard interact. A wealth constrained entrepreneur offers an investment opportunity to a venture capitalist. The project can either succeed or fail. The successful completion requires funds for the development of the project. The rate of investment controls the probability of success. A higher investment level accelerates the process of discovery. As the project continues to receive financing without achieving success, the agents change their assessment about the likelihood

---

[1]For estimates, see Poterba (1989), Sahlman (1990), Sagari and Guidotti (1991), Gompers (1995) and Amit *et al.* (1997).

[2]The staging of the funds, documented in Lerner (1994) and Gompers (1995) is perhaps the most prominent aspect of the sequential nature of venture financing.

of future success. Eventually the prospects may become too poor to warrant further investment.

The entrepreneur controls the allocation of the funds and the investment effort is unobservable to the investor. The control over the funds implies that the entrepreneur also controls the flow of information about the project. The solution of the agency conflict has to take into account the intertemporal incentives for the entrepreneur. Suppose, in any given period, the entrepreneur would consider to divert the capital flow for her private consumption. In the following period she would then be marginally more optimistic about the future of the project than the venture capitalist. The entrepreneur would know that the project did not receive any capital in the preceding period and hence could not possibly generate a success. But the venture capitalist would continue to believe that the entrepreneur did as instructed. In consequence he interprets the fact that no success has been observed as "bad news" about the project. Following a deviation, the entrepreneur will therefore keep her posterior belief about future success constant while the posterior belief of the investor is necessarily downgraded. The reward for the entrepreneur therefore consists of two components. She needs to be compensated for the foregone private benefits but also for the downgrading of her expectations about the future of the project. The longer the experimentation horizon, the larger is the option value of the diversion. In fact, the compensation could become so large as to surpass the net value of the project. In turn this implies the possibility of financial constraints in the form of an inefficient and premature end of the project.

1.2. *Results and empirical implications*

The optimal share contract and the financial constraints allow for a number of empirical implications.

First, our paper provides an analysis of the optimal evolution of the shares of entrepreneur and venture capitalist. How the parties should optimally split the prize should depend on the funding horizon and the flow of funds. Our model predicts that the share of the entrepreneur decreases towards the end. Initially, the entrepreneur's share can rise or fall, depending on the discount rate and the degree of initial optimism. The *expected* share of the entrepreneur, however, always decreases over time. Similarly, the *expected return* of the venture capitalist decreases over time and he may even make losses if the project approaches the stopping time. Empirical findings indicate that the entrepreneur is indeed penalized if the project takes too much time as her equity fraction is diluted from financing round to financing round.[3]

Second, we obtain results for the security design by extending the model to positive liquidation values in case of abandonment. The liquidation value is received by selling tangible assets or intermediate results. The venture capitalist should then either receive strip packages *combining common stock and debt* or *convertible securities.* The optimal contract should reward the entrepreneur only in the case of success. The venture capitalist should therefore keep a "hard" claim in case of failure. However, if the venture capitalist would hold exclusively hard claims, then a premature liquidation is likely.[4]

Third, a pure share or equity contract could be financed at *arm's length,* implying that it could be traded in financial markets. However an arm's length share contract may leave too much surplus to the entrepreneur and the project would be terminated too early. Our model accounts thus naturally for the observation that venture financing is typically *relationship* financing. Costly monitoring and the option to replace the entrepreneur may become desirable. Relationship specific financing permits the extension of the funding horizon.[5]

Fourth, short-term refinancing of the project can never be optimal. We call short-term financing, as distinct from staged financing, a financial policy where the entrepreneur attracts funds on a competitive basis in each period. Towards the end of the efficient investment horizon, the expected return is insufficient to cover the necessary outlays for both partners. The efficient solution can only be achieved by a long-term contract which allows for intertemporal risk-sharing. More precisely, the venture capitalist subsidizes continuation of the project toward the end in exchange for higher expected profits at the beginning. Consistent with this result, Sahlman (1990) observes that venture capitalists are protected against competition by preemptive rights and Anand and Galetovic (1997) report that venture financing is frequently supported by long-term relationships.

---

[3]As new funds are provided in exchange with stock purchase agreements, the shares of the entrepreneur become diluted, see e.g. Sahlman (1990), Lerner (1995) and Gompers (1995).

[4]The predominance of convertible preferred stock is documented by Sahlman (1990) and Trester (1998).

[5]The frequent usage of monitoring and replacement of management is documented in Gorman and Sahlman (1989), Sahlman (1990) and Lerner (1995).

We may finally remark that up-front financing as well as staged financing are consistent with our model. Since the optimal contract satisfies the intertemporal incentive constraints, the funds will be allocated by the entrepreneur as intended regardless of how the funds are provided over time.[6] In Section 7 we discuss the extension of the model to multiple signals, where stage financing arises as the unique optimal financial arrangement.

1.3. *Related literature and overview*

The theoretical research on venture finance has only recently emerged. In Hart and Moore (1994), the option of the entrepreneur to repudiate her financial obligations limits the feasible amount of outsider claims. Neher (1997) extends their approach to stage financing as an instrument to implement the optimal investment path. Admati and Pfleiderer (1994) show that a fixed fraction equity contract may give robust optimal incentives if it is efficient to allocate the control rights to the venture capitalist. Berglöf (1994) considers convertible debt in a framework of incomplete contracts to transfer control rights to the value-maximizing party. Chan *et al.* (1990) explain the optimal transition of control between entrepreneur and venture capitalist in a model with initial uncertainty about the skill of the entrepreneur. Hellmann (1996) explains the willingness of the entrepreneur to relinquish control rights by a trade-off between equity and debt induced incentives. Trester (1998) argues that the problem of an entrepreneur dissipating the firm's assets can be mitigated if the investor has no option to declare default and seize the assets. Cornelli and Yosha (1997) analyze the problem of an entrepreneur manipulating shortterm results for purposes of "window-dressing".

The paper is organized as follows. The model is presented in Section 2. The value of the project is characterized in Section 3. The structure and efficiency of short and long-term contracting is examined in Section 4. We extend the model in Section 5 by allowing for a positive liquidation value of the project and discuss issues of security design in this context. In Section 6, we analyze the relationship specific instruments such as monitoring and the occasional replacement of the entrepreneur. Section 7 discusses possible extensions of the model. Section 8 concludes.

---

[6]Sahlman (1990) discusses up-front financing, even in the presence of staging, and Gompers (1995) analyzes the structure of stage financing.

2. The Model

The venture is presented as an investment project with uncertain returns in Section 2.1. The successful realization of the venture is positively correlated with the volume of financing it receives. As the flow of investment sinks into the project, entrepreneur and investor update their assessment of the prospects of the project. The evolution of the posterior belief of eventual success represents the *learning process* of the agents, which is analyzed in Section 2.2. The moral hazard problem between the entrepreneur and the investor as well as the financing possibilities of the project are finally described in Section 2.3.

2.1. *Project with uncertain returns*

An entrepreneur owns a project with uncertain return. The project is either "good" with prior probability α_0 or "bad" with prior probability $1 - \alpha_0$. If the project is "good", then in every period t, there is a certain probability that the project is successfully completed, in which case it yields a fixed payoff R. The probability of success in period t, *conditional* on the project being good, is denoted by p_t. The probability p_t is in turn an increasing function of the investment flow in period t. Or inversely, a success probability p_t requires an investment flow of $c(p_t)$ in period t. We assume that $c(p_t)$ is a linear function of p_t:

$$c(p_t) = cp_t, \quad c > 0. \tag{2.1}$$

The maximal probability of success in each period is denoted by p (without any subscript), where $0 < p < 1$ and any probability $p_t \in [0, p]$ is feasible in each period. In other words, any investment beyond $c(p)$ does not increase the probability of success.

If the project is "bad", then it will never yield a return and the probability of success is zero independent of the capital flow. The project can receive financing over any number of periods and time is discrete and denoted by $t = 0, 1, \ldots, T$. The investment process either stops with a successful completion or the project is eventually abandoned when the likelihood of future success becomes sufficiently small.

The uncertainty of the project is resolved over time by an *experimentation process*, where the likelihood of success is positively correlated with the investment in the project. The investment problem is simple as the investment only influences the conditional probability of success in every period and independent of time. In particular, the investment flow does not influence the value of the successful realization, R, or the scrap value if the project should be abandoned. In Section 7 we shall discuss how these modifications, as well as time dependent probabilities, would enrich the predictions of our model.

2.2. *Learning*

As the experimentation process develops over time, entrepreneur and investor learn more about the prospects of the project. If success has not yet occurred at period t, then the participants in the project update their beliefs about the type of the project. We next determine the evolution of the posterior beliefs. We denote by α_{t+1} the posterior belief that the project is good, based on no discovery until and including t. The evolution of the posterior belief α_{t+1}, conditional on no success, is given by Bayes' rule as a function of the prior belief α_t and the capital flow cp_t as:

$$\alpha_{t+1} = \frac{\alpha_t(1 - p_t)}{\alpha_t(1 - p_t) + 1 - \alpha_t}. \tag{2.2}$$

The posterior belief α_{t+1} thus decreases over time when success has not been realized. The decline in the posterior belief is stronger for larger investments, as the participants in the venture become more pessimistic about the likelihood of success. The posterior belief, again conditional on no success yet, can be represented as a function of the initial belief α_0 and the sequence of investments until t, $\{cp_0, cp_1, \ldots, cp_t\}$:

$$\alpha_{t+1} = \frac{\alpha_0 \prod_{s=0}^{t}(1 - p_s)}{\alpha_0 \prod_{s=0}^{t}(1 - p_s) + 1 - \alpha_0}. \tag{2.3}$$

Under a constant investment policy $p_t = \hat{p}$, the evolution of α_t is a discrete version of a decreasing logistic function:

$$\alpha_{t+1} = \frac{\alpha_0(1 - \hat{p})^{t+1}}{\alpha_0(1 - \hat{p})^{t+1} + 1 - \alpha_0}. \tag{2.4}$$

The evolution of the posterior belief, conditional on no success, under two different constant allocation policies is displayed in Fig. 9.1.

The posterior belief changes only slowly if the participants have very precise beliefs about the nature of the project, i.e. if α_t is close to either 0 or 1. Correspondingly the event of no success is most informative if the agents have very diffuse beliefs, i.e. α_t is close to $\frac{1}{2}$. In this case the posterior beliefs change most rapidly. In any case, a higher investment level accelerates the rate at which the posteriors change over time as displayed in Fig. 9.1.

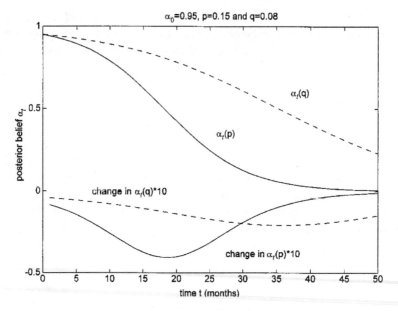

Figure 9.1.   Volume of financing *cp* and evolution of posterior belief.

2.3. *Moral hazard and financing*

The entrepreneur has no wealth initially and seeks to obtain external funds to realize the project. Financing is available from a competitive market of venture capitalists. Entrepreneur and venture capitalists have initially the same assessment about the likelihood of success, which is given by the prior belief α_0. The entrepreneur and venture capitalists are both risk-neutral and have a common discount factor $\delta \in (0,1)$.

The funds which are supplied by the venture capitalist are to be allocated by the entrepreneur to generate the desired success R. However the (correct) allocation of the funds to the project is unobservable to the venture capitalist, and thus a moral hazard problem arises between financier and entrepreneur. Indeed the entrepreneur can "shirk" and decide to (partially) withhold the investment and divert the capital flow to her private ends. An equivalent, but perhaps more classical formulation of the same moral hazard problem is following one: the efficient application of the investment requires effort, which is costly for the entrepreneur. By reducing the effort, the entrepreneur also reduces the probability of success and hence the efficiency of the employed capital. In both cases, a conflict of interest arises about the use of the funds.

Initially, the entrepreneur can suggest financial contracts to any of the venture capitalists. The selected venture capitalist then decides whether to accept or reject. The contract can be contingent on time, outcome and the capital provided by the investor. However, due to the moral hazard nature of the financing problem, the contract cannot be made contingent on the (correct) application of the funds. The design of the contract has to ensure that incentive compatibility and individual participation constraints are satisfied. The contract may contain a clause prohibiting that the entrepreneur continues the project once the contract has expired, for example by transferring ownership of the idea to the venture capitalist. If such a prohibition is not made, then the entrepreneur can again suggest financial contracts to any of the venture capitalists. For the moment, we abstractly consider contingent contracts, but in the appropriate places, we shall discuss which standard securities will be able to perform the tasks of the contingent contracts. We neglect renegotiation of contracts throughout this paper.[7]

Finally, we wish to emphasize that while there is no initial asymmetry in the information between financier and entrepreneur, the asymmetry may arise over time as the project receives funding. The source of the asymmetry is the unobservability of the allocation of funds. If entrepreneur and investor have different assessments over how the funds have been employed, then in turn they will have different posterior probabilities over the likelihood of success. Before we consider the optimal financial contract between entrepreneur and investor, we first analyze the efficient investment policy in the absence of the moral hazard problem.

3. Value of the Venture

The social value of the venture project with prior belief α_0 is maximized by an optimal investment policy and an optimal stopping point. At the stopping point the project is abandoned and no further investment is undertaken. The stopping point itself can either be characterized by the posterior belief α_T at the stopping point or the *real lime* T at which the project is abandoned. For any given investment policy $\{cp_0, \ldots, cp_T\}$ there is naturally a one-to-one relation between α_T and T through Bayes' rule as developed in Eq. (2.3).

---

[7]See Bergemann and Hege (1998) for a discussion of renegotiation in dynamic agency problems.

We denote by $V(\alpha_t)$ the value of the project with posterior belief α_t under optimal policies. The optimal policies can be obtained by standard dynamic programming arguments. Consider first the optimal stopping point α_T. Clearly, the project should receive funds as long as the expected returns from the investment exceed the costs, or

$$\alpha_T p_T R - c p_T \geqslant 0, \tag{3.1}$$

for some $p_T \in [0, p]$. Conversely, if the current expected returns do not exceed the investment cost, or

$$\alpha_T p_T R - c p_T < 0, \tag{3.2}$$

then it is optimal to abandon the project, as future returns will only decline further. It follows from Eqs. (3.1) and (3.2) that the *efficient boundary point* α^* between the investment region and the stopping region is given by

$$\alpha^* p_T R - c p_T = 0 \Leftrightarrow \alpha^* = \frac{c}{R}. \tag{3.3}$$

The posterior belief α^* at which stopping occurs decreases when either the return R increases or the marginal cost c of generating success decreases.

We notice next that if indeed the last investment occurs at α_T, then it is optimal to choose $p_T = p$ due to the linear structure of the problem. Hence we obtain the value in the terminal period:

$$V(\alpha_T) = \alpha_T p R - c p. \tag{3.4}$$

The value of the venture is then obtained recursively by the dynamic programming equation:

$$V(\alpha_t) = \max_{p_t}\{\alpha_t p_t R - c p_t + (1 - \alpha_t p_t)\delta V(\alpha_{t+1})\}, \tag{3.5}$$

where the posterior belief x_{t+1} is determined by the incoming belief α_t and the investment $c p_t$ in period t through Bayes' rule as expressed in Eq. (2.2). The value function (3.5) represents the implications of an investment policy $c p_t$ on current and future returns. An increase in $c p_t$ is costly but it increases the probability of a successful completion today and the associated expected returns $\alpha_t p_t R$. At the same time, it becomes less likely that the project will have to be continued tomorrow as $1 - \alpha_t p_t$ decreases. Finally, if success should not occur in period t even with large investment flow $c p_t$, then the posterior belief α_{t+1} will decrease and correspondingly the continuation value of the project, $V(\alpha_{t+1})$.

The value function (3.5) also indicates that the linearity in p_T which the terminal period problem (3.1) displays, is preserved in the intertemporal investment problem (3.5) as well. The optimal policy is therefore to invest maximally at the level of cp as long as the posterior belief is above α^* and stop as soon the posterior belief falls for the first time below the boundary point α^*. For transparency, we may translate this policy into a stopping time policy T^* in *real time*. In this case we ask how long can we maximally invest cp and still maintain posterior beliefs above the stopping point α^*. The optimal stopping time T^* is then given by

$$T^* = \max\left\{ T \left| \frac{\alpha_0(1-p)^T}{\alpha_0(1-p)^T + 1 - \alpha_0} \geqslant \alpha^* \right. \right\}. \tag{3.6}$$

Evidently, the optimal stopping time T^* depends on the initial belief α_0 at which the project is started, $T^* \triangleq T^*(\alpha_0)$, but we usually suppress the dependence on α_0 as a matter of convenience. The stopping time T^* then represents the time elapsed between starting at α_0 and arriving for the last time at a posterior belief exceeding α^*. The socially efficient investment policy and the value of the venture can then be obtained from the solution of the recursive problem (3.5):

Proposition 1 (Optimal investment policy):

(i) *The optimal policy is to invest maximally cp until T^*.*
(ii) *The social value of the venture is given by*

$$V(\alpha_0) = \alpha_0 p(R-c)\frac{1 - \delta^{T^*}(1-p)^{T^*}}{1 - \delta(1-p)} - (1-\alpha_0)cp\frac{1-\delta^{T^*}}{1-\delta}. \tag{3.7}$$

Proof: See Appendix A.   □

The value function $V(\alpha_0)$ presents an intuitive decomposition of the value of the project.[8] The first term in Eq. (3.7) is the expected value of the project conditional on the project being good. Notice that the value of the project is discounted at a rate which compounds the pure discount rate δ and the probability of *no* discovery $(1-p)$ which results in the factor $\delta(1-p)$. The second term captures the case that the project is bad which occurs with probability $(1 - \alpha_0)$. In this case, costly experimentation will continue until the stopping time T^* is reached and discounting occurs at

---

[8]For future reference, we denote by $V_T(\alpha_0)$ the value of the project if financing occurs at the maximal rate cp, until T, where obviously $V_{T^*}(\alpha_0) = V(\alpha_0)$.

the rate δ until the project is stopped at T^*. With the description of the socially efficient investment policy in the background, we next turn to the financial contracting between entrepreneur and venture capitalist.

4. Financial Contracting

We begin in Section 4.1 by analyzing the provision of venture capital under short-term contracts. The optimal short-term contracts are simple share contracts between entrepreneur and investor. However as a relationship governed by short-term contracts has almost no scope for intertemporal transfers, short-term contracts are generally inefficient and will lead to a premature end of the venture. Consequently, we investigate in Section 4.2 the structure of long-term contracts and in Section 4.3 their efficiency properties. The model is extended in Section 5 to allow for a positive liquidation value of the project and issues of security design appear in this context.

4.1. *Short-term contracts*

The venture capitalist offers his funds for the project in exchange against a share of the uncertain returns of the venture. Evidently the expected returns for the venture capitalist must be large enough to justify his investment. At the same time, the entrepreneur must have sufficient incentives to truthfully invest the funds in the project. As the project can only be successfully completed and yield R if the entrepreneur applies the funds correctly, it follows that the incentives provided to the entrepreneur should maximally discriminate with respect to the signal R. With the wealth constraint of the entrepreneur, the most high-powered incentive contract is obviously the following: she receives a positive share of R if the project was a success and nothing otherwise. Due to the binary nature of the project, success or failure, these contracts, which we call *share contracts*, form indeed the class of optimal contracts in this environment.

Define by S_t the share of the entrepreneur if R is realized in period t. The corresponding share of the venture capitalist is $(1 - S_t)$. In a short-term contract, the venture capitalist only promises to provide funds for a single period, and then reconsiders financing in the subsequent periods. The expected return from the current investment must at least exceed the cost of the investment and since there is competition among the venture capitalists, in equilibrium the venture capitalist will just break even. Hence the short-term contract must satisfy for any level of funding cp_t the following *participation constraint*:

$$\alpha_t p_t (1 - S_t) R = c p_t. \tag{4.1}$$

On the other hand, the remaining share S_t for the entrepreneur has to be large enough for her to invest the funds in the project and not divert them to her private ends. This forms the *incentive compatibility constraint* for the entrepreneur, which is formally stated by

$$\alpha_t p_t S_t R \geqslant c p_t. \tag{4.2}$$

By combining Eqs. (4.1) and (4.2) we infer that any financing under short-term contracts can only be continued as long as the inequality

$$\alpha_t p_t R \geqslant 2 c p_t \tag{4.3}$$

is satisfied. The critical value of the posterior belief, denoted by α^s, where short-term contracting will cease is thus given by

$$\alpha^s = \frac{2c}{R}.$$

By comparing α^s with the socially efficient stopping point α^* obtained in the previous section, we immediately obtain the following Proposition.

Proposition 2 (Short-term financing): *The venture project is stopped prematurely under short-term financing as $\alpha^s > \alpha^*$.*

The premature stopping indicated by $\alpha^s > \alpha^*$ is naturally equivalent to a funding horizon T^s which is shorter than the efficient horizon T^*, where T^s is determined as T^* in Eq. (3.6). This is a simple, but important benchmark result. It indicates that efficient financing requires some form of *intertemporal risk sharing* which can only be sustained by commitments made through long-term contracts. The necessity of intertemporal risk sharing is easy to grasp. As the posterior belief α_t deteriorates over time, the expected value which the parties expect to split decreases as well. Eventually, the competing claims emanating from the investment problem of the financier Eq. (4.1) and the agency problem of the entrepreneur Eq. (4.2) lead to a conflict. In this respect, financing a venture resembles a team problem where both investor and entrepreneur contribute. The venture capitalist must earn the equivalent of $c p_t$ to justify his investment, while the entrepreneur must earn the equivalent of $c p_t$ if she were to employ the funds towards the proper end. As α_t decreases these compensations eventually cannot be covered anymore from the *expected proceeds* in period t and it follows that short-term contracts will necessarily terminate too early.

4.2. *Long-term contracts*

In a regime of short-term financing the investor has to break even in every period as there was no commitment on either side to continue the relationship. Long-term contracts can improve the intertemporal risk-sharing by replacing the sequence of participations constraints (4.1) for every period by a single *intertemporal participation constraint*, which covers the entire funding horizon.

By offering the investor a larger share of the return in the early stages of the financing, his shares in the later stage can be lowered, and hence the project can be continued beyond α^s. But a long-term contract with the associated funding commitments offers the entrepreneur a rich set of alternative actions, many of them not desirable from the investor's point of view. For example, if the entrepreneur is promised additional funding in the next period, then she may consider to divert the funds today and bet instead on a positive realization of the project tomorrow. By implication, the incentives for the entrepreneur today have to be sufficiently strong, in particular relative to the incentives offered tomorrow. These dynamic considerations then generate a rich set of predictions about the sharing rules over time. In a *first step* we ask what the *minimal share* of the entrepreneur has to be for her to truthfully apply *any given sequence* of funds to the project. The *second step* identifies the *incentive compatible* funding policy which maximizes the value for the entrepreneur and hence is adopted in equilibrium.

The solution to the minimization problem is, again, obtained explicitly by dynamic programming methods. Consider first the final period of the contract, denoted by T. The share S_T has to be high enough for her to invest the funds in the project rather than to divert them, or formally S_T has to satisfy:

$$\min_{S_T} \alpha_T p_T S_T R \geqslant c p_T. \qquad (4.4)$$

The minimal share S_T in the ultimate period T is then given by

$$S_T = \frac{c}{\alpha_T R} \qquad (4.5)$$

and the expected value of this arrangement to the entrepreneur is denoted by $E_T(\alpha_T)$. Solving the problem recursively we obtain a sequence of value functions, denoted by $E_T(\alpha_t)$, where α_t is the current posterior belief and T is the length of the entire contract. Consequently $T - t$ is the number of remaining periods in the contract. The incentive problem in period t is given by

$$E_T(\alpha_t) \triangleq \min_{S_t}\{\alpha_t p_t S_t R + \delta(1 - \alpha_t p_t) E_T(\alpha_{t+1})\} \qquad (4.6)$$

subject to

$$E_T(\alpha_t) \geqslant \alpha_t p' S_t R + c(p_t - p') + \delta(1 - \alpha_t p') E_T(\alpha') \quad \forall p' \in [0, p_t). \quad (4.7)$$

We notice the intertemporal structure of the problem. If the entrepreneur correctly employs the funds, then with probability $\alpha_t p_t$ success occurs in period t. On the other hand, no success occurs with probability $1 - \alpha_t p_t$ in which case the project continues, but the prospects of future success will appear dimmer as α_{t+1} is given by

$$\alpha_{t+1} = \frac{\alpha_t(1 - p_t)}{1 - \alpha_t p_t} < \alpha_t.$$

The inequality Eq. (4.7) requires that for a given allocation of funds cp_t, the share S_t has to be large enough to prevent any diversion of funds. The diversion of funds, represented by $p' < p_t$, affects the payoff's for the entrepreneur in two ways. Consider first the contemporaneous effect. The likelihood of success today will be smaller as $p' < p_t$, but the entrepreneur enjoys the utility from the diverted funds $c(p_t - p')$. The second and dynamic effect is that a continuation of the contract in the next period becomes more likely. And as less funds are applied to the project today, there is less reason to change the posterior belief and clearly:

$$\alpha' = \frac{\alpha_t(1 - p')}{1 - \alpha_t p'} > \frac{\alpha_t(1 - p_t)}{1 - \alpha_t p_t} = \alpha_{t+1} \quad \text{for } p' \in [0, p_t).$$

After all, the event of no success should not surprise and lead to a smaller change in the posterior belief as less resources have been devoted to the project. In consequence, the entrepreneur will be more optimistic about *future* success as she has seen (and provided) less evidence against it, and naturally the continuation value will be higher with more optimistic beliefs:

$$\alpha' > \alpha_{t+1} \Leftrightarrow E_T(\alpha') > E_T(\alpha_{t+1}).$$

If the entrepreneur were to withhold the funds, the *private* belief α' would diverge from the *public* belief α_{t+1}, and to fend off the informational asymmetry, the entrepreneur is granted an *informational rent*. The source of the rent is the control over the conditional probability p_t and through it over the learning process, and we shall refer to it as the *learning rent*.

It is now apparent that there are two forces which help to realign the interest of the entrepreneur with the ones of the investor. First, a larger share S_t if success occurs today relative to the share S_{t+1} if success occurs tomorrow. Second, the discounting of future returns at the rate δ depresses the incentives of the entrepreneur to postpone the successful realization of the project. In consequence, the more myopic the entrepreneur is, the less binding are the intertemporal incentive constraints.

The solution S_t of the minimization problem represented by (4.6) and (4.7) delivers the expected value $E_T(\alpha_t)$ the entrepreneur receives for a given funding policy, where $E_T(\alpha_t)$ satisfies the recursive equation:

$$E_T(\alpha_t) = \alpha_t p_t S_t R + \delta(1 - \alpha_t p_t) E_T(\alpha_{t+1}). \tag{4.8}$$

In the first step, we were concerned with the minimal share the entrepreneur has to receive for any given level of funding. In the next step we ask how and when the funds should be released so as to maximize the value of the venture. As the market for venture capital is competitive, in equilibrium the net value of the project will belong entirely to the entrepreneur. Formally, the problem is then given by

$$\max_{\{cp_0, cp_1 \ldots cp_T\}} V_T(\alpha_0) \tag{4.9}$$

subject to

$$V_T(\alpha_0) \geqslant E_T(\alpha_0). \tag{4.10}$$

The constraint (4.10) incorporates the sequence of incentive constraints through $E_T(\alpha_0)$ and the participation of the investor by requiring non-negative profits in form of the inequality. The preceding analysis provides the important hints on the solution of Eqs. (4.9) and (4.10) as well. Consider any aggregate investment the investor would like to contribute to the project and the remaining issue is only how the funds should be distributed over time. If the funds are invested only slowly into the project, then the funding commitment necessarily extends over a longer horizon and the investor faces more intertemporal incentive constraints. These constraints increase the share of the entrepreneur and hence decrease the investor's share. Thus the optimal solution is to invest in each period up to the efficient level cp. The complete solution of S_t and $E_T(\alpha_t)$ based on the programs (4.9) and (4.10) is summarized as follows.

Proposition 3 (Share contract and entrepreneurial value):

(i) *The value function of the entrepreneur is given by*

$$E_T(\alpha_t) = cp\alpha_t \frac{1 - \delta^{T-t}}{1 - \delta} + cp(1 - \alpha_t) \frac{1 - \left(\frac{\delta}{1-p}\right)^{T-t}}{1 - \frac{\delta}{1-p}}. \tag{4.11}$$

(ii) *The share function of the entrepreneur is given by*

$$S_t = \frac{c(1 - p)}{\alpha_t R} + \frac{cp}{R} \frac{1 - \delta^{(T-t)}}{1 - \delta} + \frac{cp(1 - \alpha_t)}{\alpha_t R} \frac{1 - \left(\frac{\delta}{1-p}\right)^{T-t}}{1 - \frac{\delta}{1-p}}. \tag{4.12}$$

Proof: See Appendix A. □

The intertemporal contract S_t ensures that the entrepreneur employs the capital in every period towards the discovery process. The three elements in the share contract as displayed in Eq. (4.12) may seem rather inaccessible at first, but can be decomposed and traced to the different aspects of the agency problem, namely (*i*) static agency costs, (*ii*) intertemporal agency costs, and (*iii*) informational agency costs.

(*i*) If the project would be financed only for single period and hence $T = t = 0$, then the minimal share for the entrepreneur to act properly would be

$$\frac{c}{\alpha_t R}. \tag{4.13}$$

(*ii*) If the investor would like to fund the venture until time $T > 0$, but could *observe* the evolution of the posterior α_t, so that the entrepreneur would not have access to the learning rent, then the minimal share offered at time t would have to be increased by

$$\frac{cp}{R} \frac{1 - \delta^{(T-t)}}{1 - \delta}. \tag{4.14}$$

The new aspect in the intertemporal agency problem is the option to withhold financing for a single period, but continue afterwards as instructed until T. To prevent the delay in any period, the investor has to provide stronger incentives. The higher additional compensation is necessary as the deadline T, at which funding is stopped is relatively remote. As the deadline comes closer, or $T - t$ is decreasing, the need for additional incentives becomes weaker. Notice also that the Eq. (4.14) only depends on the "time to go" and is independent of the posterior belief α_t.

(*iii*) Finally the informational agency costs are represented by

$$\frac{cp(1-\alpha_t)}{\alpha_t R} \frac{1-\left(\frac{\delta}{1-p}\right)^{T-t}}{1-\frac{\delta}{1-p}}, \tag{4.15}$$

which forms the basis for the learning rent of the entrepreneur. It depends on the value of the current beliefs and the rate p at which updating of the posterior beliefs occurs relative to the rate of discounting δ. The rate of updating p is the quantity of information the entrepreneur controls at each instant of time. The informational rent is hence increasing in the quantity of information under influence of the entrepreneur, but dampened by discounting, as the value of information today is larger than tomorrow. The three elements Eqs. (4.13)–(4.15) together determine the share of the entrepreneur. An illustration of the decomposition of the sharing rule is given in Fig. 9.2.

The behavior of the shares S_t over time is thus determined by an underlying option problem. The reward implied by the shares S_t has to be equal to the value of the option of diverting funds for a single period. The value of this particular option is determined as any regular option by

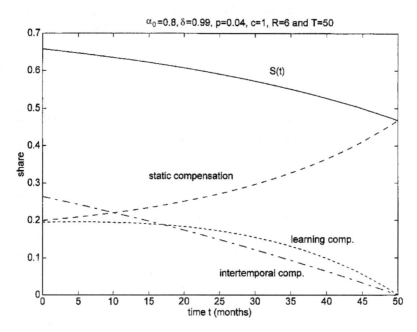

Figure 9.2. Evolution of the share $S(t)$ over time and decomposition.

the volatility of the underlying state variable and the time over which the option can be exercised. Here, the volatility is the conditional probability p at which updating occurs and the time of the option right is the remaining length of the funding, $T - t$.

The optimal (arm's length) contract is hence a time-varying *share* or *equity* contract. The time-varying share contract presents the solution to *a real option* problem, where the real option is the control of information in every period. Some implications for the intertemporal sharing of the returns are recorded next:

Proposition 4 (Evolution of shares over time):

(*i*) S_t *has at most one interior extremum in* t *and then it is a maximum.*
(*ii*) *For* δ *sufficiently close to 1,* S_t *is monotonically decreasing.*

Proof: See Appendix A.   □

Earlier in this section we identified two elements which provide incentives to allocate the funds truthfully in period t: (*a*) decreasing shares over time and (*b*) sufficiently strong discounting. Proposition 4 identifies the interplay between these two forces. If discounting does not work because the remaining time horizon is too short (*i*) or discounting is too weak (*ii*), then the shares have to fall over time. On the other hand, the shares of the entrepreneur can only increase initially (*i*), when discounting (due to the length of the funding horizon $\hat{T}$) is sufficiently strong to insure incentive compatibility.

4.3. *Equilibrium and inefficiency*

The efficient funding policy and the associated value $E_T(\alpha_0)$ to the entrepreneur for any given funding horizon T is identified in Proposition 3. The final question is to what horizon the venture capitalist is willing to extend his commitment and whether the efficient horizon T^* can be attained. The entrepreneur participates in the project until T if the expected net value of the project exceeds the expected value the entrepreneur receives, as represented by the inequality Eq. (4.10): $V_T(\alpha_0) \geqslant E_T(\alpha_0)$. The efficient horizon T^* can be achieved if at T^* we have

$$V_{T^*}(\alpha_0) \geqslant E_{T^*}(\alpha_0). \tag{4.16}$$

The equilibrium contract and the implied funding horizon is determined with the assistance of Proposition 3 and the participation constraint (4.10) of the investor. We have to distinguish two cases. If the project is not sufficiently rich to be continued until T^*, or in other words if

$$V_{T^*}(\alpha_0) < E_{T^*}(\alpha_0),$$

then the funding horizon is determined by $\hat{T}$, where $\hat{T}$ is the largest time horizon for which the value of the project exceeds the value of the compensation for the entrepreneur

$$\hat{T} = \max\{T + 1 | V_T(\alpha_0) > E_T(\alpha_0)\}. \tag{4.17}$$

In the second case, the project is sufficiently rich to allow for an efficient financing, or formally:

$$V_{T^*}(\alpha_0) \geqslant E_{T^*}(\alpha_0), \tag{4.18}$$

in which case funding is extended until T^*. In equilibrium, the net value of the project has to be allocated completely to the entrepreneur, or

$$V_{\hat{T}}(\alpha_0) = E_{\hat{T}}(\alpha_0),$$

due to competition among the venture capitalists. Hence if the inequality holds strictly in either Eqs. (4.17) and (4.18), then the remaining surplus has to be given to the entrepreneur in a way compatible with the incentive constraints. In the case of Eq. (4.17) this can be achieved by a one period continuation as a "winding-down" phase, where $S_{\hat{T}}$ is determined as in Proposition 3 but with a smaller capital flow $cp_{\hat{T}} < cp$, as cp itself would violate the participation constraint of the investor by definition of $\hat{T}$. In the case of Eq. (4.18), the project is rich enough to guarantee the entrepreneur a larger share than the one determined by S_t in Proposition 2. But as these shares also have to satisfy the sequence of incentive constraints, their intertemporal behavior will be similar to S_t. The efficiency properties of the long-term contracting are summarized in the following proposition.

Proposition 5 (Inefficiency):

    (*i*) *Long-term contracts allow for an extended funding horizon $\hat{T}$ (relative to short-term financing), but never exceed T^*.*

    (*ii*) *The funding horizon $\hat{T}$ of the optimal long-term contracts may not attain T^*.*

    (*iii*) *The funding horizon $\hat{T}$ increases in R and α_0 and decreases in c.*

Proof: See Appendix A.   □

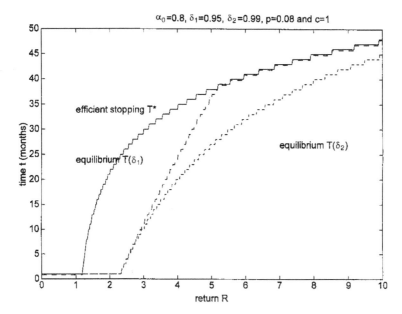

Figure 9.3.   Return R and duration of financing T.

As the funding horizon $\hat{T}$ increases with improved conditions for the project, one may wonder whether the inefficiency indicated in Proposition 5 (*ii*) will eventually disappear entirely. The answer here is rather subtle. As, to take one example, the return R increases, $\hat{T}$ increases but so does T^*. Eventually T^* may become so large that the informational rent of the entrepreneur becomes too large and the project will have to be stopped at some $\hat{T} < T^*$. The example in Fig. 9.3 illustrates that better projects (with larger R) will indeed receive longer funding commitments as Proposition 5 predicts, but the equilibrium allocation may never attain social efficiency.

5. Liquidation Value and Security Design

In this section, we introduce a liquidation value of $L_t > 0$, which is collected whenever the project ends without having succeeded. The value L_t represents intermediate outcomes of the venture and captures the proceeds from selling the remaining tangible assets if the project is liquidated. The value L_t is a deterministic function of time and does not depend on the behavior of the entrepreneur. We assume that $L_t \geqslant \delta L_{t+1}$ for all t. The liquidation value hence contains no information about the actions of the entrepreneur. It is meant to represent the value created by verifiable actions. The condition on the growth of L_t merely ensures that the project is not continued without the entrepreneur.[9]

The liquidation value enhances the social value of the project. If the project is supposed to be liquidated in T, then the expected net present value is $(1 - \alpha_0 + \alpha_0(1 - p)^T)\delta^T L_T$, which is the probability to reach T multiplied by the discounted value in T. The optimal contract should award the liquidation payoff so as to relax the financial constraint on the funding horizon $\hat{T}$. The idea of the optimal share contract under limited liability is to reward the entrepreneur if and only if she was successful. But a pure equity contract would give the entrepreneur a part of the liquidation value. This necessarily weakens the incentive structure, as by diverting funds, she would increase the likelihood of reaching the final period T, and with it an additional payoff $S_T L_T$. The optimal contract therefore needs to split the claims in the case of success from the ones in the case of liquidation. A pure equity contract cannot achieve this and a mixture between debt and common equity, or a convertible security become necessary.

Define a *debt contract* as a profile of debt claims $D_t \geqslant 0$, which are puttable at any time t and bear no coupon. Similarly, a time-varying *common stock contract* is denoted by a profile of fractions $A_t \geqslant 0$ of total equity that the entrepreneur receives if the project is terminated in t (successfully or unsuccessfully). The optimal financial contract can then be achieved by a mixture of debt and common equity, or a convertible security.

Proposition 6 (Security design):

(i) *An optimal contract is provided by a mixture of debt and common equity with*

$$D_t = \delta D_{t+1} \quad \forall t < T, \quad D_T = L_T, \quad A_t = S_t \frac{R}{R - D_t}. \tag{5.1}$$

(ii) *A convertible preferred stock held by the investor with a nominal value of D_t and converted into a share $1 - S_t$ of common stock if exercised in period t also represents an optimal contract.*

Proof: See Appendix A. □

In "strip financing" the venture capitalist retains equity *and* debt, and the debt claim increases at the rate r, where $1/(1 + r) = \delta$, until D_T

---

[9]The evolution of L_t could also be stochastic as long as $L_t \geqslant \delta \mathbb{E} L_{t+1}$.

reaches L_T. The incentive compatible equity share A_t of the entrepreneur, which would be S_t if indeed R would be distributed, has to be increased to account for the seniority claim of D_t on R in the case of success. An equivalent distribution of the payoffs can be achieved by a convertible preferred stock, as indicated by Proposition 6(ii). The time-varying *conversion price* is given by $D_t/(1 - S_t)$.[10]

Interestingly, there is some empirical evidence that debt becomes more important as a financing tool towards later stages of venture projects.[11] In our model debt has a function only at the termination date and this might provide an element to understand this pattern in the dynamic capital structure in venture financing. The preceding discussion already indicated that in many situations a pure sharing contract cannot be optimal. The corollary presents precise conditions.

Corollary 1: *A common stock contract A_t does not constitute an optimal contract if the financing ends inefficiently early:* $\hat{T} < T^*$.

A combination of debt and equity could relax the financial constraints faced by a pure equity solution. Common equity is inefficient because *equity gives too little of the liquidation value* to the venture capitalist in case the project fails. We note finally that a pure debt contract may be inefficient for exactly the opposite reason. *Debt may convey too much of the liquidation value* to the venture capitalist and encourage premature liquidation.

6. Monitoring and Job Rotation

In this section aspects of relationship financing are considered which may reduce the inefficiency indicated in Proposition 5: (i) monitoring and (ii) changing the management. Both modifications imply the transfer of substantial control rights to the venture capitalist and hence point to the *relationship aspect* of venture capital financing. The focus is on the optimal *timing* of these control instruments.

Consider first the possibility of monitoring the research effort of the venture capitalist. Monitoring is costly and the venture capitalist has to spend $mp_t > 0$ to monitor the entrepreneur in period t.[12] In return, the venture capitalist receives an accurate signal about the application

---

[10]We recall that S_t is the *minimal* incentive compatible share if the entire R is distributed.

[11]See Amit *et al.* (1997).

of the funds. The signal is verifiable and hence punishment in case of a deviation (a signal which differs from the contractually agreed effort) can be implemented. Hence, the moral hazard problem is eliminated in the periods where monitoring takes place.

The optimal timing is determined by the costs and benefits of monitoring in the intertemporal agency problem. The *expected present cost* of monitoring in period t is simply mp_t adjusted by discounting and the probability of reaching period t, as the project may be successfully completed before t. Hence the expected present cost of monitoring are decreasing with time. The benefits of monitoring come from the *static* as well as the *intertemporal* component in the share S_t. If monitoring occurs in period t then the share of entrepreneur can be set to $S_t = 0$, as there is no need to provide any incentives in period t. But monitoring in period t also affects the incentives in all preceding periods. We recall that the value of diverting funds in $t' < t$ was partially due to the possibility to have a share in the success later on. Monitoring in period t excludes this option at least for period t and hence the share of the entrepreneur can be reduced marginally in all periods *preceding* t. The direct benefits from monitoring ($S_t = 0$) are clearly decreasing in t due to discounting, but the indirect benefits increase as the number of periods which enjoy the marginal reduction increase. Efficient monitoring occurs where the reduction in the informational rent for the entrepreneur *per* unit cost of monitoring is maximized.

Proposition 7 (Monitoring policy): *The optimal policy is to monitor towards the end of the project.*

Proof: See Appendix A. □

Monitoring then occurs towards the end of the venture as the discounting of the costs and the indirect benefits dominate the direct savings due to monitoring uniformly. Notice also that in the monitoring phase all the residual gains are allocated to the investor.

The timing of the monitoring in our model is partially due to the uniform rate at which information is generated. More generally, the benefits of monitoring in period t are increasing in the amount of information

---

[12]The cost of monitoring is thus linear in the flow of funds cp_t. It can be conceived as an accounting or control system, the cost of which are increasing in the size of the operation to be monitored.

which is generated in period t. Some other implications of a more general information structure will be discussed in the next section.

Consider next the possibility of replacing the current manager, entrepreneur or not, by a new manager. The financing horizon $\hat{T}$ of the project may then be subdivided into several managerial job spells. We investigate the following extension of the model. In any period t, the entrepreneur or the incumbent manager can be replaced by a new manager. There are no differences between the original entrepreneur and successive managers, in particular concerning productivity and moral hazard. As before, the entrepreneur initially owns the project and tries to capture as much of the surplus as possible.

Managerial job rotation reduces the informational rent of each manager by restricting the duration of each individual manager. However, for the (initial) entrepreneur *and* the investor outside managers arc costly, as the founding partners have to concede shares to the future managers. The objective for the initial pair is therefore to minimize the pay-outs to succeeding managers. For this reason, the replacement of the entrepreneur should be scheduled as late as possible.

Each manager requires a compensation which is determined exactly as the shares S_t of the entrepreneur. As there is no cost in replacing the manager, the pay-out is minimized when the manager changes in every period.[13] The cost of developing the venture with a sequence of manager is then equal to $cp_t + cp_t$, where one term accounts for the cost of investment and the other for the incentive costs of the manager. Proposition 2 indicates that the efficient termination then occurs at T^s.

Proposition 8 (Replacement):

(*i*) *It is optimal to replace the entrepreneur if and only if the long-term contract would otherwise stop at $\hat{T} < T^s$.*

(*ii*) *With optimal replacement the project will be stopped at T^s. The entrepreneur remains in place until the value of the project net of compensations to the managers is disbursed to her. Thereafter, a new manager is brought in every period.*

Proof: See Appendix A.   □

---

[13] We note that the general analysis suggested here carries over to more general settings with switching costs or decreasing efficiency of subsequent managers.

The replacement of the founding entrepreneur constitutes an empirical regularity in the venture capital industry, see Gorman and Sahlman (1989). We may also add that earn outs are inefficient in this framework.[14] The rationale is essentially the same as the one exposed in Section 5. A quit payment works against the idea of making all benefits contingent on success. Thus, one insight of the present model is that earn outs, by providing insurance in the case of failure, are a costly practice in an environment with an uncertain completion date.

7. Discussion and Extensions

In this paper, the venture is characterized by a simple binary structure. In each period success is possible and the likelihood of success depends on the belief α_t and the intensity p_t at which the project is developed. The investment flow influences only the probability by which success is generated, and the prize R is constant over time. In this section we discuss how several extensions would modify our results. We consider in particular (i) time-varying returns R_t, (ii) multiple signals and staging and (iii) a time-varying information flow.

Time varying returns R_t. The value of the successful realization, R, was assumed to be constant throughout this paper. But the investment process may also have a cumulative effect on the value of a successful realization and hence lead to an increasing sequence of R_t over time. An increasing value R_t would tend to reduce the inefficiency problem documented in Proposition 5. With an increasing R_t, the share S_t of the entrepreneur could be reduced further towards the end as the incentive constraints are based on the composite $S_t R_t$. Conversely, a decreasing sequence R_t would make funding even more precarious. The positive effect of an increasing R_t points to the importance of value creation and production of tangible assets during the investment process.[15]

Multiple signals and staging. This paper portrays a simple venture which ends after a single positive signal is received. Clearly the arrival of

---

[14]See Sahlman (1990) for the observation that the founding entrepreneur often receives little protection against the threat of being put aside.

[15]Gompers (1995) presents evidence that the value of tangible assets and the length of the financing rounds increase with the duration of the venture.

good and bad news may be a more complex process. Frequently, projects are divided into various stages which are defined by the completion of certain intermediate results. In these circumstances continued financing may be conditional on the successful completion of earlier stages.

The basic model we analyzed describes the evolution of the incentives in any component of such staged projects. The entire project would simply be a sequence of such stages, each giving rise to an optimal stopping time. In each individual problem R_t would constitute the continuation value after having received an intermediate result. Each stopping problem determines how long to wait for the arrival of an intermediate signal before the project is stopped. Otherwise the previous analysis carries over and the share contract for each stage concerns the sharing of the incremental value produced in that stage.

The sequential arrival of information then supports *stage financing* as the optimal arrangement. To see this, suppose the contracting parties had the choice to contract either on a strong final signal (i.e. completion of the marketable product) or on a finer sequence of intermediate signals. The sequencing, implied by the staging, splits the horizon over which the intertemporal incentive constraints have to be compounded. In other words, the entrepreneur realizes that she must produce the intermediate result first in order to receive continued financing. This reduces her incentives to procrastinate in the intermediate periods.

Time varying information flow. The investment flow cp_t controls the conditional probability of success p_t, and via p_t, the evolution of the posterior from α_t to α_{t+1}. The marginal cost of generating success in terms of probability was assumed to be constant in the level of p_t (up to the maximum p) as well as constant over time.

The optimal investment problem with a general increasing and convex cost function $c(p_t)$ shares the same structural features as the model here, but, naturally, would not lead to an explicit solution as presented in Proposition 1 and 3. If the costs of generating success vary over time, then the information flow, associated with p_t, would be timed so as to coincide with periods where the costs are relatively low. Periods with a higher p_t would then constitute periods with a "learning boost". We saw in the previous section that the benefits of monitoring are positively related to the volatility of information. Monitoring would then tend to occur in periods where much information is produced. In the presence of sequential arrival of information, we would expect monitoring to be most prevalent in periods before the next financing stage.[16]

8. Conclusion

This paper investigated the provision of venture capital when the investment flow controls the speed at which the project is developed. As the binary outcome of the project is uncertain, the speed of development influences the (random) time at which the project yields success *and* the information which is acquired by the investment flow. The role of the entrepreneur is to control the application of the funds which are provided by the venture capitalist.

The paper provides a rationale for long-term contracting as these contracts achieve best the goal of distributing the entrepreneur's return over time in a way which maximizes the research horizon. It is further shown that the compensation of the entrepreneur is similar to an option contract, and as such depends on the length of the contract and the volatility of the information induced through her actions. The option expresses the value of the intertemporal incentive constraint. As the value of the option may become exceedingly large, relationship financing may become necessary. In consequence, the optimal timing of monitoring and replacement of the entrepreneur are analyzed.

The paper focuses on the financing of venture projects. But the interaction between investment and learning process and the incentives necessary to implement both processes is central for the financing of R&D in general. The present work may therefore be considered as a step in developing further insights into the optimal financing of innovation.

Acknowledgements

The authors would like to thank Dilip Abreu, Kyle Bagwell, James Brander, Prajit Dutta, Harald Hau, Thomas Hellmann, Phillip Leslie, Leslie Marx, Gordon Phillips, Ben Polak, Rafael Repullo, Mike Riordan, Bryan Routledge, Harald Uhlig, and Juuso Välimäki for many helpful comments. In addition, we would like to thank Joshua Lerner, the discussant, and an anonymous referee for many suggestions, which greatly improved the paper. Seminar participants at Columbia University, Tilburg University,

---

[16]See Lerner (1995) for empirical evidence on such a correlation.

Yale Law School, the Stanford University Conference on "The Economic Foundations of Venture Capital", the New York University Conference on "The Economics of Small Business Finance", and the European Summer Symposium in Financial Markets in Gerzensee provided stimulating discussions. The first author gratefully acknowledges financial support from the National Science Foundation (SBR 9709887). The second author acknowledges financial support from HEC where he was affiliated while this paper was written.

Appendix A

This appendix collects the proofs to all propositions in the paper.

Proof of Proposition 1. The value of an *arbitrary* investment policy starting at t and ending after T, denoted by $\{cp_t, cp_{t+1}, ..., cp_T\}$, can be expressed by telescoping the returns over time as

$$\sum_{s-t}^{T} \delta^{s-t} \left((\alpha_s p_s R - cp_s) \prod_{r=t}^{s-1} (1 - \alpha_r p_r) \right), \tag{A.1}$$

where, by convention, $\prod_{r=t}^{t-1}(1 - \alpha_r p_r) = 1$. From Bayes' rule in Eq. (2.2) we obtain

$$\prod_{r=t}^{s-1}(1 - \alpha_r p_r) = \frac{\alpha_t}{\alpha_s} \prod_{r=t}^{s-1}(1 - p_r) \tag{A.2}$$

and

$$\frac{\alpha_t}{\alpha_s} = \alpha_t + (1 - \alpha_t)\frac{1}{\prod_{r-t}^{s-1}(1 - p_r)}. \tag{A.3}$$

By substituting Eqs. (A.2) and (A.3) into Eq. (A.l), we obtain after collecting terms:

$$\sum_{s=t}^{T} \delta^{s-t} \left((\alpha_t p_s R - \alpha_t cp_s) \prod_{r-t}^{s-1} (1 - p_r) - cp_s(1 - \alpha_t) \right). \tag{A.4}$$

Consider now the optimal investment policy cp_t in period t. By rewriting (A.4) as

$$p_t(\alpha_t R - c) + \sum_{s=t+1}^{T} \delta^{s-t} \left(\alpha_t p_s(R - c)(1 - p_t) \prod_{r=t+1}^{s-1} (1 - p_r) - cp_s(1 - \alpha_t) \right), \tag{A.5}$$

it appears immediately that Eq. (A.5) is linear in p_t, and thus it is either optimal to allocate the capital maximally at cp or not to allocate any capital at all. Hence until the project is stopped, in each period capital is allocated at the maximal rate cp. In consequence the stopping time is given by T^* as defined in Eq. (3.6). By setting $p_s = p$ for all $s = t, t+1, \ldots, T^*$, we obtain:

$$V(\alpha_t) = \sum_{s-t}^{T^*} \delta^{s-t} (\alpha_t p (R-c)(1-p)^{s-t} - (1-\alpha_t)cp),$$

and in particular for α_0,

$$V(\alpha_0) = \alpha_0 p (R-c) \frac{1 - \delta^{T^*}(1-p)^{T^*}}{1 - \delta(1-p)} - (1-\alpha_0)cp \frac{1 - \delta^{T^*}}{1-\delta}. \qquad \square$$

The proof to Proposition 3 relies on Lemma A.l, which describes the optimal policy for the entrepreneur.

Lemma 1 (Optimal policy of the entrepreneur):

(*i*) The optimal policy for the entrepreneur is always either $p' = 0$ or $p' = p_t$.

(*ii*) If the optimal policy is $p' = p_t$ for some α_t, then p_t remains the optimal policy for all $\alpha > \alpha_t$.

Proof: (*i*) Consider first the expected value of the entrepreneur for an arbitrary assignment of shares, under the assumption that she truthfully applies the funds to the project:

$$\sum_{s=t}^{T} \delta^{s-t} \left(\alpha_s p_s S_s R \prod_{r=t}^{s-1} (1 - \alpha_r p_r) \right),$$

which can be expressed, after using Eqs. (A.2) and (A.3), as

$$\sum_{s=t}^{T} \delta^{s-t} \left(\alpha_t p_s S_s R \prod_{r=t}^{s-1} (1 - p_r) \right),$$

or, after separating the behavior in t, as

$$\alpha_t p_t S_t R + (1 - p_t) \sum_{s=t+1}^{T} \delta^{s-t} \left(\alpha_t p_s S_s R \prod_{r=t+1}^{s-1} (1 - p_r) \right).$$

If the entrepreneur would consider diverting funds in period t, then she wishes to maximize the value of the deviation p', which is given by

$$\alpha_t p' S_t R + (1 - p') \sum_{s=t+1}^{T} \delta^{s-t} \left(\alpha_t p_s S_s R \prod_{r=t+1}^{s-1} (1 - p_r) \right) + c(p_t \cdots p').$$

$$(A.6)$$

The expression Eq. (A.6) is linear in p' and the first part of the lemma follows directly.

(*ii*) By the linearity of Eq. (A.6), if it is optimal to choose $p' = p_t > 0$ under α_t, then we have

$$\alpha_t p_t S_t R - \alpha_t p_t \sum_{s=t+1}^{T} \delta^{s-t} \left(p_s S_s R \prod_{r=t+1}^{s-1} (1 - p_r) \right) \geq cp_t > 0, \qquad (A.7)$$

where the first inequality is due to the optimality and the second is due to $p_t > 0$. By monotonicity, the first inequality is preserved by any $\alpha > \alpha_t$. $\qquad \square$

Proof of Proposition 3. (*i*) Consider the value function $E_T(\alpha_0)$ of the entrepreneur in the initial period $t = 0$, obtained solving Eqs. (4.6) and (4.7) recursively. By Lemma A.1, $E_T(\alpha_0)$ has to be equal to the value generated by a maximal deviation today which is followed by compliance in all future periods:

$$E_T(\alpha_0) = cp_0 + \sum_{t=1}^{T} \delta' \left(\alpha_{t-1} p_t S_t R \prod_{s=1}^{t-1} (1 - \alpha_s p_s) \right), \qquad (A.8)$$

which in turn is equivalent to $E_T(\alpha_0) = cp_0 + (\alpha_0/\alpha_1)\delta E_T(\alpha_1)$. The general recursion of the value function then yields:

$$E_T(\alpha_t) = cp_t + \frac{\alpha_t}{\alpha_{t+1}} \delta E_T(\alpha_{t+1}), \qquad (A.9)$$

from which we obtain after recursive substitution

$$E_T(\alpha_0) = \sum_{t=0}^{T} \delta^t \frac{\alpha_0}{\alpha_t} cp_t, \qquad (A.10)$$

and rearranging by using Eq. (A.3) we get:

$$E_T(\alpha_0) = \alpha_0 c \sum_{t=0}^{T} \delta^t p_t + (1 - \alpha_0) c \sum_{t=0}^{T} \delta^t \frac{p_t}{\prod_{s=0}^{t-1}(1 - p_s)}, \qquad (A.11)$$

which in turn is equivalent to Eq. (4.11) if $p_t = p$ for all t.

(*ii*) The share contract S_t is obtained by equating Eqs. (A.9) and (4.8) which yields: $\alpha_t p_t S_t R - cp_t + (\alpha_t/\alpha_{t+1})\delta p_t E_T(\alpha_{t+1})$, and after using Eq. (A.9) again, we get

$$\alpha_t p_t S_t R = (1 - p_t)cp_t + p_t E_T(\alpha_t), \qquad (A.12)$$

which yields immediately Eq. (4.12) after replacing $E_T(\alpha_t)$ with the explicit expression obtained in Eq. (A.11) if $p_t = p$ for all t. Finally, $p_t = p$ follows immediately from Proposition 1 and the fact that the incentive compatible contract S_t only depends on the continuation value as just described.   □

Proof of Proposition 4. (*i*) Consider the share function S_t as derived in Eq. (4.12) and substitute α_t by Bayes' rule Eq. (2.4) to obtain

$$S_t = \left(1 + \frac{1 - \alpha_0}{\alpha_0(1 - p)^t}\right) \frac{c(1 - p)}{R} + \frac{cp}{R} \frac{1 - \delta^{(T-t)}}{1 - \delta}$$

$$+ \frac{1 - \alpha_0}{\alpha_0(1 - p)^t} \frac{cp}{R} \frac{1 - \left(\frac{\delta}{1-p}\right)^{T-t}}{1 - \frac{\delta}{1-p}}. \qquad (A.13)$$

It is then sufficient to prove that the share function $S_t \equiv S(t)$ as a *continuous function* of t has at most one interior extremum, and that it has to be a maximum. Based on Eq. (A.13), one obtains after some rearranging: $S'(t) \gtrless 0 \Leftrightarrow$

$$\left(\frac{1 - p}{\delta}\right)^t \left(\frac{\delta^T}{1 - \delta} + \frac{1 - \alpha_0}{\alpha_0\left(1 - \frac{\delta}{1-p}\right)} \left(\frac{\delta}{1 - p}\right)^T\right)$$

$$\lessgtr \frac{1 - \alpha_0}{\alpha_0\left(1 - \frac{\delta}{1-p}\right)} \frac{p \ln(1 - p)}{(1 - \delta)\ln\delta}, \qquad (A.14)$$

from which it follows after inspection that $S'(t)$ can cross zero at most once and only from above.

(*ii*) By inspecting again condition Eq. (A.14), we find that as $\delta \to 1$, the right-hand side of the equivalent condition eventually becomes positive and the left hand side negative, which proves the claim.   □

Proof of Proposition 5. (*i*) Consider any project with $\alpha_t > \alpha^s > \alpha_{t+1} > \alpha^*$, when financing occurs at the rate of cp. Then, by Proposition 2, there will be no financing in period $t+1$ with short-term contracts, although it would be efficient, since $\alpha_{t+1} > \alpha^*$. Under a long-term contract, the investor can be given $\alpha_t(1-S_t)pR > pc$ in period t, and $\alpha_{t+1}(1-S_{t+1})pR < pc$, and have his intertemporal participation constraint balanced over the periods. This then allows financing to proceed strictly longer than under a short-term contract and enhance the efficiency of the contractual arrangement.

(*ii*) The example associated with Fig. 9.3 verifies the claim.

(*iii*) Consider the difference $V_T(\alpha_0) - E_T(\alpha_0)$ for a given T, as given by Proposition 1 and 3. The difference is increasing in R and α_0 and decreasing with c. The equilibrium horizon $\hat{T}$ can be increased with increases in the difference $V_T(\alpha_0) - E_T(\alpha_0)$. $\qquad\square$

Proof of Proposition 6. The value of the project with liquidation value, denoted by $V_T^L(\alpha_0)$, is given by

$$V_T^L(\alpha_0) = V_T(\alpha_0) + \delta^T(\alpha_0(1-p)^T + (1-\alpha_0))L_T,$$

where $V_T(\alpha_0)$ is as defined earlier in Proposition 1. The duration $\hat{T}$ of the contract is determined by

$$\hat{T} = \max\{T + 1 | V_T^L(\alpha_0) > E_T(\alpha_0)\}, \tag{A.15}$$

where $E_T(\alpha_0)$ is as defined in Proposition 3 and the last period $\hat{T}$ is again a "winding-down" period which insures that in equilibrium:

$$V_{\hat{T}}^L(\alpha_0) = E_{\hat{T}}(\alpha_0). \tag{A.16}$$

The entrepreneur receives the loan necessary to pursue the project until $\hat{T}$ in advance.

(*i*) Since the debt claim grows at the rate r, the investor is indifferent over the particular period t at which he claims payment of the debt as long as he is assured to receive D_t in the particular period t. Moreover, in any period in which he does not claim the debt there is some probability that a success occurs. Hence he will claim the debt only if no success has been observed in period $\hat{T}$. By condition Eq. (A.16) he is willing to participate. Finally, for the entrepreneur, her share will be higher, namely A_t as specified in Eq. (5.1), to compensate her for the debt claim which has seniority. But $A_t(R - D_t)$ is just equal to the minimal compensation $S_t R$, necessary to satisfy the intertemporal incentive constraints as proven in Proposition 3. Hence the contract induces her to truthfully direct the funds to the project as well. Thus the specified mixture of debt and equity implements the optimal outcome.

(ii) The argument is almost identical to the one provided under (i). □

Proof of Corollary 1. Suppose the entrepreneur receives a share $A_T > 0$ of the liquidation proceeds if liquidation occurs in T. The modified incentive constraint in period T then becomes

$$\alpha_T p A_T R + (1 - \alpha_T p) A_T L_T \geqslant cp + A_T L_T,$$

which implies that surplus the entrepreneur can guarantee himself in T increases from cp to $cp + A_T L_T$. From Eq. (A.12) in the proof of Proposition 3, we can then infer that this in fact increases A_t in all periods. Thus the value of the entrepreneur, denoted by $E_T^L(\alpha_0)$ when she receives a share of the liquidation value, is higher than when she does not. If $\hat{T}$, defined by

$$\hat{T} = \max\{T + 1 | V_T^L(\alpha_0) > E_T^L(\alpha_0)\},$$

is indeed smaller than T^*, then we clearly have $\bar{T} > \hat{T}$, with $\bar{T}$ being defined by

$$\bar{T} = \max\{T + 1 | V_T^L(\alpha_0) > E_T(\alpha_0)\},$$

since $E_T(\alpha_0) < E_T^L(\alpha_0)$, for all T. But this implies in particular that $V_{\bar{T}}^L(\alpha_0) > V_{\hat{T}}^L(\alpha_0)$ as $\bar{T} > \hat{T}$. Since in equilibrium we have $V_{\hat{T}}^L(\alpha_0) = E_{\hat{T}}^L(\alpha_0)$ and $V_{\bar{T}}^L(\alpha_0) = E_T(\alpha_0)$, it follows that a pure sharing contract is not optimal and will not be chosen in equilibrium by the entrepreneur. □

Proof of Proposition 7. Monitoring is costly and reduces the social surplus which can be distributed between entrepreneur and venture capitalist. The benefit of monitoring is the reduction in $E(\alpha_0)$ which can then be used to extend the length of the project financing. Monitoring should therefore occur in those periods where the reduction in $E(\alpha_0)$ is maximized *per unit cost of monitoring*. The expected cost of monitoring in period t is given by $C(t) \triangleq \delta^t(\alpha_0(1-p)^t + 1 - \alpha_0)mp$. The benefit of monitoring in period t only is $B(t) \triangleq E(\alpha_0) - E^t(\alpha_0)$, where the superscript t denotes the period in which monitoring occurs. We next compute the benefit of monitoring explicitly. Suppose we would monitor in period t, then the value to the entrepreneur in t is

$$E^t(\alpha_t) = \delta(1 - \alpha_t p)E(\alpha_{t+1}) = \delta(1 - \alpha_t p) \sum_{s=t+1}^{T} \delta^{s-(t+1)} \frac{\alpha_{t+1}}{\alpha_s} cp,$$

after using Eqs. (A.9) and (A.10). Rewritten in terms of α_t, we obtain

$$E^t(\alpha_t) = \delta(1 - p) \sum_{s=t+1}^{T} \delta^{s-(t+1)} \frac{\alpha_t}{\alpha_s} cp.$$

The value function in $t - 1$, with monitoring in period t, can be obtained by backwards induction:

$$E^t(\alpha_{t-1}) = cp + \delta^2 (1 - \alpha_{t-1} p) \frac{\alpha_t}{\alpha_{t+1}} \sum_{s=t+1}^{T} \delta^{s-(t+1)} \frac{\alpha_{t+1}}{\alpha_s} cp. \qquad (A.17)$$

For the general recursion we would like to write all numerators in terms of α_{t-1}, and obtain with $\alpha_t(1 - \alpha_{t-1} p) = \alpha_{t-1}(1 - p)$, the following expression for (A.17):

$$E^t(\alpha_{t-1}) = \frac{\alpha_{t-1}}{\alpha_{t-1}} cp + \delta^2 (1 - p) \sum_{s=t+1}^{T} \delta^{s-(t+1)} \frac{\alpha_{t-1}}{\alpha_s} cp.$$

The general recursive value function is then obtained by

$$E^t(\alpha_0) = \sum_{s=0}^{t-1} \delta^s \frac{\alpha_0}{\alpha_s} cp + (1 - p) \sum_{s=t+1}^{T} \delta^s \frac{\alpha_0}{\alpha_s} cp,$$

and hence the gains from monitoring are

$$B(t) = \delta^t \frac{\alpha_0}{\alpha_t} cp + p \sum_{s=t+1}^{T} \delta^s \frac{\alpha_0}{\alpha_s} cp,$$

or equivalently, by Eqs. (A,10) and (A.12), $B(t) = \delta' \alpha_0((1-p)cp + pE(\alpha_t))$. We then need to maximize

$$\max_t \frac{B(t)}{C(t)} \Leftrightarrow \max_t \frac{(1-p)cp + pE(\alpha_t)}{(1-p)^t}. \qquad (A.18)$$

We shall show that the maximum is always achieved at $t = T$, and as t is discrete, it is sufficient to show that

$$\frac{(1-p)cp + pE(\alpha_{t+1})}{(1-p)^{t+1}} - \frac{(1-p)cp + pE(\alpha_t)}{(1-p)^t} > 0, \quad \forall t \leqslant T,$$

which is verified by substituting $E(\alpha_t)$ by $E(\alpha_{t+1})$, using Eq. (A.9). $\quad\square$

Proof of Proposition 8. (i) The social cost of continuation with a sequence of one-period managers is $cp + cp$, where one term reflects the cost of financing and the other term the incentive costs to the manager. The efficient stopping point for this arrangement is therefore α^s. The entrepreneur is only replaced by the manager when it leads to the creation of surplus, which by implication can only occur if exclusive financing with the entrepreneur would only lead to $\hat{T} < T^s$.

(ii) The compensation to the managers is minimized by replacing them in every period as they receive no intertemporal rents in this case. From the viewpoint of the entrepreneur and the venture capitalist, the compensation of new managers is a cost as similar to the monitoring cost and hence the sequential employment of entrepreneur and manager is as in the case of monitoring. $\quad\square$

References

Admati, A.R., Pfleiderer, P., 1994. Robust financial contracting and the role of venture capitalists. Journal of Finance 49, 371–402.

Amit, R., Brander, J., Zott, C., 1997. Why do venture capital firms exist? Theory and Canadian evidence. Working paper. University of British Columbia, Vancouver.

Anand, B., Galetovic, A., 1997. Small but powerful: The economics of venture capital. Working Paper, Yale School of Management.

Berglöf, E., 1994. A control theory of venture capital finance. Journal of Law, Economics and Organization 10, 247–267.

Bergemann, D., Hege, U., 1997. Dynamic Agency, Learning and the Financing of Innovation.

Chan, Y.-S., Siegel, D.R., Thakor, A.V., 1990. Learning, corporate control and performance requirements in venture capital contracts. International Economic Review 31, 365–381.

Cornelli, F., Yosha, O., 1997. Stage financing and the role of convertible debt. Working paper. London Business School.

Gompers, P.A., 1995. Optimal investment, monitoring, and the staging of venture capital. Journal of Finance 50, 1461–1489.

Gompers, P.A., Lerner, J., 1996. The use of covenants: An empirical analysis of venture partnership agreements. Journal of Law and Economics 39, 463–498.

Gorman, M., Sahlman, W.A., 1989. What do venture capitalists do? Journal of Business Venturing 4, 231–248.

Hart, O.D., Moore, J.H., 1994. A theory of debt based on the inalienability of human capital. Quarterly Journal of Economics 109, 841–879.

Hellmann, T., 1998. The allocation of control rights in venture capital contracts. Rand Journal of Economics 29, 57–76.

Jensen, M.C., Meckling, W.H., 1976. Theory of the firm: Managerial behavior, agency costs and ownership structure. Journal of Financial Economics 3, 305–360.

Lerner, J., 1994. Venture capitalists and the decision to go public. Journal of Financial Economics 35, 293–316.

Lerner, J., 1995. Venture capitalists and the oversight of private firms. Journal of Finance 50, 301–318.

Neher, D., 1997. Stage financing: An agency perspective. Working paper. Boston University.

Poterba, J.M., 1989. Venture capital and capital gains taxation. In. Summers, L.H. (Ed.). Tax Policy and the Economy, vol. 3. MIT Press, Cambridge, MA, pp. 47–67.

Sagari, S., Guidotti, G., 1991. Venture capital operations and their potential role in LDC markets. WPS 540. World Bank, Washington, DC.

Sahlman, W.A., 1990. The structure and governance of venture-capital organizations. Journal of Financial Economics 27, 473–521.

Trester, J.J., 1997. Venture capital contracting under asymmetric information. Journal of Banking and Finance 22, 675–699, this issue.

Chapter 10

The Financing of Innovation: Learning and Stopping*

Dirk Bergemann[†] and Ulrich Hege[‡]

†*Department of Economics, and Cowles Foundation of Research in Economics,*
Yale University, New Haven, CT 06520-8268, USA
dirk.bergemann@yale.edu
‡*Toulouse School of Economics, 31015 Toulouse, France*
ulrich.hege@tse-fr.eu

We consider the financing of a research project under uncertainty about the time of completion and the probability of eventual success. We distinguish between two financing modes, namely relationship financing, where the allocation decision of the entrepreneur is observable, and arm's-length financing, where it is unobservable. We find that equilibrium funding stops altogether too early relative to the efficient stopping time in both financing modes. The rate at which funding is released becomes tighter over time under relationship financing, and looser under arm's-length financing. The trade-off in the choice of financing modes is between lack of commitment with relationship financing and information rents with arm's-length financing.

1. Introduction

Motivation. Typically, when decisions are made to start an R&D project or an innovative venture, much uncertainty exists about the chances of the project and about the time and capital needed to secure success. It has been estimated that it takes about 3,000 raw ideas to eventually achieve a single major commercially successful innovation (Stevens and Burley, 1997). Research and development is tantamount to winnowing down a vast amount

---

*This chapter is reproduced from *RAND Journal of Economics*, **28**, 773–795, 1997.

of ideas and alternatives through trial and error, and it is therefore subject to considerable variance in terms of the time and money spent: it may be the 3rd or the 2,997th idea that, when tried out, produces a major success.

The research and development process for a new pharmaceutical product may serve as an illustration. The idea for a new drug is most likely based on some initial and very preliminary research, opening a vast field of possible combinations or ideas. The development itself requires substantial amounts of trials and investments before the value of the initial approach can be assessed. More information will be produced over time as to whether the project will be successful or should be abandoned due to poor results.

The uncertainty about the time and capital required is a source of potential conflict between the financiers providing the capital and the researchers or entrepreneurs carrying out the project. The purpose of the present article is to study agency problems that are directly linked to the open-endedness of the funding in R&D projects, in particular conflicts surrounding the timing of the decision to terminate a research project. Venture capitalists often refer to the decision to discontinue a project as the most important source of conflict between them and startup entrepreneurs, since entrepreneurs almost never want to abandon a project that is under way. Entrepreneurs express a strong preference for continuation regardless of present-value considerations under most circumstances, be it because they are (over)confident or because they rationally try to prolong the search, and they tend to use their discretion to (mis)represent the progress that has been made in order to secure further funding (Cornelli and Yosha, 2003).

Agency conflicts of this kind potentially occur in every situation where a researcher or entrepreneur uses external funding for her R&D efforts, as exemplified by the following three areas. First, they will affect venture capital firms financing high-tech startups. Empirical research on the venture capital industry reveals that venture capitalists are well aware of such problems and that they go to great length to build possible safeguards into their contracts.[1] Second, the optimal financing of research is also a concern for the capital budgeting for R&D expenditures process *within*

---

[1] For example, the following instruments (documented in Sahlman, 1990; Hellmann, 1998; Kaplan and Stromberg, 2002; and Gompers and Lerner, 1999); Venture capitalists retain extensive control rights, in particular rights to claim control on a contingent basis and the right to fire the founding management team; they keep hard claims in form of convertible debt or preferred stock, underpinning the right to claim control and abandon the project; and staged financing and the inclusion of explicit performance benchmarks make it possible to fine-tune the abandonment decision.

a firm. Third, the problems that we investigate arise also for governments, universities, research foundations, and other organizations that sponsor research. They need to evaluate the progress of research projects and to determine the timing for grant renewal or the decision to abandon.

In many cases, the investors in innovative projects will keep a hands-on approach on their investment. Venture capitalists are known to monitor their portfolio companies intensely, for example by soliciting reports and by visiting the company on a monthly basis, and by being involved in the decision making via board membership and other channels and control rights. Besides monitoring, venture capitalists also play an active role as advisors of startup companies, for example by getting involved in the recruitment of key employees and executives (Gompers and Lerner, 1999; Casamatta, 2003). In other words, the venture capital industry provides predominantly *relationship financing*. But not all investors in innovative startups are in fact relationship investors. Informal business angels have been characterized as being less involved in monitoring and to provide only limited advisory services (Barry, 1994; Fenn, 1997). Angel investors, who until recently channeled more money to startups than formal venture capitalists, can thus be viewed as the prototypical *arm's-length* investors in the financing of innovation.[2] Recent theory articles have cast the choice between venture capitalists and angel investors as a choice between informed and less informed investors (Chemmanur and Chen, 2003; Leshchinskii, 2003).

The distinction between relationship investors and arm's-length investors is not limited to venture capital versus angel financing. A difference between better- or worse-informed investors also arises when comparing large and experienced venture capital funds to small and young ones, close to distant, industry specialist versus generalist, independent versus corporate venture fund.

The terms relationship funding and arm's-length funding were originally introduced to distinguish between informed commercial banks and other, less-informed creditors like bondholders. In his seminal article, Rajan (1992) argues that relationship investors may use their exclusive knowledge to subsequently extract rents from successful projects. The value of relationship bank lending has been empirically confirmed (Petersen and Rajan, 1994; Berger and Udell, 1995; Degryse and Ongena, 2005). Debt in

---

[2]Chemmanur and Chen (2003) discuss the flow of funds estimation.

its various forms, including for example trade credit, is an important source of funding for startup firms (Berger and Udell, 1998), and credit markets offer a natural choice between relationship lending and credit with a greater informational distance.[3]

1.1. *Analysis*

We examine a stylized model of the funding of a research project where the merit of an idea and the time and money needed for completion are uncertain. We specifically investigate how stopping decisions are taken in the presence of agency conflicts in the form of entrepreneurial opportunism. The project will succeed with a positive probability in every period in proportion to the volume of funds provided, so that uncertainty is represented by a simple stochastic process. As continued research efforts are undertaken and no success is forthcoming, Bayesian learning will lead to a gradual downgrading of the belief in the project's prospects. The project either ends with a success or will eventually be abandoned in the light of persistent negative news. We assume that the time horizon itself is infinite, to address the essence of the uncertainty about the time to completion, but abandonment will occur in finite time.

The entrepreneur controls the allocation of the funds. She can choose to invest the funds efficiently into the project or to divert them to private ends. This agency conflict is rich because of the dynamic nature of the investment problem. When diverting the funds, the entrepreneur not only enjoys the immediate benefit from consuming the money meant for investment. She also secures the option of continued funding in the future, since nothing can be learned about the project when the funds are not invested as supposed. Thus, the entrepreneur's discretion over the funds is intimately linked to the timing of the abandonment decision.

We consider a sequence of short-term contracts that in our setting is equivalent to requiring that the contract, short or long term, can be renegotiated at all times. Thus, any decision to abandon the project after a given horizon of funding, or to reduce the speed at which funding is released, must be time-consistent. A fundamental contribution of our analysis is the

---

[3]In our model (as in countless others), debt and equity funding are indistinguishable because there is only a single positive cash flow realization. R, that is to be split between entrepreneur and investors.

fact that we embed the agency conflict about the use of resources in the context of an open-ended funding horizon, coupled with the requirement that any equilibrium be renegotiation-proof.

We model relationship and arm's-length financing by distinguishing whether the investor can or cannot observe the entrepreneur's investment decision.[4] With relationship financing, the action of the entrepreneur is *observable*, or more precisely "observable, but not verifiable" as it is usually described in the incomplete-contracts literature, and the environment is at all times one of symmetric information. With arm's-length financing, actions by the entrepreneur are *unobservable*, and we are investigating a standard moral hazard problem between investor and entrepreneur.

The basic conflict between entrepreneur and investor can be described as follows. For the entrepreneur, the project represents the possibility to win a single large prize, but also a stream of rents that she could possibly divert to her private ends. The tension between investing and diverting the funds is accentuated by the fact that the successful completion of the project automatically stops the flow of funds. The direct incentives for the entrepreneur then have to be adequate to offset the possible loss in future rents, hence they have to be increasing in the volume of future funding that the entrepreneur expects in equilibrium.

The combination of an infinite funding horizon and contract renegotiation becomes truly important in light of this fundamentally intertemporal nature of the incentive problem. The longer the funding horizon, the more valuable the entrepreneur's option to increase the probability of access to future funding by diverting funds. The natural candidate for a contractual remedy would be to declare *ex ante* that no financing will be provided after a certain funding horizon, but such a commitment would necessarily not be time-consistent by the nature of the problem presented here.

In the equilibrium analysis, we examine how entrepreneur and investor share the proceeds of the project as a function of the elapsed time, and whether funding is released at the efficient rate and until the efficient stopping point, by first considering *relationship financing* (observable actions). The information about the project is then always common for both parties, and funding renewal is negotiated under symmetric information. As funding continues and the outlook becomes less promising, the participation constraint of the investor leaves less for the direct incentive

---

[4]We would like to thank Patrick Bolton tor a suggestion to include this distinction.

of the entrepreneur. At some point, this residual will fall short of what is needed to provide incentives. The only possible solution is that the investor slows down the release in funds, which happens in the form of a reduced funding rate by the investor. This reduces the entrepreneur's option value of prolonging the project, and the incentive constraint can be met again. As time goes on and the posterior belief decreases, the slowdown in funding becomes more serious, and funding will come down to a trickle as the belief approaches the final abandonment point, which is too early relative to the efficient policy.

As we consider the case of *arm's-length financing* (actions are unobservable), we need to take into account the dynamics of the moral hazard problem. The moral hazard problem about the entrepreneur's decision in the current period translates into an adverse selection problem about beliefs in future periods. For the entrepreneur, control over the investment flow also means control over the information flow, knowing that the private beliefs of entrepreneur and investor about the project can diverge. We find that while the tension between immediate incentives and intertemporal rents remains, there is one subtle, yet important, difference in the value of a deviation for the entrepreneur. With symmetric information, the entrepreneur could renew her proposal after a deviation based on the belief held in the previous period, since nothing in the perception of the project has changed on either side. In contrast, with unobservable actions, the investor will automatically downgrade his belief after a deviation and insist in the continuation game to be compensated on the basis of his belief, which is more pessimistic than warranted. This change in the belief limits the maximal financing horizon, which relaxes the incentive constraint and facilitates funding. On the other hand, the entrepreneur commands an additional information rent since she controls the information flow.

We are then in a position to compare the overall efficiency of arm's-length and relationship financing. We identify the following basic trade-off: Under relationship financing, there is no informational asymmetry, and the information rent that compensates the entrepreneur for her control of the information flow can be saved; but under arm's-length funding, the investor is committed to stick to a finite stopping time, reducing the option value of the entrepreneur to prolong the project through deviations. We find that the second effect always dominates and arm's-length contracts allow for a higher project value.

The equilibrium is shown to be unique in both cases. We require the equilibrium to be weakly renegotiation-proof, meant to capture the inability of entrepreneur and investor to prevent recontracting or renegotiation. More precisely, we first derive the unique Markov equilibrium and then show that this equilibrium is identical to the equilibrium derived under the renegotiation-proofness assumption. We argue that our results are consistent with the typical financing cycle of startup firms where relationship financiers are gradually replaced by arm's-length sources.

In conclusion, the fundamental contracting difficulty in this article comes about by the combination of an infinite funding horizon and the possibility of renegotiation. As our analysis shows, the only way to resolve these conflicts is via *delays* in the financing of innovation — in the form of a slowdown in the release of funds. The temporal occurrence of these delays depends on the informational relationship between investor and entrepreneur: it will be frontloaded for profitable projects under arm's-length funding, and backloaded in all other cases.

1.2. *Related literature*

Besides the articles already mentioned, in particular the ones on relationship financing, our article is related to three distinct strands of the literature. First, it is linked to the literature on the financing of innovation and venture capital, in particular articles focusing on the entrepreneur's discretion to influence the stopping decision in innovative and risky projects. Qian and Xu (1998) observe that soft budget constraint problems of this kind are endemic in bureaucratic systems of R&D funding. Cornelli and Yosha (2003) address the window-dressing of performance signals to have the project continued. Dewatripont and Maskin (1995) note that having multiple investors may be a device to mitigate this problem. In the venture capital literature, moral hazard—driven stopping problems have served as the background to explain the use of such remedies as stage financing, convertible securities, and the dismissal of incumbent managers (e.g., Repullo and Suarez, 2004; Hellmann, 1998). But all contractual devices of this sort are in principle open to renegotiation. To fully account for the relevant time-consistency issues, it seems desirable to go beyond the static (two- or three-period) models employed in this literature. The question is what happens if the horizon is extended and time-consistent devices to commit to an abandonment decision are not available. This is the starting point of our article.

Second, our problem is related to a strand of the incomplete-contracts literature that investigates what is known as the strategic default problem (e.g., Hart and Moore, 1994, 1998; Bolton and Scharfstein, 1996). In this literature, the agent can threaten to default on obligations despite being solvent, and the principal's power to liquidate assets or dismiss the agent enforces payment. Our model is different in that there is only a single cash flow, with uncertain arrival time, and that the outlook of the project under consideration deteriorates over time. We are closest in spirit to three infinite-horizon models of strategic default. Gromb (1994) investigates repetitions of projects à la Bolton and Scharfstein (1996) and finds that the efficiency of the best feasible contracts deteriorates as the horizon of such repeated investments is extended. DeMarzo and Fishman (2000) address long-term contracts and the agent's ability to save. In Fluck's (1998) repeated game, only infinite-maturity outside equity can solve the agent's as well as the principal's incentive problem. Neher (1999) considers a variant of Hart and Moore (1994) where delay via staged financing enables the build-up of collateral, in contrast to our model where delay reduces the present value of future rents.

Third, our article is clearly related to literature on the advantages of arm's-length relationships in agency models. It is closest to Crémer (1995), who shows that better information about the agent's circumstances makes it more difficult for the principal to commit to sanctions. Marquez (1998) explores this idea in the context of financial contracts and relates it to competition as an alternative commitment device.

In an earlier article, Bergemann and Hege (1998) undertake a preliminary analysis of the same basic model, but the present article goes beyond the earlier one in two important dimensions. First, it thoroughly allows for renegotiation, which is entirely ignored in the earlier article. Renegotiation is at the heart of our dynamic agency problem, since the entrepreneur would like to commit *ex ante* to a finite funding horizon when she encounters financing constraint, but such a commitment would necessarily be time-inconsistent. Second, Bergemann and Hege (1998) examine only the case of unobservable actions, whereas here we account for relationship funding, which is more typical for the funding of innovative projects and puts the comparison between relationship and arm's-length financing at center stage.

The article is organized as follows. The model is formally presented in Section 2. We consider a two-period version of our model in Section 3. The equilibrium analysis begins in Section 4 with observable actions of the entrepreneur. In Section 5 we examine equilibrium financing when the allocation decision of the entrepreneur is unobservable to the investor. The structure and efficiency of the equilibria under symmetric and asymmetric information are compared in Section 6. We present some concluding remarks in Section 7. The proofs of all results are relegated to the Appendix.

2. The Model and First-Best Policy

We first describe the project, the investment technology, and the evolution of the posterior beliefs. Next, we introduce the contracting problem. Finally, we derive the efficient stopping posterior.

2.1. *Project with unknown returns*

The entrepreneur owns a project with unknown return. The project is either good with prior probability α_0 or bad with prior probability $1 - \alpha_0$. If the project is good, then the probability of success in a given period is proportional to the funds invested into the project in that period. If the project is bad, then the probability of success is zero independent of the investment flow. The project can at most generate a *single success*, which generates a fixed monetary return $R > 0$.

More precisely, if the project is good and receives an investment flow of $c\gamma$, with $c > 0$, then the probability of success is given by γ. The parameter c thus represents the constant marginal cost of increasing the success probability. We assume that $\gamma \in [0, \lambda]$ with $\lambda < 1$. In consequence, the project can never succeed with certainty in any given period. If the project succeeds, then a cash flow R is realized and distributed among the parties, and the game ends immediately. We refer to an investment flow $\gamma = \lambda$ as *full* or *maximal* funding, and to an investment $\gamma < \lambda$ as *limited* or *restricted* funding.

The uncertainty about the nature of the project is resolved over time as the flow of funds either produces a success or leads to a stopping of the project. The time horizon is discrete and infinite, time periods are denoted by $t = 0, 1, \ldots, \infty$, and the discount factor is $\delta \in (0, 1)$.

The investment process represents an experiment that produces information about the future likelihood of success. The current information is represented by the posterior belief α_t that the project is good. The evolution of the posterior belief α_t, conditional on no success in period t, is given by Bayes' rule as a function of the prior belief α_t and the investment flow γ_t:

$$\alpha_{t+1} = \frac{\alpha_t (1 - \gamma_t)}{1 - \gamma_t \alpha_t}. \tag{2.1}$$

The posterior belief α_t decreases over time if success doesn't arise. The decline in the posterior belief is stronger for larger investments flows γ_t as the agents become more pessimistic about the likelihood of future success. The posterior belief changes only slowly for very precise beliefs about the nature of the project, i.e., if α_t is close to either zero or one. Correspondingly, the event of no success is most informative with diffuse beliefs, or when α_t is close to $1/2$.

We refer to the special case of $\alpha_0 = 1$ as the "certain success project" or the shorter "certain project." With $\alpha_0 = 1$, the posterior never changes as the agents do *not* entertain the possibility that the project might be bad and α_t remains at $\alpha_t = 1$ for all $t \geq 0$. For this obvious reason, we refer to the case of $\alpha_0 = 1$ as the "certain success project" or "certain project." As a consequence of the constant posterior beliefs of $\alpha_t = 1$ for all $t \geq 0$, if funding at $\alpha_0 = 1$ can be provided at the maximal level λ, then it will be provided forever until the project succeeds, hence "certain project."

2.2. *Contracting*

The entrepreneur has initially no wealth and seeks to obtain external funds to realize the project. Financing is available from a competitive market of investors, which is represented in the model by a *single* investor who can only accept or reject contract proposals by the entrepreneur. Entrepreneur and investor share initially the same assessment about the likelihood of success represented by the prior belief α_0. The funds are supplied by the investor, and the entrepreneur controls the allocation of the funds. She can either invest the funds into the project or divert the capital flow to her private ends. We assume that the entrepreneur consumes any diverted funds immediately, i.e., she cannot accumulate funds in order to finance the project on her own in the future.[5]

The time structure in every period t is as follows. At the beginning of period t the entrepreneur can offer the investor a *share contract* s_t and a success probability $\gamma_t \geq 0$. The share s_t represents the share of the

---

[5] Alternatively, the same incentive constraints would arise if the efficient investment of the funds required costly effort by the entrepreneur. In this case the entrepreneur evidently cannot save any funds that have not been invested. We are grateful to an anonymous referee for pointing out this interpretation. In both cases, one could imagine that one unit diverted from the funds increases the entrepreneur's utility by a monetary equivalent of only $\beta < 1$ units. The smaller is β, the less attractive is diversion and hence the less acute is the agency problem that we study.

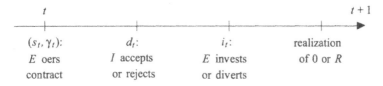

Figure 10.1.   Timeline of Events.

entrepreneur in the proceeds if the project succeeds in period t. The investor receives the remaining share $1 - s_t$. The restriction to share contracts is without loss of generality due to the binary nature of the project. After the contract proposal, the investor can decide whether to accept or reject the new contract (decision d_t). If he accepts the contract, then he provides the entrepreneur with the requested funds $c\gamma_t$ in period t to support the development of the project. If he rejects the contract, then a new proposal can be made by the entrepreneur in the subsequent period. Finally, and conditional on funding, the entrepreneur decides whether to invest the funds in the project or divert them to her private ends (decision i_t). The sequence of decisions in every period is illustrated by Figure 10.1.

2.3. *First-best policy*

The project should receive funds as long as current expected returns of the investment exceed costs:

$$\alpha_t \gamma_t R - c\gamma_t \geq 0.$$

As both return and cost are linear in γ_t, it follows that if investment is socially efficient, it should occur at the maximal level, or $\gamma_t = \lambda$. The project should receive its final investment at the lowest α_T where the current net return is positive:

$$\alpha_T \lambda R - c\lambda \geq 0,$$

which yields a socially efficient stopping point, described in terms of the posterior belief, at

$$\alpha_T \lambda R - c\lambda = 0$$

or, equivalently,

$$\alpha^* = \frac{c}{R}. \tag{2.2}$$

The social value of the project, denoted by $V(\alpha_t)$, is given by a familiar dynamic programming equation:

$$V(\alpha_t) = \max_{\gamma \in [0,\lambda]} \{\alpha_t \gamma R - c\gamma + \delta(1 - \gamma\alpha_t)V(\alpha_{t+1}(\gamma))\}.$$

The value of the program can be decomposed into the flow payoffs and continuation payoffs. The flow payoffs are the returns multiplied by the current probability of success minus the investment costs. The continuation payoffs arise conditional on no success, or with probability $(1 - \alpha_t\lambda)$, in which case the future is assessed at a new posterior, namely α_{t+1}. The efficient stopping condition (2.2) can be recovered from the dynamic programming equation at α_T by setting $V(\alpha_{T+1}) = 0$.[6]

The value of the project under the first-best policy can be determined as

$$V(\alpha_0) = \alpha_0 \lambda (R - c) \frac{1 - \delta^{T^*}(1 - \lambda)^{T^*}}{1 - \delta(1 - \lambda)} - (1 - \alpha_0)c\lambda \frac{1 - \delta^{T^*}}{1 - \delta},$$

where T^* is the maximal number of periods such that the updated belief after T^* periods of full funding is still weakly above α^*, or

$$\frac{\alpha_0(1 - \lambda)^{T^*}}{\alpha_0(1 - \lambda)^{T^*} + 1 - \alpha_0} \geq \alpha^*.$$

3. A Simple Two-Period Model

This section presents some basic insights and trade-offs in a simple two-period example with $t = 0, 1$. We compare the funding decision in the symmetric- and asymmetric-information environments.

In this finite-horizon setting, we can analyze the contracting equilibrium by backward induction. We shall start with the symmetric environment. Suppose then that in the final period, investor and entrepreneur share a common posterior α_1 that describes the probability that the project is good. If the entrepreneur makes a funding proposal (s_1, γ_1), then the investor will

---

[6]The *intertemporally* optimal stopping point is thus determined by a *static* revenue condition. The stopping point condition does not include any intertemporal element in terms of a value of information, as the posterior belief conditional on no success declines deterministically and thus there is no option value in the evolution of the posterior belief.

accept it only if his participation constraint is satisfied, or

$$\gamma_1 \alpha_1 (1 - s_1) R \geq c\gamma_1, \tag{3.1}$$

and if the incentive constraint for the entrepreneur is satisfied as well:

$$\gamma_1 \alpha_1 s_1 R \geq c\gamma_1. \tag{3.2}$$

The incentive constraint simply requires that the expected return from allocating the funds properly exceeds the value of a diversion. Due to the linear structure of the model, we can safely neglect partial diversion. In other words, inequality (3.2) also guarantees that the incentive constraints for partial diversion, or

$$\gamma_1 \alpha_1 s_1 R \geq c\gamma + (\gamma_1 - \gamma)\alpha_1 s_1 R,$$

are satisfied for all $\gamma \in [0, \gamma_t]$. From the participation and the incentive constraints, it follows that financing is provided only if

$$\gamma_1 \alpha_1 R \geq 2c\gamma_1 \Leftrightarrow \alpha_1 \geq \frac{2c}{R}.$$

We can make a first observation regarding the social efficiency of the funding decision. We just saw that equilibrium funding stops at

$$\alpha_S \triangleq \frac{2c}{R}, \tag{3.3}$$

whereas the socially efficient stopping point is given by the posterior belief

$$\alpha^* = \frac{c}{R}.$$

Hence equilibrium funding ends too early in comparison with the socially efficient funding policy. The divergence between equilibrium and social stopping arises from the rent the entrepreneur can extract due to the nonverifiability of her allocation decision.

In equilibrium, the entrepreneur offers a break-even contract to the investor that solves his participation constraint (3.1) at equality, or

$$1 - s_1 = \frac{c}{\alpha_1 R} \Leftrightarrow s_1 = \frac{\alpha_1 R - c}{\alpha_1 R}.$$

The share of the entrepreneur depends on the posterior belief about the quality of the project. Her expected profit in period $t = 1$ is given by

$$\alpha_1 \gamma_1 R - c\gamma_1 \geq 0.$$

It follows that she will always suggest a maximal funding level $\gamma_1 = \lambda$. We can then describe the expected profit of the entrepreneur as a function of the posterior belief α_1 as follows:

$$V_E(\alpha_1) = \begin{cases} \alpha_1 \lambda R - c\lambda & \text{if} \alpha_1 \geq 2c/R, \\ 0 & \text{if} \alpha_1 < 2c/R. \end{cases}$$

Going backward to period $t = 0$, we find that the participation constraint of the investor remains unchanged (except that the posterior α_1 is replaced by the prior α_0):

$$\alpha_0 \gamma_0 (1 - s_0) R \geq c\gamma_0.$$

In contrast, the incentive constraint of the entrepreneur contains an intertemporal element, or

$$\alpha_0 \gamma_0 s_0 R + (1 - \alpha_0 \gamma_0) \delta V_E(\alpha_1) \geq c\gamma_0 + \delta V_E(\alpha_0). \tag{3.4}$$

The left-hand side of inequality (3.4) represents the discounted value of the funding proposal (s_0, γ_0) on the equilibrium path. If the entrepreneur allocates the funds properly, then the project is successful with probability $\alpha_0 \gamma_0$, and an agreed share s_0 of the cash flow R is paid to the entrepreneur. With the remaining probability, namely $1 - \alpha_0 \gamma_0$, the project is not successful. In this case, there is still a chance to realize the project tomorrow. The net value of this option for the entrepreneur is given by $V_E(\alpha_1)$, taking into account the update in the belief to α_1 following the failure to succeed in period 0.

The right-hand side of the inequality (3.4) represents the value of a diversion, which now arises from two sources. First, there is a direct private benefit of $c\gamma_0$, but second, the failure to pursue the project today leads to the opportunity to realize it tomorrow. It should be noted that in contrast to the left-hand side of the inequality, the opportunity of pursuing the project tomorrow now arises with certainty, as the diversion of the funds guaranteed that the project could not be realized in period 0. Moreover, the prior α_0 is not updated, as no information regarding the project was generated in period 0.

By the same argument developed for $t = 1$, the entrepreneur will offer a break-even contract with maximal funding to the investor. Using the continuation value $V_E(\alpha_1)$, we can write the incentive constraint of the entrepreneur as follows:

$$\alpha_0 \lambda R - c\lambda + (1 - \alpha_0\lambda)\delta(\alpha_1\lambda R - \lambda c) \geq c\lambda + \delta(\alpha_0\lambda R - \lambda c).$$

The posterior α_1 is determined via Bayes' rule (see (2.1)) as

$$\alpha_1 = \frac{\alpha_0(1 - \gamma_0)}{1 - \alpha_0\gamma_0} < \alpha_0, \tag{3.5}$$

and after replacing the posterior belief α_1 in (3.5), we get the following condition on the prior for funding over two periods to be possible:

$$\alpha_0 \geq \frac{2c}{R(1 - \delta\lambda) + \delta\lambda c}.$$

We make several observations, First, for $\delta = 0$, the funding condition in period 0 is identical to the funding condition in period 1, as the entrepreneur acts purely myopically. But for all $\delta > 0$, the funding condition becomes more severe in period 0, as the denominator is a convex combination of R and c (with $R > c$). Moreover, as δ increases, more weight should be given to c and the funding condition increases in severity. In consequence, it follows that for all α_0 satisfying

$$\frac{2c}{R(1 - \delta\lambda) + \delta\lambda c} > \alpha_0 \geq \frac{2c}{R}, \tag{3.6}$$

there will be no equilibrium funding in period 0, though the project will eventually be funded in period 1. The reason for the equilibrium delay in the funding decision emerges from the incentive constraint. If the project will get funded anyhow in period 1, the entrepreneur has a strong incentive to divert the funds in period 0 and thereby guarantee himself further funding and yet maintain the possibility of success in period 1. This equilibrium delay can be prevented only if the discount factor is very low, or if the project has a very high probability of success, so that the immediate rewards outweigh the future benefits.

We consider now the asymmetric-information environment. Along the equilibrium path, entrepreneur and investor maintain symmetric information, as the investor uses his equilibrium belief about the allocation decision of the entrepreneur. The sole difference arises along the possible deviation of the entrepreneur, i.e., the off-the-equilibrium-path decision. Thus, if we consider the incentive constraint of the entrepreneur in period 0, it now reads

$$\alpha_0 \gamma_0 s_0 R + (1 - \alpha_0 \gamma_0) \delta V_E(\alpha_1) \geq c\gamma_0 + \delta \frac{\alpha_0}{\alpha_1} V_E(\alpha_1), \qquad (3.7)$$

where the only, but important, difference compared with the symmetric-information incentive constraint (3.4) arises on the right-hand side. Following a deviation by the entrepreneur in period 0, the investor observes only that the project did not succeed and continues to update his prior with the information coming from the failure to succeed. The investor therefore continues to hold his equilibrium belief α_1 formed by Bayes' rule as in (3.5). In consequence, he still wishes to be rewarded in period 1 as if the true posterior were α_1. In contrast, the entrepreneur knows that the belief α_1 is too pessimistic because conditional on the funds being diverted in period 0, no new information about the project arose, and she maintains the correct belief α_0. As the value in period 1 comes from a successful realization, the term α_0/α_1 corrects the misperception of the investor. In sum, the correct expected value of the entrepreneur following the diversion of the funds is given by $(\alpha_0/\alpha_1)V_E(\alpha_1)$. As the true probability of success is α_0, but the share is negotiated on the basis of the less optimistic belief $\alpha_1 < \alpha_0$, it follows that

$$\frac{\alpha_0}{\alpha_1} V_E(\alpha_1) < V_E(\alpha_0).$$

Since the investor's required value is larger in the arm's-length case following the investor's informational handicap, the arm's-length environment makes a deviation of the entrepreneur less attractive. This effect, which we call the *commitment effect*, is at the heart of our analysis, and we will encounter it extensively in our main analysis. Solving (3.7) in a manner similar to (3.6) leads to the condition that financing only in the second period is possible if

$$\frac{2c - \dfrac{1}{1-\lambda}\delta\lambda c}{R(1-\delta\lambda) - \dfrac{\lambda}{1-\lambda}\delta\lambda c} \geq \alpha_0 \geq \frac{2c}{R}.$$

In comparison with the symmetric environment, we find that the set of prior beliefs without financing constraints is larger in the asymmetric case, since

$$\frac{2c}{R(1-\delta\lambda) + \delta\lambda c} > \frac{2c - \dfrac{1}{1-\lambda}\delta\lambda c}{R(1-\delta\lambda) - \dfrac{\lambda}{1-\lambda}\delta\lambda c}.$$

Thus, as a consequence of the commitment effect, more projects will receive immediate funding in the asymmetric-information environment than in the symmetric-information environment. Costly delay is more frequent under symmetric information than under asymmetric information, even though it may arise in both cases.

This simple two-period model generates two important results: (i) it demonstrates the difficulties in creating efficient investment arrangements when the option of continued future funding undermines the incentives for current investment, and (ii) it shows that asymmetric information can improve the efficiency of the equilibrium by acting as a commitment device for the investor.

The two-period model contains an artificially strong discontinuity regarding the provision of incentives. In period 0, incentives to invest the funds into the project are weak, as further funding is forthcoming for sure in the period 1. In contrast, in period 1, incentives to invest the funds are strong, as by assumption no further occasions to realize the project arise. This stark contrast between the periods gave rise to the extreme nature of the equilibrium, with no funding in period 0 and maximal funding in period 1. By itself, the two-period model thus gives little guidance as to how the results carry over to a model with a more general time horizon, finite or infinite. In the remainder of the article, we shall analyze an infinite time horizon model and examine the role of the discount factor δ in the equilibrium provision of funds. The removal of the artificial final period will lead to the elimination of the discontinuity in the funding volume observed in the two-period model. We will obtain an intertemporal characterization of the funding volume that will evolve smoothly over time and display not only minimal and maximal funding levels, but typically intermediate funding at various stages of the project.

4. Relationship Financing

In this section we analyze contracting with symmetric information, that is, the entrepreneur's actions are observable (but not verifiable) for the investor. We first define the concept of Markov-perfect equilibrium. We next investigate the properties of this equilibrium and show that it coincides with the weakly renegotiation-proof equilibrium. Finally, we discuss how the contracting results would be affected if the agents could commit to long-term contracts yet could recontract in every period.

4.1. *Equilibrium*

In the environment with observable actions, the information of entrepreneur and investor is symmetric in every period. Formally, we can describe the strategies in the game as follows. Let H_t denote the set of possible public histories up to, but not including, period t. A proposal strategy by the entrepreneur is given by

$$s_t : H_t \to \mathbb{R}, \gamma_t : H_t \to [0, \lambda].$$

A decision rule by the investor is a mapping from the history and the contract proposal into a binary decision to reject ($d_t = 0$) or to accept ($d_t = 1$):

$$d_t : H_t \times \mathbb{R} \times [0, \lambda] \to \{0, 1\}.$$

Finally, an investment policy by the entrepreneur is given by

$$i_t : H_t \times \mathbb{R} \times [0, \lambda] \to \{0, \gamma_t\}.$$

The above policies all describe pure rather than mixed strategies and indeed throughout the article we focus on pure-strategy equilibria. A generic public history of the game is denoted by $h_t \in H_t$ and is simply a realized sequence of offers, funding, and investment decisions:

$$h_t = \{s_0, \ldots, s_{t-1}; \gamma_0, \ldots, \gamma_{t-1}; d_0, \ldots, d_{t-1}; i_0, \ldots, i_{t-1}\}.$$

The evolution of the posterior belief α_t is not included in the history, as it can be inferred from the sequence of public funding and investment decisions by Bayes' rule. By default, updating occurs only conditional on failure of the project, since the game ends as soon as the project succeeds and realizes the return R. Thus given any prior α_0, an arbitrary history h_t uniquely determines the current posterior belief $\alpha_t = \alpha(h_t)$. For a given quadruple $\{s_t, \gamma_t, d_t, i_t\}_{t=0}^{\infty}$ of strategies, denote the value function of the entrepreneur at the beginning of period t by $V_E(h_t)$ and the value function of the investor by $V_I(h_t)$.

We restrict our attention initially to Markovian equilibria where strategies are allowed to depend only on the payoff-relevant part of the history of the game, which in this model is fully represented by a single state variable, the posterior belief α_t in every period t. The Markov equilibrium outcome is subsequently shown to be identical to a (weakly) renegotiation-proof equilibrium outcome. Formally, a Markov-perfect equilibrium following Maskin and Tirole (2001) is defined as

Definition 1 (Markov-perfect equilibrium): A Markov-perfect equilibrium is a subgame-perfect equilibrium

$$\{s_t^*, \gamma_t^*, d_t^*, i_t^*\}_{t=0}^{\infty},$$

such that the sequence of policies satisfies $\forall h_t \in H_t, \forall h_{t'}' \in H_{t'}, \forall s_t, s_{t'}', \forall \gamma_t,$ $\gamma_{t'}', \forall d_t, d_{t'}'$:

$$\alpha(h_t) = \alpha(h_{t'}')$$
$$\Rightarrow s_t^*(h_t) = s_{t'}^*(h_{t'}'), \gamma_t^*(h_t) = \gamma_{t'}^*(h_{t'}');$$
$$\alpha(h_t) = \alpha(h_{t'}'), \quad s_t = s_{t'}', \gamma_t = \gamma_{t'}'$$
$$\Rightarrow d_t^*(h_t, s_t, \gamma_t) = d_{t'}^*(h_{t'}', s_{t'}', \gamma_{t'}');$$
$$\alpha(h_t) = \alpha(h_{t'}'), \quad s_t = s_{t'}', \gamma_t = \gamma_{t'}', d_t = d_{t'}'$$
$$\Rightarrow i_t^*(h_t, s_t, \gamma_t, d_t) = i_{t'}^*(h_{t'}', s_{t'}', \gamma_{t'}', d_{t'}'). \tag{4.1}$$

Thus, a Markov-perfect equilibrium imposes the requirement that the continuation play be identical after any two histories h_t and $h_{t'}'$ with an identical belief, $\alpha(h_t) = \alpha(h_{t'}')$, noting that the particular history h_t may differ from $h_{t'}'$ in its date $t \neq t'$, and/or in its past actions. Since the moves in any period are sequential, the relevant state for the investor is not only the belief about α_t but must also include the entrepreneur's contract offer, and similarly for the entrepreneur's final capital allocation decision. In the Markovian setup, we can write the entrepreneur's and the investor's value functions simply as a function of the current belief, $V_E(\alpha_t)$ and $V_I(\alpha_t)$, respectively.

4.2. *Analysis*

Consider the situation of the investor at an arbitrary point of time. He receives a proposal by the entrepreneur to fund a project for the current period in exchange for shares in the proceeds of the project should it succeed in the current period. As the current contract commits neither investor nor entrepreneur to any future course of action, the investor is willing to accept the proposal (s_t, γ_t) as long as the expected returns are nonnegative, or

$$\alpha_t(1 - s_t)\gamma_t R \geq c\gamma_t. \tag{4.2}$$

The inequality then represents the participation constraint of the investor. However, the expected returns can materialize only if the entrepreneur decides to put the funds to work in the project rather than divert them to her private ends. This is the incentive problem of the entrepreneur.

Consider first the final period where the entrepreneur receives funding in equilibrium. This final period will arise when the belief $\alpha = \alpha_t$ has deteriorated so much that it will be impossible to solicit any future funds. In that final period the entrepreneur has to choose between investing and diverting, or

$$\alpha_t s_t \gamma_t R \geq c\gamma_t. \tag{4.3}$$

Exactly as shown in the two-period model above, jointly the inequalities (4.2) and (4.3) imply that for any funding to occur in equilibrium, the expected flow return from the investment must cover both the cost of the funds for the investor and the opportunity costs for the entrepreneur,

$$\alpha_t \gamma_t R \geq 2c\gamma_t.$$

The critical posterior belief at which funding will certainly cease is therefore given by α_S defined by the identity (3.3) above, where α_S is twice as large as the efficient stopping belief, $\alpha_S = 2\alpha^*$.

In all preceding periods, the incentive constraint for the entrepreneur has to take into account her future opportunities. As in the two-period model (see equation (3.4)), the constraint can be represented in terms of her value function:

$$\alpha_t \gamma_t s_t R + (1 - \alpha_t \gamma_t) \delta V_E(\alpha_{t+1}) \geq c\gamma_t + \delta V_E(\alpha_t). \tag{4.4}$$

For any sharing rule s_t, the entrepreneur can either invest the funds (left-hand side) or divert them (right-hand side). She benefits from complying with the contract via two different sources. Either the project succeeds in period t, giving her a share of the return $s_t R$, or it does not succeed, in which case she has access to future rounds of funding. With a funding flow of γ_t, the probabilities of these events are $\alpha_t \gamma_t$ and $1 - \alpha_t \gamma_t$ respectively. If the project fails in the current period, then the posterior belief declines to α_{t+1}. The alternative action for the entrepreneur is to simply divert the funds today and then face a similar problem tomorrow as the state of the project remains unchanged. Therefore, the equilibrium can be characterized by a sequence of participation constraints for the investor (as in (4.2)) and a sequence of incentive constraints for the entrepreneur (as in (4.4)).

In equilibrium, the entrepreneur will never leave the investor with more net value than is necessary to obtain the funding. The equilibrium share s_t^* is therefore determined by the exact fulfillment of the participation constraint. The investor receives zero net utility when the participation constraint (4.2) is binding and we can solve for the equilibrium sharing rule:

$$s_t^* = \frac{\alpha_t R - c}{\alpha_t R}. \tag{4.5}$$

We refer to contracts that leave the investor with zero net utility as break-even contracts and observe that the break-even share is independent of the funding flow.[7] Using (4.5), we may rewrite the incentive constraint (4.4) as

$$\alpha_t \gamma_t R - c\gamma_t + (1 - \alpha_t \gamma_t) \delta V_E(\alpha_{t+1}) \geq c\gamma_t + \delta V_E(\alpha_t). \tag{4.6}$$

The dynamic incentive constraint shows that the return R has to be sufficiently high to cover the static as well as the dynamic incentive costs. The static part simply says that the gross return $\alpha_t \gamma_t R$ has to be sufficiently large to compensate the investor for his cost $c\gamma_t$ as well as dissuade the entrepreneur from diverting the current flow, an additional $c\gamma_t$. The

---

[7]We prove the result that only break-even contracts will be offered in equilibrium formally in Lemma 2 as a property that holds for all subgame-perfect equilibria, and not only for Markovian equilibria.

dynamic part accentuates the incentive problem. By rewriting (4.6) as

$$\alpha_t \gamma_t (R - \delta V_E(\alpha_{t+1})) \geq 2c\gamma_t + \delta(V_E(\alpha_t) - V_E(\alpha_{t+1})), \qquad (4.7)$$

it says that the net return for the entrepreneur after paying for the static cost is $R - \delta V_E(\alpha_{t+1})$ rather than R itself. This is natural, as the success today leads to an end of the project and preempts future payouts to the entrepreneur. On the other hand, even if the project is funded yet without success, then the future value of the project is determined by α_{t+1} rather than α_t. By diverting the funds, the entrepreneur could escape the downgrade that constitutes the dynamic part of incentive costs.

A lower level of funding γ_t certainly affects both sides of the inequality (4.7), but at different rates. A lower equilibrium level reduces the gains on the left-hand side but only with the weight $\alpha_t < 1$, whereas on the right-hand side it affects the current costs of funding and, what is more important, the intertemporal incentive costs. A low funding level γ_t reduces the current value function, $V_E(\alpha_1)$, but a lower γ_t also decreases the difference between α_t and α_{t+1} and with it the difference of the value functions $V_E(\alpha_t) - V_E(\alpha_{t+1})$. This argument suggests that a marginal decrease in γ_t always leads to a larger decrease in the right-hand side of the incentive constraint than the left-hand side of the incentive constraint.

Given the impact of the funding rate on the intertemporal incentive constraints, we might then ask whether it is conceivable that the project receives full funding with $\gamma_t = \lambda$ until the last period $\alpha_T = \alpha_S$. With period T being the last period of funding, the continuation value at α_{t+1} would be zero, $V_E(\alpha_{T+1}) = 0$, and the value function at α_T would be given by

$$V_E(\alpha_T) = \alpha_T \lambda R - c\lambda.$$

But if we insert these two continuation valuations into the incentive constraint (4.6), we are led to a contradiction because we obtain, after rearranging,

$$\alpha_T \lambda R \geq 2c\lambda + \delta(\alpha_T \lambda R - c\lambda).$$

The inequality clearly cannot be satisfied at $\alpha_T = \alpha_S$, as we already have $\alpha_S \lambda R = 2c\lambda$ and $\alpha_S \lambda R - c\lambda = c\lambda$. This argument already indicates that in equilibrium, funding has to (eventually) slow down from the maximal level λ to a lower level $\gamma_t < \lambda$, which can be sustained by the incentive constraint (4.6).

In light of this finding, we might ask whether full funding with $\gamma_t = \lambda$ will ever occur. To answer this question, it is helpful to consider the limit case of the incentive constraint (4.7) with $\alpha_0 = 1$, the case of the "certain project." Using the fact that with the certain project, $\alpha_0 = 1$, the posterior beliefs remain constant and equal to the prior beliefs, the intertemporal incentive constraint (4.6) can then be written as

$$\lambda R - c\lambda + (1 - \lambda)\delta V_E(1) \geq c\lambda + \delta V_E(1). \tag{4.8}$$

In equilibrium, the value function of the entrepreneur is the discounted and risk-adjusted sum of the per-period returns:

$$V_E(1) = \frac{\lambda R - \lambda c}{1 - \delta(1 - \lambda)}.$$

We then insert $V_E(1)$ into the incentive constraint (4.8),

$$\lambda R - c\lambda + (1 - \lambda)\delta \frac{\lambda R - \lambda c}{1 - \delta(1 - \lambda)} \geq c\lambda + \delta \frac{\lambda R - \lambda c}{1 - \delta(1 - \lambda)}, \tag{4.9}$$

and obtain a necessary and sufficient condition on R for full funding to occur:

$$R \geq 2c + \lambda c \frac{\delta}{1 - \delta}. \tag{A}$$

By considering the deviation option of the entrepreneur, the sources of the determination for the critical value R become more transparent. If we consider $\alpha_0 = 1$, then if full funding is to occur in equilibrium, we know that the entrepreneur can always guarantee herself at least $c\lambda$ in every period, since a diversion would simply lead to renewed attempts of funding in next period. With $\alpha_0 = 1$ and $V_E(\alpha_t) = V_E(\alpha_{t+1})$, we can rearrange the incentive constraint (4.8) to read

$$\lambda R \geq 2\lambda c + \delta \lambda V_E(1). \tag{4.10}$$

Now the entrepreneur has the option to divert the funds in every period, which secures him at least a perpetuity of $c\lambda$, hence $V_E(1) = \lambda c/(1 - \delta)$. The inequality (4.10) then states that the return from the project, R, has to cover at least $c + c$, which are the current costs for entrepreneur and investor, and the increment in the perpetual rent that is at the discretion of the entrepreneur via his option to deviate and thereby to increase by λ the probability of getting access to the perpetual rent. The rent of the entrepreneur thus has a contemporaneous and an intertemporal component.

Condition (A) turns out to be a key condition in our analysis. A project where the payoff R is large enough relative to the marginal cost c of success, the contemporaneous rent c, and the increment in the perpetual rent $\lambda c[\delta/(1-\delta)]$ so as to satisfy (4.10) is termed a "high-return" project (in incentive-adjusted terms), as opposed to a "low-return" project where the condition is violated.

Definition 2: (*low- and high-return projects*). The project is a *low-return* project if $R < 2c + \lambda c[\delta/(1-\delta)]$, and it is a *high-return* project if $R \geq 2c + \lambda c[\delta/(1-\delta)]$.

We can now ask what happens to equilibrium funding when the critical inequality (A) is violated. Staying with the special case of the certain project and hence $\alpha_0 = 1$, the solution is almost apparent from the analysis of the incentive constraint (4.8). For funding to occur, the incentive constraint has to be reestablished. This requires that the rent arising from a diversion is lowered, which can only mean that the funding level is lowered to an appropriate level $\gamma_t^* < \lambda$. In fact, we can obtain the equilibrium funding $\gamma_t = \gamma_t^*$ by solving the incentive constraint (4.9) as an equality. The solution γ_t^* to the equality (4.9) is also the unique equilibrium funding level. While it is by now clear that a funding level above γ_t^* could not be sustained in equilibrium, we shall now argue that any funding level strictly lower than γ_t^* could not form an equilibrium either. We observe first that if the funding level is set below γ_t^*, then the incentive constraint (4.8) would again hold as a strict inequality. Because the entrepreneur then has slack in his incentive constraint, he could ask the investor for a higher funding level, even λ, by "bribing" the investor and offering him a slightly larger share of the surplus than the break-even contract. The investor would agree, since he would be offered a strictly positive net surplus, and given the continuation values, he would be assured that the entrepreneur's incentive constraint still holds. In consequence, an interior level of funding $\gamma \in (0,1)$ can be sustained in equilibrium only if the incentive constraint of the entrepreneur is met as an equality. The insight that an interior level of funding is always associated with a binding incentive constraint is of course not restricted to $\alpha_t = 1$, but is valid more generally for all $\alpha_t \leq 1$. We now summarize the equilibrium funding decisions for the certain project.

Theorem 1: (*certain project*).

(i) *If the certain project has high returns, then it receives full funding in all periods.*

(ii) *If the certain project has low returns, then it receives restricted funding in all periods:*

$$\gamma_t^* = \frac{1-\delta}{\delta c}(R - 2c) < \lambda, \forall t.$$

The equilibrium funding level of the low-return project is increasing in the final value R and decreasing in cost c. An increase in the discount factor δ increases the value of the option to divert, and hence the investor responds in equilibrium by decelerating the flow of funds as it becomes more difficult to satisfy the incentive constraint.

The equilibrium in the general case of an evolving α_t can now almost be conceived by replacing the constant value R by the dynamically evolving value $\alpha_t R$. As long as $\alpha_t R$ is sufficiently large, unrestricted funding will be possible, yet as $\alpha_t R$ decreases, funding will have to decrease as well so as to maintain the incentive constraint of the entrepreneur.

The critical value of the posterior belief, denoted by $\overline{\alpha}$, at which funding will become restricted can in fact be easily obtained. Under the hypothesis that the value function of the entrepreneur is just equal to, but not larger than, the perpetual rent that the entrepreneur can secure by deviating forever, the incentive constraint allows us to solve for the value function as

$$V_E(\alpha_t) = V_E(\alpha_{t+1}) = \frac{\lambda c}{1-\delta}.$$

The critical posterior belief is then computed by solving (4.6) with the continuation values given by the perpetual rents, or

$$\overline{\alpha}\lambda R - \lambda c + (1 - \overline{\alpha}\lambda)\delta \frac{\lambda c}{1-\delta} = \lambda c + \lambda c \frac{\delta}{1-\delta},$$

from which we can infer the threshold to be

$$\overline{\alpha} = \frac{2c}{R - \lambda c \dfrac{\delta}{1-\delta}}. \tag{4.11}$$

Full funding at α_t is possible if and only if $\alpha_t \geq \overline{\alpha}$. We can summarize our findings as follows.

Theorem 2 (Relationship funding): *The Markov-perfect equilibrium is unique, and funding is always provided until $\alpha_T = \alpha_S$.*

(i) *If the project has low returns, then it receives restricted funding in all periods.*

(ii) *If the project has high returns, then it receives full funding for all $\alpha_t \geq \overline{\alpha}$ and restricted funding for all $\alpha_t < \overline{\alpha}$.*

(iii) *If funding is restricted, then γ_t is strictly decreasing in t.*

The sharing rule associated with the equilibrium is given by (4.5). For high-return projects, there is a critical value $\overline{\alpha}$ such that the project will receive maximal funding as long as $\alpha_t \geq \overline{\alpha}$. Low-return projects that have insufficient returns to cover current costs and perpetual rents, even at $\alpha_t = 1$, are then always subject to restricted funding. In both cases, the volume of funding will decrease over time with the deterioration in the expected returns $\alpha_t R$.

4.3. *Renegotiation-proof equilibrium*

The notion of a Markov equilibrium imposes a stationarity requirement on the offer and acceptance decisions of the agents. In the context of our model, the Markovian assumption has a natural interpretation as a consistency requirement on the process of (re)negotiation between the two parties; namely, the Markovian condition requires that entrepreneur and investor find an arrangement mutually acceptable whenever they have found the same arrangement acceptable in the past and absent any new information about the nature of the project.

We now strengthen this intuition by considering arbitrary history-dependent policies instead. However, we impose a condition that the policies must be *time-consistent* in the sense that if the players can coordinate on a certain policy in a subgame, they are also able to coordinate on the same policy in any other subgame where the circumstances are the same, that is, if they share the same belief about α_t. In other words, we assume that they are able to avoid any Pareto-inferior outcome under exactly the same circumstances. To this end, we invoke the refinement of weakly renegotiation-proof equilibrium first suggested by Farrell and Maskin (1989) for repeated games. The adaptation of the equilibrium notion to dynamic games is straightforward.

Definition 3 (weakly renegotiation-proof): A subgame-perfect equilibrium $\{s_t^*, \gamma_t^*, d_t^*, i_t^*\}_{t=0}^{\infty}$ is weakly renegotiation-proof if there do not exist continuation equilibria at some h_t and $h_{t'}'$ with $\alpha(h) = \alpha(h_{t'}')$ and $h_t \neq h_{t'}'$ such that $V_E(h_t) \geq V_E(h_{t'}')$ and $V_I(h_t) \geq V_I(h_{t'}')$, with at least one strict inequality.

The renegotiation considered here occurs between time periods. It is conceptually different from renegotiation in static principal-agent models as considered by Fudenberg and Tirole (1990) or Hermalin and Katz (1991). The notion of weakly renegotiation-proof is often interpreted as an internal consistency requirement. Indeed, Farrell and Maskin (1989) suggested a strengthening of the notion by defining as strongly renegotiation-proof any weakly renegotiation-proof profile with none of its continuation equilibria being strictly Pareto dominated by another weakly renegotiation-proof profile. This distinction is immaterial to our argument, as they all coincide in this sequential move game with symmetric information.

Theorem 3 (equivalence): *The unique Markov-perfect equilibrium is identical to the unique weakly renegotiation-proof equilibrium.*

The equivalence can be illustrated by the following simple example of equilibrium strategy profiles that form a subgame-perfect, but not renegotiation-proof, equilibrium. The example also shows that renegotiation-proofness indeed imposes restrictions on the equilibrium set.

(i) *The entrepreneur offers in each period break-even contracts and invests funds if her private value to invest exceeds her private value to divert. If the investor has observed no deviations in the past, then the investor provides maximal funding if he breaks at least even and can expect the entrepreneur to invest. He rejects any contract proposal that doesn't meet the above conditions.*

(ii) *If there were any deviations in the past, then entrepreneur and investor pursue the stationary equilibrium strategies as described earlier.*

Consider these strategy profiles for a certain project, $\alpha_0 = 1$, with low returns, $R < 2c + \lambda c[\delta/(1 - \delta)]$. By Theorem 1, the Markov-perfect equilibrium permits only restricted funding with

$$\gamma^* = \frac{1 - \delta}{\delta c}(R - 2c),$$

and the resulting equilibrium value for the entrepreneur is

$$V_E(1) = \frac{1}{\delta}(R - 2c).$$

In contrast, suppose part (i) of the strategy profile indeed forms a subgame-perfect equilibrium. Then the value for the entrepreneur would be

$$\hat{V}_E(1) = \frac{\lambda R - \lambda c}{1 - \delta(1 - \lambda)}.$$

As it is immediately verified that offer and acceptance strategies in (i) have the best-response property if the entrepreneur subsequently invests, it remains to verify her incentive constraint, which can be written as

$$\frac{\lambda R - \lambda c}{1 - \delta(1 - \lambda)} \geq c\lambda + (R - 2c),$$

which leads after the obvious cancellations to

$$R \leq 2c + \lambda c \frac{\delta}{1 - \delta}.$$

This is precisely the condition of low returns that we imposed for this example. Thus, the outlined strategy profile would allow full funding everywhere along the equilibrium path by relying on the stationary equilibrium as an off-the-equilibrium punishment path. The strategy profiles rely in an obvious way on continuation plays that are not renegotiation-proof. As the investor receives zero utility on and off the equilibrium path, it is sufficient to note that the entrepreneur receives different values on and off the equilibrium path to find that the strategy profile is not weakly renegotiation-proof.

4.4. *Bargaining and long-term contracts*

We have so far imposed two strong assumptions on the structure of contracts, namely: (i) that all the bargaining power rests with the entrepreneur and (ii) that only short-term contracts are possible. Here, we briefly discuss the robustness of the results if we relax either of the assumptions.

Bargaining power. Consider first a change in the bargaining power. Suppose that the investor now makes all contract offers and the entrepreneur accepts or rejects all proposals. Still, the participation constraint of the investor and the incentive constraint of the entrepreneur have to hold in any equilibrium. As long as both constraints are binding, they uniquely determine the equilibrium. In the model, both constraints were binding in the region $\alpha_t < \overline{\alpha}$, where only reduced funding was feasible. Therefore, nothing would change in this region with the redistribution of the bargaining power: The pattern of funding and the distribution of the surplus remain the same.

A change in the allocation can arise only if one of the inequalities is not binding any more, and the change would then pertain to the distribution of the surplus. Now in the benchmark model, the incentive constraint was slack only in the region of optimistic posterior beliefs $\alpha_t \geq \overline{\alpha}$. But there, maximal funding was guaranteed anyhow, and a shift in bargaining power would not alter that. It follows that the funding pattern in equilibrium would remain unaffected by a change in the bargaining power.

Long-term contracts. In this article we analyze short-term contracts in which the participation constraint of the investor has to hold in every period. Consider then an extension of the contracting space to allow for long-term contracts that are valid for any arbitrary horizon of T periods. We maintain our requirement that existing contracts can be renegotiated or new contracts be concluded in every future period. Formally, this allows us to substitute the sequence of participation constraints that had to be met in every period by a single intertemporal participation constraint that has to hold only at the time of entry into the contract. In contrast, the sequence of period-by-period incentive constraints needs to be maintained, as they guarantee the proper allocation of investment funds in every period.

The advantages of a long-term contract reside naturally with a possible intertemporal smoothing of the entrepreneur's expected payoffs. More precisely, it is then possible to reallocate the entrepreneur's payoff stream over time so as to make it coincide with the stream that is necessary to guarantee incentives. Thus, in every moment where the project's current net cash flow $(\alpha_t R - c)\lambda$ exceeds what is needed to maintain the entrepreneur's incentives, or as long as $\alpha_t > \overline{\alpha}$, there is a surplus that can be reallocated. The entrepreneur concedes a larger share to the investor today in exchange for receiving a larger share herself in the future. Conversely, the investor makes profits initially in return for a commitment to *subsidize* the project later on, when α_t falls below $\overline{\alpha}$.

Hence, for high-return projects with an optimistic prior belief, $\alpha_0 > \overline{\alpha}$, long-term contract can strictly improve upon the allocation of short-term contracts. The region where full funding is provided can then be extended beyond the threshold $\overline{\alpha}$. The funding pattern, however, would remain as before, insofar as the project would receive full funding initially and then switch to lower funding levels. By contrast, we observe that as soon as funding is reduced, our previous argument of the intertemporal smoothing effect of long-term contracting never applies and long-term contracts can do no better than short-term contracts. Therefore, if $\alpha_0 \leq \overline{\alpha}$, which is always the case for low-return projects, there is no role for long-term contracts and the equilibrium is unaffected by the larger set of feasible contracts. The reason is that then the project in no instance has high-enough returns to generate surplus beyond participation and incentive constraints.

5. Arm's-Length Financing

In this section we assume that the investment decision by the entrepreneur is unobservable by the investor. We first consider Markovian equilibria, to maintain consistent equilibrium conditions across different informational structures. Below we define a Markov sequential equilibrium and then present the equilibrium analysis. We then show that the Markovian restriction is immaterial, as the unique Markov sequential equilibrium coincides with the unique sequential equilibrium. Finally, we again discuss robustness when changes in the bargaining power or long-term contracts are introduced.

5.1. *Equilibrium*

As we consider the contracting problem with unobservable actions by the entrepreneur, the observable history of the game begins to differ for entrepreneur and investor. The entrepreneur still observes all past realizations of the strategic choices, and a private history h_t for her is still given by

$$h_t = \{s_0, \ldots, s_{t-1}; \gamma_0, \ldots, \gamma_{t-1}; d_0, \ldots, d_{t-1}; i_0, \ldots, i_{t-1}\}.$$

The investor, however, is not able to observe the action of the entrepreneur anymore. Along any arbitrary sample path without success, the observable history to him is given by

$$\hat{h}_t = \{s_0, \ldots, s_{t-1}; \gamma_0, \ldots, \gamma_{t-1}; d_0, \ldots, d_{t-1}\}.$$

Denote by $\widehat{H}_t$ the set of all possible such histories. In consequence, the evolution of the posterior belief may differ for entrepreneur and investor. We continue to denote by α_t the entrepreneur's posterior belief based on the history $h_t, \alpha_t \triangleq \alpha(h_t)$. We refer to the belief that the investor holds at time t and after observing the restricted public history $\hat{h}_t$ as $\widehat{\alpha}_t \triangleq \widehat{\alpha}(\hat{h}_t)$, which will depend on the observed history $\hat{h}_t$ as well as on the investor's belief about the entrepreneur's past investment behavior, $\{\hat{i}_0, \ldots, \hat{i}_{t-1}\}$. By Bayes' law there is a one-to-one relationship between the estimate regarding the entrepreneur's past investments $\{\hat{i}_0, \ldots, \hat{i}_{t-1}\}$ and the belief about $\widehat{\alpha}(\hat{h}_t)$. The estimate regarding $\{\hat{i}_0, \ldots, \hat{i}_{t-1}\}$ depends on the incentives provided through the past and future share contracts $\{s_0, \ldots, s_t, \ldots\}$. As before, updating occurs only conditional on current failure of the project, since the game ends as soon as the project succeeds.

Because entrepreneur and investor observe different histories, the payoff-relevant part of the history is now represented by two state variables, the two (possibly different) posterior beliefs about the likelihood of success, α_t and $\widehat{\alpha}_t$. The suitably adapted Markovian equilibrium concept can then be stated as follows.

Definition 4: (*Markov sequential equilibrium*). A Markov sequential equilibrium is a sequential equilibrium

$$\{s_t^*, \gamma_t^*, d_t^*, i_t^*\}_{t=0}^{\infty}$$

if $\forall h_t \in H_t, \forall h_{t'}' \in H_{t'}, \forall \widehat{h}_t \in \widehat{H}_t, \forall \widehat{h}_{t'}' \in \widehat{H}_{t'}$ and $\forall s_t, s_{t'}', \forall \gamma_t, \gamma_{t'}', \forall d_t, d_{t'}'$:

$$\alpha(h_t) = \alpha(h_{t'}')$$
$$\Rightarrow s_t^*(h_t) = s_{t'}^*(h_{t'}'); \gamma_t^*(h_t) = \gamma_{t'}^*(h_{t'}');$$
$$\widehat{\alpha}(h_t) = \widehat{\alpha}(h_{t'}'), \quad s_t = s_{t'}', \gamma_t = \gamma_{t'}';$$
$$\Rightarrow d_t^*(\widehat{h}_t, s_t, \gamma_t) = d_{t'}^*(\widehat{h}_{t'}', s_{t'}', \gamma_{t'}');$$
$$\alpha(h_t) = \alpha(h_{t'}'), \quad s_t = s_{t'}', \gamma_t = \gamma_{t'}', d_t = d_{t'}'$$
$$\Rightarrow i_t^*(h_t, \gamma_t, s_t, d_t) = i_{t'}^*(h_{t'}', \gamma_{t'}', s_{t'}', d_{t'}'). \tag{5.1}$$

The Markovian sequential equilibrium ensures that the continuation strategies are time-consistent and identical after any history with an identical pair of rationally updated beliefs, α_t and $\widehat{\alpha}_t$. The Markovian restrictions contained in (5.1) are equivalent to the ones formulated earlier in (4.1), with the exception being that the underlying histories and beliefs differ for entrepreneur and investor.

5.2. *Analysis*

Before we go to the details of the analysis, it might be useful to describe intuitively where the differences in the equilibrium incentives arise and how they matter for the equilibrium funding. Conditional on receiving the funds, the entrepreneur still has the option to either invest or divert the funds. The differences arise in how entrepreneur and investor evaluate these different options. Clearly, the investor is willing to provide the funds only if he is convinced that the funds will be directed to the project. Consider then the counterfactual of a diversion of the funds by the entrepreneur. Following a deviation, the entrepreneur would know that the funds didn't benefit the project and hence a failure of the project to succeed in this period will not surprise her at all. In contrast, for the investor, a deviation remains a counterfactual and thus he is downgrading his beliefs about the future value of the project, as the current failure induces a downward change in his beliefs. Thus a deviation, as an off-the-equilibrium behavior by the entrepreneur, leads to a divergence in the posterior about the future likelihood of success. More precisely, the entrepreneur maintains her estimate $\alpha_{t+1} = \alpha_t$, whereas the investor continues to update his belief to a lower value $\widehat{\alpha}_{t+1} < \widehat{\alpha}_t$. Such a divergence of beliefs per se could not arise in the environment with observable actions.

How does the possibility of divergent beliefs influence the equilibrium incentives? Ultimately the divergence imposes more discipline on the funding decisions of the investor and therefore tends to ease the funding problem. Because a deviation will still lead to a lowering in the posterior belief of the investor, he will ask for a larger share of the return R. This leads directly to a higher cost of obtaining funds from the point of view of the entrepreneur. The option of delaying the investment decision until the next period thus becomes less attractive.

We examine next how these changes will be reflected in the participation and incentive constraints. The participation constraint of the investor remains unchanged at

$$\widehat{\alpha}_t s_t \gamma_t R \geq \gamma_t c,$$

with the exception that it is evaluated at $\widehat{\alpha}_t$ rather than α_t. The modification is immaterial along the equilibrium path, as $\alpha_t = \widehat{\alpha}_t$. However, the incentive constraint of the entrepreneur changes to reflect the divergence of the beliefs off the equilibrium path. Formally, the incentive constraint is given by

$$\alpha_t \gamma_t s_t R + (1 - \alpha_t \gamma_t)\delta V_E(\alpha_{t+1}) \geq c\gamma_t + \delta V_E(\alpha'_{t+1}, \widehat{\alpha}_{t+1}),$$

where, momentarily, we express the value function as determined by both beliefs. We observe that off the equilibrium path, the posterior belief of entrepreneur and investor diverge, and hence the value function off the equilibrium path depends on the specific beliefs of the entrepreneur, α'_{t+1}, and the investor, $\widehat{\alpha}_{t+1}$. Along the equilibrium path, the entrepreneur invests the funds into the project, and a diversion occurs off the equilibrium path. The continuation value conditional on a diversion is therefore described by two different beliefs, the correct belief α'_{t+1} of the entrepreneur and the incorrect belief $\hat{\alpha}_{t+1}$ of the investor. After a one-period deviation, the off-the-equilibrium-path belief of the entrepreneur, α'_{t+1}, is given simply by $\alpha'_{t+1} = \alpha_t$, whereas the investor holds the "equilibrium" belief $\hat{\alpha}_{t+1} = \alpha_{t+1}$. Hence, off the equilibrium path, the investor will accept only contracts that will break even under his posterior belief $\hat{\alpha}_{t+1}$ and all subsequent updates of his posterior. How then does this affect the continuation value of the entrepreneur off the equilibrium path? The answer is rather straightforward. It will be as if her continuation value would indeed be determined by the belief of the investor $\hat{\alpha}_{t+1} = \alpha_{t+1}$, but because the entrepreneur privately knows that the true posterior, conditional on diversion, is still given by α_t, she simply exchanges the posterior belief $\hat{\alpha}_{t+1} = \alpha_{t+1}$ of the investor for her own, $\alpha'_{t+1} = \alpha_t$. This allows us to relate the off-the-equilibrium-path value function to the on-the-equilibrium-path value function as follows:

$$V_E(\alpha'_{t+1} = \alpha_t, \hat{\alpha}_{t+1} = \alpha_{t+1}) = \frac{\alpha_t}{\alpha_{t+1}} V_E(\alpha_{t+1}). \tag{5.2}$$

After all, the value function of the entrepreneur, on and off the equilibrium path, is simply the discounted sum of success probabilities, or

$$V_E(\alpha_{t+1}) = R(\alpha_{t+1}s_{t+1}\gamma_{t+1} + \delta(1 - \gamma_{t+1}\alpha_{t+1})\alpha_{t+2}s_{t+2}\gamma_{t+2}$$
$$+ \delta^2(1 - \gamma_{t+2}\alpha_{t+2})\alpha_{t+3}s_{t+3}\gamma_{t+3} + \cdots). \tag{5.3}$$

After replacing the conditional success probabilities $\alpha_{t+2}\gamma_{t+2}, \alpha_{t+3}\gamma_{t+3}, \cdots$ with the unconditional success probabilities viewed from α_{t+1}, and making repeated use of Bayes' formula,

$$\alpha_{t+2} = \frac{\alpha_{t+1}(1 - \gamma_{t+1})}{1 - \gamma_{t+1}\alpha_{t+2}},$$

we can rewrite the value function (5.3) as a sequence of unconditional probabilities,

$$V_E(\alpha_{t+1}) = \alpha_{t+1}R(\gamma_{t+1}s_{t+1} + \delta(1 - \gamma_{t+1})\gamma_{t+2}s_{t+2}$$
$$+ \delta^2(1 - \gamma_{t+1})(1 - \gamma_{t+2})\gamma_{t+3}s_{t+3} + \cdots), \tag{5.4}$$

that only invoke the current belief α_{t+1} and the current and future flow probabilities $\gamma_{t+1}, \gamma_{t+2}, \gamma_{t+3}, \ldots$. This shows how the identity (5.2) arises. The terms of the contract, namely $s_{t+1}, s_{t+2}, \ldots$, are conditional on a deviation of the entrepreneur and they are determined by the belief of the investor. In consequence, the accepted shares conditional on the deviation will be identical to the shares the parties would agree upon on the equilibrium path as $\widehat{\alpha}_{t+1} = \alpha_{t+1}$, hence the term $V_E(\alpha_{t+1})$. But privately, the entrepreneur knows that the true probability of future success is α_t rather than α_{t+1}. Hence the ratio term α_t/α_{t+1} corrects for the fact that the value function $V_E(\alpha_{t+1})$ underestimates the true probability of success when $\alpha'_{t+1} = \alpha_t$.

We can therefore rewrite the incentive constraint in the asymmetric information environment as follows:

$$\alpha_t \gamma_t R - \gamma_t c + (1 - \alpha_t \gamma_t)\delta V_E(\alpha_{t+1}) \geq c\gamma_t + \frac{\alpha_t}{\alpha_{t+1}}\delta V_E(\alpha_{t+1}). \qquad (5.5)$$

The reader may realize that the left-hand side of the inequality, which represents the on-the-equilibrium-path behavior, remains identical to the one in the observable environment (see expression (4.4)). The change occurs on the right-hand side of the inequality, or the off-the-equilibrium-path behavior. The flow value of a diversion still contains the immediate benefit of $c\gamma_t$. But as the investor continues to believe that an investment occurred, he will only accept future proposals as if an investment today had indeed occurred. In consequence, the value function of the entrepreneur will have to evolve (almost) as if the current failure had to be attributed to the project rather than the diversion of the entrepreneur. There is one benefit, however, for the entrepreneur from the continued updating. She will know that the true probability is still α_t rather than α_{t+1}. Thus instead of multiplying the future probability of success with α_{t+1}, she is certain that it is indeed α_t.

The intertemporal incentive constraint (5.5) may be rewritten, after canceling the obvious terms, as

$$\alpha_t \gamma_t R - c\gamma_t \geq c\gamma_t + \gamma_t + \frac{\alpha_t}{\alpha_{t+1}}\delta V_E(\alpha_{t+1}). \qquad (5.6)$$

In general, then, funding toward the end of the lifetime of the project will become easier with an arm's-length relationship. But now a complementary problem may arise at the beginning of the project. If indeed funding will be generous close to the end of the project, then the entrepreneur may have less incentive at the beginning of the project to invest funds, as the future will offer plenty of opportunities to generate success. Thus an easing of the incentive constraint near the end of the project may tighten the incentive constraint at the beginning of the project, when the assessment in terms of the beliefs α_t is still very positive. This indicates that the monotonicity in the funding volume may indeed be reversed with unobservable actions. We first state the results and then comment on some of the equilibrium properties. The threshold posterior belief at which the funding policy will switch is denoted by $\bar{\bar{\alpha}}$ and given by

$$\bar{\bar{\alpha}} = \frac{2c - \dfrac{2c\lambda}{1-\delta} + \dfrac{\lambda\delta c}{1-\delta}}{R - \dfrac{2c\lambda}{1-\delta}}.$$

Theorem 4 (arm's-length funding): *The Markov sequential equilibrium is unique, and funding stops at $\alpha_T = \alpha_S$.*

(i) *If the discount factor is low, $\delta < (2 - 2\lambda)/(2 - \lambda)$, then*

   (a) *a low-return project receives restricted funding at all times, and*
   (b) *a high-return project receives full funding for $\alpha_t \geq \bar{\bar{\alpha}}$ and restricted funding for $\alpha_t < \bar{\bar{\alpha}}$.*

(ii) *If the discount factor is high, $\delta \geq (2 - 2\lambda)/(2 - \lambda)$, then*

   (a) *a low-return project receives restricted funding for $\alpha_t > \bar{\bar{\alpha}}$ and full funding for $\alpha_t < \bar{\bar{\alpha}}$, and*
   (b) *a high-return project receives full funding at all times.*

The difference in the equilibrium funding policies between arm's-length and relationship financing are now easily discussed. For low discount factors, or $\delta < (2 - 2\lambda)/(2 - \lambda)$, the equilibrium funding over time displays exactly the same dynamics under arm's-length and relationship funding. Yet a difference emerges for high discount factors, or $\delta \geq (2 - 2\lambda)/(2 - \lambda)$. The absence of a commitment problem with arm's-length funding completely restores the efficiency of the funding decision until $\alpha_T = \alpha_S$ for a high-return project. For a project with low returns, it at least allows the reestablishment of funding efficiency close to α_S. The above condition on the discount factor can be restated symmetrically as a condition on the winning probability as

$$\delta \geq \frac{2 - 2\lambda}{2 - \lambda} \Leftrightarrow \lambda \geq \frac{2 - 2\delta}{2 - \delta}. \tag{B}$$

For convenience, we shall henceforth refer to large and small discount factors depending on whether δ does or does not satisfy condition (B).

Definition 5 (small and large discount factors): The discount factor is said to be *small* if $\delta < (2 - 2\lambda)/(2 - \lambda)$, and it is said to be *large* if $\delta \geq (2 - 2\lambda)/(2 - \lambda)$.

We observe that an increase in the winning probability λ leads to a lower bound on the discount factor and vice versa. The role of the discount factor, or for that matter of the success probability, should not come entirely as surprise given our earlier discussion on the inefficiencies in relationship funding. The lack of commitment became especially damaging to the social efficiency of the equilibrium when the discount factor was high; the intertemporal rent of the entrepreneur was then high as well, and the only possible equilibrium resolution of this conflict required the investor to slow down the release of the funds. As the informational asymmetry allows the investor to overcome this lack of commitment, the discrepancy between arm's-length and relationship funding arises precisely where arm's-length funding was most affected by the lack of commitment. The efficiency is thus reestablished for high-return projects and improves the funding volume for low-return projects with high discount factors.

Corollary 1 (funding evolution):

(i) *If the discount factor is small, then the funding volume is decreasing over time.*

(ii) *If the discount factor is large, then the funding volume is increasing over time.*

Whether funding will eventually become unrestricted as α_t is sufficiently close to one is again determined by the high-return condition, $R \geq 2c + [\delta\lambda/(1 - \delta)]c$, that we encountered earlier in the symmetric environment. The reappearance of the condition is plausible because for α_t sufficiently close to one, the differences in the beliefs of entrepreneur and investor after a deviation become arbitrarily small, as a current failure barely changes the very optimistic view of the investor. More precisely, for any fixed funding flow $\gamma_t > 0$, the difference in the belief before and after a single unsuccessful investment γ_t,

$$\lim_{\alpha_t \to 1} (\alpha_t - \alpha_{t+1}) = \lim_{\alpha_t \to 1} \left(\alpha_t - \frac{\alpha_t(1 - \gamma_t)}{1 - \alpha_t\gamma_t} \right) = 0,$$

converges to zero when the initial belief at α_t is arbitrarily optimistic about the likelihood of eventual success. The asymmetry in the information between entrepreneur and investor is thus arbitrarily small when the incoming belief α_t is close to one, and in consequence the asymmetric contracting problem becomes arbitrarily close to the symmetric contracting problem. These arguments can be retraced formally by comparing the two incentive constraints, the symmetric incentive constraint (see (4.6)),

$$\alpha_t \gamma_t R - c\gamma_t + (1 - \alpha_t \gamma_t)\delta V_E(\alpha_{t+1}) \geq c\gamma_t + \delta V_E(\alpha_t),$$

and the asymmetric incentive constraint (see (5.5)),

$$\alpha_t \gamma_t R - c\gamma_t + (1 - \alpha_t \gamma_t)\delta V_E(\alpha_{t+1}) \geq c\gamma_t + \frac{\alpha_t}{\alpha_{t+1}}\delta V_E(\alpha_{t+1}).$$

We now observe that as $\alpha_t \to 1$, we have

$$\lim_{\alpha_t \to 1}(\alpha_t - \alpha_{t+1}) = 0 \quad \text{and} \quad \lim_{\alpha_t \to 1}\frac{\alpha_t}{\alpha_{t+1}} = 1.$$

By continuity of the value function, it then follows from the above that

$$\lim_{\alpha_t \to 1}(V_E(\alpha_t) - V_E(\alpha_{t+1})) = 0,$$

and hence symmetric and asymmetric incentive constraints become identical as $\alpha_t \to 1$. Moreover, as $\alpha_t \to 1$, the posterior beliefs will change for a long period of time only very slowly as the funds fail to generate a success. This means that for a long period of time, symmetric and asymmetric incentive conditions will almost be identical and hence the values generated through them will be very close as the more distant events matter less due to discounting.

5.3. *Sequential equilibrium*

The characterization of the equilibrium seemed to rely strongly on the Markovian assumption. In particular, we represented the incentive problem of the entrepreneur through a Bellman equation. But there is one crucial difference to relationship financing: As the investor continues to lower his belief every time he provided funds yet did not observe success, he reaches the posterior belief α_S after finitely many positive funding decisions. This is true on the equilibrium path as well as off. Thus, in contrast to the symmetric environment, the horizon of the game effectively becomes finite. This allows us to analyze the game by backward induction over a finite horizon. As the (static) equilibrium in any final period where $\alpha_T \geq \alpha_S$, yet $\alpha_{T+1} < \alpha_S$, is unique, we can then construct the equilibrium recursively. Moreover, the stage game has a unique equilibrium for any given continuation payoff. In a sequential equilibrium, the investor's beliefs $\alpha(\widehat{h}_t)$ are tied down according to Bayes' rule after all possible histories, including off-the-equilibrium-path histories, which is sufficient to guarantee the uniqueness of the continuation equilibrium everywhere. It follows that backward induction leads to a unique sequential equilibrium independent of the Markov assumption.[8]

The construction of the equilibrium in Theorem 4 is thus in fact constructing the unique sequential equilibrium, where the posterior belief α_t merely serves to summarize the beliefs of the players for a given history, but not as a restriction on the conditioning of the strategies.

Corollary 2: *The unique Markov sequential equilibrium is the unique sequential equilibrium.*

Thus, since the equilibrium play in a sequential equilibrium always follows a finite-horizon logic in the environment with unobservable actions, there is no need to refer to any formal concept of renegotiation-proofness in order to make sure that the outcome is the one that we have in mind, where it is impossible throughout to find a Pareto-improving continuation play by rescinding the equilibrium contracts.

5.4. *Bargaining and long-term contracts*

As before, we may ask how sensitive the equilibrium results are to the specifics of the contracting model, in particular the distribution of bargaining power and the restriction to short-term contracts.

Bargaining power. Suppose now that the investor makes all the offers and the entrepreneur can respond only with acceptance or rejection. With a small discount factor, the equilibrium funding pattern is comparable to the one under symmetric information, and changes in bargaining structure do not at all affect the funding volume. With a large discount factor, the funding pattern remains in its qualitative properties but the equilibrium displays fewer inefficiencies. The reason is that whenever funding is unrestricted, the project's expected cash flow leaves some free surplus after participation and incentive constraints are satisfied. The question is then whether a better overall allocation is achieved if this surplus is distributed to the investor rather than the entrepreneur. Giving the surplus to the investor means a lower expected equilibrium payoff for the entrepreneur. Recall that the minimum value of the entrepreneur that guarantees incentive compatibility is recursively constructed. Thus, a lower expected compensation in the future (since the free surplus is given to the investor) translates into a lower

---

[8] Perfect Bayesian equilibrium cannot be used here, since adverse selection is a consequence of the entrepreneur's unobservable actions, not of chance moves of nature.

option value of diverting and hence into a lower minimum compensation today. The incentive problem of the entrepreneur in the current period is eased. In consequence, a change in the bargaining power would allow an increase of the area where maximal funding is provided and would increase the volume of funding over the entire horizon.

Long-term contracts. The reasons why there can be benefits from adopting (renegotiation-proof) long-term contracts are closely related. As long-term contracts replace the flow participation constraint of the investor with a single initial constraint, intertemporal smoothing is possible. With a large discount factor, the project is initially constrained, and a free surplus arises toward the end of the relationship. As discussed for changes in the bargaining power, allocating this surplus to the investor lowers the entrepreneur's expected future value, and hence eases the current incentive problem. Moreover, in return for making expected profits toward the end, the investor can agree to *subsidize* the project elsewhere, i.e., to provide full funding while accepting a current share $(1 - s_t)\alpha_t \lambda R$ that falls short of the investment flow $c\lambda$. The question is then when to schedule this subsidy phase. The answer is that this subsidy phase should be scheduled as soon as possible, but the requirement that the equilibrium be immune to renegotiation is an effective constraint on this. As a consequence, if the project has low returns, the intertemporal smoothing arrangement will allow an early start and an extension of the final phase where full funding can be provided, but only limited funding is possible initially. If the project has high returns, then full funding is possible from the start and can be continued even beyond α_S.

By contrast, with a small discount factor, the dynamics of the funding pattern are reversed and resemble roughly the picture with observable actions. The project is constrained toward the end, necessitating a slowdown in the release of funds. The intertemporal smoothing option of long-term contracts allows to prolong the initial full-funding phase. But as soon as the surplus is exhausted, the optimal contract reverts back to the sequence of contracts described above, with the same funding volume. For low-return projects or if α_0 is so small that short-term contracts never allow for full funding, then there is never a surplus to redistribute intertemporally, and long-term contracting cannot improve upon short-term contracts.

6. Observability and the Commitment to Stop

In the previous two sections, we gave separate accounts of the environment with observable actions and with unobservable actions. We provide a comparison of the two cases in this section that we interpret to reflect the initial choice between relationship financing and arm's-length financing when the project is set up.

We will conduct this comparison by maintaining the assumption that once the financing mode is chosen, the investor is committed to the informational environment throughout. The source of this commitment is not explained in the model, and we will informally discuss possible transition from one funding mode to the other at the end of this section.

The immediate benefit of relationship funding is the absence of private information during the development of the relationship. It means in particular that the design of the contract does not have to account for the extraction of private information. It thus circumvents the learning rent that is associated with the private information. We have shown above that under relationship financing, three different components of rents must be awarded to the entrepreneur to make her willing to invest and risk early success, namely the contemporaneous rent equal to the immediate gain in consumption that a deviation affords, the intertemporal rent to compensate for the option to receive sure continued financing when deviating, and finally the learning rent driven by the fact that only the entrepreneur knows whether or not something has actually been learned about α_t. By contrast, only two of these components were present in the case of relationship financing, since there was no need for the learning rent.

The (implicit) cost of the relationship funding resides with the ability of the entrepreneur to restart the relationship after she diverted funds in previous periods. As the investor can't commit himself to refuse a contract with positive net payoffs, the entrepreneur was essentially able to extract a rent equivalent to an infinite stream of funds λc, worth $\lambda c[\delta/(1-\delta)]$. In contrast, the asymmetry in the arm's-length relationship reduces the ability of the entrepreneur to renegotiate at favorable terms and hence weakens the incentives for the entrepreneur to delay investment into the project.

With this basic trade-off between the two funding modes, we find that the possible cost of an arm's-length relationship, namely the learning rent, is small in comparison to the benefit from commitment. Therefore, we arrive at the following result.

Theorem 5 (Comparison): *For all posterior beliefs α_t the funding volume is larger under arm's-length than under relationship financing.*

To gain more insight into this result, it is helpful to discuss separately the case of a low and high discount factor. With a small discount factor, we showed earlier that the funding pattern decreases in both informational environments. We then show that for all posterior beliefs α_t, the funding volume is higher with arm's-length funding. With a large discount factor, we have shown that the evolution of the funding levels displays opposite signs: γ_t is (weakly) decreasing in t with observable actions, but it is increasing in t with unobservable actions. Here, in order to show that arm's-length funding always occurs at a higher volume than with relationship funding, it is sufficient to compare the initial equilibrium funding volume, which again can be shown to be higher with arm's-length funding.

The clear Pareto ranking between the two financing modes is a rather striking result. From a naive point of view it may appear counterintuitive, since it says that the financing mode with an informational asymmetry separating financier and entrepreneur is more efficient. While it is well known that asymmetric information may offer advantages in a principal-agent model (see Crémer, 1995), our analysis shows that this argument is particularly prevalent when the agency relationship is open-ended and the agent has the option to extend it over a very long horizon.

The dynamic model shows that the relative advantage of arm's-length funding increases over time. In the beginning, there may be no difference between the speed with which funds can be released, especially for high-return projects. But as projects become protracted and prospects become relatively poor, arm's-length funding eventually offers an increasing advantage compared with relationship funding, which does not offer a commitment to stop at a given time and therefore allows the entrepreneur to threaten an infinite series of deviations even when only few profitable rounds of experimentation are left. As bargaining power shifts to the investor, the advantage of arm's-length contracts toward the end of the projects increases even further.

The typical financing cycle of business startups involves close relationships with financiers early on, and this is exemplified by the activity of venture capitalists who not only provide capital, but also monitor the projects very closely and get involved as advisors (see Gompers and Lerner. 1999; Casamatta, 2003). There is evidence that the value of relationship funding decreases over the typical financial cycle of an innovative firm: for example, in venture funded projects, syndicates tend to grow and to include more uninformed investors later on (Hege, Palomino, and Schwienbacher, 2003). There are also empirical findings that in banking relationships, financiers assume a more passive role over time (e.g., Ongena and Smith, 2001). As projects mature and require more capital, while having more tangible assets and research results to offer, their funding sources tend to become more diverse: funding typically starts with equity-dominated venture financing and adds more and more debt-like instruments overtime as the firm grows and its funding needs expand (Berger and Udell, 1998). Thus, even before a successful technology startup reaches financial maturity and is funded by genuine outside investors, such as dispersed shareholders or bondholders, many ventures already go through a process of gradually decreasing the reliance on relationship financiers. This pattern is consistent with the comparison of the two funding modes in our model, which lends support to the notion that the financial cycle of innovative projects evolves from more-informed to less-informed investors.

7. Conclusion

In this article we present a dynamic agency model in which time and outcome of the project were uncertain. The model prominently features three aspects that together are defining elements for a wide class of agency problems of research and development activities: (i) the eventual returns from the project are uncertain, (ii) more information about the likelihood of success arrives with investment into the project, and (iii) investor and entrepreneur (innovator) cannot commit to future actions. The analysis focuses on Markovian equilibria, but we showed that this is a rather mild or even immaterial restriction in the context of the model. The equilibrium analysis proceeds sequentially, starting with symmetric information and ending with asymmetric information. The funding level was determined endogenously and depends on the returns of the project, the discount factor, and the informational asymmetry between entrepreneur and investor.

The impatience of the entrepreneur is an important determinant in the volume of funding, as the severity of the incentive constraint increased with the discount factor. This is in contrast to the results in the theory of repeated moral hazard games, where discount factors close enough to one often allow the equilibrium set to reach the efficiency frontier. In addition, we showed that the recursive structure of the incentive constraint leads to distinct funding dynamics under asymmetric information, where with large discount factors, the incentive constraint tends to actually relax over time and allow a larger funding rate as the project approaches its terminal period.

The basic trade-off between arm's-length and relationship financing revealed in this article is that arm's-length financing offers the advantage that the investor is implicitly committed to a finite stopping horizon, while relationship financing saves up on the learning rent because investor and entrepreneur update beliefs symmetrically.

Finally, some possible extensions of our model should be mentioned. First, a worthwhile extension is to consider the equilibrium behavior when there are *competing projects*, formed by different entrepreneurs. As competition may limit the rent of each entrepreneur, parallel research for an identical objective might be an arrangement that improves efficiency despite the inevitable duplication of R&D efforts. In a *winner-takes-all* competition, the threat of preemption by a competitor will limit the intertemporal rent of each entrepreneur. Similarly, launching competing research teams may increase the *ex ante* value for an organization despite the multiplication of research efforts.

Second, it is conceivable that the entrepreneur may initially own some, perhaps small, investment funds. We then might ask how inside and outside funds are optimally mixed over time. We are confident that a delayed use of the entrepreneur's equity can be shown to be optimal in some cases. This should notably be the case if the entrepreneur's funds help alleviate financing constraints when the promise of the project deteriorates, as is typically the case under relationship financing. The open-horizon principal-agent model developed here should allow us to analyze the relative merits of these different incentive tools and their role in mitigating the contracting problems of compounded information rents.

Appendix

Proofs of Theorems 2–4 and Corollary 2 follow.

Lemma A1: *In every subgame-perfect equilibrium no funding occurs for* $\alpha_t < \alpha_S$.

Proof: The proof is by contradiction. Suppose there exists a subgame-perfect equilibrium with funding in period t, or

$$\alpha_t \gamma_t s_t R + (1 - \alpha_t \gamma_t)\delta V_E(h_{t+1}) \geq c\gamma_t + \delta V_E(h'_{t+1}),$$

or

$$\alpha_t \gamma_t s_t R - \gamma_t c + (1 - \alpha_t \gamma_t)\delta V_E(h_{t+1}) \geq \delta V_E(h'_{t+1}), \qquad (A.1)$$

and

$$\alpha_t \gamma_t (1 - s_t) R - \gamma_t c + (1 - \alpha_t \gamma_t)\delta V_1(h_{t+1}) \geq \delta V_1(h''_{t+1}), \qquad (A.2)$$

yet

$$\alpha_t \gamma_t R - 2c\gamma_t < 0. \tag{A.3}$$

As entrepreneur and investor can always guarantee themselves at least a zero lifetime utility by offering contracts without funding (i.e., $\gamma_t = 0$) and by refusing all other contracts, respectively, it follows that $V_E(h_t) \geq 0, V_I(h_t) \geq 0$ for all histories h_t and all periods t. A necessary condition for the validity of (A.1) and (A.2) is therefore

$$\alpha_t \gamma_t s_t R - \gamma_t c + (1 - \alpha_t \gamma_t)\delta V_E(h_{t+1}) \geq 0, \tag{A.4}$$

and

$$\alpha_t \gamma_t (1 - s_t)R - \gamma_t c + (1 - \alpha_t \gamma_t)\delta V_1(h_{t+1}) \geq 0. \tag{A.5}$$

Under the hypothesis of (A.3), it follows that at least one of the agents, entrepreneur or investor, must incur a loss in period t in exchange to a strictly positive continuation utility in period $t + 1$, and from (A.4) and (A.5) we can infer that

$$V_E(h_{t+1}) + V_1(h_{t+1}) \geq \frac{2c\gamma_t - \alpha_t \gamma_t R}{\delta(1 - \alpha_t \gamma_t)} > 0. \tag{A.6}$$

It follows that entrepreneur and investor jointly expect to be compensated for the current loss in the future. Yet due to discounting and the possibility of a positive realization, the current loss translates to higher present value gains starting from tomorrow. But future gains can only be generated from the value of the project, yet since α_t is decreasing over time, (A.3) implies that $\alpha_{t+1}\gamma_{t+1}R - 2c\gamma_{t+1} < 0$, and a repetition of the same argument allows us to infer that by forwarding (A.6) by one period, we obtain the condition

$$V_E(h_{t+2}) + V_I(h_{t+2}) \geq \frac{1}{\delta(1 - \alpha_{t+1}\gamma_{t+1})}$$

$$\times \left(2c\gamma_{t+1} - \alpha_{t+1}\gamma_{t+1}R + \frac{2c\gamma_t - \alpha_t \gamma_t R}{\delta(1 - \alpha_t \gamma_t)} \right) > 0,$$

and by induction on t, we then come to the conclusion that the value functions of the entrepreneur and investor are growing without bounds as $t \to \infty$, which delivers the desired contradiction because the sum of the value functions has to be finite because the value of the project is finite. $\qquad\square$

Lemma A2: *In every subgame-perfect equilibrium only break-even contracts have a positive probability of being accepted.*

Proof: Suppose in equilibrium a contract (s_t, γ_t) is offered and accepted. Then it has to satisfy

$$\alpha_t \gamma_t s_t R + (1 - \alpha_t \gamma_t) \delta V_E(h_{t+1}) \geq c\gamma_t + \delta V_E(h'_{t+1})$$

and

$$\alpha_t \gamma_t (1 - s_t) R - \gamma_t c + (1 - \alpha_t \gamma_t) \delta V_I(h_{t+1}) \geq \delta V_I(h''_{t+1}). \tag{A.7}$$

Denote for the purpose of this proof a break-even contract by $\bar{s}_t$, where $\bar{s}_t$ is given by $\bar{s}_t = (\alpha_t R - c)/\alpha_t R$. We first show that in every equilibrium and at every t, $s_t \geq \bar{s}_t$. The proof is by contradiction. Suppose that the equilibrium contract is given by $s_t < \bar{s}_t$. It then follows that every other contract s'_t with $s_t < s'_t$ that is more advantageous for the entrepreneur must be rejected by the investor. It follows that his outside option, which is given by $V_I(h''_{t+1})$, must satisfy

$$0 < \frac{\alpha_t \gamma_t (1 - s_t) R - \gamma_t c}{\delta} \leq V_I(h''_{t+1}),$$

as the value function along the equilibrium path satisfies $V_I(h_{t+1}) \geq 0$. Consider then the continuation equilibrium starting at h''_{t+1}. It follows that starting at $t + 1$, the investor must be offered some contracts with strictly positive net value to him to generate the strictly positive continuation payoff. Yet, again for him to reject all lower offers by the entrepreneur, it must be that his outside option, represented by a decision to reject a lower offer, must be sufficiently high, and in fact since

$$V_I(h''_{t+1}) = \alpha_{t+1}\gamma_{t+1}(1 - s_{t+1})R - \gamma_{t+1}c + (1 - \alpha_{t+1}\gamma_{t+1})\delta V_I(h_{t+2})$$

$$= \delta V_I(h''_{t+2}),$$

it follows that

$$0 < \frac{\alpha_t \gamma_t (1 - s_t) R - \gamma_t c}{\delta^2} \leq V_I(h''_{t+1}),$$

and thus by induction we find a sequence of continuation games in which the equilibrium value of the investor grows without bound, which leads to the desired contradiction because the value of the game is finite and the value of the entrepreneur is guaranteed to be nonnegative.

It remains to discuss the case of $s_t \geq \bar{s}_t$. As $s_t \geq \bar{s}_t$ for all t, it follows that $V_I(h_t) = 0$ for all h_t. In this case, the intertemporal participation constraint (A.7) of the investor becomes $\alpha_t \gamma_t (1 - s_t) R - \gamma_t c \geq 0$, which can only be satisfied with $s_t = \bar{s}_t$ for all t, which is the desired conclusion.

$\square$

Proof of Theorem 2, parts (i) and (ii): The incentive constraint for the entrepreneur in period t is given by $\alpha_t s_t \gamma_t R + \delta(1 - \alpha_t \gamma_t) V(\alpha_{t+1}) \geq c\gamma_t + \delta V(\alpha_t)$. The participation constraint for the investor is given by

$$\alpha_t(1 - s_t)\gamma_t R \geq c\gamma_t. \tag{A.8}$$

The participation constraint is always binding, and the sharing rule is given by $s_t = 1 - c/(\alpha_t R)$. The incentive constraint for the entrepreneur becomes $\alpha_t \gamma_t R - c\gamma_t + \delta(1 - \alpha_t \gamma_t) V(\alpha_{t+1}) \geq c\gamma_t + \delta V(\alpha_t)$, and if funding is constrained in α_t, the incentive constraint is binding with

$$\alpha_t \gamma_t R - c\gamma_t + \delta(1 - \alpha_t \gamma_t) V(\alpha_{t+1}) = c\gamma_t + \delta V(\alpha_t). \tag{A.9}$$

The equilibrium value for the entrepreneur can then be expressed by $V(\alpha_t) = c\gamma_t/(1 - \delta)$, and the indifference condition leads to a difference equation determining the equilibrium funding γ_t:

$$\alpha_t \gamma_t R - c\gamma_t + \delta(1 - \alpha_t \gamma_t) \frac{c\gamma_{t+1}}{1 - \delta} = c\gamma_t + \delta \frac{c\gamma_t}{1 - \delta}. \tag{A.10}$$

Suppose initially that $\gamma_t = \lambda$ and $\gamma_{t+1} < \lambda$. Then it must be that $\alpha_t \lambda R - \lambda c + (1 - \alpha_t \lambda)\delta V(\alpha_{t+1}) \geq \lambda c + \delta V(\alpha_t)$ and in consequence $V(\alpha_t) \geq c\lambda/(1 - \delta)$, whereas $V(\alpha_{t+1}) = c\gamma_{t+1}/(1 - \delta) < c\lambda/(1 - \delta)$. It follows that a necessary and sufficient condition for maximal funding is given by $\alpha_t \lambda R - \lambda c + (1 - \alpha_t \lambda)\delta[c\lambda/(1 - \delta)] \geq \lambda c + \delta[c\lambda/(1 - \delta)]$, or $\alpha_t \geq \bar{\alpha} = 2c/[R - [\delta c\lambda/(1 - \delta)]]$. As the critical posterior beliefs $\bar{\alpha}$ is a probability, $\bar{\alpha} = 2c/[R - [c\lambda\delta/(1 - \delta)]] \leq 1 \Leftrightarrow 2c + [c\lambda\delta/(1 - \delta)] \leq R$, we obtain the distinction between low-return and high-return projects. It follows that for all $\alpha_t < \bar{\alpha}, \lambda_t < \lambda$, as the funding volume has to be lowered. By the same argument, there is maximal funding for $\alpha_t \geq \bar{\alpha}$.

Part (iii). We proceed by contradiction in two steps. We first show that if $\gamma(\alpha)$ were to be decreasing in α on some segment, then it can only occur for

$$R < 2c + \frac{c\lambda\delta}{1-\delta}. \tag{A.11}$$

We then argue by contradiction that even if R satisfies inequality (A.11), $\gamma(a)$ has to be increasing. To this end, we rewrite the difference equation (A.10), using the fact that $\alpha_t = \alpha_{t+1}/(1 - \gamma_t + \alpha_{t+1}\gamma_t)$ to get

$$(\gamma_{t+1} - \gamma_t)\frac{1-\gamma_t}{\gamma_t} = \alpha_{t+1}\left(\gamma_t\left(\frac{2-\delta}{\delta}\right) - \frac{(1-\delta)R}{\delta c}\right) + 2\frac{1-\delta}{\delta}(1-\gamma_t). \tag{A.12}$$

For an arbitrary and fixed γ_{t+1} and α_{t+1}, we then investigate the nature of the solution for γ_t. The right-hand side of the equality (A.12) is linear in γ_t. The left-hand side is a convex function of γ_t, initially decreasing, zero at $\gamma_t = \gamma_{t+1}$, displaying a minimum at $\gamma_t = \sqrt{\gamma_{t+1}}$, and remaining negative thereafter. It follows that a necessary condition for $\gamma_{t+1} \geq \gamma_t$, is that

$$\alpha_{t+1}\left(\lambda\left(\frac{2-\delta}{\delta}\right) - \frac{(1-\delta)R}{\delta c}\right) + 2\frac{1-\delta}{\delta}(1-\lambda) \geq 0. \tag{A.13}$$

By a previous part of the theorem, $\alpha_{t+1} \leq 2c/[R - [c\lambda\delta/(1-\delta)]]$, the inequality (A.13) can be written as

$$\frac{2c}{R - \frac{c\lambda\delta}{1-\delta}}\left(\lambda\left(\frac{2-\delta}{\delta}\right) - \frac{(1-\delta)R}{\delta c}\right) + 2\frac{1-\delta}{\delta}(1-\lambda) \geq 0,$$

and after canceling the obvious terms, we have $2c[(1-\delta)/\delta)] - R[(1-\delta)/\delta] + c\lambda \geq 0$. From this we can infer that $\gamma_{t+1} > \gamma_t$ requires

$$R < 2c + \frac{c\lambda\delta}{1-\delta}. \tag{A.14}$$

We proceed by contradiction using again the difference equation (A.10), which is now valid everywhere (for all α_t) by the earlier argument. Consider first the right-hand side of (A.10). We first show that for $\alpha_t > \alpha_{t+1}$ and for all equilibrium values $\gamma_t \in [0, \lambda]$,

$$\alpha_t\left(\gamma_t - \frac{1-\delta}{\delta}\frac{R}{c}\right) + 2\frac{1-\delta}{\delta} < \alpha_{t+1}\left(\gamma_t - \frac{1-\delta}{\delta}\frac{R}{c}\right) + 2\frac{1-\delta}{\delta},$$

or

$$\gamma_t < \frac{1-\delta}{\delta}\frac{R}{c},$$

which has to hold as $c\gamma_t\delta/(1-\delta) < R$, as the agent cannot receive more than the gross value of the project. Next suppose there is an increasing segment, or $\gamma_t \leq \gamma_{t+1}$. Then it follows from the above property and the left-hand side of the difference equation that $\gamma_{t+1} < \gamma_{t+2}$. It follows that if funding is weakly increasing at some time segment t and $t+1$, it will be strictly increasing thereafter. Further, if we view the difference equation as a function expressing γ_{t+1} in dependence of γ_t, then for a fixed α_t it is a convex function and hence $\gamma_{t+2} - \gamma_{t+1} > \gamma_{t+1} - \gamma_t$. As α_t is decreasing over time, this makes the difference equation even more increasing. It follows that if the function is not monotonic decreasing, then it must be increasing at an increasing rate, but this leads to the desired contradiction because funding has to be lower than $\lambda < 1$ everywhere by virtue of (A.14).   □

Proof of Theorem 3: ($\Rightarrow$) It is a direct implication of the definition of the Markov-perfect equilibrium that the equilibrium value functions of the players depend only on the payoff-relevant state of the game. The set of equilibrium values at any α is therefore a singleton for every player, and it follows that any Markov-perfect equilibrium is also a weakly renegotiation-proof equilibrium.

($\Leftarrow$) We first show that every weakly renegotiation-proof equilibrium has to be a Markov-perfect equilibrium. The uniqueness of the weakly renegotiation-proof equilibrium then follows from the uniqueness of the Markov-perfect equilibrium. We first observe that every weakly renegotiation-proof equilibrium is a subgame-perfect equilibrium. By Lemma A2, it is then sufficient to consider break-even contracts. This implies that the value function of the investor is equal to zero along every continuation path, or $V_I(h_t) = 0$ for all h_t. By Definition 3 of the weakly renegotiation-proof equilibrium, it then follows that the equilibrium value function of the entrepreneur has to take on the same value for any two histories, h_t and h_s, which generate the same posterior belief. In other words, for all h_t and h_s, we have

$$\alpha(h_t) = \alpha(h_s) \Rightarrow V_E(h_t) = V_E(h_s). \tag{A.15}$$

Consider then the incentive constraint of the entrepreneur in any subgame-perfect equilibrium in period t : $\alpha_t\gamma_t s_t R + (1-\alpha_t\gamma_t)\delta V_E(h_{t+1}) \geq c\gamma_t + \delta V_E(h'_{t+1})$. By Lemma A2, we can restrict our attention to break-even contracts, which leads to $\alpha_t \gamma_t R - \gamma_t c + (1 - \alpha_t \gamma_t)\delta V_E(h_{t+1}) \geq c\gamma_t + \delta V_E(h'_{t+1})$. By the earlier argument, represented by the implication (A.15), the posterior belief α has to be a sufficient statistic for the history h_t with respect to the value function of the entrepreneur, and hence

$$\alpha_t \gamma_t R - \gamma_t c + (1 - \alpha_t \gamma_t)\delta V_E(\alpha_{t+1}) \geq c\gamma_t + \delta V_E(\alpha_t). \tag{A.16}$$

It is immediate from here that even weakly renegotiation-proof equilibrium must also be a Markov-perfect equilibrium, as it satisfies the equilibrium conditions of the Markov-perfect equilibrium. But since by Theorem 2 the Markov-perfect equilibrium is unique, it follows that the weakly renegotiation-proof equilibrium is unique as well. $\qquad\square$

Proof of Theorem 4: We characterize the equilibrium funding through a sequence of lemmas. Lemma A3 establishes necessary and sufficient conditions for restricted and unrestricted funding at $\alpha_T = \alpha_S$. Lemma A4 establishes the switching point from restricted to unrestricted funding and establishes the difference equation that governs the equilibrium funding as it is restricted. Lemma A5 establishes properties of the fixed point and thereby the areas where funding is restricted and unrestricted. Lemma A6 establishes the monotonicity of the funding volume as a function of time.

Lemma A3: *The equilibrium funding volume at $\alpha_T = \alpha_S$ is given by*

$$\gamma_T = \lambda \Leftrightarrow \delta \geq \frac{2 - \lambda}{2 - 2\lambda} \tag{A.17}$$

and

$$\gamma_T < \lambda \Leftrightarrow \delta < \frac{2 - \lambda}{2 - 2\lambda}. \tag{A.18}$$

Proof: We consider the ultimate α_T and penultimate prior beliefs α_{T-1} with $\alpha_T = \alpha_S$. We begin with (A.17). Consider the incentive constraint (5.6) evaluated at the penultimate period, $T - 1$:

$$\alpha_{T-1}\lambda R - c\lambda \geq c\lambda + \lambda \frac{\alpha_{T-1}}{\alpha_T}\delta V_E(\alpha_T). \tag{A.19}$$

We can express α_{T-1} in terms of λ and α_T, $\alpha_{T-1} = \alpha_T/[1 - \lambda(1 - \alpha_T)]$, and rewrite (A.19) as

$$\frac{\alpha_T}{1 - \lambda(1 - \alpha_T)}\lambda R \geq 2c\lambda + \lambda\frac{1}{1 - \lambda(1 - \alpha_T)}\delta V_E(\alpha_T). \tag{A.20}$$

By hypothesis $\alpha_T = \alpha_S$, and hence

$$V_E(\alpha_T) = \alpha_t \gamma_T R - c\gamma_T = c\gamma_T. \tag{A.21}$$

Inserting the value (A.21) into the incentive constraint (A.20), we obtain, after rewriting,

$$\lambda(1 - \alpha_T)2c \geq \delta c\gamma_T. \tag{A.22}$$

The incentive constraint (A.22) is the most difficult to satisfy if γ_T is chosen maximally, i.e., $\gamma_T = \lambda$. Using the fact that $\alpha_T = \alpha_S = 2c/R$, we get $\lambda(R - 2c)2 \geq \delta R\lambda$, which can be written as

$$R(2 - \delta) \geq 4c. \tag{A.23}$$

Using the fact that we distinguish between low- and high-return projects, we can express the inequality (A.23) in terms of δ and λ exclusively. For suppose that $R < 2c + c\lambda\delta/(1 - \delta)$. Then for $R(2 - \delta) \geq 4c$, it follows that $2c + c\lambda\delta/(1 - \delta) > 4c/(2 - \delta)$ or $\delta > (2 - \lambda)/(2 - 2\lambda)$. On the other hand, for $R \geq 2c + c\lambda\delta/(1 - \delta)$ and $R(2 - \delta) < 4c$, we necessarily have $4c/(2 - \delta) > 2c + c\lambda\delta/(1 - \delta)$ or $\delta < (2 - \lambda)/(2 - 2\lambda)$. Thus, it follows for $R \geq 2c + [c\lambda\delta/(1 - \delta)]$ as well as for $R < 2c + [c\lambda\delta/(1 - \delta)]$ that $\delta \geq (2 - \lambda)/(2 - 2\lambda)$ is a necessary and sufficient condition for full funding in the terminal period. It further follows that if

$$\delta < \frac{2 - \lambda}{2 - 2\lambda},$$

then for (A.22) to hold, $\gamma_T < \lambda$.   □

Lemma A4: *The equilibrium switching point is given by*

$$\bar{\bar{\alpha}} = \frac{2c - 2\dfrac{c\lambda}{1 - \delta} + c\dfrac{\lambda\delta}{1 - \delta}}{R - 2\dfrac{c\lambda}{1 - \delta}}.$$

Proof: We first derive the difference equation for the investor's funding decision, provided that the incentive constraint of the entrepreneur is binding. The beliefs of entrepreneur and investor are symmetric along the equilibrium path. In consequence, the contracts on the equilibrium path are the break-even contracts and satisfy

$$\alpha_t \gamma_t s_t R = \alpha_t \gamma_t R - c\gamma_t. \tag{A.24}$$

The value of the entrepreneur along the equilibrium path can then be represented as

$$V_E(\alpha_t) = \alpha_t \gamma_t R - c\gamma_t + \delta(1 - \alpha_t \gamma_t) V_E(\alpha_{t+1}). \tag{A.25}$$

We can now directly consider the recursive incentive problem of the entrepreneur. The incentive constraint of the entrepreneur changes to reflect the divergence of the beliefs off the equilibrium path. It is given by

$$\alpha_t \gamma_t R - c\gamma_t + \delta(1 - \alpha_t \gamma_t) V_E(\alpha_{t+1}) \geq c\gamma_t + \delta \frac{\alpha_t}{\alpha_{t+1}} V_E(\alpha_{t+1}).$$

Since $1 - \alpha_t \gamma_t = (\alpha_t / \alpha_{t+1})(1 - \gamma_t)$, the incentive constraint may be rewritten as

$$\alpha_t \gamma_t R - 2c\gamma_t \geq \delta \gamma_t \frac{\alpha_t}{\alpha_{t+1}} V_E(\alpha_{t+1}). \tag{A.26}$$

When funding is restricted, (A.26) must hold with equality, and hence it can be written as

$$V_E(\alpha_{t+1}) = \frac{1}{\delta} \left(\alpha_{t+1} R - \frac{\alpha_{t+1}}{\alpha_t} 2c \right). \tag{A.27}$$

Substituting (A.27) back into (A.25) and using $\alpha_{t+1}/\alpha_t = (1 - \gamma_t)/(1 - \alpha_t \gamma_t)$, we get

$$V_E(\alpha_t) = (\alpha_t R - 2c) + c\gamma_t. \tag{A.28}$$

Forwarding (A.28) for one period and equating to (A.27) yields

$$(\alpha_{t+1} R - 2c) + c\gamma_{t+1} = \frac{1}{\delta} \left(\alpha_{t+1} R - \frac{\alpha_{t+1}}{\alpha_t} 2c \right).$$

Solving, substituting for α_{t+1}/α_t leads to the difference equation

$$\gamma_{t+1} = 2 + \left(\frac{\alpha_t(1 - \gamma_t)}{(1 - \alpha_t \gamma_t)} \frac{R}{c} \frac{1 - \delta}{\delta} - \frac{1}{\delta} \left(\frac{(1 - \gamma_t)}{1 - \alpha_t \gamma_t} 2 \right) \right).$$

Making use of the backward expression of the belief ratios, $\alpha_t/\alpha_{t+1} = 1/(\alpha_{t+1}\gamma_t + (1-\gamma_t))$, and solving, we get the backward difference equation of the form $\gamma_t = f(\gamma_{t+1}, \alpha_{t+1})$,

$$\gamma_t = \frac{1}{(1-\alpha_{t+1})}\left(\frac{1}{2}\gamma_{t+1}\delta - (1-\delta)\left(\frac{\alpha_{t+1}R}{2c}-1\right)\right), \tag{A.29}$$

where we observe that γ_t is a linear increasing function of γ_{t+1}.

The equilibrium switching point $\bar{\alpha}$ is given by the unique α that results in a fixed point of the difference equation $\lambda = f(\lambda, \alpha)$ at full funding level,

$$\bar{\alpha} = \frac{2c - 2\dfrac{c\lambda}{1-\delta} + \dfrac{c\lambda\delta}{1-\delta}}{R - 2\dfrac{c\lambda}{1-\delta}},$$

which completes this lemma. $\qquad\square$

Lemma A5:

(i) The switching point $\bar{\alpha}$ satisfies $\bar{\alpha} \in (\alpha_s, 1)$ if either

    (a) $R \geq 2c + [c\lambda\delta/(1-\delta)]$ and $\delta < (2-2\lambda)/(2-\lambda)$, or
    (b) $R < 2c + [c\lambda\delta/(1-\delta)]$ and $\delta \geq (2-2\lambda)/(2-\lambda)$.

(ii) The switching point satisfies $\bar{\alpha} \notin (\alpha_s, 1)$ otherwise.

Proof: The proof is omitted, as it simplex requires the algebraic verification of the conditions stated in Lemma A5 applied to the switching point $\bar{\alpha}$. $\qquad\square$

Lemma A6:

(i) If $\delta < (2-\lambda)/(2-2\lambda)$, the funding volume γ_t is decreasing over time.
(ii) If $\delta \geq (2-\lambda)/(2-2\lambda)$, the funding volume γ_t is increasing in time.
(iii) The monotonicity is strict in either case provided that $\gamma_t < \lambda$.

Proof: To describe the monotonicity properties of the difference equation, it is useful to analyze the fixed point of the mapping, $\gamma = f(\gamma, \alpha)$ for all

$\gamma < \lambda$. It is given by

$$\gamma(\alpha) = \frac{(\alpha R - 2c)^{\frac{1-\delta}{\delta}}}{(2\alpha - 2^{\frac{1-\delta}{\delta}}(1-\alpha) - 1)c}.$$

The fixed point $\gamma(\alpha)$ has a derivative that leads to

$$\gamma'(\alpha) > 0 \Leftrightarrow \delta < \frac{2-\lambda}{2-2\lambda} \tag{A.30}$$

and

$$\gamma'(\alpha) < 0 \Leftrightarrow \delta > \frac{2-\lambda}{2-2\lambda}. \tag{A.31}$$

We start with $\delta < (2-\lambda)/(2-2\lambda)$ and show that the difference equation γ_t must be strictly decreasing in time t. The argument is by contradiction. Suppose then there exists t and $t+1$ such that $\gamma_t \leq \gamma_{t+1}$. Then it follows by the property of the fixed point as a function of α, as displayed in (A.30), and the fact that γ_t is linear increasing in γ_{t+1}, as displayed in (A.29), that $\gamma_{s-1} < \gamma_s$ for all $s < t$. The funding flow in the period $t = 0$ is then determined, again using (A.29), by

$$\gamma_0 = \frac{1}{1-\alpha_0}\left(\frac{1}{2}\delta\gamma_1 - (1-\delta)\left(\frac{\alpha_0 R}{2c} - 1\right)\right). \tag{A.32}$$

Consider then the limit $\alpha_0 \to 1$. Since

$$\lim_{\alpha_0 \to 1} \frac{1}{1-\alpha_0} = \infty,$$

and γ_0 is bounded by $\gamma_0 \in [0, \lambda]$, it follows that the expression in brackets in (A.32) has to go to zero as $\alpha_0 \to 1$, or

$$\gamma_1 = 2\frac{1-\delta}{\delta}\left(\frac{R}{2c} - 1\right). \tag{A.33}$$

Because γ_t is supposed to be a local maximum, it follows from (A.29) that an upper bound for γ_t is obtained by setting $\gamma_{t+1} = \gamma_t$, or

$$\gamma_1 \leq \frac{(R\alpha_{t+1} - 2c)^{\frac{1-\delta}{\delta}}}{(2\alpha_{t+1} - 2\frac{1-\delta}{\delta}(1-\alpha_{t+1}) - 1)c}.$$

Yet we find that

$$\frac{(R\alpha_{t+1} - 2c)\frac{1-\delta}{\delta}}{(2\alpha_{t+1} - 2\frac{1-\delta}{\delta}(1 - \alpha_{t+1}) - 1)c} < 2\frac{1-\delta}{\delta}\left(\frac{R}{2c} - 1\right)$$

provided that $\delta \le (2-\lambda)/(2-2\lambda)$, which leads to the desired contradiction. Thus there is no local maximum either, and hence, $\gamma_t \ge \gamma_{t+1}$ for all t.

Consider next the case of $\delta \ge (2 - \lambda)/(2 - 2\lambda)$. We argue as above by contradiction. Suppose there is a segment t and $t + 1$ such that $\gamma_t > \gamma_{t+1}$. Then it follows again from the fixed-point property, (A.31), that for all $s \le t, \gamma_{s-1} \ge \gamma_s$. As $\bar{\bar{\alpha}} \in (\alpha_s, 1)$, it further follows that there must be at least one local minimum, say at γ_t. For the local minimum at γ_t, we obtain a lower bound by setting as above $\gamma_{t+1} = \gamma_t$,

$$\gamma_t > \frac{(R\alpha_{t+1} - 2c)\frac{1-\delta}{\delta}}{(2\alpha_{t+1} - 2\frac{1-\delta}{\delta}(1 - \alpha_{t+1}) - 1)c}. \tag{A.34}$$

The right-hand side presents the value of the fixed point $\gamma(\alpha)$, which is decreasing by (A.30). The lower bound is therefore lowest for $\alpha_{t+1} = 1$, from which it follows that the local minimum γ_t must satisfy

$$\gamma_t > \frac{(R - 2c)\frac{1-\delta}{\delta}}{(2 - 1)c}, \tag{A.35}$$

yet because it is a local minimum, it has to satisfy

$$\gamma_t \le \gamma_1 = 2\frac{1-\delta}{\delta}\left(\frac{R}{2c} - 1\right), \tag{A.36}$$

where γ_1 was computed earlier at (A.33). But for $\delta \ge (2 - \lambda)/(2 - 2\lambda)$, the conditions (A.35) and (A.36) lead to a contradiction. $\qquad\square$

Proof of Corollary 2: It is immediately verified that the derivation of the Markov sequential equilibrium above relied only on a backward-induction argument. In the construction of the equilibrium the belief α_t merely served to summarize the information of the players but never to restrict the history contingency of the strategies employed by the agents. $\qquad\square$

Proof of Theorem 5: Denote by γ_t^s and γ_t^α the funding volume in the symmetric and asymmetric case, respectively. Consider first the case of

$\delta \geq (2 - \lambda)/(2 - 2\lambda)$. If, in addition, $R \geq 2c + [\lambda c \delta/(1 - \delta)]$, then $\gamma_t^\alpha = \lambda$ everywhere, and thus the condition $\gamma_t^\alpha \geq \gamma_t^s$ for all α_t, with strict inequality in the final periods, is satisfied. For $R < 2c + [\lambda c \delta/(1 - \delta)]$, we know by Theorem 2 that the funding volume γ_t^s is increasing in α and by Theorem 4 that the funding volume γ_t^α is (weakly) decreasing in α. A necessary and sufficient condition for $\gamma_t^\alpha \geq \gamma_t^s$ is therefore given by

$$\lim_{\alpha_0 \to 1} (\gamma_t^\alpha - \gamma_t^s) \geq 0. \tag{A.37}$$

For the symmetric-information case, condition (A.10) in the proof of Theorem 2 can be solved as the following difference equation:

$$\frac{\gamma_{t+1} - \gamma_t}{\gamma_t} = \alpha_t \left(\gamma_{t+1} - \frac{1 - \delta}{\delta} \frac{R}{c} \right) + 2 \frac{1 - \delta}{\delta}. \tag{A.38}$$

As we solve the respective difference equations, (A.38) and (A.29) term the proof of theorem 4, for γ_0 as $\alpha_0 \to 1$, we find

$$\lim_{\alpha_0 \to 1} \gamma_0^s = \lim_{\alpha_0 \to 1} \gamma_0^\alpha = \frac{1 - \delta}{\delta} \left(\frac{R}{c} - 2 \right),$$

which establishes the validity of (A.37).

Consider next the case of

$$\delta < \frac{2 - \lambda}{2 - 2\lambda},$$

where by Theorems 2 and 4 γ_t^α and γ_t^s are both increasing in α. We proceed by establishing a lower bound on γ_t^α and an upper bound on γ_t^s, denoted by $\underline{\gamma}_t^\alpha$ and $\bar{\gamma}_t^s$, respectively. We then show that $\underline{\gamma}_t^\alpha \geq \bar{\gamma}_t^s$, completing the result. As $\gamma_t^\alpha \leq \gamma_{t-1}^\alpha$ in this case, a lower bound $\underline{\gamma}_t^\alpha$ is established by looking for the fixed point $\gamma_{t-1}^\alpha = \gamma_t^\alpha$ in the difference equation (A.29),

$$\underline{\gamma}_t^\alpha = \frac{\dfrac{1 - \delta}{\delta} \left(\dfrac{\alpha_t R}{2c} - 1 \right)}{\dfrac{1}{2} - (1 - \alpha_t) \dfrac{1}{\delta}}. \tag{A.39}$$

Considering then the difference equation of the symmetric environment (A.38) and using the fact that by hypothesis $\gamma_{t-1}^s \geq \gamma_t^s$ in (A.38), we obtain an upper bound:

$$\bar{\gamma}_t^s = \frac{1-\delta}{\delta} \left(\frac{R}{c} - \frac{2}{\alpha_t} \right). \tag{A.40}$$

Comparing (A.39) and (A.40) and requiring that $\underline{\gamma}_t^a \geq \bar{\gamma}_t^s$ leads to, after multiplying and canceling the obvious terms,

$$\left(\frac{R}{2c} - \frac{1}{\alpha_t} \right) + \left(\frac{R}{c} - \frac{2}{\alpha_t} \right) \frac{1-\delta}{\delta} (1 - \alpha_t) \geq \alpha_t \left(\frac{R}{2c} - \frac{1}{\alpha_t} \right),$$

which holds for all $\alpha_t \geq \alpha_s$. ☐

Acknowledgments

The authors would like to thank the Editor, Raymond Deneckere, and three referees for detailed comments and suggestions. The authors benefited from discussions with Dilip Abreu, Patrick Bolton, Amil Dasgupta, Thomas Hellmann, Godfrey Keller, Dima Leshchinskii, Farshad Mashayekhi, Georg Noldeke, Kjell Nyborg, Ted O'Donoghue, David Pearce, Paul Pfleiderer, Rafael Repullo, and Luca Rigotti. Seminar participants at Basel, Budapest, LSE, Mannheim, Princeton, Rotterdam, Stockholm, Tilburg, Econometric Society Winter Meetings 1998 and EFA 1999 provided stimulating discussions. The first author gratefully acknowledges financial support from the National Science Foundation (grant no. SBR 9709887) and an A.P. Sloan Research Fellowship, the second author from a TMR grant of the European Commission.

References

BARRY, C.B. "New Directions in Research on Venture capital Finance." *Financial Management*, Vol. 23 (1994), pp. 3–15.

BERGEMANN, D. AND HEGE, U. "Venture Capital Financing, Moral Hazard and Learning." *Journal of Banking and Finance*, Vol. 22 (1998), pp. 703–755.

BERGER, A.N. AND UDELL, G. "Relationship Lending and Lines of Credit in Small Firm Finance." *Journal of Business,* Vol. 68 (1995), pp. 351–382.

——— AND ——— "The Economics of Small Business Finance: The Roles of Private Equity and Debt Markets in the Financial Growth Cycle." *Journal of Banking and Finance*, Vol. 22 (1998), pp. 613–673.

BOLTON, P. AND SCHARFSTEIN, D. "Optimal Debt Structure and the Number of Creditors." *Journal of Political Economy*, Vol. 104 (1996), pp. 1–25.

CASAMATTA, C. "Financing and Advising: Optimal Financial Contracts with Venture Capitalists." *Journal of Finance*, Vol. 58 (2003), pp. 2058–2086.

CHEMMANUR, T. AND CHEN, Z. "Angels, Venture Capitalists, and Entrepreneurs: A Dynamic Model of Private Equity Financing." Mimeo, Carroll School of Management, Boston College, 2003.

CORNELLI, F. AND YOSHA, O. "Stage Financing and the Role of Convertible Securities." *Review of Economic Studies*, Vol. 70 (2003), pp. 1–32.

CREMER, J. "Arm's Length Relationships." *Quartely Journal of Economics*, Vol. 110 (1995), pp. 275–295.

DEGRYSE, H. AND ONENA, S. "Distance, Lending Relationships, and Competition." *Journal of Finance*, Vol. 60 (2005), pp. 231–266.

DEMARZO, P. AND FISHMAN, M.J. "Optimal Long-Term Financial Contracting with Privately Observed Cash-Flows." Mimeo, Stanford University Graduate School of Business, and Northwestern University Kellogg Graduate School of Management, 2000.

DEWATRIPONT, M. AND MASKIN, E. "Credit and Efficiency in Centralized and Decentralized Economies" *Review of Economic Studies*, Vol. 62(1995) pp. 541–555.

FARREL, J. AND MASKIN, E. "Renegotiation in Repealed Games." *Games and Economic Behavior*, Vol. 1 (1989), pp. 327–360.

FLUCK, Z. "Optimal Financial Contracting: Debt Versus Outside Equity." *Review of Financial Studies*, Vol. 11 (1998), pp. 383–418.

FUDENBERG, D. AND TIROLE, J. "Moral Hazard and Renegotiation in Agency Contracts." *Econometrica*, Vol. 58 (1990), pp. 1279–1319.

GOMPERS, P. AND LERNER, J. *The Venture Capital Cycle*. Cambridge, Mass.: MIT Press, 1999.

GROMB, D. "Renegotiation in Debt Contracts." Mimeo, MIT, 1994.

HART, O.D. AND MOORE, J.H. "A Theory of Debt Based on the Inalienability of Human Capital." *Quarterly Journal of Economics*, Vol. 109 (1994), pp. 841–879.

—— AND ——. "Default and Renegotiation: A Dynamic Model of Debt." *Quarterly Journal of Economics*, Vol. 113 (1998), pp. 1–41.

HEGE, U., PALOMINO, F., AND SCHWIENBACHER, A. "The Determinants of Venture Capital Performance: Europe and United States." Mimeo, HEC School of Management, 2003.

HELLMANN, T. "The Allocation of Control Rights in Venture Capital Contracts." *RAND Journal of Economics*, Vol. 29 (1998), pp. 57–76.

HERMALIN, B. AND KATZ, M. "Moral Hazard and Verifiability: The Effects of Renegotiation in Agency." *Econometrica*, Vol. 59 (1991), pp. 1735–1753.

KAPLAN, S. AND STROMBERG, P. "Financial Contracting Theory Meets the Real World: An Empirical Analysis of Venture Capital Contracts." *Review of Economic Studies*, Vol. 70 (2003), pp. 281–315.

LESHCHINSKII, D. "Project Externalities and Moral Hazard." Mimeo, HEC School of Management, 2003.

MARQUEZ, R. "Competition and the Choice Between Informed and Uninformed Finance." Mimeo, R.H. Smith School of Business, University of Maryland, 1998.

MASKIN, E. AND TIROLE, J. "Markov Perfect Equilibrium." *Journal of Economic Theory,* Vol. 100 (2001), pp. 191–219.

NEHER, D. "Stage Financing: An Agency Perspective." *Review of Economic Studies,* Vol. 66 (1999), pp. 255–274.

ONGENA, S. AND SMITH, D.C. "The Duration of Bank Relationships." *Journal of Financial Economics,* Vol. 61 (2001), pp. 449–475.

PETERSEN, M.A. AND RAJAN, R.G. "The Benefits of Lending Relationships: Evidence from Small Business Data." *Journal of Finance,* Vol. 49 (1994), pp. 3–37.

QIAN, Y. AND XU, C. "Innovation and Bureaucracy Under Soft and Hard Budget Constraints." *Review of Economic Studies,* Vol. 65 (1998), pp. 151–164.

RAJAN, R. "Insiders and Outsiders: The Choice Between Informed and Arm's Length Debt." *Journal of Finance,* Vol. 47 (1992), pp. 1367–1400.

REPULLO, R. AND SUAREZ, J. "Venture Capital Finance: A Security Design Approach." Mimeo, CEMFI, 1998.

SAHLMAN, W.A. "The Structure and Governance of Venture-Capital Organizations." *Journal of Financial Economics,* Vol. 27 (1990), pp. 473–521.

STEVENS, G. AND BURLEY, J. "3000 Raw Ideas = 1 Commercial Success." *Research-Technology Management,* Vol. 40 (1997), pp. 16–27.

Chapter 11

Learning About the Arrival of Sales*

Robin Mason[†,1] and Juuso Välimäki[‡]

† University of Birmingham, Birmingham B15 2TT, United Kingdom
r.a.mason@bham.ac.uk
‡ Department of Economics, Aalto University School of Business,
02150 Espoo, Finland
juuso.valimaki@aalto.fi

We analyze optimal stopping when the economic environment changes because of learning. A primary application is optimal selling of an asset when demand is uncertain. The seller learns about the arrival rate of buyers. As time passes without a sale, the seller becomes more pessimistic about the arrival rate. When the arrival of buyers is not observed, the rate at which the seller revises her beliefs is affected by the price she sets. Learning leads to a higher posted price by the seller. When the seller does observe the arrival of buyers, she sets an even higher price. ©2011 Elsevier Inc. All rights reserved.

JEL classification: D42; D81; D82

Keywords: Optimal stopping, learning, uncertain demand.

1. Introduction

Models of optimal stopping capture the trade-off between known current payoffs and the opportunity cost of future potentially superior possibilities.

---

*This chapter is reproduced from *Journal of Economic Theory*, **146**(4), 1699–1711, 2011.
[1]Robin Mason acknowledges financial support from the ESRC under Research Grant RES-062-23-0925.

339

340 Learning and Intertemporal Incentives

In many of their economic applications, the environment in which the stopping decisions are taken is assumed to be stationary. In the standard job search model, for example, a rejected offer is followed by another draw from the same distribution of offers. See [19] for a survey. The predictions of these models are at odds with empirical observations. Prices of houses decline as a function of time on the market; see e.g. [17]. Reservation wages of unemployed workers decline in the duration of the unemployment spell: see e.g. [14].

We propose a simple and tractable model of optimal stopping where the economic environment changes as a result of learning. Rather than using the deterministic model of optimal stopping typically used in optimal search models, we generalize the other canonical model of stopping in economic theory, where stopping occurs probabilistically (as in e.g. models of R&D).

We cast the model in the language of the classic problem of how to sell an asset [10]. (In the concluding section, we discuss alternative interpretations of the model.) A seller posts a take-it-or-leave-it price; potential buyers arrive according to a Poisson process and observe the posted price. They buy if and only if their valuation exceeds the posted price. The seller is uncertain about the arrival rate of buyers. The seller does not observe whether a buyer is present: she sees only whether a sale occurs or not. The seller updates her beliefs about the arrival rate in a Bayesian fashion, becoming more pessimistic about future arrivals after each period when a sale does not occur. The rate at which beliefs decline depends on the current posted price. Hence, when choosing the current price of her asset, the seller controls her immediate expected profit, conditional on a sale, as well as the beliefs about future demand if no sale occurs.

Even though our model is a standard Bayesian learning model, the belief dynamics are a little unfamiliar. When we describe the evolution of beliefs over time, we are implicitly conditioning on the event that no sale occurred in the current period. Since this event is bad news about the prospect of a future sale, the seller becomes more pessimistic over time. The seller can control the downward drift of her beliefs through her choice of price.

We identify two key effects. A more pessimistic future implies a lower current value of being a seller and hence a lower current price. We call this feature the *controlled stopping effect*. By setting a lower price, the seller can cause her beliefs to fall further in the event of no sale occurring. A more pessimistic seller has a lower value than a more optimistic seller — there is a capital loss from learning. Holding fixed the probability of a sale occurring, the seller has an incentive to increase her price in order to reduce this capital loss. We call this feature the *controlled learning effect*.

Which effect dominates? Does learning cause the seller to raise or lower her price, relative to the case when no learning occurs? The benchmark is a model where the rate of contact between the seller and buyers is fixed at her current expected level. We show that in our model, the controlled learning effect dominates: the optimal posted price exceeds the price posted in the equivalent model with no learning.

We gain further insight into the effect of learning by looking at the case where the seller does observe the arrival of buyers. In this case, the updating of the seller's beliefs is independent of the price that she sets. We show that the seller optimally sets an even higher price than when she cannot observe arrivals. The key difference when arrivals are observed is that now the seller can raise her price after positive news i.e. after a buyer arrival, even when there is no sale. We therefore obtain an upper bound for the price increase resulting from controlled learning, when arrivals are not observed: it is no larger than the price increase resulting from the observability of arrivals.

A number of papers have studied the problem of a monopolist who learns about the demand it faces through its pricing experience: see [20, 16, 9, 1, 18].[2] The general difficulty in analyzing this type of situation is the updating that occurs when the seller observes that the object has failed to sell at a particular price. There are certain valuation distributions for which updating can be done analytically. Even for these cases, the amount of progress that can be made is minimal.

A notable exception is Trefler [23], who provides a characterization of the expected value of information and the direction of experimentation in several cases of learning by a monopoly seller. There are two main differences between our work and Trefler's. First, we consider the sale of a single item; as a result, the monopolist faces a stopping problem (albeit one where stopping occurs probabilistically). In Trefler's model, the true (but unknown) economic environment remains stationary across periods, and the model is thus one of repeated sales. Secondly, in out set-up, the Bellman equations are first-order differential equations; they are therefore particularly easy to analyze. This allows us to give more characterization than Trefler of the seller's optimal policy.

---

[2]In these papers, the demand curve is fixed and the learning process narrows down the initial uncertainty about it. In other papers, the object of the learning is not constant, but changes over time in some random way. See, for example, [3, 21, 12].

Our paper is also related to the literature on search with learning. Burdett and Vishwanath [6], Van den Berg [24] and Smith [22] analyze an optimal stopping problem when wage offers come from exogenous sources. Anderson and Smith [2] analyze a matching model where each match results in a current immediate payoff and reveals new information about the productivity of the partners. In all these papers, beliefs change i.e. the environment is non-stationary; but the change in beliefs is not affected by the action chosen by the economic agent. We develop a tractable framework in which agents affect their learning through their actions.

Finally, there are other papers, from a number of different fields, that involve learning about a Poisson process. Amongst others, see [8, 15, 13, 4, 5].

The paper is organized as follows. Section 2 presents the model. In Section 3, we look at a couple of benchmarks without learning, which are useful when assessing the effects of learning in the model. Section 4 characterizes the optimal pricing policy when learning takes place. Section 5 looks at the case when the seller can observe the arrival of buyers to the market. Section 6 discusses alternative interpretations of the model and concludes. Appendix A contains longer proofs.

2. The model

Consider a seller setting a sequence of take-it-or-leave-it prices over time for a single unit of a good of known quality. Time is discrete. Time periods are denoted by $t = 0, 1, \ldots, \infty$; each time interval is of length $dt > 0$, which we take to be arbitrarily short. It is commonly known that the seller's valuation is 0. The seller announces at the start of each period the price for that period. If a buyer announces that she is willing to buy at that price, then the seller sells and the problem stops. The seller is infinitely lived and discounts the future at rate r.

The probability that the seller encounters a buyer in a given period depends on two factors: whether a buyer is present during the period; and whether that buyer is willing to pay the seller's posted price.

Buyers arrive randomly to the market according to a Poisson process. With an arrival rate λ, in any given time interval $(t, t + dt]$, there is a probability λdt that a buyer arrives to the market; and a probability of $O((dt)^2)$ that more than one buyer arrives to the market. Since we take the continuous-time limit, the realization of two buyers arriving is ignored. There are two states. In the low state L, the Poisson arrival rate is $\lambda_L > 0$.

In the high state H, the Poisson arrival rate is $\lambda_H > \lambda_L$. Let $\Delta\lambda := \lambda_H - \lambda_L$. Each buyer is present (if at all) for one period, and then disappears.

Buyers' valuations are drawn independently from a distribution $F(\cdot)$ with a density function $f(\cdot)$. Given that the buyers are short-lived, they purchase the good immediately if and only if their valuation exceeds the price. Conditional on a buyer being present, the probability of this is $1 - F(p)$. We assume that $F(p)$ has an increasing hazard rate.[3]

The seller does not observe whether a buyer is present: it observes only when a sale occurs. (We comment further on this below.) So in the seller's view, the probability of a sale in any given period, given an arrival rate $\lambda \in \{\lambda_L, \lambda_H\}$, is $\lambda dt(1 - F(p))$.

The state variable in this problem is the seller's scalar belief $\pi \in [0, 1]$ that the state of the world is high. The only event that is relevant for updating the seller's beliefs is when no sale has occurred in a period. In the continuous time limit $dt \to 0$,

$$d\pi(t) := \pi(t + dt) - \pi(t) = -\pi(t)(1 - \pi(t))\Delta\lambda(1 - F(p))dt \leqslant 0. \quad (2.1)$$

So the seller becomes more pessimistic about the state of the world when no sale occurs. The seller affects the updating of beliefs through the price that it sets: a high price implies a higher probability of no sale occurring, and a higher posterior.

One modeling assumption deserves further comment: the seller observes only sales and not the arrivals of buyers. (For example, the seller places an advertisement in a newspaper each period; the seller cannot tell if a buyer has seen the advertisement.) This assumption ensures two things. First, the probability of stopping, $\lambda dt(1 - F(p))$, is of order dt. Secondly, the change in beliefs $d\pi$ also is of order dt. In the continuous-time limit, as $dt \to 0$, this ensures that the dynamic programming problem yields a differential, rather than a difference equation; this makes the analysis much simpler. In Section 5, we compare this model to the one where arrivals are observed.

3. A Non-Learning Benchmark

We start with a benchmark intertemporal model where the seller's belief π does not change over time. Since the problem remains the same in all periods, the dynamic programming problem of the seller is particularly easy

---

[3]We could allow the overall arrival rate of buyers to respond to the seller's price, with little change to the analysis.

to solve. This solution will serve as a natural point of comparison to the model that incorporates learning i.e. with π changing over time. We refer to this case as the *repeated* problem. (See also [11]; our repeated benchmark corresponds to his "naive learner" case.)

For the repeated problem, the seller's Bellman equation is

$$V_R(\pi) = \max_p V_R(p, \ \pi) \tag{3.2}$$

where

$$V_R(p, \ \pi) := \left\{ \frac{\lambda(\pi)(1 - F(p))p}{r + \lambda(\pi)(1 - F(p))} \right\}.$$

The first-order condition for the repeated problem is

$$p - \frac{1 - F(p)}{f(p)} = V_R(\pi). \tag{3.3}$$

For any π, Eq. (3.3) has a unique solution, because $F(\cdot)$ has an increasing hazard rate. Denote this solution as $pR(\pi)$.

It will be helpful for later arguments to establish a couple of facts about $pR(\pi)$. First, denote the static price as ps; it solves $pS - (1 - F(pS))/f(pS) = 0$. Since $V_R(\cdot) \geqslant 0$, the following result is immediate.

Proposition 1: *The repeated price is greater than the static price: $pR(\pi) \geqslant pS$ for all $\pi \in [0, \ 1]$.*

The intuition is straightforward: the prospect of continuation in the repeated problem acts as an opportunity cost to a current sale; see Eq. (3.3). This opportunity cost leads the seller to set a higher price.

Next, we establish the monotonicity of the repeated price $pR(\pi)$.

Proposition 2: *The repeated price $pR(\pi)$ is non-decreasing in the level of the seller's belief π for all $\pi \in [0, \ 1]$.*

Proof: Total differentiation of Eq. (3.3) means that $pR(\pi)$ is non-decreasing in π iff $V_R(\pi)$ is non-decreasing in π. Since the value in any program is realized only at the stopping moment, the value of any fixed policy must be increasing in the arrival rate and hence π. Hence the optimal policy starting at π yields a higher value when started at state $\pi' > \pi$ and as a consequence $V_R(\pi') > V_R(\pi)$. $\qquad\square$

The proof relies on two facts: the prospect of continuation acts as an opportunity cost to a current sale; and the value of continuation increases with the belief π. Hence the repeated price increases in π.

4. The Effect of Learning

Consider next the model with learning: the *dynamic* problem. The crucial difference compared to the repeated problem is that now π evolves over time according to Eq. (2.1). The seller anticipates this and adjusts the optimal price to control both the current probability of a sale occurring and the future value of sales.

We show in Appendix A (see Lemma A.1) that $V_D(\pi)$ is convex; as a result, $V_D(\pi)$ is differentiable almost everywhere. Hence we can use a Taylor series expansion for $V_D(\pi + d\pi)$, and take the continuous-time limit $dt \to 0$, to write the Bellman equation as

$$V_D(\pi) = \max_p \left\{ V_R(p, \ \pi) - \frac{\Delta\lambda(1 - F(p))\pi(1 - \pi)V_D'(\pi)}{r + \lambda(\pi)(1 - F(p))} \right\}. \qquad (4.4)$$

(See Appendix A for a derivation of this equation.) Eq. (4.4) shows that the dynamic value function is equal to a repeated value, $V_R(pD, \ \pi)$ (evaluated at the optimal dynamic price, pD), minus a term that arises due to learning. The learning term is the present value of the infinite stream of the capital losses due to learning, using the effective discount rate $r + \lambda(\pi)(1 - F(p))$.

The capital loss from learning has two parts. $\pi(1 - \pi)V_D'(\pi)$ represents the impact of a unit of learning on future profits. It is unaffected by the current choice of p and hence we ignore its effect for the moment. The term

$$-\frac{\Delta\lambda(1 - F(p))}{r + \lambda(\pi)(1 - F(p))}$$

is affected by p in two ways. The first way measures the effect of p on the effective discount rate $r + \lambda(\pi)(1 - F(p))$:

$$-\frac{\Delta\lambda\lambda(\pi)f(p)(1 - F(p))}{(r + \lambda(\pi))(1 - F(p)))^2} \leqslant 0.$$

We call this effect the *controlled stopping effect*. It captures the increasing pessimism of the seller. Since it has a negative sign, this effect on its own leads to a lower posted price by the seller.

The second way is related to the change in the seller's posterior, $\pi(1 - \pi)\Delta\lambda f(p)$, resulting from a change in p, normalized by the effective discount rate $r + \lambda(\pi)(1 - F(p))$:

$$\frac{\Delta\lambda f(p)}{r + \lambda(\pi)(1 - F(p))} \geqslant 0.$$

We call this effect the *controlled learning effect*. This term is positive, and so on its own leads to a higher posted price by the seller.

In this model, the sum of these two effects is

$$\frac{r\Delta\lambda f(P)}{(r + \lambda(\pi)(1 - F(p)))^2} \leqslant 0.$$

That is, the controlled learning effect dominates the controlled stopping effect.[4]

Hence the first order condition for the dynamic problem is

$$p - \frac{1 - F(p)}{f(p)} = V_R(p, \pi) + \frac{\Delta\lambda f(p)}{r + \lambda(\pi)(1 - F(p))}\pi(1 - \pi)V_D'(\pi). \quad (4.5)$$

Eq. (4.5) can be contrasted with Eq. (3.3). Since $V_D(\pi)$ is non-decreasing (from the same argument used in the proof of Proposition 2), the price posted by the seller who learns about the state of the world is *above* the price of the non-learning seller. As a result, the probability that the learning seller makes a sale is below that of the non-learning seller.

We summarize this discussion in the following theorem.

Theorem 1: *In the case of learning about the arrival rate of buyers, the learning price $pD(\pi)$ is greater than the repeated price $pR(\pi)$ for all $\pi \in [0, 1]$.*

We can compare this result to Trefler's [23]. By Theorem 4 of Trefler, the seller moves her price in the direction that is more informative, in the Blackwell sense, relative to the static price pS. In our model, the expected

---

[4]This will not be the case in all models. In an earlier version of this paper, we considered also the case where the seller learns about the distribution from which buyers' valuations are drawn. We showed that either effect may then dominate.

continuation value of information, [7] is

$$I(p, \pi) := E[V_D(\pi + d\pi)] - V_D(\pi)$$

$$= -\lambda(\pi)(1 - F(p)) \left(V(\pi) + \frac{\Delta\lambda}{\lambda(\pi)} \pi(1 - \pi) V'(\pi) \right) dt,$$

ignoring terms in $(dt)^2$. Notice that this expression conditions on no sales in the period — an event which is bad news for the seller. So, with this definition, the expected continuation value of information is negative. Hence

$$\frac{\partial I(p, \pi)}{\partial p} = \lambda(\pi) f(p) \left(V(\pi) + \frac{\Delta\lambda}{\lambda(\pi)} \pi(1 - \pi) V'(\pi) \right) dt \geqslant 0.$$

This means that a higher price is more informative, in the Blackwell sense. Trefler's result therefore implies that $pD(\pi) \geqslant pS$. This is indeed what we find. We know from Proposition 1 that the repeated price is greater than the static price: $pR(\pi) \geqslant pS$; Theorem 1 tells us that the dynamic price is higher still. The comparison between the repeated and dynamic price does not appear in Trefler, since he does not consider the sale of a single asset.

The next two results give some intuitive properties of the seller's optimal decision.

Property 1 (Monotonicity): *The dynamic learning price, $pD(\pi)$, and the probability of sale, $\lambda(\pi)(1 - F(pD(\pi)))$, are non-decreasing in the seller's belief π.*

Proof: See Appendix A. □

Property 2 (Discounting): *The dynamic learning price $pD(\pi)$ is decreasing in the discount rate r.*

Proof: Eq. (A.1) implies that $pD(\pi)$ is decreasing in r iff $rV_D(\pi)$ is increasing in r. To show the latter, differentiate the Bellman equation (4.4), using the Envelope Theorem:

$$\frac{\partial}{\partial r}(rV_D(\pi)) = \max_p \left\{ -\lambda(\pi)(1 - F(p)) \frac{\partial V_D(\pi)}{\partial r} - \pi(1 - \pi) \right.$$

$$\left. \Delta\lambda(1 - F(p)) \frac{\partial^2 V_D(\pi)}{\partial\pi\partial r} \right\}.$$

Basic arguments establish that $\partial V_D(\pi)/\partial r \leqslant 0$ and $\partial^2 V_D(\pi)/\partial\pi\partial r \leqslant 0$. The result follows.   □

An infinitely patient seller ($r = 0$) will set his price at the upper bound of the support of the valuation distribution. In contrast, a myopic seller ($r = \infty$) will set his price at the static level, which is clearly lower. The dynamic price moves between these two extremes monotonically in r.[5]

4.1. *Two-point example*

We can illustrate these results in a simple case where buyers' valuations can take two values, $\bar{v} > \underline{v} > 0$, with probabilities $\beta \in (0, 1)$ and $1 - \beta$ respectively. For the problem to be interesting, assume that $\beta\bar{v} < \underline{v}$, so that in the static problem, the seller would set a price equal to $\underline{v}$. Also assume that if the seller knows that $\lambda = \lambda_H$, her optimal price is $\bar{v}$; but if the seller knows that $\lambda = \lambda_L$, her optimal price is $\underline{v}$.

By standard arguments, the seller's price strategy then takes a simple form: set the price at $\bar{v}$ when the posterior is above some cut-off; and $\underline{v}$ for posteriors below this level. Let the cut-off level of the posterior be $\pi^* \in (0, 1)$. We illustrate the seller's value function in Fig. 11.1; in Appendix A, we show how to determine the seller's value function and the cut-off π^*. The value function is a strictly convex function of π in the region $\pi \geqslant \pi^*$, while for $\pi < \pi^*$, it is a linear function.

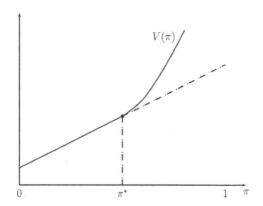

Figure 11.1.   The value function with a 2-point valuation distribution.

---

[5]Perhaps surprisingly, the dynamic price is not monotonic in the arrival rate parameters λ_H and λ_L, a fact confirmed in numerical examples.

For a seller who is certain of the arrival rate, the optimal price is $\bar{v}$ for all $\lambda \geqslant \lambda_0$, and $\underline{v}$ for $\lambda < \lambda_0$, for some λ_0. (See Appendix A for a derivation of λ_0.) Then define π_0 to be such that $\pi_0 \lambda_H + (1 - \pi_0)\lambda_L := \lambda_0$. We show in Appendix A that $\pi^* \leqslant \pi_0$: that is, learning causes the seller to set a higher price, as Theorem 1 predicts.

5. Observed Arrivals

Now suppose that the state of the world i.e. λ is not known, but that the seller observes when buyers arrive to the market. The seller's beliefs are updated according to Bayes' rule. Given an initial belief π, the posterior after an arrival is

$$\pi + d\pi_1 = \frac{\pi \lambda_H dt}{\pi \lambda_H dt + (1 - \pi)\lambda_L dt} = \frac{\lambda_H}{\lambda(\pi)}\pi. \tag{5.6}$$

Notice that $\pi + d\pi_1 > \pi$: the seller becomes more optimistic after an arrival. In fact,

$$d\pi_1 = \pi(1 - \pi)\frac{\Delta\lambda}{\lambda(\pi)}.$$

The posterior after no arrival is

$$\pi + d\pi_0 = \frac{\pi(1 - \lambda_H dt)}{\pi(1 - \lambda_H dt) + (1 - \pi)(1 - \lambda_L dt)} = \frac{1 - \lambda_H dt}{1 - \lambda(\pi)dt}\pi. \tag{5.7}$$

Since $\pi + \Delta\pi_0 < \pi$ in this case, the seller becomes more pessimistic in the event of no arrival:

$$d\pi_0 = -\pi(1 - \pi)\frac{\Delta\lambda dt}{1 - \lambda(\pi)dt}.$$

Note two things about the seller's beliefs in this case. First, since beliefs must be a martingale, the expected posterior equals the prior: $\lambda(\pi)dt(\pi + d\pi_1) + (1 - \lambda(\pi)dt)(\pi + d\pi_0) = \pi$. Secondly, beliefs are unaffected by the price that the seller sets.

After ignoring higher order terms in dt, the seller's Bellman equation can be written as

$$(1 + rdt)U(\pi) = \max_p \{\lambda(\pi)dt(p(1 - F(p)) + F(p)U(\pi + d\pi_1))\}$$

$$+ (1 - \lambda(\pi)dt)U(\pi + d\pi_0),$$

where $U(\cdot)$ is the value function when the seller observes the arrivals of buyers.

Notice that the term in the second line of this expression does not depend on the seller's price p. Hence the first-order condition (which is necessary and sufficient for an interior optimum) is

$$p - \frac{1 - F(p)}{f(p)} = U(\pi + d\pi_1). \tag{5.8}$$

Denote the solution to this first-order condition $pO(\pi)$. In the next proposition, we compare Eqs. (4.5) and (5.8) to conclude that the seller's price is higher when it can observe the arrivals of buyers.

Proposition 3: $pO(\pi) \geqslant pD(\pi)$: *the seller's price is higher when it observes the arrival of buyers.*

Proof: First note that $U(\pi) \geqslant V_D(\pi)$ for all $\pi \in [0, 1]$ (with strict equality when $\pi \in \{0, 1\}$): the seller's value when it observes buyers' arrivals (i.e. has more information) is never less than when it cannot. Secondly, recall that $V_D(\pi)$ is a convex function. The first-order condition for the dynamic problem with unobserved arrivals may be alternatively written as

$$pD - \frac{1 - F(pD)}{f(pD)} = V_D(\pi) + d\pi_1 V_D'(\pi),$$

where the expression for $d\pi_1$ has been substituted. Using the first two facts stated at the start of the proof, it follows that $V_D(\pi) + \Delta\pi_1 V_D'(\pi) \leqslant U(\pi + d\pi_1)$. This proves the proposition. $\qquad\square$

The proof of the proposition is illustrated in Fig. 11.2. The first-order condition (5.8) shows the intuitive fact that, when the seller observes arrivals, the price is based on the opportunity cost of a sale when an arrival occurs. (The price when no arrival occurs is irrelevant.) In contrast, when the seller does not observe arrivals, the price is based on an opportunity cost conditional on no sale occurring. This involves a capital loss from learning, $d\pi_1 V_D'(\pi)$: the marginal change in the seller's value arising from the change in the seller's beliefs. Bayes' rule tells us that the downward drift in beliefs following no sale, when arrivals are not observed, is proportional to the upward jump in beliefs when an arrival is observed. The factor relating the two is the probability of a sale $\lambda(\pi)dt(1 - F(p))$ with unobserved arrivals.

This result establishes an upper bound for the price that the seller charges when arrivals are unobserved. The extent to which learning is

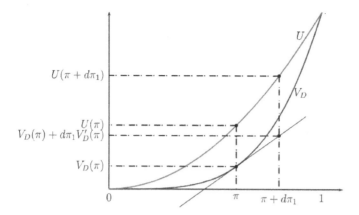

Figure 11.2. The proof of Proposition 3.

controlled when arrivals are not observed is bounded above by the price increase resulting from full observability of arrivals.

6. Conclusion

We have written the model in terms of a seller learning about demand for her good. The model can easily be generalized to accommodate other standard economic models. For example, we can interpret the model as a job search problem when an unemployed worker sets his reservation wage while learning about employment opportunities. By introducing a flow cost for actions, it is possible to transform the model of learning about arrival rates into a model of R&D where the true success probability is initially unknown. Furthermore, there is no particularly compelling reason to concentrate only on Bayesian learning models. The change in posteriors could equally well reflect accumulation of past accumulated actions. Hence this model can be used as a basis for a non-Poisson model of R&D.[6]

The paper is written in terms of ever more pessimistic continuation beliefs. Obviously we could have taken an application where continuation beliefs drift upwards conditional on no stopping. As an example, consider the maintenance problem of a machine whose durability is uncertain. (Or the problem of finding the optimal effort level to demand from an agent or seller.) As long as the machine does not break down, beliefs about its

---

[6]A previous version of this paper includes such an application and is available from the authors upon request.

longevity become more optimistic. Apart from the obvious change in signs, the controlled stopping and learning effects are present in such models as well.

Appendix A

Lemma A.1: $V_D(\pi)$ *is convex in* π.

Proof: Consider two points, π and π' and let $\pi^\alpha := \alpha\pi + (1 - \alpha)\pi'$. We want to show that for all π, π' and π^α, we have $V_D(\pi^\alpha) \leqslant \alpha V_D(\pi) + (1 - \alpha)V_D(\pi')$. Denote by $\mathbf{p}^\alpha$ the path of optimal prices starting from π^α. Denote by $W^H(\mathbf{p})$ the expected profit from an arbitrary path of prices $\mathbf{p}$ conditional on the true state being $\lambda = \lambda_H$ and similarly for $W^L(\mathbf{p})$. Then we have $V_D(\pi^\alpha) = \pi^\alpha W^H(\mathbf{p}^\alpha) + (1 - \pi^\alpha)W^L(\mathbf{p}^\alpha)$. We also know that

$$V_D(\pi) \geqslant \pi W^H(\mathbf{p}^\alpha) + (1 - \pi)W^L(\mathbf{p}^\alpha),$$
$$V_D(\pi') \geqslant \pi' W^H(\mathbf{p}^\alpha) + (1 - \pi')W^L(\mathbf{p}^\alpha)$$

since $\mathbf{p}^\alpha$ is a feasible price path.
  Hence

$$\begin{aligned}
\alpha V_D(\pi) &+ (1 - \alpha)V_D(\pi') \\
&\geqslant \alpha(\pi W^H(\mathbf{p}^\alpha) + (1 - \pi)W^L(\mathbf{p}^\alpha)) + (1 - \alpha)(\pi' W^H(\mathbf{p}^\alpha) \\
&\quad + (1 - \pi')W^L(\mathbf{p}^\alpha)) = (\alpha\pi + (1 - \alpha)\pi')\, W^H(\mathbf{p}^\alpha) \\
&\quad + (1 - \alpha\pi - (1 - \alpha)\pi')\, W^L(\mathbf{p}^\alpha) = \pi^\alpha W^H(\mathbf{p}^\alpha) \\
&\quad + (1 - \pi^\alpha)W^L(\mathbf{p}^\alpha) = V_D(\pi^\alpha).
\end{aligned}$$

This proves the claim. $\square$

A.1. *The continuous-time Bellman equation*

In Section 4, we state the continuous-time Bellman equation (4.4). We now show how this Bellman equation arises as the limit of a discrete-time

Bellman equation. In discrete time, the Bellman equation is

$$V_D(\pi) = \max_p \{\lambda(\pi)dt(1 - F(p))p + (1 - rdt)(1 - \lambda(\pi)$$

$$dt(1 - F(p)))V_D(\pi + d\pi)\}.$$

Noting that $V_D(\cdot)$ is differentiable, write

$$V_D(\pi + d\pi) = V_D(\pi) + d\pi V_D'(\pi) + (d\pi)^2 V_D''(\pi) + \cdots$$

Next note that $d\pi$ is of order dt. Since we take the limit $dt \to 0$, ignore terms in $(dt)^2$ or higher. The Bellman equation is then

$$V_D(\pi) = \max_p \{\lambda(\pi)dt(1 - F(p))p + (1 - rdt - \lambda(\pi)dt(1 - F(p)))V_D(\pi)$$

$$+ d\pi V_D'(\pi).$$

Substituting from Eq. (2.1) for $d\pi$ and re-arranging, we arrive at Eq. (4.4).

A.2. *Proof of Property 1*

To prove the first part, rewrite the first-order condition (4.5) as

$$\frac{(1 - F(p))^2}{f(p)} = \frac{rV_D(\pi)}{\lambda(\pi)}. \tag{A.1}$$

Eq. (A.1) implies that $pD(\pi)$ is non-decreasing in π iff $rV_D(\pi)/\lambda(\pi)$ is non-decreasing in π.
But

$$\frac{\partial}{\partial \pi}\left(\frac{rV_D(\pi)}{\lambda(\pi)}\right) = \frac{r}{\lambda(\pi)^2}\left(\pi\Delta\lambda\left(V_D'(\pi) - \frac{V_D(\pi)}{\pi}\right) + \lambda_L V'(\pi)\right);$$

by the facts that V_D is increasing and convex in π, this is non-negative and the result follows.

To prove the second part, consider the Bellman equation:

$$rV_D(\pi) = \max_p \left\{ \lambda(\pi)(1 - F(p)) \left(p - V_D(\pi) - \pi(1 - \pi)\frac{\Delta\lambda}{\lambda(\pi)}V_D'(\pi) \right) \right\}$$

$$\equiv P(\pi)\Lambda(\pi)$$

where

$$P(\pi) \equiv \lambda(\pi)(1 - F(pD(\pi))), \quad \Lambda(\pi) \equiv pD(\pi) - V_D(\pi) - \pi(1-\pi)\frac{\Delta\lambda}{\lambda(\pi)}V_D'(\pi).$$

Then $P_D'(\pi) = (rV_D'(\pi) - P_D(\pi)\Lambda'(\pi))/\Lambda(\pi)$, assuming that $\Lambda \neq 0$. From the first-order condition, $\Lambda(\pi) = (1 - F(pD(\pi)))/f(pD(\pi))$. By assumption, $(1 - F(p))/f(p)$ is decreasing in p. From the first part of the property, $p_D'(\pi) \geqslant 0$. This implies that $\Lambda'(\pi) \leqslant 0$.

A.3. The numerical example in Section 4.1

We assume that if the seller knows that $\lambda = \lambda_H$, its optimal price is $\bar{v}$; but if the seller knows that $\lambda = \lambda_L$, its optimal price is $\underline{v}$. This is equivalent to

$$\frac{\beta\lambda_H\bar{v}}{r + \beta\lambda_H} \geqslant \frac{\lambda_H\underline{v}}{r + \lambda_H} > \frac{\lambda_L\underline{v}}{r + \lambda_L} \geqslant \frac{\beta\lambda_L\bar{v}}{r + \beta\lambda_L}. \tag{A.2}$$

For $\pi \geqslant \pi^*$, the seller's value function is the solution to the ODE

$$\pi(1 - \pi)\Delta\lambda V'(\pi) + \left(\frac{r}{\beta} + \lambda(\pi)\right)V(\pi) - \bar{v}\lambda(\pi) = 0.$$

The general solution to this ODE is of the form

$$V(\pi) = k_1 \left(\frac{1 - \pi}{\pi}\right)^{\frac{r+\beta\lambda_H}{\beta\Delta\lambda}} \pi + \frac{\beta\lambda_L\bar{v}}{r + \beta\lambda_L} + \frac{r\beta\Delta\lambda\bar{v}}{(r + \beta\lambda_L)(r + \beta\lambda_H)}\pi$$

for $\pi \geqslant \pi^* > 0$. The first term represents the option value from learning to the seller. (k_1 is a constant of integration that will be determined below.) It is a convex function of π.

For $\pi < \pi^*$, the seller sets its price at $\underline{v}$; its value function is then given by the solution to a second ODE: $\pi(1 - \pi)\Delta\lambda V'(\pi) + (r + \lambda(\pi))V(\pi) - \underline{v}\lambda(\pi) = 0$. The general solution to this ODE is of the form

$$V(\pi) = k_2 \left(\frac{1 - \pi}{\pi}\right)^{\frac{r+\lambda_H}{\Delta\lambda}} \pi + \frac{\lambda_L \underline{v}}{r + \lambda_L} + \frac{r\Delta\lambda\underline{v}}{(r + \lambda_L)(r + \lambda_H)}\pi.$$

The first term represents the option value from learning to the seller. It is unbounded as π approaches zero; hence the constant of integration k_2 must equal zero. In words, the seller has no option value when $\pi \leqslant \pi^*$. The reason is that there is no prospect of the seller's posterior increasing above π^*; hence the seller's price is constant (equal to $\underline{v}$) once the posterior falls below π^*. There is then no option for the seller to value. So, for $\pi < \pi^*$,

$$V(\pi) = \frac{\lambda_L \underline{v}}{r + \lambda_L} + \frac{r\Delta\lambda\underline{v}}{(r + \lambda_L)(r + \lambda_H)}\pi.$$

There are two remaining parameters to determine: k_1 and the optimal boundary π^*. Since π^* is chosen optimally, value matching and smooth pasting apply; these two boundary conditions are sufficient to determine the remaining parameters. In particular,

$$\pi^* = \frac{(-(\underline{v} - \beta\bar{b})r^2 + (\beta(\bar{v} - \underline{v})\lambda_L - (\underline{v} - \beta\bar{v})\lambda_H)r + \beta(\bar{v} - \underline{v})\lambda_L\lambda_H)\lambda_L}{(-(\underline{v} - \beta\bar{v})r^2 + \beta(\bar{v} - \underline{v})(\lambda_L + \lambda_H)r + \beta(\bar{v} - \underline{v})\lambda_L\lambda_H)\Delta\lambda}.$$

Let λ_0 be such that

$$\frac{\beta\lambda_0\bar{v}}{r + \beta\lambda_0} = \frac{\lambda_0\underline{v}}{r + \lambda_0}, \quad \text{i.e. } \lambda_0 = \frac{r(\underline{v} - \beta\bar{v})}{\beta(\bar{v} - \underline{v})}.$$

For a seller who is certain of the arrival rate, the optimal price is $\bar{v}$ for all $\lambda \geqslant \lambda_0$, and $\underline{v}$ for $\lambda < \lambda_0$. Then define π_0 to be such that $\pi_0\lambda_H + (1 - \pi_0)\lambda_L = \lambda_0$. If $\pi^* \leqslant \pi_0$, then uncertainty causes the seller to set a higher price in this case with a two-point valuation distribution. Lengthy but straightforward calculation shows that this inequality holds if

$$\frac{\underline{v}}{r + \lambda_H} < \frac{\beta\bar{v}}{r + \beta\lambda_H}.$$

which was assumed at the outset: see the inequalities in (A.2). In summary: learning about an unknown arrival rate causes the seller to set a higher price.

References

[1] P. Aghion, P. Bolton, C. Harris, B. Jullien, Optimal learning by experimentation, Rev. Econ. Stud. 58 (1991) 621–654.

[2] A. Anderson, L. Smith, Explosive convexity of the value function in learning models, May 21, 2006.

[3] R.J. Balvers, T.F. Cosimano, Actively learning about demand and the dynamics of price adjustment, Econ. J. 100 (1990) 882–898.

[4] D. Bergemann, U. Hege, The financing of innovation: learning and stopping, RAND J. Econ. 36 (4) (2005) 719–752.

[5] A. Bonatti, J. Horner, Collaborating, Amer. Econ. Rev. (2010), forthcoming, available at SSRN: http://ssrn.com/abstract= 1396261.

[6] K. Burdett, T. Vishwanath, Declining reservation wages and learning, Rev. Econ. Stud. 55 (1988) 655–665.

[7] M. DeGroot, Uncertainty, information and sequential experiments, Ann. Math. Stat. 33 (1962) 404–419.

[8] J. Driffill, M. Miller, Learning and inflation convergence in the ERM, Econ. J. 103 (1993) 369–378.

[9] D. Easley, N. Kiefer, Controlling a stochastic process with unknown parameters, Econometrica 56 (1988) 1045–1064.

[10] S. Karlin, Stochastic models and optimal policy for selling an asset, in: K.J. Arrow, S. Karlin, H. Scarf (Eds.), Studies in Applied Probability and Management Science, Stanford University Press, 1962, pp. 148–158.

[11] G. Keller, Passive learning: a critique by example, Econ. Theory 33 (2007) 263–269.

[12] G. Keller, S. Rady, Optimal experimentation in a changing environment, Rev. Econ. Stud. 66 (1999) 475–507.

[13] G. Keller, S. Rady, M. Cripps, Strategic experimentation with exponential bandits, Econometrica 73 (2005) 39–68.

[14] T. Lancaster, A. Chesher, An econometric analysis of reservation wages, Econometrica 51 (1983) 1661–1676.

[15] D.A. Malueg, S.O. Tsutsui, Dynamic R&D competition with learning, RAND J. Econ. 28 (1997) 751–772.

[16] A. McLennan, Price dispersion and incomplete learning in the long run, J. Econ. Dynam. Control 7 (1984) 331–347.

[17] A. Merlo, F. Ortalo-Magne, Bargaining over residential real estate: Evidence from England, J. Urban Econ. 56 (2004) 192–216.

[18] L. Mirman, L. Samuelson, A. Urbano, Monopoly experimentation, Int. Econ. Rev. 34 (1993) 549–563.

[19] R. Rogerson, R. Shimer, R. Wright, Search-theoretic models of the labor market: A survey, J. Econ. Lit. XLIII (2005) 959–988.

[20] M. Rothschild, A two-armed bandit theory of market pricing, J. Econ. Theory 9 (1974) 185–202.

[21] A. Rustichini, A. Wolinsky, Learning about variable demand in the long run, J. Econ. Dynam. Control 19 (1995) 1283–1292.

[22] L. Smith, Optimal job search in a changing world, Math. Soc. Sci. 38 (1999) 1–9.

[23] D. Trefler, The ignorant monopolist: Optimal learning with endogenous information, Int. Econ. Rev. 34 (1993) 565–581.

[24] G. Van den Berg, Nonstationarity in job search theory, Rev. Econ. Stud. 57 (1990) 255–277.

Chapter 12

Learning and Information Aggregation in an Exit Game[*]

Pauli Murto[†] and Juuso Välimäki[‡]

*Department of Economics, Aalto University School of Business,
02150 Espoo, Finland*
[†]*pauli.murto@aalto.fi*
[‡]*juuso.valimaki@aalto.fi*

We analyze information aggregation in a stopping game with uncertain pay-offs that are correlated across players. Players learn from their own private experiences as well as by observing the actions of other players. We give a full characterization of the symmetric mixed strategy equilibrium, and show that information aggregates in randomly occurring exit waves. Observational learning induces the players to stay in the game longer. The equilibria display aggregate randomness even for large numbers of players.

Keywords: Learning, optimal stopping, dynamic games.

JEL Codes: C73, D81, D82, D83

1. Introduction

Learning in dynamic decision problems comes in two different forms. Players learn from their own individual, and often private, observations about the fundamentals of their economic environment. At the same time, they may learn by observing the behavior of other players in analogous situations. In this paper, we analyze the interplay of private and observational learning in a timing game with pure informational externalities. We show that even though private information accumulates steadily over time in our model, it is transmitted across players in occasional bursts.

---

[*]This chapter is reproduced from *Review of Economic Studies*, **78**, 1426–1461, 2011.

There are many economic environments where both forms of learning are important. Learning about the quality of a service, the profitability of a new technology, or the size of a new market are examples of this type. Another aspect that is relevant in such contexts is the degree to which the uncertainty is common across players. Often it is reasonable to assume that part of the uncertainty is common to all agents, while part is idiosyncratic. For example, demand may be high or low. For a population of monopolistically competing firms, the market is profitable to a larger fraction of firms if demand is high. Learning from others is useful to the extent that it can be used to determine the overall demand level. It is not sufficient, however, as it may be that the product of an individual firm does not appeal to the consumers even when demand is high.

To represent private learning, we use a standard discounted single-player experimentation model in discrete time. Players do not know their type at the beginning of the game. Over time, they learn by observing signals that are correlated with their true payoff type. We assume binary types. Good types gain in expected terms by staying in the game, while bad types gain by exiting the game. We assume that information accumulates according to a particularly simple form. Good types observe a perfectly informative signal with a constant probability in each period that they stay in the game, while bad types never see any signals.[1] Uninformed players become more pessimistic as time passes and their optimal strategy is to exit the game once a threshold level of pessimism is reached.

Our timing game involves many players that face the experimentation problem outlined above and whose initially unknown types are correlated. The correlation in the players' types means that observing others in the game yields additional information. The basic mechanism is that the more pessimistic players are more likely to exit, and therefore, observing another player exit is bad news, while observing another player stay is good news. Uninformed players gain from this additional information, and this creates an incentive to wait as in Chamley and Gale (1994). But in contrast to Chamley and Gale (1994), private learning prevents the players from waiting indefinitely. Our model strikes a balance between the benefits from delaying in order to learn more from others and the costs from increased pessimism as a result of private learning.

---

[1]The actual form of information revelation is not very important for the logic of our model. The important assumption is that it takes some time for any player to become so pessimistic that exiting is optimal.

We show that the game has a unique symmetric equilibrium in mixed strategies. In order to highlight the effects of learning and waiting, we eliminate the observation lags by reducing the time interval between consecutive decision moments. We show that the symmetric equilibrium can be characterized by two modes of behavior that we label the *flow modes* and the *exit waves*. For most of the time, the game is in the flow mode. In that mode, the bad news from receiving no informative private signals is balanced by the good news from the observation that no other player exits. Exits are infrequent and prior to any exit, the beliefs of the uninformed players evolve smoothly.

Whenever some player exits, the beliefs of the other players become more pessimistic. This means that from the perspective of an individual uninformed player, it is suddenly optimal to exit as well. However, if all the uninformed players were to exit immediately, this would reveal so much information that an individual player would find it optimal to wait (since the cost of delay is small for frequent periods). As a result, the equilibrium must be in mixed strategies that balance the incentives to exit and wait. These randomizations are likely to result in further exits. If that is the case, then pessimism persists and yet another round of randomizations is called for. We call this phase of consecutive exits an *exit wave*.[2] Only when there is a period with no exits, a sufficient level of optimism is restored and the exit wave ends. An exit wave thus ends either in a collapse of the game where the last uninformed player exits or in a reversion to the flow mode. In the symmetric equilibrium, the play fluctuates randomly between flow modes and exit waves until a collapse ends the game. In Section 7, we argue that exit decisions are bunched together also in similar models where we allow for heterogeneity between the uninformed players.[3]

When the number of players is increased toward infinity, the pooled information on the aggregate state becomes accurate. One might conjecture that, conditional on the state, aggregate randomness would vanish by the law of large numbers. We show that this is not the case. Even in large games, transitions between the modes remain random. The size of an individual exit wave as measured by the total number of exits during the

---

[2] Other models that display waves of action that resemble our exit waves include Bulow and Klemperer (1994) and Toxvaerd (2008). However, these models depend on the direct payoffs externalities arising from scarcity, whereas our waves are purely informational.
[3] Without observational learning, exit decisions would then vary smoothly as a function of the private information of the players.

wave also remains random. Information is thus aggregated during quick random bursts. We compute the exit probabilities during exit waves and the hazard rate for their occurrence when the number of players is large.

We show that even though in the large game limit, the aggregate state must eventually be revealed, this can never happen earlier than the efficient stopping time conditional on the true state. In particular, this means that information is aggregated efficiently in the high state: almost all uninformed players exit as if they knew the true state. But if the state is low, information aggregation fails: the players learn the state too late. In terms of the pay-offs, the possibility to observe other players' behavior helps the good types by making them less likely to exit the game. At the same time, observational learning hurts the bad types as they are encouraged to stay in the game too long.

1.1. *Related literature*

This paper is related to the literature on herding and observational learning where players have private information about a common state variable at the beginning of the game. Early papers in this literature assumed an exogenously given order of moves for the players, e.g. Banerjee (1992), Bikhchandani, Hirshleifer and Welch (1992), and Smith and Sorensen (2000). A number of later papers have endogenized the timing of action choices. Among those, the most closely related to ours is Chamley and Gale (1994).[4] In that paper, a number of privately informed players consider investing in a market of uncertain aggregate profitability. The model exhibits herding with positive probability: the players' beliefs may get trapped in a region with no investment even if the market is profitable. In our model, private learning during the game prevents the beliefs from getting trapped. The difference between the models is best seen by eliminating observation lags, i.e. letting period length go to zero. In Chamley and Gale, information aggregates incompletely in a single burst at the start of the game. In our model, information is revealed eventually but at a slow rate.

Caplin and Leahy (1994) and Rosenberg *et al.* (2007) consider models with gradual private learning about common values. While these papers are close to ours in their motivation, each makes a crucial modeling assumption that leads to qualitatively different information aggregation properties

---

[4]See also a more general model (Chamley, 2004). An early contribution along these lines is Mariotti (1992).

from ours. Caplin and Leahy assume a continuum of players. This implies that the actions of the players either reveal no information or result in full information revelation. In contrast, a key feature of our model is that a large number of players reveal information at a moderate rate, a possibility hence ruled out by Caplin and Leahy. Rosenberg *et al.* (2007) assume a finite number of players like we do, but they assume signals that may make a player so pessimistic after one period that exiting is the dominant strategy right away. As a result, when the number of players is increased towards infinity, the exit behavior after the first period reveals the state by the law of large numbers. Due to these modeling assumptions, the aggregate behavior in the large game limit is essentially deterministic conditional on state both in Caplin and Leahy (1994) and Rosenberg *et al.* (2007).[5] Our model displays a qualitatively different form of endogenous information transmission between the players, where private information is accumulated gradually and is revealed in randomly occurring exit waves. In Section 7, we discuss some extensions of our main model where the pattern of information aggregation remains similar to the main model.

Another difference to the literature mentioned above is that by combining common and idiosyncratic uncertainty, our paper relaxes the assumption of perfect payoff correlation across players made in Chamley and Gale (1994), Caplin and Leahy (1994), and Rosenberg *et al.* (2007). This makes it possible to have a richer analysis of the welfare consequences of social learning. The pure common values case is obtained in our model as a special case.

Our paper is also related to the literature on strategic experimentation. That literature focuses on the private provision of public information rather than aggregation of privately held information. Examples of such models are Bolton and Harris (1999) and Keller *et al.* (2005). The key difference is that in those models, the signals of all players are publicly observable, whereas in our model, the players see only each other's actions.

The paper is organized as follows. Section 2 sets up the discrete-time model and Section 2.1 presents an alternative interpretation for the model as a model of irreversible investment. Section 3 describes the flow of information in the game, and Section 4 provides the analysis of the symmetric equilibrium. In Section 5, we discuss information aggregation in large games. In Section 6, we characterize the symmetric equilibrium in

---

[5]The main emphasis in Rosenberg *et al.* (2007) is on the characterization of the symmetric equilibrium in games with a small number of players.

the continuous time limit. Section 7 presents a number of extensions of the basic model and some suggestions for future work. Appendix A contains the proofs for all results given in Sections 3, 4, and 6. The proofs for Section 5 are longer and they are given in Appendix B.

2. Model

The model is in discrete time with periods $t = 0, 1, \ldots, \infty$. The discount factor per period is $\delta = e^{-r\Delta}$, where Δ is the period length and $r > 0$ is the pure rate of time preference. The set of players is $\{1, \ldots, N\}$.

Before the game starts, nature chooses an (aggregate) state randomly from two alternatives: $\theta \in \{H, L\}$. Let q^0 denote the common prior $q^0 = \Pr\{\theta = H\}$. After choosing the state, nature chooses randomly and independently the type of each player. Each player is either good or bad, but the players do not know their type at the beginning of the game. If $\theta = H$, the probability of being good is ρ^H, while if $\theta = L$, the probability of being good is ρ^L, where $0 \leq \rho^L < \rho^H \leq 1$. In the special case, where $\rho^H = 1$ and $\rho^L = 0$, the players' types are perfectly correlated and the game is one of pure common values. Conditional on the state, the player types are drawn independently for all players. All types are initially unobservable to all players, but the parameters q^0, ρ^H, and ρ^L are common knowledge.

The information about the aggregate state and the individual types arrives gradually during the game. As long as a player stays in the game, she receives a random signal $\zeta \in \{0, 1\}$ in each period. Signals have two functions: they generate payoffs and transmit information. For a bad-type player, $\zeta = 0$ with Probability 1. For a good player, $\Pr\{\zeta = 1\} = \lambda\Delta$, where $\lambda > 0$ is a commonly known parameter.[6] Upon receiving the signal, the players collect a payoff $\zeta \cdot v$, where v is the value of the good signal $\zeta = 1$ for the players. Note that a good signal occurs with a probability that depends linearly on the period length, and as a result, information arrives at a constant rate in real time. The signal realizations across periods and players (conditional on the state and the type) are assumed to be independent. Each player observes only her own signals. We use the terms *informed* and *uninformed* to refer to the players' knowledge of their own type: players who have had a good signal are informed, other players are uninformed.

---

[6]We assume throughout that Δ is so small that $\lambda\Delta \ll 1$.

At the beginning of each period t, all active players i make a binary decision a_i^t. They either exit, $a_i^t = 0$, or continue, $a_i^t = 1$. Exiting is costless but irreversible: once a player exits, she becomes inactive and receives the outside option payoff 0. Hence, we require that whenever $a_i^t = 0$, then $a_i^s = 0$ for all $s > t$. We call player i active in period t if she has stayed in the game up to that point in time. We denote by $\mathcal{N}$ the set of active players, and we let n denote their number. If the player continues in the game, she pays the (opportunity) cost $c \cdot \Delta$. The cost c and the benefit v are parameters for which we assume $0 < c < \lambda v$. The expected payoff per period is $(\lambda v - c)\Delta > 0$ for a good player and $-c\Delta < 0$ for a bad player. This means that if the players knew their types, bad types would exit immediately, and good types would never exit.

At the beginning of each period, the players choose their actions simultaneously. Actions are publicly observed. At the end of the period, the players observe their own private signals. Hence, the past actions are common knowledge, but the players do not know each others' informational types, i.e. whether the other players are informed.

The history of player i consists of her private history of own past signals, and the public history consisting of the actions of all players. Since a good signal reveals fully the player's type, the uninformed have never observed the good signal. Conditional on a good signal, it is a dominant strategy to stay in the game forever. It is therefore sufficient to specify the exit behavior of the uninformed players. For the uninformed players, all relevant information is contained in the public history of past actions, and therefore, we call this public information simply the *history*. Formally, a history h^t in period t is a sequence of actions:

$$h^t = \{a^0, a^1, \ldots, a^{t-1}\},$$

where $a^t = (a_1^t, \ldots, a_N^t)$. Denote by H^t the set of all such histories up to t and let $H = \bigcup_{t=0}^{\infty} H^t$. A history $h^\infty = \{a^t\}_{t=0}^\infty$ gives a sequence of action profiles for the entire game.

A (behavior) strategy for an uninformed player i is a mapping

$$\sigma_i : H \to [0, 1]$$

that maps all histories to an exit probability.[7] A strategy profile in the game is a vector $\boldsymbol{\sigma} = (\sigma_1, \ldots, \sigma_N)$.

---

[7]Since exit is irreversible, we require that $\sigma_i(h^t) = 1$ for a player that has already exited at some earlier period $s < t$.

Each player maximizes her expected discounted sum of payoffs as estimated on the basis of her own signal history, observations of the other players' behavior, and the initial prior probability assessment q^0. By equilibrium, we mean a perfect Bayesian equilibrium of the above game. In an equilibrium, all actions in the support of $\sigma_i(h^t)$ are best responses to σ_{-i} for all i and for all h^t.

2.1. *Interpretation as an investment game*

We can interpret the game as an investment model where a number of firms have the option of undertaking an irreversible investment. The project qualities are correlated across firms. A good project yields $c\Delta > 0$ per period, whereas a bad project yields $(c - \lambda v)\Delta < 0$ per period. The fixed investment cost is normalized to zero. Before undertaking the project, each firm learns about the quality of her individual potential project as follows. In each period of this prior experimentation phase, a "failure" signal occurs with probability $\lambda\Delta$ if and only if the project is bad.[8] The firms do not, however, earn the proceeds $c\Delta$ before the actual investment has been made.

To see that this is equivalent to our exit game, consider the capitalized value of undertaking the action "exit" in our original model. If the player type is bad, then by exiting the player avoids the fixed cost $c\Delta$ from today to eternity. Therefore, the capitalized value of exit is equal to the value of investing in a "good" project in the investment model. A "good" type in the original model avoids the cost $c\Delta$ by exiting, but at the same time, she forgoes the expected payoffs $\lambda v\Delta$ per period. The net capitalized value of exit is then equal to the value of investing in a "bad" project. This shows that the two models are isomorphic (with "good" types interpreted as "bad" projects and "bad" types as "good" projects).

3. Beliefs

In this section, we describe the two different forms of learning in our model. First, as long as a player stays in the game, she receives in every period a direct signal ζ^t on her own private type. The strength of this signal is exogenously given, and Bayesian updating resulting from such signals has been studied extensively in the literature. We let p_t denote the player's

---

[8]See Décamps and Mariotti (2004) for another investment model with this kind of learning.

belief on the event that her type is good given that she is uninformed after t periods. By p_t^θ, we denote the probability of being good type conditional on being uninformed after t periods and conditional on state θ. Section 3.1 below describes the evolution of these beliefs, which is based on private signals only.

The second form of learning depends on the publicly observed actions and is endogenous in our model. Only uninformed players exit in our game and the number of uninformed players depends on the state. As a result, players learn about the state by observing the other players' actions. Since each player's own type is also correlated with the state, this information is payoff relevant.

We denote by $\pi_i^\theta(h^t)$ the posterior probability with which player i is uninformed given history h^t, conditional on state θ. Using this notation, we can define the exit probability of player i conditional on state θ as

$$\zeta_i^\theta(h^t) := \sigma_i(h^t)\pi_i^\theta(h^t). \tag{3.1}$$

As long as $\xi_i^H(h^t) \neq \xi_i^L(h^t)$, the other players learn about the true aggregate state by observing the exit decisions of player i. Note that the speed of observational learning depends on σ_i through two channels. The direct effect is through current exit probabilities $\sigma_i(h^t)$, while the indirect effect summarizes the observations from all past periods in the statistic $\pi_i^\theta(h^t)$. We denote by $\widehat{q}(h^t)$ the probability of the event that the state is high given a public history h^t. Section 3.2 describes the evolution of $\widehat{q}(h^t)$, which is based only on the public history.

Section 3.3 combines the two forms of learning to derive the beliefs of an uninformed player in our model. Since p_t^θ does not depend on observational learning and since $\widehat{q}(h^t)$ is by definition independent of private observations, this step simply puts together the information contained in p_t^θ and $\widehat{q}(h^t)$. We denote by $p(h^t)$ a player's belief that her type is good given that she is uninformed and given the public history h^t. By $q(h^t)$, we denote her belief that the state is high. We summarize the different forms of updating in the table below:

(Section 3.1) $\qquad\qquad\qquad p_t^\theta \xrightarrow{\zeta^t} p_{t+1}^\theta$

$\qquad\qquad\qquad\qquad\qquad\qquad\qquad\searrow$

$\qquad\qquad\qquad\qquad\qquad\qquad\qquad\qquad (p(h^{(t+1)}), q(h^{(t+1)}))$ (Section 3.3)

$\qquad\qquad\qquad\qquad\qquad\qquad\qquad\nearrow$

(Section 3.2) $\quad \widehat{q}(h^t) \xrightarrow{\sigma(h^t), \pi^\theta(h^t)} \widehat{q}(h^{t+1})$

3.1. *Private learning*

We start with the analysis of an isolated player that can only learn from her own signals. Denote by p_t the current belief of an uninformed player about her type, i.e.

$$p_t := \Pr\{\text{"type of player } i \text{ is good"} \mid \text{"}i \text{ is uninformed in period } t\text{".}\}.$$

If the player continues for another period and receives a bad signal $\zeta = 0$, the new posterior p_{t+1} is obtained by Bayes' rule:

$$p_{t+1} = \frac{p_t(1 - \lambda\Delta)}{p_t(1 - \lambda\Delta) + 1 - p_t}. \tag{3.2}$$

The updating formula is essentially the same if the player knows the true aggregate state. We let p_t^θ denote the player's belief on her own type conditional on state θ. Using $p_0^\theta = p^\theta$, equation (3.2) gives us the formula for p_t^θ:

$$p_t^\theta = \frac{\rho^\theta(1 - \lambda\Delta)^t}{\rho^\theta(1 - \lambda\Delta)^t + (1 - \rho^\theta)}. \tag{3.3}$$

Note that p_t^θ is a strictly decreasing function of t and since it conditions on the state of the world, it will not be affected by learning from others.

3.2. *Observational learning*

To describe observational learning in our model, we consider for the moment how player i learns from the behavior of players $j \neq i$ if she ignores her private signals. We denote by $\widehat{q}_i(h^t)$ the belief of i on the aggregate state, when learning is based only on the behavior of the other players. Alternatively, we may think of $\widehat{q}_i(h^t)$ as the belief of player i as an *outside observer* to the game.

Recall that $\zeta_j^\theta(h^t)$ denotes the probability with which an active player $j \in \mathcal{N}(h^t)$ exits at history h^t. If $\zeta_j^L(h^t) > \zeta_j^H(h^t) > 0$ and j does not exit, then i believes that j is more likely to be informed, and that state H is relatively more likely. To describe the belief updating, we denote by $A_{-i}(h^t)$ the random vector containing the actions of all active players, excluding i, at history h^t. The probability of a given exit vector a_{-i}^t is then

$$P_\theta(A_{-i}(h^t) = a_{-i}^t) = \prod_{\substack{j \neq i \\ j \in \mathcal{N}(h^t)}} ((1 - a_j^t)\zeta_j^\theta(h^t) + a_j^t(1 - \zeta_j^\theta(h^t))), \tag{3.4}$$

where we use shorthand notation P_θ to denote probability conditional on state:

$$P_H(\cdot) := \Pr(\cdot | \theta = H), \ P_L(\cdot) := \Pr(\cdot | \theta = L).$$

After observing the exit vector a_{-i}^t, player i updates her belief $\widehat{q}_i(h^t)$ according to Bayes' rule as follows:

$$\widehat{q}_i(h^{t+1}) = \frac{\widehat{q}_i(h^t) P_H(A_{-i}(h^t) = a_{-i}^t)}{\widehat{q}_i(h^t) P_H(A_{-i}(h^t) = a_{-i}^t) + (1 - \widehat{q}_i(h^t)) P_L(A_{-i}(h^t) = a_{-i}^t)}.$$
(3.5)

Note that $\widehat{q}(h^{t+1})$ depends on $\zeta_j^\theta(h^t)$ through equation (3.4), which in turn depends on $\pi_j^\theta(h^t)$ through equation (3.1). Therefore, in order to complete the description of observational learning in our model, we must also specify the evolution of $\pi_i^\theta(h^t)$, $i = 1, \ldots, N$, for the fixed strategy profile σ. These posteriors change for two reasons within each period. First, at the beginning of each period, exit decisions are realized. If player i continues, then the other players believe that i is more likely to be informed and update their beliefs using Bayes' rule as follows:

$$\pi_i'^\theta(h^t) = \frac{\pi_i^\theta(h^t)(1 - \sigma_i(h^t))}{1 - \sigma_i(h^t)\pi_i^\theta(h^t)}, \ \theta \in \{H, L\}.$$
(3.6)

Second, uninformed players become informed within the current period with Probability $1 - p_t^\theta \lambda \Delta$ (conditional on not exiting). Combining these two steps, the updated belief after history h^{t+1} is

$$\pi_i^\theta(h^{t+1}) = \frac{\pi_i^\theta(h^t)(1 - \sigma_i(h^t))}{1 - \sigma_i(h^t)\pi_i^\theta(h^t)} \cdot (1 - p_t^\theta \lambda \Delta), \ \theta \in \{H, L\}.$$
(3.7)

3.3. *Combined learning*

The remaining task is to combine the two forms of learning to derive the beliefs of the uninformed players in the game. Recall that $\widehat{q}_i(h^t)$ denotes i's belief on aggregate state if she ignores all of her private signals. Therefore, the actual belief of an uninformed player i is obtained by combining the information contained in $\widehat{q}_i(h^t)$ with the information contained in her private history (i.e. with the fact that i remains uninformed).

Let π_t^θ denote the *ex ante* probability with which a player (that stays in the game with Probability 1) is uninformed in period t, conditional on state θ. The player is of a bad type with probability $(1 - \rho^\theta)$, and all bad types remain uninformed with Probability 1. The player is a good type with probability ρ^θ, and good types remain uninformed with probability $(1 - \lambda\Delta)$ in each period. Hence, we have

$$\pi_t^\theta = (1 - \rho^\theta) + \rho^\theta (1 - \lambda\Delta)^t. \tag{3.8}$$

We denote by $q_i(h^t)$ the belief of an uninformed player i on the aggregate state. This belief differs from $\widehat{q}_i(h^t)$ only to the extent that the private history of i affects her belief, and therefore the relationship between the two is given by Bayes' rule as follows:

$$q_i(h^t) = \frac{\widehat{q}_i(h^t)\pi_t^H}{\widehat{q}_i(h^t)\pi_t^H + (1 - \widehat{q}_i(h^t))\pi_t^L}, \tag{3.9}$$

where π_t^L and π_t^H are given by equation (3.8).

We denote by $p_i(h^t)$ the belief of an uninformed player i on her own type. This belief is tightly linked to $q_i(h^t)$. By the law of iterated expectation, we have

$$p_i(h^t) = q_i(h^t)p_t^H + (1 - q_i(h^t))p_t^L, \tag{3.10}$$

where p_t^L and p_t^H are given by equation (3.3). Inserting equation (3.9) in this equation, we have

$$p_i(h^t) = \frac{\widehat{q}_i(h^t)\pi_t^H p_t^H + (1 - \widehat{q}_i(h^t))\pi_t^L p_t^L}{\widehat{q}_i(h^t)\pi_t^H + (1 - \widehat{q}_i(h^t))\pi_t^L}. \tag{3.11}$$

We end this section with two propositions that characterize learning in our model. The first proposition establishes that whatever the strategy profile, the players are always more likely to exit in state L than in state H. In particular, the likelihood ratio across states is bounded away from 1, which guarantees that an exit is always informative about the aggregate state.

To state the result, note that equation (3.8) implies that the *ex ante* likelihood ratio across states of being uninformed changes monotonically over time:

$$\frac{\pi_t^L}{\pi_t^H} > \frac{\pi_{t-1}^L}{\pi_{t-1}^H} > \cdots > \frac{\pi_1^L}{\pi_1^H} = \frac{1 - \rho^L \lambda\Delta}{1 - \rho^H \lambda\Delta} > 1. \tag{3.12}$$

With this observation at hand, we can prove our first result on the informativeness of exits.

Proposition 1: *For any strategy profile σ, we have*

$$\frac{\zeta_i^L(h^t)}{\zeta_i^H(h^t)} \geq \frac{\pi_t^L}{\pi_t^H} > 1$$

for all $t > 0$ and h^t such that $\sigma_i(h^t) > 0$.

The second proposition ranks strategy profiles according to their informativeness. For a profile σ and a history h^t, we let the random variable $P_i^{t+1}(h^t, \sigma(h^t))$ denote the posterior of player i on her own type at the beginning of period $t+1$, assuming that she is uninformed at the beginning of period t. The randomness in the posterior arises from i's private signal realization and the realized exit decisions of the players other than i. The following proposition shows that higher exit probabilities by other players induce a mean preserving spread (in the sense of Rothschild and Stiglitz (1970)) on the posterior.

Proposition 2: *Take an arbitrary history h^t with $t > 0$ and two strategy profiles σ and σ' with $\sigma(h^s) = \sigma'(h^s)$ for $s = 0, \ldots, t - 1$. Then $P_i^{t+1}(h^t, \sigma(h^t))$ dominates $P_i^{t+1}(h^t, \sigma'(h^t))$ in the sense of second-order stochastic dominance if $\sigma_j'(h^t) \geq \sigma_j(h^t)$ for all $j \neq i$ and $\sigma_j'(h^t) > \sigma_j(h^t)$ for some $j \neq i$.*

The economic content of this proposition is rather immediate. Since all players' types are correlated with the state of the world θ, having maximal information on the state is also maximal information on an individual type. The total amount of information available to the players is captured by the vector of information types for the players, i.e. an enumeration of all players that are informed. A pure strategy profile $\sigma(h^t) = 1$ transmits all this information since under this strategy, players exit if and only if they are uninformed. The profile with $\sigma(h^t) = 0$ conveys no information. Any intermediate exit probability can be seen as a convex combination of these two signal structures, and it is to be expected that the combination with a higher weight on the informative signal is more informative with respect to the true informational state of a player.

4. Equilibrium Analysis

4.1. *Isolated player*

It is again useful to start with the case of an isolated player. The decision problem of the isolated player is to choose whether to continue or exit at period t. Standard arguments show that the problem is a Markovian optimal stopping problem with the posterior probability $p = p_t$ as the state variable. We let $V_m(p)$ denote the value function of the isolated player. Stopping at posterior p yields a payoff of 0. If the player continues, she pays the cost $c\Delta$ and gets a good signal $\zeta = 1$ with probability $p\lambda\Delta$. In this case, the player learns that her expected payoff per period is $(\lambda v - c) \cdot \Delta$, and thus the value function jumps to

$$V^+(\Delta) := \frac{(\lambda v - c) \cdot \Delta}{1 - \delta}.$$

If the signal is $\zeta = 0$, then p falls to p_{t+1} according to equation (3.2). The Bellman equation for the optimal stopping problem can thus be written as

$$V_m(p) = \max \left\{ 0, \ -c\Delta + p\lambda\Delta(v + \delta V^+(\Delta)) \right.$$

$$\left. + (1 - p\lambda\Delta)\delta V_m \left(\frac{p(1 - \lambda\Delta)}{p(1 - \lambda\Delta) + (1 - p)} \right) \right\}. \qquad (4.1)$$

The optimal policy is to stop as soon as p falls below a threshold level, which we denote by $p^*(\Delta)$. Standard arguments establish that the value function $V_m(p)$ is increasing, convex and continuous in p. The threshold $p^*(\Delta)$ is obtained from equation (4.1) by setting $V_m(p^*(\Delta)) = 0$:

$$p^*(\Delta) = \frac{c}{\lambda(v + \delta V + (\Delta))}. \qquad (4.2)$$

We shall see that $p^*(\Delta)$ plays a crucial role also in the model with many players. Denote by $t^*(\Delta)$ the period in which p falls below $p^*(\Delta)$ if all signals so far have been bad:

$$t^*(\Delta) := \min\{t \in \mathbb{N} | p_t \leq p^*(\Delta)\}.$$

We denote the optimal strategy of the isolated player by

$$a_m(p_t) = \begin{cases} 1 & \text{if } p_t > p^*(\Delta), \\ 0 & \text{if } p_t \leq p^*(\Delta). \end{cases}$$

4.2. *Symmetric equilibrium*

In this section, we show that the exit game with observational learning has a unique symmetric equilibrium.[9] Furthermore, the equilibrium value

---

[9]The game has also asymmetric equilibria, where the players act in a predetermined order conditioning their actions on the outcomes of the previous moves by the other players.

functions of the individual players can be written as functions of their belief on their own type only. With symmetric strategies, all uninformed players have identical beliefs, and therefore we drop the subscripts i from the beliefs and the strategies of the uninformed players. In particular, we let $\sigma(h^t)$ denote the probability with which each uninformed player exits at history h^t in symmetric equilibrium, and we let $p(h^t)$ denote the belief of an uninformed player on her own type at history h^t.

We start by showing that if a symmetric equilibrium exists for the stopping game, then the equilibrium value function is closely related to the value function of the isolated player. Let $V(h^t)$ denote the equilibrium value of an uninformed player at history h^t. Lemma 1 says that the equilibrium value is equal to the value of an isolated player evaluated at the current belief. The key observation for this result is that as long as $\sigma(h^t) = 0$, there is no observational learning and thus the information available in the game is identical to the information available to the isolated player. On the other hand, whenever $\sigma(h^t) > 0$, the players can learn from each other, but then their value must be zero since they choose to exit with a positive probability.[10]

Lemma 1: *For any symmetric equilibrium of the exit game,*

$$V(h^t) = V_m(p(h^t)).$$

With the help of Lemma 1, we can derive a symmetric equilibrium strategy profile recursively. To see this, note that if a symmetric equilibrium is given for periods $0, \ldots, t - 1$, we can calculate the beliefs of uninformed players at history h^t as explained in Section 3. Consider then exit probabilities at history h^t. By Lemma 1, the payoff for the next period is given by $V(h^{t+1}) = V_m(p(h^{t+1}))$, and therefore, all we have to do is to find an exit probability $\sigma(h^t)$ that induces a probability distribution for $p(h^{t+1})$ that makes the players indifferent between exiting and staying. This indifference condition must equate the discounted expected value for the next period with the cost of staying for one period, so we can write it as

$$\delta\mathbb{E}V_m(P^{t+1}(h^t; \sigma(h^t))) = c(h^t) \cdot \Delta,$$

Since the properties of such equilibria are essentially similar to the herding models with exogenous order of moves, we do not discuss them further (details about asymmetric equilibria are available from authors upon request).

[10] In Section 7, we discuss how this result is modified in extensions of the current model where the active players may have heterogeneous private histories.

where we use notation $c(h^t)$ to denote cost of staying net of expected payoff per time unit:

$$c(h^t) := c - p(h^t)\lambda v.$$

The next lemma shows that increasing the exit probabilities for the current period increases the players' incentive to stay. This result follows from two observations. First, by Proposition 2, increasing the exit probability for the current period induces a mean preserving spread for the next period belief $p^{t+1}(h^t, \sigma(h^t))$. Second, we know from Section 4.1 that the isolated player's value function V_m is convex. The lemma also guarantees that this monotonicity property is strict at the point of indifference, which is essential for the uniqueness of symmetric equilibrium.

Lemma 2: *The expected continuation payoff* $\mathbb{E}V_m(P^{t+1}(h^t; \sigma(h^t)))$ *is weakly increasing in* $\sigma(h^t)$. *Furthermore, for each* h^t *such that* $p(h^t) < p^*(\Delta)$, *there is at most one exit probability* $\sigma(h^t)$ *satisfying*

$$\delta\mathbb{E}V_m(P^{t+1}(h^t; \sigma(h^t))) = c(h^t) \cdot \Delta.$$

Lemma 2 guarantees that for each h^t (except for the knife-edge case $p(h^t) = p^*(\Delta)$), at most one exit probability can make the players indifferent between exiting and staying.[11] However, if the players are so pessimistic that even the information transmitted by the pure strategy profile $\sigma(h^t) = (1, \ldots, 1)$ is not sufficient to compensate for the loss $c(h^t)\Delta$ of waiting for one more period, then it is a dominant action for all the uninformed players to exit with Probability 1. When this happens, we say that the game *collapses*.

With these preliminaries, we are ready to prove the existence and the uniqueness of a symmetric equilibrium.[12]

---

[11] For the non-generic case where $p(h^t) = p^*(\Delta)$ after some history h^t, all randomizations $\sigma(h^t)$ that result in posteriors $p(h^{t+1}) \leq p^*(\Delta)$ for all a^t are compatible with equilibrium.

[12] The uniqueness is modulo the multiplicity at $p(h^t) = p^*$ as explained in the previous footnote.

Theorem 1: *The stopping game has a unique symmetric equilibrium where the exit probability at history h^t is given by*

$$
\sigma(h^t) = \begin{cases} 0 & \text{if } \delta\mathbb{E}V_m(P^{t+1}(h^t;0)) > c(h^t)\Delta, \\ \sigma^*(h^t) \in [0,1] & \text{if } \delta\mathbb{E}V_m(P^{t+1}(h^t;0)) \\ & \quad \leq c(h^t)\Delta \leq \delta\mathbb{E}V_m(P^{t+1}(h^t;1)), \\ 1 & \text{if } \delta\mathbb{E}V_m(P^{t+1}(h^t;1)) < c(h^t)\Delta, \end{cases}
$$

where $\sigma^*(h^t)$ solves

$$
\delta\mathbb{E}V_m(P^{t+1}(h^t;\sigma^*(h^t))) = c(h^t)\Delta.
$$

The symmetric equilibrium has a simple structure. Whenever the players' beliefs on their own type are above the threshold of the isolated player, i.e. $p(h^t) > p^*(\Delta)$, then $\delta\mathbb{E}V_m(P^{t+1}(h^t;0)) > c(h^t)\Delta$ and thus the equilibrium actions coincide with those prescribed by the optimal decision rule of the isolated player (i.e. stay). On the other hand, if the players are very pessimistic, then $\delta\mathbb{E}V_m(P^{t+1}(h^t;1)) < c(h^t)\Delta$, and again equilibrium actions coincide with the isolated player (i.e. exit with probability one). With intermediate beliefs equilibrium behavior differs from isolated player: in equilibrium, the exits take place with a probability that exactly balances the players' incentives to exit and wait, whereas an isolated player would exit with probability one.

We will obtain a sharper characterization of the symmetric equilibrium when we decrease the time lag between successive periods towards zero. This will be done in Section 6.

5. Information Aggregation in Large Games

In this section, we analyze information aggregation in the symmetric equilibrium of the game as the number of players grows without bound. As a benchmark case for comparison, we use the case where information is pooled, i.e. the players share all the information with each other. If the number of players is large in this benchmark, the (weak) law of large numbers implies that the players can determine the true aggregate state with arbitrarily high accuracy. As a result, for large games, the efficient benchmark in terms of information aggregation is simply the one where all the players know the aggregate state θ. Nevertheless, idiosyncratic uncertainty about player types remains also in the efficient benchmark: conditional on the state, each player is still uncertain about her own type. In state θ, an uninformed player believes that she is a good type with probability p_t^θ, and therefore it is optimal for her to exit as soon as p_t^θ falls below $p^*(\Delta)$. Hence, the efficient exit period in state θ is given by $t_\theta^*(\Delta)$ defined as

$$t_\theta^*(\Delta) := \min\{t : p_t^\theta \leq p^*(\Delta)\}, \ \theta = H, L,$$

where p_t^θ is given by equation (3.3). Since $p_t^L < p_t^H$, we have $t_L^*(\Delta) < t^*(\Delta) < t_H^*(\Delta)$.[13] That is, it is efficient to experiment longer in state $\theta = H$ than in state $\theta = L$.

The main result of this section is Theorem 2, which says that by decreasing the period length, we eliminate the possibility that a large number of players exit too early relative to this efficient benchmark. This means that for large games, information is aggregated efficiently in state $\theta = H$ because $t_H^*(\Delta)$ is an upper bound for all exit times of the uninformed players. However, if $\theta = L$, information aggregation fails: all players exit too late in expectation.

Since we vary the number of players N and the period length Δ while keeping all the other parameters of the model fixed, we denote by $\Gamma(\Delta, N)$ the game parameterized by Δ and N. We denote by $X(h^t)$ the number of players that exit the game at history h^t in the unique symmetric equilibrium of the game:

$$X(h^t) := n(h^t) - n(h^{t+1}).$$

As a first step towards Theorem 2, we consider the effect of a large number of exits on the beliefs. We already showed in Proposition 1 that individual exit probabilities are different across the two states, which allows the players to make inferences based on observed exits. It is therefore natural to expect that if a large number of players exit, then all the remaining players should have accurate beliefs on the aggregate state. Proposition 3 shows that this must indeed be the case with a high probability. The idea in the proof is to follow the belief of an outside observer to the game and to show that this belief must converge to truth at a high rate as the number of exits grows. The argument uses techniques developed in Fudenberg and Levine (1992).

---

[13]Unless ρ^L and ρ^H are very close to each other and Δ is large in which case we may have $t_L^*(\Delta) = t_H^*(\Delta)$. In that case, the optimal action does not depend on state, and observational learning is irrelevant.

Proposition 3: *For all $\varepsilon > 0$, there is some $K \in \mathbb{N}$ such that*

$$P_H\{h^\infty : n(h^t) \leq N - K \text{ and } q(h^t) < 1 - \varepsilon \text{ for some } h^t \in h^\infty\} < \varepsilon,$$
(5.1)

$$P_L\{h^\infty : n(h^t) \leq N - K \text{ and } q(h^t) > \varepsilon \text{ for some } h^t \in h^\infty\} < \varepsilon, \quad (5.2)$$

for any game $\Gamma(\Delta, N)$.

A couple of remarks are in order. First, the bound K for the number of exits in the Proposition is independent of Δ and N. Hence, by increasing N, we can make sure that the state is revealed if an arbitrarily small fraction of players exit. Second, although we assume a symmetric equilibrium profile in this section, Proposition 3 actually holds for all strategy profiles σ (equilibrium or not) as long as some private information has been accumulated before the first exits.

Proposition 3 implies that once a large number of players have exited and $\theta = H$, then with a high probability all the remaining players are so convinced of the true state that no further exits take place before the efficient exit period $t_H^*(\Delta)$. This would suggest that the total number of suboptimally early exits must be bounded. However, we must also consider the possibility that an arbitrarily large number of players exit within *a single period, before* they have learnt the true state. Our second step towards Theorem 2 is to show that by reducing period length Δ towards zero, we can eliminate this possibility. This is established in Proposition 4 below.

We need some notation to keep track of the passage of real time as we vary Δ.[14] Let τ_θ denote the efficient exit time corresponding to state θ in the limit $\Delta \to 0$:

$$\tau_\theta := \lim_{\Delta \to 0} [t_\theta^*(\Delta) \cdot \Delta], \quad \theta = H, L. \tag{5.3}$$

To link real time to the corresponding period of a discrete-time model, we define $t(\tau, \Delta)$ as the last period before an arbitrary real time τ:

$$t(\tau, \Delta) := \max\{t : t \cdot \Delta \leq \tau\}. \tag{5.4}$$

---

[14]This notation will also be useful in the following section where the continuous time limit of the model is considered.

Proposition 4: *For all $\tau < \tau_H$ and $\varepsilon > 0$, there are constants $\overline{\Delta} \in \mathbb{R}^+$ and $K \in \mathbb{N}$ such that*

$$P_H\{h^\infty : X(h^t) > K \text{ for some } t \leq t(\tau, \Delta)\} < \varepsilon,$$

for any game $\Gamma(\Delta, N)$ with $\Delta < \overline{\Delta}$.

The proof of Proposition 4 is lengthy but the intuition is straightforward. If the players were to adopt a strategy that induces a large number of exits with a non-negligible probability within a single period, then this would generate a very informative signal about the state. For all $\tau < \tau_H$, the value of such a signal is positive. If the waiting cost is low enough (i.e. Δ is small enough), then all the players would have a strict preference to observe the signal rather than exit contradicting the hypothesized positive probability of exits.

Combining Propositions 3 and 4 gives us the theorem that bounds the total number of sub-optimally early exits in the game. This result implies that in the double limit where we increase the number of players and reduce the period length, the fraction of players that exit suboptimally early shrinks to zero.

Theorem 2: *For all $\tau < \tau_H$ and $\varepsilon > 0$, there are constants $\overline{\Delta} \in \mathbb{R}^+$ and $K \in \mathbb{N}$ such that*

$$P_H\left\{ h^\infty : \sum_{t=0}^{t(\tau,\Delta)} X(h^t) > K \right\} < \varepsilon,$$

for any game $\Gamma(\Delta, N)$ with $\Delta < \overline{\Delta}$.

This theorem is important for us in two respects. First, as we will see in the next section, it allows us to compute explicitly the statistical properties of the equilibrium path in the limit where Δ is small and N is large. Second, it carries a central message about information aggregation in our model: information cannot be fully aggregated before the efficient exit time conditional on the true state.

Perhaps the most illuminating special case is the case of perfectly correlated private types (i.e. $\rho^H = 1$ and $\rho^L = 0$) where full information on the aggregate state is sufficient to determine all players' payoff types. In this case $\tau_H = \infty$ and Theorem 2 implies that in state H almost all the players remain in the game forever. In state L, all the players exit the game eventually, but they do so later than they would in the absence of observational learning.

Although we have assumed symmetric strategies throughout this section, the results would go through for asymmetric equilibria as well. The proof of Proposition 3 is valid for any asymmetric equilibrium strategy profile as such. The proof of Proposition 4 uses symmetry in two lemmas (Lemmas 6 and 8 in Appendix B). However, even there symmetry is used merely for convenience (the number of exits within a period is binomially distributed, which leads more easily to the desired results).

6. Exit Waves

In this section, we characterize the symmetric equilibrium in the limit as $\Delta \downarrow 0$. We have several reasons for this. The first reason is substantive. In a model with endogenous timing decisions, it is important to know if the results depend on an exogenously imposed reaction lag Δ. Second, it turns out that the inherent dynamics of the model are best displayed in the limit: information aggregation happens in randomly occurring bursts of sudden activity. We call these bursts of activity *exit waves*. Third, when we also let $N \to \infty$, we can compute the statistical properties of the equilibrium path in an explicit form.

We may view the public history h^∞ generated by the symmetric equilibrium $\sigma(\Delta, N)$ in the game $\Gamma(\Delta, N)$ from a slightly different angle. Suppose that the players are to be treated anonymously. Then the vector $t(\Delta, N) = (t_1(\Delta, N), \ldots, t_N(\Delta, N))$, where $t_k(\Delta, N)$ indicates the period in which kth exit took place gives a full account h^∞. The profile $\sigma(\Delta, N)$ induces a probability distribution on $\mathbb{R}^N$ on instants of exit measured in continuous time τ (i.e. the kth exit takes place at time $\tau_k = t_k(\Delta, N) \cdot \Delta$). We denote this distribution, conditional on state θ, by $F_{\Delta,N}^\theta(\tau)$. We investigate the limiting distribution

$$F_N^\theta(\tau) = \lim_{\Delta \downarrow 0} F_{\Delta,N}^\theta(\tau),$$

where the convergence is taken to be in the sense of weak convergence. Observational learning then results from the differences between $F_N^H(\tau)$ and $F_N^L(\tau)$.

In Section 6.1, we keep the number of players fixed at N. We show that when there was no exit in the previous period, the probability of an exit within the current period is of the order Δ. This means that exits arrive according to a well-defined hazard rate, and we say that the game is in the *flow mode*. On the other hand, if there was an exit in the previous period, then the probability of an exit in the current period is bounded away from zero, and we say that the game is in an *exit wave*.

In Section 6.2, we consider the limiting distributions

$$F^\theta(\tau) = \lim_{N \to \infty} F_N^\theta(\tau)$$

defined on the set of sequences of exit times $\{\tau_k\}_{k=1}^\infty$. In particular, we compute the distributions for the first K exit instants and we also calculate the probability of the event that the game collapses by time instant τ, i.e. the probability of the event $\{\tau_k \leq \tau$ for all $k\}$. We make use of Poisson approximations and Theorem 2 when computing the size of the exit events and the probability of a collapse given that an exit event started.

6.1. *The structure of the symmetric equilibrium*

In this section, we keep the number of players N fixed. Since we are interested in the limit $\Delta \to 0$, we parameterize the game and its histories with the period length Δ. We say that the game is in the *flow mode* at history h^t if no players exited at history h^{t-1}, i.e. if $X(h^{t-1}) = 0$. The game is in an *exit wave* at history h^t if $X(h^{t-1}) > 0$. Finally, we say that the game *collapses* at history h^t if $\sigma(h^t) = 1$. The collapse is an absorbing state: since all uninformed players exit, the game is effectively over, and $\sigma(h^s) = 1$ for all $s > t$. This means that for a game with a given Δ, we have three mutually exclusive sets of histories, corresponding to the flow mode, exit wave, and collapse, respectively,

$$H^f(\Delta) := \{h^t : X(h^{t-1}) = 0 \text{ and } \sigma(h^t) < 1\},$$

$$H^w(\Delta) := \{h^t : X(h^{t-1}) > 0 \text{ and } \sigma(h^t) < 1\},$$

$$H^c(\Delta) := \{h^t : \sigma(h^t) = 1\}.$$

In order to relate the discrete decision periods to real time instants, we define

$$p^* := \lim_{\Delta \downarrow 0} p^*(\Delta),$$

$$\tau^* := \lim_{\Delta \downarrow 0} t^*(\Delta) \cdot \Delta$$

where $p^*(\Delta)$ and $t^*(\Delta)$ denote the belief threshold and the corresponding exit time as defined in Section 4.1.

We start by showing that the beliefs of the uninformed players are qualitatively different in the two active modes. When the game is in the flow mode and Δ is small, beliefs stay close to p^* (as long as $t \geq t^*(\Delta)$, i.e. players have started to randomize). In contrast, in an exit wave, the beliefs are bounded away from p^*.

Lemma 3: (*i*) *For all* $\varepsilon > 0$, *there is a* $\overline{\Delta} > 0$ *such that*

$$p(h^t) \in (p^* - \varepsilon, p^* + \varepsilon)$$

for all $h^t \in H^f(\Delta)$, $t \geq t^*(\Delta)$, $\Delta < \overline{\Delta}$.
(*ii*) *There is a* $v > 0$ *and a* $\overline{\Delta} > 0$ *such that*

$$p(h^t) < p^* - v$$

for all $h^t \in H^w(\Delta)$, $\Delta < \overline{\Delta}$.

The following proposition shows that the active players also behave differently in the two modes. In the flow mode, the probability with which any exits take place within a period is at most proportional to the period length. In contrast, in an exit wave, the corresponding probability is bounded away from zero even in the limit $\Delta \to 0$.

Proposition 5: (*i*) *There is a* $\kappa > 0$ *such that*

$$P_H(X(h^t) > 0) < P_L(X(h^t) > 0) < \kappa\Delta \text{ for all } h^t \in H^f(\Delta), \Delta > 0.$$

(*ii*) *There is a* $\underline{p} > 0$ *and a* $\overline{\Delta} > 0$ *such that*

$$P_L(X(h^t) > 0) > P_H(X(h^t) > 0) > \underline{p} \text{ for all } h^t \in H^w(\Delta), 0 < \Delta < \overline{\Delta}.$$

The first claim in the above proposition justifies our use of the term *flow mode*. The flow mode comes to an end with a well-defined hazard rate. The actual computation of the equilibrium hazard rate is not hard in principle. Nevertheless, the formula will depend on the evolution of $\pi^\theta(h^t)$ and it is not possible to give a closed-form solution for the continuous-time limit of the updating formula (3.7).[15] In the following section, we compute the explicit hazard rate in the limit as $N \to \infty$.

The second point to note is that since at least one player must exit at each period for the exit wave to continue, the total number of players N gives an upper bound for the periods within an exit wave. Therefore, the real time duration of an exit wave is bounded from above by ΔN and hence vanishes as $\Delta \to 0$. As a result, we may view the limit exit waves as

---

[15]The computation is somewhat complicated because the updating of $\pi^\theta(h^t)$ depends on the equilibrium randomization probabilities $\sigma(h^t)$.

(rather complicated) randomization events between the flow mode and the collapse.

Finally, every exit wave results in a collapse with a strictly positive probability. To see this, recall that in an exit wave, $p(h^t)$ must be bounded away from p^* by part (ii) of Lemma 3, whereas by part (i) of the same lemma, the belief must be back within a small neighborhood of p^* once the game returns to the flow mode. Since this belief is a martingale, we can conclude that a return to the flow mode cannot happen with Probability 1, hence the collapse must take place with a positive probability.[16] Note that if the number of players is small, then the first exit starting a wave may in fact lead to an immediate collapse. Then the exit wave lasts only one period and ends up in a collapse with probability one. If this is not the case, then a similar martingale argument establishes that the game must return to the flow mode with a strictly positive probability.

6.2. *Exit events in large games*

The large game limit $N \to \infty$ simplifies the computations for a number of reasons. First, we can use Poisson approximations of the Binomial distribution for the number of exits within each period of an exit wave. Second, as long as the state has not been fully revealed, we know that the probability with which an individual player has exited is negligible. To see this, note that if each player would exit with a non-negligible probability, then in the limit $N \to \infty$ this would mean that the total number of exits explodes, which by Proposition 3 implies full sate revelation. The simplification that we obtain from this observation is that we can use the continuous time limit of equation (3.8) instead of equation (3.7) to compute the conditional probabilities for the players to be uninformed. Third, we can apply Theorem 2, which implies that the game collapses in state $\theta = H$ before the efficient stopping time τ_H with a vanishing probability as $N \to \infty$.

Let $p^\theta(\tau)$ and $\pi^\theta(\tau)$ denote the continuous time limits of equations (3.3) and (3.8):

$$p^\theta(\tau) := \frac{\rho^\theta e^{-\lambda\tau}}{(1-\rho^\theta) + \rho^\theta e^{-\lambda\tau}},$$

$$\pi^\theta(\tau) := (1-\rho^\theta) + \rho^\theta e^{-\lambda\tau}.$$

---

[16]The argument also utilizes the observation that an exit wave takes only a vanishing amount of real time in the limit $\Delta \to 0$. This guarantees that the possibility that the belief of a player jumps upwards during the wave *due to a private signal* can be ignored.

We compute first the hazard rate of exits in the flow mode. In particular, assume that k players have exited the game at real times $\tau_1, \ldots, \tau_k$, and the game is in the flow mode at real time τ. Using the fact that the likelihood ratio of exit across the states is given by $\pi^L(\tau)/\pi^H(\tau)$, and the fact that the belief of an uninformed player must stay close to p^* as long as no player exits (as required by Lemma 3), we can calculate the hazard rate with which an additional player exits:

Proposition 6: *In the limit $N \to \infty$, the instantaneous hazard rate of $k + 1$st exit at some $\tau \in (\tau_k, \tau_H)$, conditional on the first k exit times $\tau_1, \ldots, \tau_k$, is given by*

$$\frac{f^\theta_{k+1}(\tau | \tau_1, \ldots \ldots, \tau_k)}{1 - F^\theta_{k+1}(\tau | \tau_1, \ldots, \tau_k)}$$

$$= \pi^\theta(\tau) \lambda \frac{p^*(1 - p^*)(p^H(\tau) - p^L(\tau))}{(p^* - p^L(\tau))(p^H(\tau) - p^*)(\pi^L(\tau) - \pi^H(\tau))}. \qquad (6.1)$$

Note that the right-hand side does not depend on $\tau_1, \ldots, \tau_k$, and therefore, the hazard rate of $k + 1$st exit depends only on τ. Furthermore, note that equation (6.1) applies also in the case where no player has yet exited. Hence, irrespective of the number of players that have already exited, equation (6.1) gives the hazard rate with which an exit wave starts at time τ, $\tau > \tau^*$, conditional on the game being in the flow mode.

Every exit wave leads either to a collapse or a return to the flow mode. With a large number of players, it is easy to compute the probabilities with which either possibility occurs. To see this, let us again utilize the fact that in the flow mode an uninformed player must have belief p^* on her own type (by Lemma 3). Since the beliefs on one's own type and on the aggregate state are linked by equation (3.10), we can equivalently express this by requiring the belief on the aggregate state at time τ to be given by

$$q^*(\tau) := \frac{p^* - p^L(\tau)}{p^H(\tau) - p^L(\tau)}, \tau^* \leq \tau \leq \tau_H.$$

Note that $q^*(\tau)$ is strictly increasing within $[\tau_L, \tau_H]$ and $q^*(\tau_L) = 0$ and $q^*(\tau_H) = 1$.

Consider now the posterior after the first exit of an exit wave that takes place at real time τ. Since the belief just before the exit is given by $q^*(\tau)$ and the exit reveals one player to be uninformed, the posterior after the first exit is given by Bayes' rule:

$$q^-(\tau) = \frac{\pi^H(\tau)q^*(\tau)}{\pi^H(\tau)q^*(\tau) + \pi^L(\tau)(1 - q^*(\tau))} < q^*(\tau).$$

By Theorem 2, the game returns to the flow mode with a probability that converges to 1 as $N \to \infty$ in state $\theta = H$. Therefore, if the game collapses, $q(\tau+) = 0$. On the other hand, we know from Lemma 3 that if the game returns to the flow mode, we have $q(\tau+) = q^*(\tau)$. Let $\phi^\theta(\tau)$ denote the probability of the collapse given an exit event at $\tau < \tau_H$, conditional on state θ. By Theorem 2,

$$\phi^H(\tau) = 0,$$

so the probability of the collapse estimated by a player with belief $q^-(\tau)$ is $(1 - q^-(\tau))\phi^L(\tau)$. Therefore, by the martingale property of the belief of this player we have:[17]

$$q^-(\tau) = (1 - (1 - q^-(\tau))\phi^L(\tau))q^*(\tau),$$

which gives

$$\phi^L(\tau) = \frac{q^*(\tau) - q^-(\tau)}{q^*(\tau)(1 - q^-(\tau))} = \frac{\pi^L(\tau) - \pi^H(\tau)}{\pi^L(\tau)}. \tag{6.2}$$

Since (6.1) gives the hazard rate with which an exit wave starts, and equation (6.2) gives the probability with which a given exit wave leads to collapse, we get the hazard rate of collapse by multiplying them

Corollary 1: *In the limit $N \to \infty$, the instantaneous hazard rate of collapse at time $\tau \in (\tau^*, \tau_H)$, conditional on state, and conditional on being in the flow mode at τ, is*

$$\chi^H(\tau) = 0, \tag{6.3}$$

$$\chi^L(\tau) = \lambda \frac{p^*(1 - p^*)(p^H(\tau) - p^L(\tau))}{(p^* - p^L(\tau))(p^H(\tau) - p^*)}. \tag{6.4}$$

Note that the equations (6.3) and (6.4) tell us how an outside observer learns from the actions of the players. Since the game collapses only in state L, an outside observer becomes gradually more optimistic over time

---

[17]If the game has $N < \infty$ players, then collapse will take place at a posterior $q^C > 0$ and as a consequence, the probability of a collapse is higher than in equation (6.2).

about the aggregate state if there is no collapse. By Bayes' rule, the belief dynamics of the outside observer is given by

$$\frac{d\widehat{q}(\tau)}{d\tau} = \widehat{q}(\tau)(1 - \widehat{q}(\tau))\chi^L(\tau).$$

If $\rho^H < 1$, $\rho^H(\tau)$ falls to p^* at $\tau = \tau_H$. We see from equation (6.4) that

$$\lim_{\tau \to \tau_H} \chi^L(\tau) = \infty.$$

In words, the hazard rate of collapse in state L explodes at τ_H. This is to be expected because τ_H is an upper bound for all exit times. As a result, the state is eventually fully revealed to the outside observer: if the game collapses strictly before τ_H, then $\theta = L$, otherwise $\theta = H$.

Corollary 1 applies also to the special case of perfectly correlated types (i.e. $\rho^H = 1$ and $\rho^L = 0$). In that case, $p^H(\tau) \equiv 1$ and $p^L(\tau) \equiv 0$, and hence equation (6.4) reduces to $\chi^L(\tau) = \lambda$. To understand this simple formula, note that an uninformed player learns fully her own type from either receiving a good private signal (in which case she must be a good type) or by observing the game collapse (which in the case of perfectly correlated types means that she must be a bad type). Lemma 3 requires that her belief stay constant as long as none of these two events occur. This is possible only if both events arrive at the same rate conditional on state. Note that in this special case $\tau_H = \infty$, and hence if $\theta = H$, the game never collapses and the outside observer learns only asymptotically the true state.

We end this section by describing the sequence of events *within* a given exit wave that takes place at real time τ. We use index $s = 1, 2, \ldots$ to refer to the consecutive periods within the exit wave. Let q_s denote the belief on the aggregate state in the sth period of the wave, and let X_s denote the number of exits at that period. Note that since we are considering the limit $\Delta \to 0$, the duration of the exit wave in real time is zero in the limit.

Fix a period s and the corresponding belief q_s. Lemma 3 implies that we must have $q_s < q^*(\tau)$. On the other hand, the same lemma implies that if s is the last period of the exit wave (i.e. no player exits), then we must have $q_{s+1} = q^*(\tau)$.

Proposition 7: *Consider period s of an exit wave taking place at time τ. As $N \to \infty$, X_s^θ converges in distribution to a Poisson random variable* *with parameter:*

$$\frac{\pi^\theta(\tau)}{(\pi^L(\tau) - \pi^H(\tau))} \log\left(\frac{q^*(\tau)}{(1 - q^*(\tau))} \cdot \frac{(1 - q_s)}{q_s}\right) \quad for \ \theta \in \{H, L\}. \quad (6.5)$$

If the realized number of exits is $X_s = k$, *the next period belief is*

$$q_{s+1}(k) := \frac{(\pi^H(\tau))^k q^*(\tau)}{(\pi^H(\tau))^k q^*(\tau) + (\pi^L(\tau))^k (1 - q^*(\tau))}. \quad (6.6)$$

Note from equation (6.6) that the number of exits in the previous stage is a sufficient statistic for the belief in the current stage, which in turn by equation (6.5) is a sufficient statistic for current stage randomization probabilities. Hence, the number of exits at period s is a random variable distributed according to the Poisson distribution with a parameter that depends on the number of exits at period $s - 1$.

The exit event taking place at real time instant τ reverses to the flow mode at the first s such that $X_s = 0$. Hence, the probability of the exit wave ending at period s depends on the number of exits at period $s - 1$ and can be calculated from a Poisson distribution with the parameter given by equation (6.5).

7. Discussion

We analyzed a stopping game where the players base their decisions on their privately acquired information and on the behavior of the other players in a similar situation. The possibility to observe the actions of others makes the players more likely to postpone their actions. But this, in turn, reduces the informativeness of their actions, thus reducing the incentives to wait. The symmetric equilibrium balances these effects and leads to aggregate delays and randomly arriving exit waves. We showed that even when the number of players gets large, aggregate uncertainty persists in equilibrium. Information is aggregated gradually until a sudden collapse leads to full revelation of the aggregate state.

We kept the model as simple as possible in order to highlight the interplay between individual and social learning. In the following subsections, we discuss a number of extensions of the model.

7.1. *Heterogeneous private values*

In the main model, all players have an identical opportunity cost c and an identical value from the signal v. We could have equally well assumed that c_i is private information and drawn from a common distribution $F(c_i)$ at the beginning of the game. With this specification, the game has a pure Perfect Bayesian equilibrium in symmetric strategies. An isolated player has an optimal policy characterized by a stopping threshold $p^*(\Delta; c_i)$ and an optimal value given by $V_m(p(h^t); c_i)$. With heterogeneous private values, the conclusion of Lemma 1 holds only for players with high c_i. For each history (h^t, c_i), there exists a cut-off cost type $c(h^t)$ such that the symmetric equilibrium value $V(h^t; c_i)$ is equal to the expected payoffs of the isolated player for all $c_i > c(h^t)$.

Adjusting the cut-off level $c(h^t)$ plays the same role as changing the equilibrium randomization probabilities $\sigma(h^t)$ in the main model. It is no longer true that all players get the same payoffs as they would as isolated players (the players with lower c benefit from observational learning). Nevertheless, the characterization of the equilibrium path remains almost unchanged.

It is perhaps worth noticing that with heterogeneous costs, the isolated players' optimal stopping times are continuously distributed as a function of c_i. In the case where the individual types are perfectly correlated (i.e. $\rho^H = 1$ and $\rho^L = 0$) and players observe each others' actions, almost all players exit the game at the same time if the number of players is large.

7.2. *Bounded signals*

We stressed the importance of our assumption that the players do not become unboundedly pessimistic as a result of their private learning. Yet we allow for the fact that signals are perfectly informative for the good types. In this section, we argue that this assumption is purely for notational simplicity.

A more general model would allow for two Poisson rates of signal arrivals, λ^G if the player is a good type, and λ^B for a bad type. The players' private histories are now summed by k_i^t, the number of signals that player i has observed up to period t. The optimal stopping policy of the isolated player is still characterized by a cut-off rule $p^*(\Delta)$. Denote the symmetric equilibrium payoff in this case by $V(h^t, k_i^t)$. In analogy to Lemma 1, we can show that for all h^t such that $p(h^t, k_i^t) < p^*(\Delta)$ for some active players i, there is a cut-off level of private histories $k(h^t)$ such that all players with $k_i^t < k(h^t)$ earn the same expected continuation payoff as an isolated player with the same belief of her own type.

The equilibrium randomization probabilities are determined through the requirement that the players with exactly $k(h^t)$ signals must be indifferent between continuing and exiting. Those players with fewer than $k(h^t)$ signals exit with probability one if they have not exited already.

In the modified model, all players are *ex ante* identical, but they become heterogeneous due to their different private experiences. Since only the most pessimistic players exit with positive probability, the *ex ante* equilibrium value in the model with observational learning exceeds the *ex ante* expected payoffs of the isolated player. Furthermore, with perfectly correlated private types, almost all players exit the game at the same time when the number of players is large.

7.3. *Payoff externalities*

Finally, a more challenging extension would be to incorporate payoff externalities in the model. The payoff could e.g. depend on the number of players present in the market. It seems to us that beyond the two-player case, quite different analytical techniques would be needed to cover this case. Lemma 1 has no analogue in this extension, and as a result, the analysis will have to be quite different. In our view, this is an interesting and challenging direction for further research.

Appendix A: Shorter Proofs

Proof of Proposition 1. Since $\zeta_i^\theta(h^t) = \sigma_i(h^t)\pi_i^\theta(h^t)$, all we have to do is to show that

$$\frac{\pi_i^L(h^t)}{\pi_i^H(h^t)} \geq \frac{\pi_t^L}{\pi_t^H} \qquad (A.1)$$

for all $t > 0$ (note that $\pi_t^L/\pi_t^H > 1$ follows from equation (3.12)).

We use induction. As an induction hypothesis, assume that equation (A.1) holds for some $t \geq 0$. Using equations (3.7) and (A.1), we then have

$$\frac{\pi_i^L(h^{t+1})}{\pi_i^H(h^{t+1})} = \left(\frac{\pi_i^L(h^t)}{\pi_i^H(h^t)}\right)\left(\frac{1 - \sigma_i(h^t)\pi_i^H(h^t)}{1 - \sigma_i(h^t)\pi_i^L(h^t)}\right)\left(\frac{1 - p_t^L\lambda\Delta}{1 - p_t^H\lambda\Delta}\right)$$

$$\geq \frac{(1 - p_t^L\lambda\Delta)}{(1 - p_t^H\lambda\Delta)}\frac{\pi_t^L}{\pi_t^H}. \qquad (A.2)$$

On the other hand, using equations (3.3) and (3.8), we have

$$\pi_{t+1}^\theta = (1 - p_t^\theta\lambda\Delta)\pi_t^\theta, \ \theta = H, L.$$

Combining this with equation (A.2) gives us the induction step:

$$\frac{\pi_i^L(h^{t+1})}{\pi_i^H(h^{t+1})} \geq \frac{\pi_{t+1}^L}{\pi_{t+1}^H}.$$

Noting that $\pi_i^L(h^0) = \pi_i^H(h^0) = \pi_0^L = \pi_0^H = 1$ gives us:

$$\frac{\pi_i^L(h^0)}{\pi_i^H(h^0)} \geq \frac{\pi_0^L}{\pi_0^H},$$

and therefore, the proof by induction is complete. $\square$

Proof of Proposition 2. Construct an experiment X_i on $\Theta = \{H, L\}$ with outcomes in $S^{X_i} = \{0, 1\}$. The joint probabilities on the states and outcomes are given by the following stochastic matrix P^{X_i}:

P^{X_i}	$\theta = H$	$\theta = L$
$s^{X_i} = 1$	$1 - \sigma_i(h^t)\pi_i^H(h^t)$	$1 - \sigma_i(h^t)\pi_i^L(h^t)$
$s^{X_i} = 0$	$\sigma_i(h^t)\pi_i^H(h^t)$	$\sigma_i(h^t)\pi_i^L(h^t)$

We interpret the event $\{\theta = H\}$ as the event that the state is good and the event $\{s^{X_i} = 1\}$ as the decision of player i to stay in the game. The joint probability over (θ, s^{X_i}) simply reflects the conditional exit probabilities given strategy σ.

Consider next another experiment Y_i on Θ with outcomes in $S^{Y_i} = \{0, 1\}$ and the associated stochastic matrix P^{Y_i}

P^{Y_i}	$\theta = H$	$\theta = L$
$s^{Y_i} = 1$	$1 - \sigma_i'(h^t)\pi_i^H(h^t)$	$1 - \sigma_i'(h^t)\pi_i^L(h^t)$
$s^{Y_i} = 0$	$\sigma_i'(h^t)\pi_i^H(h^t)$	$\sigma_i'(h^t)\pi_i^L(h^t)$

with $\sigma_i'(h^t) > \sigma_i(h^t)$. Then we can write

$$P^{X_i} = \Phi P^{Y_i},$$

where the stochastic matrix Φ is given by

Φ	$s^{Y_i} = 1$	$s^{Y_i} = 0$
$s^{X_i} = 1$	1	$\frac{\sigma_i'(h^t) - \sigma_i(h^t)}{\sigma_i'(h^t)}$
$s^{X_i} = 0$	0	$\frac{\sigma_i(h^t)}{\sigma_i'(h^t)}.$

Since Φ is a stochastic matrix that is independent of θ, X_i is a garbling of Y_i, and therefore Y_i is sufficient for X_i.

Since the individual exit decisions X_i are independent (conditional on the informational status of the players), the same argument as above applies for the joint experiments $X := \times_{i=1}^{n(h^t)} X_i$ and $Y = \times_{i=1}^{n(h^t)} Y_i$.

Finally, consider two experiments $X^\omega = (X, Z)$ and $Y^\omega = (Y, Z)$ on $\Omega = \{G, B\}$, where X and Y are as above and Z is an experiment with outcomes in $S^Z = \{0, 1\}$. Since θ is correlated with ω, the information contained in X and Y is also information on Ω. We interpret Z as the individual learning experiment on own type and hence the matrix of conditional probabilities for that experiment is given by P^Z:

P^Z	$\omega = G$	$\omega = B$
$s^Z = 1$	$\lambda\Delta$	0
$s^Z = 0$	$1 - \lambda\Delta$	1

Since (X, Z) is a garbling of (Y, Z) by the argument above, we know that (Y, Z) is sufficient for (X, Z) with respect to Ω. The assertion that $P_i^{t+1}(h^t, \sigma(h^t))$ second-order stochastically dominates $P_i^{t+1}(h^t, \sigma'(h^t))$ follows from Blackwell's theorem. $\qquad\square$

Proof of Lemma 1. (i) Assume that σ is a symmetric equilibrium profile and $p(h^t) > p^*(\Delta)$. If i continues for one period and then exits, her payoff is

$$p(h^t)\lambda\Delta(v + \delta V^+(\Delta)) - c\Delta > 0,$$

where the inequality follows from equation (4.2). This implies that $\sigma(h^t) = 0$ if $p(h^t) > p^*(\Delta)$.

(ii) Assume that $p(h^t) < p^*(\Delta)$. Note from equation (4.2) that for such a belief we have

$$p(h^t)\lambda(v + \delta V^+(\Delta)) < c. \qquad (A.3)$$

We show that $\sigma(h^t) > 0$. Assume on the contrary that $\sigma(h^t) = 0$. Let h^τ be the first continuation history of h^t on the equilibrium path such that $\sigma(h^\tau) > 0$. In other words, $h^\tau = (h^t, a^t, \ldots, a^{\tau-1})$, where $a^s = (1, \ldots, 1)$ and $\sigma(h^s) = 0$ for all $s \in \{t, t+1, \ldots, \tau - 1\}$ and $\sigma(h^\tau) > 0$.

First, note that if no such h^τ exists, then the players never exit, and their value calculated at history h^t is given by

$$V(h^t) = \frac{p(h^t)\lambda v\Delta - c\Delta}{1 - \delta} < 0,$$

where the inequality follows from equation (A.3). This is a contradiction because equilibrium value must be at least zero due to the exit option.

Since exiting is in the support of the equilibrium strategy at h^τ, we have

$$V(h^\tau) = 0,$$

and hence the value in period $\tau - 1$ is given by

$$V(h^{\tau-1}) = -c\Delta + p(h^{\tau-1})\lambda\Delta(v + \delta V^+(\Delta)) < 0,$$

which contradicts optimal behavior at $h^{\tau-1}$. It follows that $\sigma(h^t) > 0$ if $p(h^t) < p^*(\Delta)$. This implies that $V(h^t) = 0$ if $p(h^t) < p^*(\Delta)$. If $p(h^t) = p^*(\Delta)$, we have

$$V(h^t) = \max\{0, p^*(\Delta)\lambda\Delta(v + \delta V^+(\Delta)) - c\Delta\} = 0.$$

(iii) Since $V(h^t) = 0$ whenever $p(h^t) \leq p^*$, and since $\sigma(h^t) = 0$ for $p(h^t) > p^*$, the pure strategy $a_m(p(h^t))$ is a best response for each player after each history h^t, given the strategy profile $\boldsymbol{\sigma}$. For any h^t such that $p(h^t) > p^*(\Delta)$, $p(h^{t+1})$ is updated on the equilibrium path using the isolated player's belief updating formula in equation (3.2) since $\sigma(h^t) = 0$. Therefore, $V(h^t) = V_m(p(h^t))$ for each h^t. □

Proof of Lemma 2. By Lemma 1,

$$\mathbb{E}V(h^{t+1}) = \mathbb{E}V_m(P^{t+1}(h^t; \sigma(h^t))).$$

Furthermore, $V_m(p)$ is convex in p and $P^{t+1}(h^t; \sigma(h^t))$ is second-order stochastically decreasing in $\sigma(h^t)$ by Proposition 2 and hence the first claim follows.

To prove the second claim, suppose that $p(h^t) < p^*(\Delta)$ and that there exists a $\sigma(h^t)$ such that

$$\delta\mathbb{E}V_m(P^{t+1}(h^t; \sigma(h^t))) = c(h^t)\Delta. \tag{A.4}$$

We claim that for all $\sigma'(h^t) > \sigma(h^t)$,

$$\mathbb{E}V_m(P^{t+1}(h^t; \sigma'(h^t))) > c(h^t)\Delta. \tag{A.5}$$

To see this, consider the exit decision of player i when all players use the symmetric strategy $\sigma(h^t)$ and equation (A.4) holds. Since $p(h^t) < p^*(\Delta)$, there must be an exit decision vector $\widehat{a}^t_{-i}$ for players other than i that makes player i's exit decision a^t_i pivotal in the following sense. In what follows, we use the notation $p(h^t; \sigma(h^t))$ and $V(h^t; \sigma(h^t))$ to emphasize the dependence of these quantities on the equilibrium strategies. For $h_0^{t+1} := (h^t, (\hat{a}_{-i}^t, 0))$,

$$p(h_0^{t+1}; \sigma(h^t)) < p^*(\Delta), \tag{A.6}$$

and for $h_1^{t+1} := (h^t, (\hat{a}_{-i}^t, 1))$,

$$p(h_1^{t+1}; \sigma(h^t)) > p^*(\Delta). \tag{A.7}$$

Furthermore,

$$\Pr\{A_{-i}(h^t) = \hat{a}_{-i}^t\} > 0.$$

Suppose next that player i exits with probability $\sigma_i' > \sigma(h^t)$ and all other players exit with probability $\sigma(h^t)$ after history h_t. We consider the beliefs of an arbitrary player $j \neq i$ following this change in the strategy profile at history h^t.

Denote the profile where all players but i exit with probability $\sigma(h^t)$ and i exits with probability σ_i' by (σ_{-i}, σ_i'). By Proposition 2, and by the convexity of $V_m(p)$, we know that for every a_{-i}^t

$$\delta \mathbb{E}_{a_i} V((h^t, (a_{-i}^t, a_i)); (\sigma_{-i}, \sigma_i')) \geq \delta \mathbb{E}_{a_i} V((h^t, (a_{-i}^t, a_i)); \sigma(h^t)).$$

Therefore, the payoff of players other than i is strictly increasing in σ_i', if for the exit vector $\hat{a}_{-i}^t$, the previous inequality is strict, i.e.

$$\delta \mathbb{E}_{a_i} V((h^t, (\hat{a}_{-i}^t, a_i)); (\sigma_{-i}, \sigma_i')) > \delta \mathbb{E}_{a_i} V((h^t, (\hat{a}_{-i}^t, a_i)); \sigma(h^t)).$$

But this follows immediately from equations (A.6) and (A.7) and the strict convexity of the isolated player value function V_m in the neighborhood of $p^*(\Delta)$. To see this, note that

$$p(h_0^{t+1}; (\sigma_{-i}, \sigma_i')) = p(h_0^{t+1}; \sigma(h^t)) < p^*(\Delta),$$
$$p(h_1^{t+1}; (\sigma_{-i}, \sigma_i')) > p(h_1^{t+1}; \sigma(h^t)) > p^*(\Delta),$$

and:

$$0 = \frac{\partial^- V_m(p(h_0^{t+1}; \sigma(h^t)))}{\partial p} < \frac{\partial^- V_m(p(h_1^{t+1}; \sigma(h^t)))}{\partial p},$$

where $\frac{\partial^- V_m(p)}{\partial p}$ denotes the derivative from the left (which exists by the convexity of $V_m(p)$) of V_m at p.

Starting with the strategy profile (σ_{-i}, σ_i'), change the exit probability of all players $j \neq i$ to $\sigma_j' = \sigma_i'$ and denote the resulting symmetric profile by $\sigma'(h^t)$. By Proposition 2, the payoff to all players is weakly increased. Therefore for all j,

$$\mathbb{E}V_j(h^{t+1}; \sigma'(h^t)) = \mathbb{E}V_m(P^{t+1}(h^t; (\sigma'_{-i}, \sigma'_i))) \geq \mathbb{E}V_m(P^{t+1}(h^t; (\sigma_{-i}, \sigma'_i)))$$
$$> \mathbb{E}V_m(P^{t+1}(h^t; \sigma(h^t))) = \mathbb{E}V_j(h^{t+1}; \sigma(h^t)). \qquad \square$$

Proof of Theorem 1. All we have to do is to check that $\sigma(h^t)$ is optimal for all players under all three cases given in the Theorem, and that this is the only symmetric exit probability with this property.

Lemma 1 implies that it is optimal to stay (exit) at h^t iff $\sigma(h^t)$ satisfies

$$\delta\mathbb{E}V_m(P^{t+1}(h^t; \sigma(h^t))) \geq (\leq)c(h^t)\Delta. \qquad (A.8)$$

Consider now cases (i)–(iii) below. These cases cover all possibilities and are mutually exclusive because $\mathbb{E}V_m(P^{t+1}(h^t; \sigma(h^t)))$ is increasing in $\sigma(h^t)$ by Lemma 2.

(i) Assume that

$$\delta\mathbb{E}V_m(P^{t+1}(h^t; 0)) > c(h^t)\Delta.$$

Then it is strictly optimal for all the players to stay in the game if $\sigma(h^t) = 0$. Moreover, by Lemma 2, $\delta\mathbb{E}V_m(P^{t+1}(h^t; x)) > c(h^t)\Delta$ for all $x \geq 0$, so $\sigma(h^t) = 0$ is the unique symmetric equilibrium action in that case.

(ii) Assume that

$$\delta\mathbb{E}V_m(P^{t+1}(h^t; 0)) \leq c(h^t)\Delta \leq \delta\mathbb{E}V_m(P^{t+1}(h^t; 1)).$$

First note that $\mathbb{E}V_m(P^{t+1}(h^t; \sigma(h^t)))$ is continuous in $\sigma(h^t)$ as a result of the continuity of the Bayes' rule in $\sigma(h^t)$. Lemma 2 implies that there is a unique value $\sigma^*(h^t)$ for which

$$\delta\mathbb{E}V_m(P^{t+1}(h^t; \sigma^*(h^t))) = c(h^t)\Delta.$$

Moreover, for all $\sigma(h^t) < \sigma^*(h^t)$ the strictly optimal action is to exit, and for all $\sigma(h^t) > \sigma^*(h^t)$ the strictly optimal action is to stay. Thus, $\sigma^*(h^t)$ is the unique symmetric equilibrium action in this case.

(iii) Assume that

$$\delta\mathbb{E}V_m(P^{t+1}(h^t; 1)) < c(h^t)\Delta.$$

Then it is strictly optimal for all the players to exit if $\sigma(h^t) = 1$. Moreover, by Lemma 2, $\delta\mathbb{E}V_m(P^{t+1}(h^t; x)) < c(h^t)\Delta$ for all $x \leq 1$, so $\sigma(h^t) = 1$ is the unique symmetric equilibrium action in that case. $\qquad \square$

Proof of Lemma 3. (i) We claim first that for each $\varepsilon > 0$ there exists a $\overline{\Delta} > 0$ such that for all $\Delta < \overline{\Delta}$,

$$p(h^t) < p^*(\Delta) + \varepsilon \text{ for all } h^t \text{ such that } t \geq t^*(\Delta).$$

Fix an $\varepsilon > 0$ and suppose that the claim does not hold. Then, with an arbitrarily small Δ, we must be able to find some history h^t such that if there are no exits, the belief jumps above $p^*(\Delta) + \varepsilon$:

$$p(h^t, 1) > p^*(\Delta) + \varepsilon, \tag{A.9}$$

where we denote by $(h^t, 1)$ the history at $t+1$ with no exits at h^t.

Consider the continuation value of an arbitrary uninformed player i at $V(h^t)$. Clearly, this value is bounded from below by the payoff of the event that no player exits at h^t weighted by the probability of that event, which gives us:

$$V(h^t) \geq -c\Delta + \Pr\{A^t_{-i} = 1\}e^{-r\Delta}V_m(p_{t+1}(h_t, 1)), \tag{A.10}$$

where $\Pr\{A^t_{-i} = 1\}$ is the probability that i assigns to the event that no players $j \neq i$ exits at h^t. But since the players are willing to randomize at h^t, we must have $V(h^t) = 0$. Using this fact and equation (A.9), we can write equation (A.10) as

$$c\Delta > \Pr\{A^t_{-i} = 1\}e^{-r\Delta}V_m(p^*(\Delta) + \varepsilon).$$

Since $V_m(p)$ is strictly increasing for $p > p^*(\Delta)$, there is an $\eta > 0$ such that

$$e^{-r\Delta}V_m(p^*(\Delta) + \varepsilon) > \eta.$$

Hence, we have:

$$\Pr\{A^t_{-i} = 1\} < \frac{c\Delta}{e^{-r\Delta}V_m(p^*(\Delta) + \varepsilon)} < \frac{c}{\eta}\Delta. \tag{A.11}$$

On the other hand, a natural lower bound for $\Pr\{A^t_{-i} = 1\}$ is given by

$$\Pr\{A^t_{-i} = 1\} \geq \Pr\{\text{all players } j \in \mathcal{N}(h^t)\backslash i \text{ are informed}\}.$$

To evaluate this lower bound, note that for an arbitrary $j \neq i$,

$$\Pr\{j \text{ is informed}\} \geq 1 - \pi^L_t,$$

and therefore,

$$\Pr\{A^t_{-i} = 1\} \geq (1 - \pi^L_t)^{n(h^t)-1}. \tag{A.12}$$

Since $\pi_t^L < 1$ and $n(h^t) \leq N$, equation (A.12) contradicts equation (A.11) for small enough Δ, and the claim is established.

It remains to show that if $h^t \in H^f(\Delta)$, then $p(h^t) > p^*(\Delta) - \varepsilon$ for sufficiently small Δ. Note that $h^t \in H^f(\Delta)$ means that $h^t = (h^{t-1}, 1)$. If $p(h^{t-1}) < p^*(\Delta)$, then $p(h^{t-1}, 1) > p^*(\Delta)$, otherwise the players would strictly prefer to exit at h^{t-1}. If $p(h^{t-1}) \geq p^*(\Delta)$, then the players stay with probability one and $p(h^{t-1}, 1) > p^*(\Delta)(1 - \lambda\Delta)$ by equation (3.2). Hence, there exists an $\eta > 0$ such that

$$p(h^t) > p^*(\Delta) - \eta\Delta. \tag{A.13}$$

Together with the above claim, this establishes part (i) of the Lemma.

(ii) Let

$$h^t(k) = (h^{t-1}, a^t(k)),$$

where $a^t(k)$ is a vector of exit decisions where exactly k active players exit at history h^{t-1}. By Bayes' rule, we know that

$$\frac{1 - q(h^t(k))}{q(h^t(k))} = \frac{1 - q(h^t(0))}{q(h^t(0))} \left(\frac{\pi^L(h^{t-1})}{\pi^H(h^{t-1})} \right)^k.$$

By Proposition 1, there is an $\eta > 0$ such that $\frac{\pi^L(h^{t-1})}{\pi^H(h^{t-1})} > 1 + \eta$. Therefore, for all $k \geq 1$, there is an η' such that

$$q(h^t(k)) < q(h^t(0)) - \eta'.$$

By equation (3.11), there exists a $v > 0$ such that

$$p(h^t(k)) < p(h^t(0)) - v.$$

By part (i), for all $\varepsilon > 0$, there exists a $\overline{\Delta} > 0$ such that for all $\Delta < \overline{\Delta}$,

$$p(h^t(0)) < p^* + \varepsilon.$$

Since $\varepsilon > 0$ can be chosen arbitrarily small, the claim follows. $\square$

Proof of Proposition 5. (i) Take an arbitrary $h^t \in H^f(\Delta)$. If $p(h^t) > p^*(\Delta)$, then no player wants to exit and we have $P_L(X(h^t) > 0) = 0$. Suppose therefore that $p(h^t) \leq p^*(\Delta)$. Since $h^t \in H^f(\Delta)$, we know from equation (A.13) in the proof of the previous lemma that

$$p(h^t) > p^*(\Delta) - \eta\Delta \tag{A.14}$$

for some η.

Since $p(h^t) \leq p^*(\Delta)$ and $h^t \in H^f(\Delta)$, the players must be indifferent between staying and exiting. In the absence of observational learning, the loss from staying in the game for an additional period to an uninformed player is

$$c\Delta - \lambda \Delta p(h^t)(v + e^{-r\Delta}V^+(\Delta))$$
$$= \Delta[c - \lambda p(h^t)(v + e^{-r\Delta}V^+(\Delta))]$$
$$< \Delta[c - \lambda(p^*(\Delta) - \eta\Delta)(v + e^{-r\Delta}V^+(\Delta))]$$
$$= \Delta\lambda\eta\Delta(v + e^{-r\Delta}V^+(\Delta)) := \widehat{\kappa}(\Delta)^2, \qquad (A.15)$$

where the inequality uses equation (A.14) and the last line uses equation (4.2). Hence, the loss is at most quadratic in Δ. This loss must be compensated by the gain from observational learning, which by the same logic as in part (i) of the previous lemma, is bounded from below by

$$\Pr\{A^t_{-i} - 1\}e^{-r\Delta}V_m(p(h^t_n, 1)), \qquad (A.16)$$

where $p(h^t, 1)$ is the belief at $t + 1$ with no exits at h^t, and $\Pr\{A^t = 1\}$ is the probability of no exits at h^t as estimated by an arbitrary uninformed player. A lower bound for the value at at $p(h^t, 1)$ is given by

$$V_m(p(h^t, 1)) \geq (p(h^t, 1) - p^*(\Delta))\lambda\Delta(v + e^{-r\Delta}V^+(\Delta)). \qquad (A.17)$$

Combining equations (A.15), (A.16), and (A.17), we have

$$p(h^t, 1) - p^*(\Delta) < \frac{\widehat{\kappa}(\Delta)^2}{\lambda\Delta(v + e^{-r\Delta}V^+(\Delta))\Pr\{A^t_{-i} = 1\}e^{-r\Delta}}.$$

Noting that $\Pr\{A^t_{-i} = 1\}$ is bounded away from 0 for the same reason as in the proof of Lemma 3, we can conclude that there is a $\eta' > 0$ such that

$$p(h^t, 1) < p^*(\Delta) + \eta'\Delta. \qquad (A.18)$$

Combining equations (A.14) and (A.18), we can conclude that

$$\frac{p(h^t, 1)}{1 - p(h^t, 1)} \cdot \frac{1 - p(h^t)}{p(h^t)} < \frac{p^*(\Delta) + \eta'\Delta}{1 - p^*(\Delta) - \eta'\Delta} \cdot \frac{1 - p^*(\Delta) + \eta\Delta}{p^*(\Delta) - \eta\Delta} < 1 + \eta''\Delta$$
$$(A.19)$$

for some $\eta'' > 0$.

Let $\xi^\theta(h^t)$ denote the probability with which an arbitrary player exits at h^t, conditional on θ. Bayes' rule gives:

$$\frac{p(h^t,1)}{1-p(h^t,1)} = \frac{(1-\xi^H(h^t))^{n(h^t)-1}}{(1-\xi^L(h^t))^{n(h^t)-1}} \frac{p(h^t)}{1-p(h^t)} > \frac{1-\xi^H(h^t)}{1-\xi^L(h^t)} \frac{p(h^t)}{1-p(h^t)}.$$
$$(A.20)$$

By Proposition 1, there is some $\alpha < 1$ such that

$$\xi^H(h^t) < \alpha \xi^L(h^t),$$

and therefore by equation (A.20),

$$\frac{p(h^t,1)}{1-p(h^t,1)} \cdot \frac{1-p(h^t)}{p(h^t)} > \frac{1-\alpha\xi^L(h^t)}{1-\xi^L(h^t)}. \qquad (A.21)$$

Combining equations (A.19) and (A.21) gives:

$$\xi^L(h_n^t) < \frac{\eta''}{2(1-\alpha)}\Delta.$$

Therefore,

$$P_L(X(h^t) > 0) < N\xi^L(h^t) < \frac{N\eta''}{2(1-\alpha)}\Delta := \kappa\Delta.$$

(ii) From part (ii) of Lemma 3, we know that there is a $\overline{\Delta} > 0$ and $v > 0$ such that $p(h^t) < p^*(\Delta) - v$ for all $h^t \in H^w(\Delta), \Delta < \overline{\Delta}$. Indifference requires that $p(h^{t+1},1) > p^*(\Delta)$. The result follows then from Bayes' rule. □

Proof of Proposition 6. By Proposition 5, the probability of exit per period in the flow mode is at most proportional to Δ. Therefore, we know that in the limit $\Delta \to 0$, there is a well-defined hazard rate for the exits. Moreover, by equation (3.1), the probability of exit for an individual player is proportional to the probability with which that player is uninformed, and therefore in the limit $N \to \infty$, the hazard rate of exit conditional on state must be proportional to $\pi^\theta(\tau)$:

$$\frac{f_{k+1}^0(\tau|\tau_1,\ldots,\tau_k)}{1-F_{k+1}^\theta(\tau|\tau_1,\ldots,\tau_k)} = \pi^\theta(\tau)K,$$

where K is a number that we will next determine using Lemma 3.

Consider the change in beliefs within short $d\tau$ induced by a given exit intensity K. Denote by $q(\tau)$ the belief of an uninformed player on the aggregate state at real time instant τ. Within $d\tau$, this belief changes for two reasons. First, a good signal arrives with probability $p^\theta(\tau)\lambda d\tau$, and second, an exit takes place with probability $\pi^\theta(\tau)Kd\tau$. Therefore, Bayes' rule gives:

$$q(\tau + d\tau) = \frac{q(\tau)(1 - p^H(\tau)\lambda d\tau)(1 - \pi^H(\tau)Kd\tau)}{\begin{array}{c}q(\tau)(1 - p^H(\tau)\lambda d\tau)(1 - \pi^H(\tau)Kd\tau)\\ +(1 - q(\tau))q(\tau)(1 - p^L(\tau)\lambda d\tau)3(1 - \pi^L(\tau)Kd\tau)\end{array}}.$$

With $d\tau$ small, this can be expressed as

$$\frac{dq(\tau)}{d\tau} = q(\tau)(1 - q(\tau))[(\pi^L(\tau) - \pi^H(\tau))K + (p^L(\tau) - p^H(\tau))\lambda]. \quad (A.22)$$

On the other hand, we know by Lemma 3 that when $\Delta \to 0$, the belief of an uninformed player on aggregate state must be given by

$$q(\tau) = \frac{p^* - p^L(\tau)}{p^H(\tau) - p^L(\tau)}. \qquad (A.23)$$

Differentiating this with respect to τ gives us:

$$\frac{dq(\tau)}{d\tau} = \lambda \frac{p^H(\tau)(1 - p^H(\tau))(p^* - p^L(\tau)) + p^L(\tau)(1 - p^L(\lambda))(p^H(\tau) - p^*)}{(p^H(\tau) - p^L(\tau))^2},$$

$$(A.24)$$

where we have used:

$$\frac{dp^\theta(\tau)}{d\tau} = -\lambda p^\theta(\tau)(1 - p^\theta(\tau)).$$

Equating equations (A.22) and (A.24), using equation (A.23) and solving for K gives us:

$$K = \lambda \frac{p^*(1 - p^*)(p^H(\tau) - p^L(\tau))}{(p^* - p^L(\tau))(p^H(\tau) - p^*)(\pi^L(\tau) - \pi^H(\tau))}. \qquad \square$$

Proof of Proposition 7. Consider an arbitrary period of an exit wave, where belief is given by q_s. The number of exits at this period is binomially distributed with parameters $\sigma_n \pi^\theta(\theta)$ and n, where n denotes the number of active players in the game and σ_n denotes the exit probability of an individual uninformed player. In case there is no exit, the next period belief is given by Bayes' rule by

$$q_{s+1}(0) = \frac{(1 - \sigma_n \pi^H(\tau))^{n-1} q_s}{(1 - \sigma_n \pi^H(\tau))^{n-1} q_s + (1 - \sigma_n \pi^L(\tau))^{n-1}(1 - q_s)}.$$

Lemma 3 requires that

$$\lim_{n \to \infty} \frac{(1 - \sigma_n \pi^H(\tau))^{n-1} q_s}{(1 - \sigma_n \pi^H(\tau))^{n-1} q_s + (1 - \sigma_n \pi^L(\tau))^{n-1}(1 - q_s)} = q^*(\tau),$$

which gives us

$$\frac{q^*(\tau)}{1 - q^*(\tau)} = \frac{q_s}{1 - q_s} \lim_{n \to \infty} \frac{(1 - \sigma_n \pi^H(\tau))^{n-1}}{(1 - \sigma_n \pi^L(\tau))^{n-1}}.$$

Taking logarithm on both sides and evaluating limits gives us

$$\lim_{n \to \infty} (n-1)\sigma_n = \frac{1}{\pi^L(\tau) - \pi^H(\tau)} \log\left(\frac{q^*(\tau)}{1 - q^*(\tau)} \frac{1 - q_s}{q_s}\right).$$

Therefore,

$$\lim_{n \to \infty} n \sigma_n \pi^\theta(\theta) = \frac{\pi^\theta}{\pi^L(\tau) - \pi^H(\tau)} \log\left(\frac{q^*(\tau)}{1 - q^*(\tau)} \frac{1 - q_s}{q_s}\right),$$

and the result is given by the Poisson approximation of Binomial distribution.

The second claim is an immediate consequence of the Bayes' rule with k exits. $\qquad\square$

Appendix B: Proofs for Section 5

This appendix contains the proofs for Propositions 3, 4, and Theorem 2. Our arguments rely on the convergence of the outside observer's belief $\hat{q}(h^t)$, which we define as the posterior belief of the event $\{\theta = H\}$ at public history h^t:

$$\hat{q}(h^t) = \frac{q_0 P_H\{h^t\}}{q_0 P_H\{h^t\} + (1 - q_0)P_L\{h^t\}}.$$

Note that $\hat{q}(h^t)$ differs slightly from $\hat{q}_i(h^t)$ that was defined in Section 3.2: $\hat{q}(h^t)$ is based on actions of all players (and represents therefore a true *outside* observer), while $\hat{q}_i(h^t)$ is based on actions of players other than i.

To link $\hat{q}(h^t)$ to the beliefs of actual players in the game, note that player i's belief $q_i(h^t)$ differs from $\hat{q}(h^t)$ only to the extent that i's private information affects her belief. Lemma 4 below guarantees that player i's private history cannot overwhelm a sufficiently strong public history:[18]

Lemma 4: *Suppose that $\rho^H < 1$. Then for all $\varepsilon > 0$, there is some $\delta > 0$ such that the following implications hold for all i:*

$$\widehat{q}(h^t) \geq 1 - \delta \Rightarrow q_i(h^t) \geq 1 - \varepsilon \text{ and} \tag{B.1}$$

$$\widehat{q}(h^t) \leq \delta \Rightarrow q_i(h^t) \leq \varepsilon. \tag{B.2}$$

Proof: Recall from Section 3.2 the definition of $\widehat{q}_i(h^t)$ as the belief based on public histories of all players other than i. In addition to this, $\widehat{q}(h^t)$ also conditions on actions of i. Consider the effect of this additional information. The most favorable piece of evidence in terms of state $\theta = H$ that could ever be obtained form i's actions is the one that fully reveals i to be informed. The likelihood ratio of being informed across the states is $(1-\pi_t^H)/(1-\pi_t^L)$, so Bayesian rule gives us an upper bound for $\widehat{q}(h^t)$ as expressed in terms of $\widehat{q}_i(h^t)$:

$$\frac{\widehat{q}(h^t)}{1 - \widehat{q}(h^t)} \leq \frac{1 - \pi_t^H}{1 - \pi_t^L} \frac{\widehat{q}_i(h^t)}{1 - \widehat{q}_i(h^t)}. \tag{B.3}$$

On the other hand, we can write the relationship between $q_i(h^t)$ and $\widehat{q}_i(h^t)$ using equation (3.9):

$$\frac{q_i(h^t)}{1 - q_i(h^t)} = \frac{\pi_t^H}{\pi_t^L} \frac{\widehat{q}(h^t)}{1 - \widehat{q}(h^t)}. \tag{B.4}$$

Combining equations (B.3) and (B.4) gives us:

$$\frac{q_i(h^t)}{1 - q_i(h^t)} \geq \frac{\pi_t^H}{\pi_t^L} \frac{1 - \pi_t^L}{1 - \pi_t^H} \frac{\widehat{q}(h^t)}{1 - \widehat{q}(h^t)}. \tag{B.5}$$

By equation (3.8), π_t^H and $1 - \pi_t^L$ are bounded away from zero, and therefore the first equation of Lemma 4 follows directly from equation (B.5). The second equation follows from the fact that $q_i(h^t) \leq \widehat{q}(h^t)$, which in turn follows directly from equations (3.8) and (3.9). □

---

[18]In the pure common values case, where $\rho^H = 1$, the ratio $\frac{\pi_t^H}{\pi_t^L} \to 0$ as $t \to \infty$. In that case, the statement below holds for all t up to an arbitrary, fixed $\bar{t}$. This modification is not essential for any of our results.

Proof of Proposition 3

Our proof strategy is to follow the evolution of the outside observer's belief along a filtration that samples the players' actions sequentially one player at a time. We show that this belief must converge to truth as the number of exits increases. Furthermore, this implies the convergence of the actual players' beliefs in the original filtration where all actions within a period are sampled simultaneously. The key step in the argument is Lemma 5 below, which implies that the belief process that we consider has a strong drift towards truth when sampled at the points where any player exits. With this Lemma at hand, the rest of the argument is a relatively straightforward application of Theorem A.1. of Fudenberg and Levine (1992).

We use index $s \in \mathbb{N}$ to track the moments of observation starting from period $t^*(\Delta)$ in the following way. At $s = 1$, the action of Player 1 in period $t^*(\Delta)$ is observed. At $s = 2$, the action of Player 2 in period $t^*(\Delta)$ is observed, and so on. Once the decisions of all N players in period $t^*(\Delta)$ have been sampled, the process moves to the next time period. At $s = N+1$, Player 1's action in period $t^*(\Delta) + 1$ is observed, and so on. This means that we map every $s \in \mathbb{N}$ to the corresponding period $t(s)$ and player $i(s)$ as follows:

$$t(s) := \left\lfloor \frac{s}{N} \right\rfloor + t^*(\Delta),$$

$$i(s) := s - N \cdot \left\lfloor \frac{s}{N} \right\rfloor.$$

Let ξ_s^θ denote the exit probability of Player $i(s)$ in period $t(s)$ with equilibrium strategy profile σ (nothing in the proof requires this to be symmetric):

$$\xi_s^\theta := \xi_{i(s)}^\theta (h^{t(s)}),$$

where we set $\xi_i^\theta(h^t) = 0$ if $a_{i(s)}^{t(s)-1} = 0$ (i.e. probability of exit is zero for a player that has already exited). We use $x_s \in \{0, 1\}$ as an indicator for Player $i(s)$ exiting in period $t(s)$:

$$x_s = \begin{cases} 1 & \text{if } a_{i(s)}^{t(s)-1} = 1 \text{ and } a_{i(s)}^{t(s)} = 0, \\ 0 & \text{otherwise.} \end{cases}$$

We use notation h_s to refer to the history of exits up to s:

$$h_s = \left(a_{i(1)}^{t(1)}, \dots, a_{i(s)}^{t(s)} \right),$$

and we denote by $\widehat{q}_s$ the belief process of the outside observer, who observes the players sequentially:

$$\widehat{q}_s = \Pr\{\theta = H | h_s\}, \quad s \in \mathbb{N}.$$

By Bayes' rule, this belief evolves according to

$$\widehat{q}_0 = q_0,$$

$$\widehat{q}_s = \begin{cases} \frac{\widehat{q}_{s-1}\xi_s^H}{\widehat{q}_{s-1}\xi_s^H+(1-\widehat{q}_{s-1})\xi_s^L} & \text{if } x_s = 1 \\ \frac{\widehat{q}_{s-1}(1-\xi_s^H)}{\widehat{q}_{s-1}(1-\xi_s^H)+(1-\widehat{q}_{s-1})(1-\xi_s^L)} & \text{if } x_s = 0 \end{cases}, s = 1, 2, \ldots \quad (B.6)$$

Note that for all $s = t \cdot N$, $t \in \mathbb{N}$, the belief $\widehat{q}_s$ coincides with the belief at period t of an outside observer that observes all players simultaneously at each period:

$$\widehat{q}_s = \widehat{q}(h^t), \quad s = t \cdot N, \quad t \in \mathbb{N}.$$

For all other values of s, $\widehat{q}_s$ is the belief of an outside observer who has observed only a subset of players in the last period.

Let X_∞ denote the total number of players that exit the game:

$$X_\infty := \sum_{t=0}^{\infty} X(h^t).$$

We next define an increasing sequence of natural numbers $\{s(k)\}_{k=1}^{X_\infty}$ as follows:

$$s(0) = 0,$$

$$s(k) = \min\{s > s(k-1) | x_s = 1\}, \quad k = 1, \ldots, X_\infty.$$

Hence, $\{\widehat{q}_{s(k)}\}_{k=1}^{X_\infty}$ is a subset of $\{\widehat{q}_s\}_{s=1}^{\infty}$ sequence, that samples the beliefs immediately after realized exits.

Define:

$$L_k := \begin{cases} \frac{1-\widehat{q}_{s(k)}}{\widehat{q}_{s(k)}} & \text{for } k = 1, \ldots, X_\infty. \\ 0 & \text{for } k = X_\infty + 1, \ldots \end{cases} \quad (B.7)$$

In words, L_k is the likelihood ratio for the event $\{\theta = L\}$ sampled after realized exits. It is clear that under the event $\{\theta = H\}$, this process is a martingale. The next lemma is the key to our argument, and it states that this process is an *active supermartingale*, as defined in Fudenberg and Levine (1992).

Lemma 5: *There exists an $\eta > 0$ such that*

$$P_H\big(|L_{k+1}/L_k - 1| > \eta \,|\, h_{s(k)}\big) > \eta \qquad (\text{B.8})$$

for all $L_k > 0$.

Proof: Note first that

$$\{|L_{k+1}/L_k - 1| \le \eta\} \Leftrightarrow (1-\eta)L_k \le L_{k+1} \le (1+\eta)L_k. \qquad (\text{B.9})$$

By Proposition 1 and equation (3.12), there is some $\gamma > 0$ such that

$$\frac{\xi_s^L}{\xi_s^H} > 1 + \gamma \qquad (\text{B.10})$$

for all $s \in \mathbb{N}$. We choose some $\eta > 0$ small enough to ensure that

$$\frac{(1+\eta)}{(1-\eta)^2} < 1 + \gamma. \qquad (\text{B.11})$$

Write

$$\widetilde{L}_s := \frac{1 - \widehat{q}_s}{\widehat{q}_s}, \quad s \in \mathbb{N}.$$

Note that $L_k = \widetilde{L}_{s(k)}$ for $k = 1, \ldots, X_\infty$. Using equations (B.6) and (B.10), we have

$$\widetilde{L}_s = \begin{cases} \frac{\xi_s^L}{\xi_s^H} \widetilde{L}_{s-1} > (1+\gamma)\widetilde{L}_{s-1} & \text{if } x_s = 1, \\[2mm] \frac{(1-\xi_s^L)}{(1-\xi_s^H)} \widetilde{L}_{s-1} < \widetilde{L}_{s-1} & \text{if } x_s = 0. \end{cases}$$

By definition of $s(k)$, we have $x_{s(k+1)} = 1$, and therefore, we have

$$\widetilde{L}_{s(k+1)} > (1+\gamma)\widetilde{L}_{s(k+1)-1}. \qquad (\text{B.12})$$

Noting that $L_{k+1} = \widetilde{L}_{s(k+1)}$ and $L_k = \widetilde{L}_{s(k)}$, and using equations (B.9) and (B.12), we have

$$\{|L_{k+1}/L_k - 1| \le \eta\} \Rightarrow \left\{ \widetilde{L}_{s(k+1)-1} < \frac{1+\eta}{1+\gamma} \widetilde{L}_{s(k)} \right\}. \qquad (\text{B.13})$$

Let $\bar{s}$ be the first observation point after $s(k)$ at which $\widetilde{L}_{\bar{s}}$ is below $\frac{1+\eta}{1+\gamma}\widetilde{L}_{s(k)}$ in case there are no exits:

$$\bar{s} := \min\left\{ s' > s(k) : \left(\prod_{j=s(k)+1}^{s'} \frac{1-\xi_j^L}{1-\xi_j^H} \right) \widetilde{L}_{s(k)} < \frac{1+\eta}{1+\gamma}\widetilde{L}_{s(k)} \right\}.$$

Then it follows from equations (B.13) and (B.9) that:

$$\{|L_{k+1}/L_k - 1| \le \eta\}$$
$$\Rightarrow \left\{ x_s = 0 \forall s = s(k) + 1, \ldots, \bar{s} \text{ and } \widetilde{L}_{s(k+1)} > \frac{1-\eta}{1+\eta}(1+\gamma)\widetilde{L}_{\bar{s}} \right\}.$$
$$(B.14)$$

But, since $\widetilde{L}_s$ is a supermartingale under $\theta = H$, we have

$$\mathbb{E}(\widetilde{L}_{s(k+1)}|h^{\bar{s}}, \theta = H) < \widetilde{L}_{\bar{s}},$$

which implies the following (using the fact that $\widetilde{L}_{s(k+1)}$ is bounded from below by 0):

$$P_H\left(\widetilde{L}_{s(k+1)} < \frac{1-\eta}{1+\eta}(1+\gamma)\widetilde{L}_{\bar{s}}|h^{\bar{s}}\right) \ge 1 - \frac{(1+\eta)}{(1-\eta)(1+\gamma)} > \eta,$$

where the last inequality follows from equation (B.11). Combining this with equation (B.14), we note that

$$P_H(|L_{k+1}/L_k - 1| > \eta|h_{s(k)}) > \eta. \qquad \square$$

Lemma 5 says that L_k, $k \in \mathbb{N}$ is an *active supermartingale* with activity η, as defined in Fudenberg and Levine (1992). We need this property to apply Theorem A.1. of Fudenberg and Levine (1992), which we restate here for convenience:

Theorem B1 (Fudenberg and Levine): *Let $l_0 > 0$, $\varepsilon > 0$, and $\eta \in (0,1)$ be given. For each $\underline{L}$, $0 < \underline{L} < l_0$, there is some $K < \infty$ such that*

$$\Pr\left(\sup_{k > K} L_k \le \underline{L}\right) \ge 1 - \varepsilon$$

for every active supermartingale L with $L_0 = l_0$ and activity η.

With these preliminaries at hand, we are ready to finish the proof of Proposition 3:

Proof of Proposition 3. Fix an $\varepsilon > 0$. Consider the stochastic process L_k, $k \in \mathbb{N}$, defined in equation (B.7). Note from equation (B.7) that

$$L_k \leq \underline{L} \Leftrightarrow \left(\left\{ \widehat{q}_{s(k)} \geq \frac{1}{1 + \underline{L}} \right\} \text{ or } \{k > X_\infty\} \right).$$

We set $\underline{L}$ small enough to guarantee:

$$L_k \leq \underline{L} \Rightarrow (\{\widehat{q}_{s(k)} > 1 - \varepsilon\} \text{ or } \{k > X_\infty\}). \tag{B.15}$$

By Lemma 5, we know that L_k is an active supermartingale with activity η. By Theorem B1, we can therefore set K high enough to guarantee that

$$P_H \left\{ h^\infty : \sup_{k > K} L_k \leq \underline{L} \right\} \geq 1 - \varepsilon. \tag{B.16}$$

Combining this with equation (B.15), we have

$$P_H \{ h^\infty : n(h^t) \leq N - K \text{ and } \widehat{q}(h^t) < 1 - \varepsilon \text{ for some } h^t \in h^\infty \} < \varepsilon. \tag{B.17}$$

We have now proved the Proposition as regards equation (5.1). Knowing this, the part concerning equation (5.2) follows from Bayes' rule as follows. Define the following event:

$$A(K, \varepsilon) := \{ h^\infty : n(h^t) < N - K \text{ and } \varepsilon < \widehat{q}(h^t) < 1 - \varepsilon \text{ for some } h^t \in h^\infty \}.$$

Then, by the definition of $A(K, \varepsilon)$, the posterior of $\{\theta = H\}$ conditional on reaching $A(K, \varepsilon)$ must be between ε and $1 - \varepsilon$:

$$\varepsilon < \frac{q_0 P_H(A(K, \varepsilon))}{q_0 P_H(A(K, \varepsilon)) + (1 - q) P_L(A(K, \varepsilon))} < 1 - \varepsilon. \tag{B.18}$$

Since equation (B.17) holds for any ε given large enough K, we know that $P_H(A(K, \varepsilon))$ can be made arbitrarily small by increasing K. Therefore, for equation (B.18) to hold, also $P_L(A(K, \varepsilon))$ must go to zero as K is increased, which implies that for any $\varepsilon > 0$, we can find K large enough to ensure that

$$P_L \{ h^\infty : n(h^t) \leq N - K \text{ and } \widehat{q}(h^t) < \varepsilon \text{ for some } h^t \in h^\infty \} < \varepsilon. \qquad \square$$

Proof of Proposition 4

We work through a number of lemmas. First, we formalize the intuitive fact that whenever the probability that a large number of players exit within the current period is non-negligible, the realized actions generate a precise signal about the state of the world. In particular, if the true state is $\theta = H$, then the beliefs of all players must be very close to one after that period.

Lemma 6: *For all $\varepsilon > 0$ and $q > 0$, there is some $K \in \mathbb{N}$ such that*

$$P_L(X(h^t) > K) > \frac{1}{2} \Rightarrow P_H(q(h^{t+1}) > 1 - \varepsilon) > 1 - \varepsilon,$$

whenever $q(h^t) > q$ and $t \geq t^*(\Delta)$.

Proof: Denote

$$\mu_\theta := \mathbb{E}[X(h^t)|\theta.] = n(h^t)\xi^\theta(h^t).$$

Since $X(h^t)$ is a random variable that can only take positive values, the following must hold:

$$P_L(X(h^t) > K) > \frac{1}{2} \Rightarrow \mu_L > \frac{1}{2}K. \tag{B.19}$$

By Proposition 1, we know that there is some $\gamma > 0$ such that

$$\frac{\xi^L(h^t)}{\xi^H(h^t)} > 1 + \gamma$$

for all $t \geq t^*(\Delta)$. Consider the random variable

$$Z(h^t) := \frac{X(h^t)}{\mu_L}.$$

We have

$$\mathbb{E}[Z(h^t)|\theta = H] = \frac{n(h^t)\xi^H(h^t)}{n(h^t)\xi^L(h^t)} < \frac{1}{1+\gamma}, \tag{B.20}$$

$$\mathbb{E}[Z(h^t)|\theta = L] = \frac{n(h^t)\xi^L(h^t)}{n(h^t)\xi^L(h^t)} = 1, \tag{B.21}$$

$$var[Z(h^t)|\theta = H] = \frac{n(h^t)\xi^H(h^t)(1 - \xi^H(h^t))}{(n(h^t)\xi^L(h^t))^2} < \frac{1}{n(h^t)\xi^L(h^t)} = \frac{1}{\mu_L}, \tag{B.22}$$

$$var[Z(h^t)|\theta = L] = \frac{n(h^t)\xi^L(h^t)(1 - \xi^L(h^t))}{(n(h^t)\xi^L(h^t))^2} < \frac{1}{n(h^t)\xi^L(h^t)} = \frac{1}{\mu_L}. \tag{B.23}$$

Consider the event

$$A = \left(Z(h^t) \leq \frac{1 + \frac{1}{2}\gamma}{1 + \gamma} \right).$$

The formulas (B.20–B.23) imply that

$$\lim_{\mu_L \to \infty} P_H(A) = 1 \quad \text{and} \quad \lim_{\mu_L \to \infty} P_L(A) = 0.$$

By equation (B.19), assuming $P_L(X(h^t) > K) > \frac{1}{2}$ and increasing K will increase μ_L without bound. Hence, the result follows from Bayes' rule by considering the likelihood ratio across states of event A as K is increased. □

The next step in the proof, Lemma 7, bounds the probability with which a large number of players may exit within an arbitrary single period. By Lemma 6, a random experiment that induces a large number of players to exit with a non-negligible probability is very informative on the aggregate state. Any uninformed player would like to stay in the game until τ_H if she knew the state to be H. Suppose next that the probability of state H is bounded away from zero. As the period length is reduced towards zero, the players would rather wait and observe the result of an informative experiment than exit immediately. Lemma 7 formalizes this argument.

Lemma 7: *For all $\tau < \tau_H$ and $q > 0$, there exist a $K \in \mathbb{N}$ and a $\overline{\Delta} \in \mathbb{R}^+$ such that*

$$q(h^t) > q \Rightarrow P_L(X(h^t) > K) < \frac{1}{2}$$

whenever $\Delta < \overline{\Delta}$ and $t \leq t(\tau, \Delta)$.

Proof: Fix a $\tau < \tau_H$ and a $q > 0$. Lemma 6 implies the existence of a function $\phi : \mathbb{N} \to \mathbb{R}^+$ with

$$\lim_{K \to \infty} \phi(K) = 0,$$

such that for all h^t, $t \leq t(\tau, \Delta)$, and $q(h^t) > q$:

$$P_L(X(h^t) > K) > \frac{1}{2} \Rightarrow P_H(q(h^{t+1}) > 1 - \phi(K)) > 1 - \phi(K). \quad \text{(B.24)}$$

Recall the definition of p_t^H in equation (3.3), i.e. the belief of a player on her own type conditional on state H. If $\tau < \tau_H$, we can choose an $\eta > 0$ and a $\Delta' > 0$ such that $p_{t(\tau,\Delta)+1}^H > p^*(\Delta) + \eta$ for all $\Delta < \Delta'$. This follows directly from the continuity of $p^*(\Delta)$ and the definition of τ_H. This means that we can choose a K high enough so that

$$q(h^{t+1}) > 1 - \phi(K) \Rightarrow p(h^{t+1}) > p^*(\Delta) + \eta. \quad \text{(B.25)}$$

We choose a K such that equations (B.24) and (B.25) hold for all h^t, $t \leq t(\tau, \Delta)$, for which $q(h^t) > q$. Take any such history and assume that $P_L(X(h^t) > K) > \frac{1}{2}$. Consider next the expected payoff that an uninformed player would get by staying in the game with probability one at that history. We want to find a lower bound for that payoff. Since $q(h^t) > q$, the posterior for $\theta = H$ is bounded from below by q. By equations (B.24) and (B.25), $1 - \phi(K)$ is a lower bound for the probability that $p(h^{t+1}) > p^*(\Delta) + \eta$, conditional on $\theta = H$. Finally, $V_m(p^*(\Delta) + \eta) > 0$ is the value of the isolated player at belief $p^*(\Delta) + \eta$. Therefore, the continuation payoff for a player that stays is bounded from below by

$$V(h^t) \geq -c\Delta + e^{-r\Delta} \cdot q \cdot (1 - \phi(K)) \cdot V_m(p^*(\Delta) + \eta). \tag{B.26}$$

We see from equation (B.26) that we guarantee $V(h^t) > 0$ by setting Δ small enough and K large enough. Since then it is strictly optimal for any individual player to stay in the game, this contradicts the presumption that $P_L(X(h^t) > K) > \frac{1}{2}$. We thus conclude that for high enough $K \in \mathbb{N}$ and small enough $\overline{\Delta} \in \mathbb{R}^+$ the implication

$$q(h^t) > q \Rightarrow P_L(X(h^t) > K) < \frac{1}{2}$$

holds whenever $\Delta < \overline{\Delta}$ and $t \leq t(\tau, \Delta)$. $\qquad\qquad\square$

Lemma 8 shows that if a large number or players exit within a period, then the belief of an uninformed player falls to a very low level.

Lemma 8: *For all $\tau < \tau_H$ and $q > 0$, there exist a $K \in \mathbb{N}$ and a $\overline{\Delta} \in \mathbb{R}^+$ such that the following implication holds on the equilibrium path of any game $\Gamma(\Delta, N)$ with $\Delta < \overline{\Delta}$:*

$$\{t \leq t(\tau, \Delta) \wedge q(h^t) > q \wedge X(h^t) > K\} \Rightarrow \{q(h^{t+1}) < q\}.$$

Proof: Fix a $\tau < \tau_H$ and a $q > 0$. By Lemma 7, fix $K' \in \mathbb{N}$ and Δ' such that

$$P_L(X(h^t) > K') < \frac{1}{2} \tag{B.27}$$

whenever $\Delta < \Delta'$, $q(h^t) > q$, $t \leq t(\tau, \Delta)$. Since $\tau < \tau_H$, the same logic that led to equation (B.25) allows us to fix $\Delta'' > 0$ and $q^* < 1$ such that whenever $t \leq t(\tau, \Delta)$ and $\Delta < \Delta''$, the following implication holds:

$$q(h^t) > q^* \Rightarrow p(h^t) > p^*(\Delta). \tag{B.28}$$

Define $\overline{\Delta} = \min(\Delta', \Delta'')$. For the rest of the proof, we assume that $\Delta < \overline{\Delta}$, and we take an arbitrary history h^t such that $t \leq t(\tau, \Delta), q(h^t) > q$, and $\xi^H(h^t) > 0$. Our goal is to find a K such that $X(h^t) > K$ would imply $q(h^{t+1}) < q$.

Consider the expression for the probability of k exits:

$$P_\theta(X(h^t) = k) = \binom{n}{k} (\xi^\theta(h^t))^k (1 - \xi^\theta(h^t))^{n(h^t)-k}. \tag{B.29}$$

Since $\xi^H(h^t) < \xi^L(h^t)$, it follows by straightforward algebra from equation (B.29) that

$$\frac{P_H(X(h^t) = k)}{P_L(X(h^t) = k)} > \frac{P_H(X(h^t) = k')}{P_L(X(h^t) = k')} \quad \text{for } k < k'. \tag{B.30}$$

It then also follows that

$$\frac{P_H(X(h^t) = k')}{P_L(X(h^t) = k')} < 2. \tag{B.31}$$

To see why, assume the contrary. Then, we have

$$P_H(X(h^t) \leq K') = \sum_{k=0}^{K'} P_H(X(h^t) = k) > 2 \cdot \sum_{k=0}^{K'} P_L(X(h^t) = k) > 2 \cdot \frac{1}{2} = 1,$$

where the first inequality uses equation (B.30) and the presumption that equation (B.31) does not hold, whereas the second inequality follows from equation (B.27). But a probability of an event cannot be greater than one, so equation (B.31) must hold.

Consider next the following expression:

$$\frac{P_H(X(h^t) = K' + K'')}{P_L(X(h^t) = K' + K'')}$$

$$= \frac{\binom{n}{K' + K''} (\xi^H(h^t))^{K'+K''} (1 - \xi^H(h^t))^{n(h^t)-K'-K''}}{\binom{n}{K' + K''} (\xi^L(h^t))^{K'+K''} (1 - \xi^L(h^t))^{n(h^t)-K'-K''}}$$

$$= \left(\frac{\xi^H(h^t)}{\xi^L(h^t)}\right)^{K'} \left(\frac{1-\xi^H(h^t)}{1-\xi^L(h^t)}\right)^{n(h^t)-K'}$$

$$\cdot \left(\frac{\xi^H(h^t)}{\xi^L(h^t)}\right)^{K''} \left(\frac{1-\xi^L(h^t)}{1-\xi^H(h^t)}\right)^{K''}$$

$$= \frac{P_H(X(h^t)=K')}{P_L(X(h^t)=K')} \cdot \left(\frac{\xi^H(h^t)}{\xi^L(h^t)}\right)^{K''} \cdot \left(\frac{1-\xi^L(h^t)}{1-\xi^H(h^t)}\right)^{K''}.$$

By equation (B.31),

$$\frac{P_H(X(h^t)=K')}{P_L(X(h^t)=K')} < 2.$$

Also, since $\xi^L(h^t) > \xi^H(h^t)$, we have

$$\left(\frac{1-\xi^L(h^t)}{1-\xi^H(h^t)}\right)^{K''} < 1.$$

By Proposition 1, we have

$$\lim_{K''\to\infty} \left(\frac{\xi^H(h^t)}{\xi^L(h^t)}\right)^{K''} = 0,$$

and therefore, we can set K'' high enough to ensure

$$\frac{P_H(X(h^t)=K'+K'')}{P_L(X(h^t)=K'+K'')} < \frac{1-q^*}{q^*}q. \tag{B.32}$$

Since $\xi^H(h^t) > 0$, we know from equation (B.28) that $q(h^t) < q^*$ (otherwise no player would want to exit). Therefore, Bayes' rule and simple algebra leads to

$$q(h^{t+1}|X(h^t)=K'+K''.)$$

$$= \frac{q(h^t)P_H(X(h^t)=K'+K'')}{q(h^t)P_H(X(h^t)=K'+K'') + (1-q(h^t))P_L(X(h^t)=K'+K'')}$$

$$< \frac{q(h^t)}{1-q(h^t)} \frac{P_H(X(h^t)=K'+K'')}{P_L(X(h^t)=K'+K'')}$$

$$\leq \frac{q^*}{1-q^*} \frac{P_H(X(h^t)=K'+K'')}{P_L(X(h^t)=K'+K'')} < q,$$

where the last inequality follows from equation (B.32). By equation (B.30), this means that

$$q(h^{t+1}|X(h^t) = k) < q$$

for any $k > K$, where we have set $K : K' + K''$. □

Finally, we state a lemma that limits the probability with which an outside observer's belief $\widehat{q}(h^t)$ could ever get small values if $\theta = H$. This result is simply a formalization of the notion that a Bayesian observer is not likely to be convinced of the untrue state.

Lemma 9: *For all $\varepsilon > 0$, there is a $q > 0$ such that*

$$P_H\{h^\infty : \widehat{q}(h^t) \leq q \text{ for some } h^t \in h^\infty\} < \varepsilon.$$

Proof: Consider the event

$$A = \{h^\infty : \widehat{q}(h^t) \leq q \text{ for some } h^t \in h^\infty\}.$$

The posterior probability of $\theta = H$ conditional on reaching A is

$$\frac{q_0 P_H(A)}{q_0 P_H(A) + (1 - q_0)P_L(A)} \leq q$$

by the definition of the event A. Since $P_L(A) \leq 1$, we have:

$$P_H(A) \leq \frac{(1 - q_0)q}{q_0(1 - q)},$$

which can be made arbitrarily small by decreasing q. □

With Lemmas 8 and 9 at hand, it is now easy to finish the proof of Proposition 4.

Proof of Proposition 4. Fix a $\tau < \tau_H$ and an $\varepsilon > 0$. Using Lemma 9 and noting that the divergence of the uninformed players' belief from outside observer's belief is bounded by Lemma 4, we can choose a $q > 0$ such that

$$P_H\{h^\infty : q(h^t) \leq q \text{ for some } h^t \in h^\infty\} < \varepsilon. \tag{B.33}$$

Next, by Lemma 8, we can choose a $K \in \mathbb{N}$ and a $\overline{\Delta} \in \mathbb{R}^+$ such that

$$\{t \leq t(\tau, \Delta), q(h^t) > q, X(h^t) > K\} \Rightarrow \{q(h^{t+1}) < q\},$$

whenever $\Delta < \overline{\Delta}$. Thus, if there is some $h^t, t \leq t(\tau, \Delta)$, for which $X(h^t) > K$, we must have either $q(h^t) \leq q$ or $q(h^{t+1}) \leq q$. But by equation (B.33) this cannot happen with probability greater than ε, and as a result, we have

$$P_H\{h^\infty : X(h^t) > K \text{ for some } t \leq t(\tau, \Delta)\} < \varepsilon$$

if $\Delta < \overline{\Delta}$. □

Proof of Theorem 2

Proof of Theorem 2. Fix a $\tau < \tau_H$ and an $\varepsilon > 0$. Then, by Lemma 4 and equation (3.11), we can choose a $q' < 1$ and a $\Delta' > 0$ such that whenever $\Delta < \Delta'$ and $t \leq t(\tau, \Delta)$, the following holds:

$$\widehat{q}(h^t) \geq q' \Rightarrow p(h^t) > p^*(\Delta). \tag{B.34}$$

Next, by Proposition 3, we can choose a K' such that

$$P_H\{h^\infty : n(h^t) \leq N - K' \text{ and } \widehat{q}(h^t) < q' \text{ for some } h^t \in h^\infty\} < \frac{\varepsilon}{2}. \tag{B.35}$$

Assume that $\sum_{t=0}^{t(\tau,\Delta)} X(h^t) \geq K'$ (if not, then there is nothing to prove), and denote by $t_{K'}$ the first period with fewer than $N - K'$ active players left in the game:

$$t_{K'} := \min\{t : n(h^t) \leq N - K'\}. \tag{B.36}$$

Since $\sum_{t=0}^{t(\tau,\Delta)} X(h^t) \geq K'$, we must have

$$t_{K'} \leq t(\tau, \Delta) + 1. \tag{B.37}$$

Equations (B.34) and (B.35) mean that the probability that any player exits in $[t_{K'}, \ldots, t(\tau, \Delta)]$ is less than $\frac{\varepsilon}{2}$:

$$P_H\left\{h^\infty : \sum_{t=t_{K'}}^{t(\tau,\Delta)} X(h^t) > 0\right\} < \frac{\varepsilon}{2}. \tag{B.38}$$

By the definition of $t_{K'}$ in equation (B.36), we know that

$$\sum_{t=0}^{t_{K'}-2} X(h^t) < K'. \tag{B.39}$$

Finally, by equation (B.37) and Proposition 4, we can find a $\Delta'' \in \mathbb{R}^+$ and a K'' such that

$$P_H\{h^\infty : X(h^{t_{K'}-1}) > K''\} < \frac{\varepsilon}{2}. \tag{B.40}$$

Noting that equation (B.38) holds when $\Delta < \Delta'$ and equation (B.40) holds when $\Delta < \Delta''$, we may set $K := K' + K''$ and $\overline{\Delta} := \min(\Delta', \Delta'')$ and combine equations (B.38–B.40) to get:

$$P_H \left\{ h^\infty : \sum_{t=0}^{t(\tau, \Delta)} X(h^t) > K \right\} < \varepsilon,$$

whenever $\Delta < \overline{\Delta}$.   □

Acknowledgment

We would like to thank numerous seminar audiences and, in particular the editor, Bruno Biais, three anonymous referees, Dirk Bergemann, Hikmet Gunay, Johannes Hörner, Godfrey Keller, Elan Pavlov, Sven Rady, Larry Samuelson, and Peter Sorensen for useful comments. An earlier version of this paper was called 'Learning in a Model of Exit'. Murto thanks the Academy of Finland, and both authors thank Yrjö Jahnsson's foundation for financial support during the writing of this paper.

References

BANERJEE, A.V. (1992), "A Simple Model of Herd Behavior", *Quarterly Journal of Economics*, **107**, 797–817.

BIKHCHANDANI, S., HIRSHLEIFER, D. and WELCH, I. (1992), "A Theory of Fads, Fashion, Custom, and Cultural Change as Informational Cascades", *Journal of Political Economy*, **100**, 992–1026.

BOLTON, P. and HARRIS, C. (1999), "Strategic Experimentation", *Econometrica*, **67**, 349–374.

BULOW, J. and KLEMPERER, P. (1994), "Rational Frenzies and Crashes", *Journal of Political Economy*, **102**, 1–23.

CAPLIN, A. and LEAHY, J. (1994), "Business as Usual, Market Crashes, and Wisdom After the Fact", *American Economic Review*, **84**, 548–565.

CHAMLEY, C. (2004), "Delays and Equilibria with Large and Small Information in Social Learning", *European Economic Review*, **48**, 477–501.

CHAMLEY, C. and GALE, D. (1994), "Information Revelation and Strategic Delay in a Model of Investment", *Econometrica*, **62**, 1065–1086.

DÉCAMPS, J.-P. and MARIOTTI, T. (2004), "Investment Timing and Learning Externalities", *Journal of Economic Theory*, **118**, 80–102.

FUDENBERG, D. and LEVINE, D. (1992), "Maintaining a Reputation When Strategies are Imperfectly Observed", *Review of Economic Studies*, **59**, 561–579.

KELLER, G., RADY, S. and CRIPPS, M. (2005), "Strategic Experimentation with Exponential Bandits", *Econometrica*, **73**, 39–68.

MARIOTTI, M. (1992), "Unused Innovations", *Economics Letters*, **38**, 367–371.

ROSENBERG, D., SOLAN, E. and VIEILLE, N. (2007), "Social Learning in One Arm Bandit Problems", *Econometrica*, **75**, 1591–1611.

ROTHSCHILD, M. and STIGLITZ, J. (1970), "Increasing Risk I: Definition", *Journal of Economic Theory*, **2**, 225–243.

SMITH, L. and SORENSEN, P. (2000), "Pathological Outcomes of Observational Learning", *Econometrica*, **68**, 371–398.

TOXVAERD, F. (2008), "Strategic Merger Waves: A Theory of Musical Chairs", *Journal of Economic Theory*, **140**, 1–26.

Chapter 13

Information Acquisition and Efficient Mechanism Design*

Dirk Bergemann[†] and Juuso Välimäki[‡]

[†] *Department of Economics, and Cowles Foundation of Research in Economics, Yale University, New Haven, CT 06520-8268, USA*
dirk.bergemann@yale.edu
[‡] *Department of Economics, Aalto University School of Business, 02150 Espoo, Finland*
juuso.valimaki@aalto.fi

We consider a general mechanism design setting where each agent can acquire (covert) information before participating in the mechanism. The central question is whether a mechanism exists that provides the efficient incentives for information acquisition ex-ante and implements the efficient allocation conditional on the private information ex-post.

It is shown that in every private value environment the Vickrey-Clark-Groves mechanism guarantees both ex-ante as well as ex-post efficiency. In contrast, with common values, ex-ante and ex-post efficiency cannot be reconciled in general. Sufficient conditions in terms of sub- and supermodularity are provided when (all) ex-post efficient mechanisms lead to private under- or over-acquisition of information.

Keywords: Auctions, mechanism design, information acquisition, ex-ante and ex-post efficiency.

*This chapter is reproduced from *Econometrica*, **70**, 1007–1034, 2002.

1. Introduction

1.1. *Motivation*

In most of the literature on mechanism design, the model assumes that a number of economic agents possess a piece of information that is relevant for the efficient allocation of resources. The task of the mechanism designer is to find a game form that induces the agents to reveal their private information. An efficient mechanism is one where the final allocation is efficient given all the private information available in the economy.

In this paper, we take this analysis one step further. We assume that before participating in the mechanism each agent can covertly obtain additional private information at a cost. After the information has been acquired, the mechanism is executed. Hence the primitive notion in our model is an information gathering technology rather than a fixed informational type for each player. It is clear that the properties of the mechanism to be played in the second stage affect the players' incentives to acquire information in the ex-ante stage.

The main results in this paper characterize information acquisition in ex-post efficient mechanisms. Efficiency of a mechanism in this paper is understood in the same sense as in the original contributions by Vickrey, Clarke, and Groves. In particular, we do not impose balanced budget or individual rationality constraints on the mechanism designer. In the independent private values case, we show that the Vickrey-Clarke-Groves (henceforth VCG) mechanism induces efficient information acquisition at the ex-ante stage.

The common values case is much less straightforward to analyze. In light of the recent results by Dasgupta and Maskin (2000) and Jehiel and Moldovanu (2001), it is in general impossible to find mechanisms that would induce ex-post efficient allocations. Adding an ex-ante stage of information acquisition does not alleviate this problem. The two basic requirements for incentive compatibility of the efficient allocation rule are that the signals to the agents be single dimensional and that the allocation rule be monotonic in the signals. Even when these two conditions are met, we show that any efficient ex-post mechanism does not result in ex-ante efficient information acquisition. We use ex-post equilibrium as our solution concept. An attractive feature of this concept for problems with endogenously determined information is that the mechanisms do not depend on the distributions of the signals. By the revenue equivalence theorem, any allocation rule that can be supported in an ex-post equilibrium results in the same expected payoffs to all of the players as the VCG mechanism, provided that the lowest type receives the same utility in the mechanisms. But the defining characteristic of the VCG mechanism is that an agent's payoff changes only when the allocation changes due to his announcement of the signal. As a result, the payoffs cannot reflect the direct informational effects on other agents, and hence the private and social incentives will differ in general.

We also investigate the direction in which the incentives to acquire information are distorted. We restrict our attention to the case where the efficient allocation rule can be implemented in an ex-post equilibrium and derive new necessary and sufficient conditions for the ex-post implementability. It turns out that under our sufficient conditions for implementability, the information acquisition problem also satisfies the conditions for the appropriate multi-agent generalization of a monotone environment as defined in Karlin and Rubin (1956) and Lehmann (1988). As a result, we can expand the scope of our theory beyond signal structures that satisfy Blackwell's order of informativeness to the much larger class of signals ordered according to their effectiveness as defined in Lehmann (1988). We show that in settings with conflicting interests between agent i and all other agents, as expressed by their marginal utilities, every ex-post efficient mechanism results in excessive information acquisition by agent i. With congruent interests between agent i and agents $-i$, there is too little investment in information by agent i at the ex-ante stage.

The paper is organized as follows. The model is laid out in the next section. Section 3 presents the case of a single unit auction as an example of the general theory. The analysis of the independent private values case is given in Section 4. Results on efficient ex-post implementation are presented in Section 5. Section 6 deals with ex-ante efficiency in the common values case and Section 7 concludes.

1.2. *Literature*

This paper is related to two strands of literature in mechanism design. It extends the ideas of efficient mechanism design pioneered by Vickrey (1961), Clarke (1971), and Groves (1973) in an environment with fixed private information to an environment with information acquisition.

Our results on ex-post efficient mechanisms in common values environmentss complement recent work by Dasgupta and Maskin (2000) and Jehiel and Moldovanu (2001). Dasgupta and Maskin (2000) suggest a generalization of the VCG mechanism to obtain an efficient allocation in the context of multi-unit auctions with common values. Jehiel and Moldovanu analyze the efficient design in a linear setting with multidimensional signals and interdependent allocations. We give necessary conditions as well as weaker sufficient conditions for the efficient design in a general nonlinear environment. The results here are valid for general allocation problems and not only for single or multi-unit auctions.

The existing literature on information acquisition in mechanism design is restricted almost entirely to the study of auctions.[1] In the private values setting, Tan (1992) considers a procurement model where firms invest in R&D expenditure prior to the bidding stage. In the symmetric equilibrium with decreasing returns to scale, he observes that revenue equivalence holds between first and second price auction. Stegeman (1996) shows that the second price auction induces efficient information acquisition in the single unit independent private values case. Our results in the private values case can thus be seen as extensions of these earlier results. We show that analogous results hold for a much larger class of models and that as long as the conditions of the revenue equivalence theorem are satisfied, there is no need to analyze separately different indirect mechanisms that result in efficient allocation. Matthews (1977 and 1984) considers endogenous information acquisition in a pure common values auction and analyzes the convergence of the winning bid to the true value of the object when the number of bidders increases. Those papers are different from our papers in at least two respects. First, Matthews compares given auction forms for a single unit auction rather than taking a mechanism design approach to general allocation problems. Second, as Matthews considers the pure common value model, the efficient level of information acquisition is always identical to zero. Persico (2000) compares the equilibrium incentives of the bidders to acquire information in first and second price auctions within a model of affiliated values. Persico (2000) also uses the same notion of informativeness of information structures as we use. Again the main difference between his approach and ours is that we take the mechanism design approach rather than compare given auction formats for a single-unit auction.

---

[1] The notable exceptions are Rogerson (1992), who analyzes an n-person investment and subsequent allocation problem in Bayesian-Nash equilibrium; Crémer, Khalil, and Rochet (1998a and 1998b), who study information acquisition in a Baron-Myerson adverse selection model; and Auriol and Gary-Bobo (1999), who consider decentralized sampling in a collective decision model for a public good.

2. Model

2.1. *Payoffs*

Consider a setting with I agents, indexed by $i \in \mathscr{I} = \{1, \ldots, I\}$. The agents have to make a collective choice x from a compact set X of possible alternatives. Uncertainty is represented by a set of possible states of the world, $\Omega = \times_{i=l}^{l} \Omega_i$, where Ω_i is assumed to be a finite set for every i.[2] An element $\omega \in \Omega$ is a vector $\omega = (\omega_i, \omega_{-i}) = (\omega_1, \ldots, \omega_i, \ldots, \omega_l)$. The prior distribution $q(\omega)$ is common knowledge among the players. The marginal distribution over ω_i is denoted by $q_i(\omega_i)$ and we assume that the prior distribution $q(\omega)$ satisfies independence across i, or

$$q(\omega) = \prod_{i=1}^{I} q_i(\omega_i)$$

We assume that agent i's preferences depend on the choice x, the state of the world ω, and a transfer payment t_i, in a quasilinear manner:

$$u_i(x, \omega) - t_i.$$

We also assume that u_i is continuous for all i. The mechanism designer is denoted with a subscript 0, and her utility is assumed to be

$$\sum_{i=1}^{I} t_i + u_0(x).$$

The model is said to be a *private value model* if, for all ω, ω',

$$\omega_i = \omega_i' \Rightarrow u_i(x, \omega) = u_i(x, \omega'). \tag{2.1}$$

If condition (2.1) is violated, then the model displays *common values*.

2.2. *Signals and posteriors*

Agent i can acquire additional information by receiving a noisy signal about the true state of the world. Let S_i be a compact set of possible signal realizations that agent i may observe. Agent i acquires information by

---

[2]The extension to a compact, but not necessarily finite, state space would only change sums to integrals in the appropriate formulae.

choosing a distribution from a family of joint distributions over the space $S_i \times \Omega_i$:

$$\{F^{\alpha_i}(S_i, \omega_i)\}_{\alpha_i \in A_i}, \tag{2.2}$$

parameterized by $\alpha_i \in A_i$. We refer to $F^{\alpha_i}(s_i, \omega_i)$ as the signal and s_i as the signal realization. Since the conditional distribution of s_i depends only on Ω_i and since the prior on Ω satisfies independence across i, s_i is independent of s_j for all $i \neq j$.[3] Each A_i is assumed to be a compact interval in $\mathbb{R}$. The cost of information acquisition is captured by a cost function $c_i(\alpha_i)$ and $c_i(\cdot)$ is assumed to be continuous in α_i for all i. We endow $\Delta(S_i \times \Omega_i)$ with the topology of weak convergence and assume that $F^{\alpha_i}(s_i, \omega_i)$ is continuous in α_i in that topology. This ensures that the marginal distributions on S_i are continuous in α_i as well.

Agent i acquires information by choosing α_i. Each fixed α_i corresponds to a statistical experiment, and observing a signal realization $s_i \in S_i$ leads agent i to update his prior belief on ω_i according to Bayes' rule. The resulting posterior belief, $p_i(\omega|s_i, \alpha_i)$ summarizes the information contained in the signal realization s_i, with

$$p_i(\omega_i|s_i, \alpha_i) = \frac{f^{\alpha_i}(s_i, \omega_i)}{\sum_{\omega_i' \in \Omega_i} f^{\alpha_i}(s_i, \omega_i')}.$$

Considered as a family of distributions on Ω_i parameterized by s_i, we assume that $p_i(\omega_i|s_i, \alpha_i)$ is continuous in s_i in the weak topology on Ω_i.[4]

A profile of signal realizations $s = (s_1, \ldots, s_I)$ leads to a posterior belief $p(\omega|s, \alpha)$, which can be written by the independence of the prior belief and the signals as

$$p(\omega|s, \alpha) = \prod_{i=1}^{I} p_i(\omega_i|s_i, \alpha_i).$$

---

[3]Since we focus on ex-post equilibria, this independence is not needed for the results on efficient implementation. If we wanted to extend the analysis to Bayesian implementation, then this assumption would have real strength. The independence assumptions allow us to give conditions on the economic fundamentals that lead to over- and underacquisition of information in the ex-ante stage.

[4]The continuity and compactness assumptions made above are sufficient to guarantee that the choice set of each agent is compact and that the objective function is continuous in the choice variable.

In many instances, it is convenient to let the signal realization s_i be directly a posterior belief $p_i(.)$. The experiment α_i can then be represented directly by a joint distribution over ω_i and p_i.

2.3. *Efficiency*

The ex-ante efficient allocation requires each individual agent i to acquire the efficient amount of information and the allocation x to be optimal conditional on the posterior beliefs of all agents. Since the model has quasilinear utilities, Pareto efficiency is equivalent to surplus maximization.[5] The social utility is defined by

$$u(x,\omega) \triangleq \sum_{i=0}^{I} u_i(x,\omega).$$

The expected social surplus of an allocation x conditional on the posterior belief $p(\omega)$ is given by

$$u(x,p) \triangleq \sum_{\omega \in \Omega} u(x,\omega)p(\omega). \tag{2.3}$$

The *ex-post efficient* allocation $x(p)$ maximizes $u(x,p)$ for a given p. Given the assumptions made in the previous subsection, it is clear that a maximizer exists for all p.

Similarly, denote by p_{-i} the information held by all agents but i, with $p_{-i}(\omega) = q_i(\omega_i)\Pi_{j\neq i}p_j(\omega_j)$ and let $x_{-i}(p_{-i})$ be the allocation that maximizes the expected social value $u_{-i}(x,p_{-i})$ of all agents excluding i, with

$$u_{-i}(x,\omega) \triangleq \sum_{j\neq i} u_j(x,\omega) \tag{2.4}$$

and

$$u_{-i}(x,p_{-i}) \triangleq \sum_{\omega \in \Omega} u_{-i}(x,\omega)p_{-i}(\omega). \tag{2.5}$$

Let $F^\alpha(p)$ be the distribution induced on posteriors by the vector of experiments, where $\alpha = (\alpha_1,\ldots,\alpha_I)$ and let $c(\alpha) = \sum_i c_i(\alpha_i)$. An *ex-ante*

---

[5]Recall that the mechanism designer collects all the payments and receives utility from them.

efficient allocation is a vector of experiments, α^*, and an ex-post efficient allocation $x(p)$, such that α^* solves

$$\max_{\alpha \in A} \int u(x(p), p) dF^\alpha(p) - c(\alpha). \tag{2.6}$$

Observe that since we have used the posterior probabilities as arguments in the choice rule, the optimal allocation $x(p)$ does not depend on α. Again, given the continuity and compactness assumptions made in the previous subsection, a solution is guaranteed to exist.

3. Illustrating Examples

In this section, we present an example of a single unit auction with two bidders. It is meant to introduce the basic arguments for the private and common values results and to indicate how to extend the logic of the arguments to any number of agents and allocations. A similar example is discussed in Maskin (1992) with a signal space but without an underlying state space. After presenting the example, we briefly discuss the role of the independence assumptions of ω_i across i by arguing how it could arise in the auction setting and then in a different environment, namely procurement.

3.1. *Information acquisition in an auction*

The set of allocations is the set of possible assignments of the object to bidders, or $X = \{x_1, x_2\}$, where x_i denotes the decision to allocate the object to bidder $i \in \{1, 2\}$. The state space of agent i is given by $\Omega_i = \{0, 1\}$. We begin with a private value model, where the value of the object for bidder i is $u_i(x_i, \omega) = 2\omega_i$ and $u_i(x, \omega) = 0$ for $x \neq x_i$. We let the signal of agent i be simply his posterior belief $p_i = Pr(\omega_i = 1)$. The expected (ex-post) utility for agent i depends on p_i and p_j: $u_i(x_i, p_i, p_j) = 2p_i$.[6] The direct VCG mechanism in this setting is the second price auction where bidder i pays the reported valuation of bidder j conditional on obtaining the object. Ex-post efficiency implies that i gets the object if $u_i(x_i, p_i, p_j) \geq u_j(x_j, p_i, p_j)$, i.e. if $p_i \geq p_j$. It follows that the equilibrium utility of bidder

---

[6]The notation in this section is in minor conflict with the general notation presented in the previous section to take advantage of the binary structure of the example: (i) p_i in this section is simply a scalar rather than a probability distribution and (ii) the expected gross utility is written as a function of p_i and p_j rather than the implied probability vector p over the state space Ω, which is here simply: $\Omega = \{0, 1\} \times \{0, 1\}$.

i, conditional on obtaining the object, is $u_i(x_i, p_i, p_j) - u_j(x_j, p_i, p_j)$. For an arbitrary fixed realization $p_j = \hat{p}$, the valuations by i and j are depicted in Figure 13.1a as functions of p_i. The equilibrium net utility of bidder i has the same slope in p_i as the social utility, as displayed in Figure 13.1b.

Consider next information acquisition within this auction. With a binary state structure, a signal is more informative if the posteriors are more concentrated around 0 and 1. Around $\hat{p}$, a local increase in informativeness can be represented as a lottery (with equal probability) over $\hat{p} - \varepsilon$ and $\hat{p} + \varepsilon$ for some $\varepsilon > 0$. The convexity of the equilibrium net utility (see Figure 13.1b) implies that information has a positive value. More importantly, the private marginal value of the lottery coincides with the social marginal value. As a result each agent acquires the socially efficient level of information. The logic of this argument extends to all private value problems as the utility $u_{-i}(x, p)$ of all agents but i is constant in p_i.

To extend the example to a common values environment, let $u_i(x_i, \omega) = 2\omega_i + \omega_j$. The expected valuation is then $u_i(x_i, p_i, p_j) = 2p_i + p_j$ and under an efficient allocation rule i gets the object when $p_i \geq p_j$. For a given $p_j = \hat{p}$, the utilities are displayed as functions of p_i in Figure 13.2a. The valuation of bidder j now varies with p_i, even though it is less responsive to p_i than the valuation of i. The valuations therefore satisfy a familiar single-crossing condition. However, as the valuation of bidder j varies with p_i, the original VCG mechanism does not induce truth telling in ex-post equilibrium. If we were to apply the mechanism, the equilibrium utility of agent i would be $u_i(x_i, p_i, \hat{p}) - u_j(x_j, p_i, \hat{p})$, but for any $p_i > \hat{p}$, there is an $\varepsilon > 0$ such that bidder i could lower his report to $p_i - \varepsilon$, still get the object, but receive $u_i(x_i, p_i, \hat{p}) - u_j(x_j, p_i - \varepsilon, \hat{p}) > u_i(x_i, p_i, \hat{p}) - u_i(x_i, p_i, \hat{p})$. The above argument remains valid until $p_i = \hat{p}$, where a lower report would induce an undesirable change in the allocation. Thus by asking bidder i to pay $u_j(x_j, \hat{p}, \hat{p})$, incentive compatibility is preserved. The equilibrium utility of agent i is then $u_i(x_i, p_i, \hat{p}) - u_j(x_j, \hat{p}, \hat{p})$. When we now compare the slopes of individual payoffs and social payoffs locally at $\hat{p}$, we find that the equilibrium utility of agent i has a sharper kink than the social utility as depicted in Figure 13.2b.

As before, more information can be represented locally as a randomization over posteriors around $\hat{p}$. In equilibrium bidder i has excessive incentives to acquire information relative to the socially optimal level as his objective function is (locally) more convex. Conversely, agent i has insufficient incentives to acquire information if $\partial u_j(x_j, p_i, p_j)/\partial p_i < 0$.

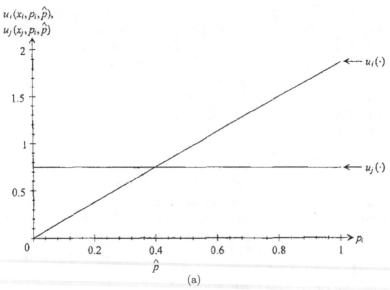

(a)

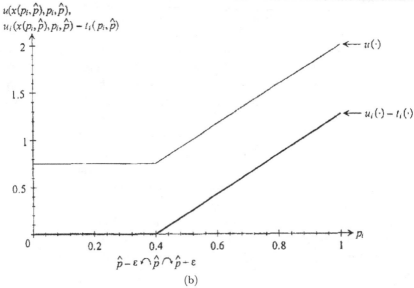

(b)

Figure 13.1.   (a) Private value utilities $u_i(x_i, p_i, \hat{p})$ and $u_j(x_j, p_i, \hat{p})$ for $p_j = \hat{p} = 3/8$.
(b) Social value $u(x(p_i, \hat{p}), p_i, \hat{p})$ and equilibrium utility $u_i(x(p_i, \hat{p}), p_i, \hat{p}) - t_i(p_i, \hat{p})$.

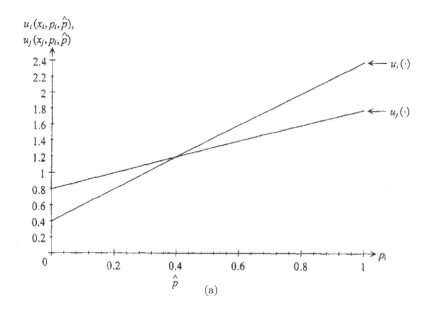

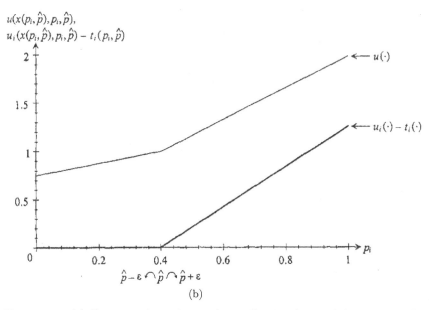

Figure 13.2. (a) Common value utilities $u_i(x_i, p_i, \hat{p})$ and $u_j(x_j, p_i, \hat{p})$ for $p_j = \hat{p} = 3/8$. (b) Social value $u(x(p_i, \hat{p}), p_i, \hat{p})$ and equilibrium utility $u_i(x(p_i, \hat{p}), p_i, \hat{p}) - t_i(p_i, \hat{p})$.

Next we briefly sketch how these insights generalize beyond the current example. The (single) crossing of the utilities at $p_i = \hat{p}$ has two important implications.

First, it indicates that it is socially efficient to change the assignment from agent j to agent i at $p_i = \hat{p}$. Consequently the social utility satisfies at $p_i = p_j = \hat{p}$:

$$\frac{\partial u(x_i, p_i, p_j)}{\partial p_i} - \frac{\partial u(x_j, p_i, p_j)}{\partial p_i} \geq 0. \qquad (3.1)$$

Consider an order $\succ$ on X, such that $x_i \succ x_j$. If the local condition (3.1) holds for all $p_i, p_j \in [0, 1]$, then $u(x, p_i, p_j)$ is supermodular in (x, p_i). The second implication of the single crossing condition is that at $p_i = p_j = \hat{p}$

$$\frac{\partial u_i(x_i, p_i, p_j)}{\partial p_i} - \frac{\partial u_i(x_j, p_i, p_j)}{\partial p_i} \geq 0, \qquad (3.2)$$

where the latter condition is necessary for truth telling by agent i. (The partial derivative of the second term is naturally equal to zero for all (p_i, p_j) in auctions without externalities.) If $u_i(x, p_i, p_j)$ is supermodular in (x_i, p_i), i.e., if we require (3.2) to hold globally for all $p_i, p_j \in [0, 1]$, we obtain a sufficient condition for truth telling by agent i. Finally, the condition for under- or overacquisition of information by bidder i was related to the responsiveness of the utility of agent j, $u_j(x, p_i, p_j)$ to p_i. We can now restate the conditions for overacquisition of information by agent i in terms of

$$\frac{\partial u_j(x_i, p_i, p_j)}{\partial p_i} - \frac{\partial u_j(x_j, p_i, p_j)}{\partial p_i} \leq 0, \qquad (3.3)$$

or equivalently that $u_j(x, p_i, p_j)$ is *submodular* in (x, p_i). We shall see in the subsequent sections that the supermodularity conditions for $u_i(x, p_i, p_j)$ and $u_i(x, p_i, p_j)$ are sufficient and almost necessary conditions for efficient implementation and that sub- or supermodularity conditions for $u_{-i}(x, p_i, p_{-i})$, provide sufficient conditions for over- and underacquisition of information in ex-post efficient mechanism.

3.2. *Independence*

The example just discussed is a linear version of an auction model introduced by Maskin (1992) and Dasgupta and Maskin (2000) to analyze privatization and related asset sale problems. A very similar model with common values and bidders with independent private information appears in Bulow, Huang, and Klemperer (1999) to analyze takeovers (with toeholds). The sale of a company is a fine example to see how the independence of ω_i across i might arise in substantive economic problems. Consider the sale of a company whose primary value is its client list. The state ω_i would then describe the extent to which the client list of the target company overlaps with the acquiring firm. The acquiring firm could either value differing client lists (to gain access to new clients) or overlapping client lists (to enhance the cross-selling of products). For competitive reasons, it also matters for firm j how the client list of the target compares with the client list of firm j, which is represented by ω_j. This introduces the common value aspect into the takeover contest. The single crossing condition in this environment simply states that the marginal value of information about the extent of the intersection between the client list of firm i and the target is larger for firm i than for firm j. The independence of ω_i across i then amounts to assuming that the clients are distributed independently across the acquiring firms i and j.

The general interpretation in the context of an asset sale, where the value of the asset could arise from a proprietary technology, a marketing strategy, a specific product or market niche or alike, is then that value of the asset for the acquiring firm is determined by the match of the asset with the characteristics of the acquiring firm. The independence assumption requires that the characteristics of the acquiring firms are distributed independently across firms.

Yet a different environment where independence of the state variable ω_i arises quite naturally, is in the context of procurement and R&D. The following is a stylized version of a model recently analyzed by D'Aspremont, Bhattacharya, and Gerard-Varet (2000). Their analysis focuses on bargaining with information sharing, whereas we adopt a procurement interpretation. Suppose there are two firms who compete for a contract by the government, regarding a project, say a weapons or software system, which has a social value v. The two firms, $i \in \{1, 2\}$, are pursuing the realization of the same project, but follow different design routes or approaches. Uncertainty in the model is described by, for simplicity, a binary state space $\Omega_i = \{0, 1\}$, where ω_i, represents the probability that the research of firm i will eventually be successful. If there are many possible ways of pursuing the same goal, it is sensible to assume that the priors on Ω_i satisfy independence. Denote the (expected) cost of firm i of completing the project, independent of eventual failure or success, by γ_i. The government agency designs a game form that decides which of the firms should pursue the project. The space of possible decisions is then $X = \{0, 1, 2, 12\}$ where

$x = i$ stands for the case where firm i continues with its research, $x = 12$ denotes the case where both firms continue in the race, and $x = 0$ denotes the case where neither firm continues with research in the second stage. In the first stage, each firm can obtains information about its research project by observing a signal realization s_i generated by an information structure α_i. Each choice of α_i and realization of s_i result in a posterior belief $p_i(\omega_i|s_i, \alpha_i)$. We may assume that if $\omega_1 = \omega_2 = 1$, then firm i wins the race with probability π_i and firm j with probability π_j. Assuming quasilinearity of the payoffs in the cost, we can write the ex-post utilities $u_i(x, \omega_i, \omega_j)$ for the case that both projects receive support by

$$u_i(12, 1, 0) = v - \gamma_i, \quad u_i(12, 1, 1) = \pi_i v - \gamma_i,$$
$$u_i(12, 0, 0) = u_i(12, 0, 1) = -\gamma_i,$$

and for the case that only a single project receives support by

$$u_i(i, \omega_i, \cdot) = \omega_i v - \gamma_i,$$
$$u_i(j, \cdot, \cdot) = 0.$$

Since $u_i(12, \omega_i, \omega_j)$ depends nontrivially on ω_j when $\omega_i = 1$, this model is one with common values. Let $p_i \triangleq p_i(\omega_i = 1|s_i, \alpha_i)$ and we observe that the socially optimal decision (conditional on the posterior beliefs) is to have firm i engage in the race (possibly jointly with firm j) whenever

$$p_i v - \gamma_i > 0 \quad \text{and} \quad p_i v - \gamma_i > p_j v - \gamma_j.$$

It is efficient to have both firms doing research in the second stage if

$$p_i(1 - p_j)v > \gamma_i \quad \text{and} \quad p_j(1 - p_i)v > \gamma_j.$$

In words, only the superior firm should continue with the research if its success is sufficiently certain (this is to avoid duplication of costly effort). If the posterior probability of success is in the intermediate range for both firms, then they should engage in further research jointly.[7] The independence here reflects the fact that each firm is pursuing an independent research program.

---

[7]It can also be verified that the ex-post efficient allocation satisfies the required single crossing properties and hence our results apply to this example as well.

4. Private Values

This section considers information acquisition in the context of independent private values. For this environment Vickrey (1961), Clarke (1971), and Groves (1973) showed in increasing generality that the ex-post efficient allocation can be implemented in a direct revelation mechanism.

Definition 1: A *direct revelation mechanism* is defined by a pair (x, t), where x is an outcome function, $x : S \to X$, and t is a transfer scheme, $t : S \to \mathbb{R}^I$.

In the private value environment we may consider without loss of generality the set of signal realizations S to be the probability simplex Δ over the state space Ω. The efficient allocation is implemented in dominant strategies if the transfer function has the following form: For all $i \in \mathscr{I}$,

$$t_i(p) = h_i(p_{-i}) - u_{-i}(x(p), p_{-i}), \tag{4.1}$$

where $h_i(p_{-i})$ is an arbitrary function of p_{-i}. We refer to the class of mechanisms that implement the efficient allocation with a transfer function of the form displayed in (4.1) as *the Vickrey-Clark-Groves (VCG) mechanism.*

Definition 2: A vector of experiments, α, is a *local social optimum* if for every i, α_i solves

$$\alpha_i \in \arg\max_{\alpha_i' \in A_i} \left\{ \int u(x(p), p) dF^{(\alpha_i', \alpha_{-i})}(p) - c(\alpha_i', \alpha_{-i}) \right\}.$$

Notice that local here refers to the property that α solves the maximization problem for each agent separately, or Nash locality. In consequence, a local social optimum may not necessarily be a solution to the problem when the experiments of all agents are jointly maximized.

Theorem 1 (Private Values): *With independent private values, every local social optimum can be achieved by the VCG mechanism.*

Proof: See Appendix. $\square$

With the VCG mechanism the equilibrium net utility of agent i behaves as the social utility up to $h_i(p_{-i})$, which does not depend on p_i. It therefore follows that the decision problem of agent i with respect to the information acquisition in terms of the posterior belief p_i is equivalent to the problem faced by the social planner. An immediate consequence of Theorem 1 is the following corollary.

Corollary 1:   *The ex-ante efficient allocation can be implemented by the VCG mechanism.*

Proof: See Appendix.   □

The ex-ante stage has many equilibria if there are multiple local social optima.[8] It follows that the VCG mechanism uniquely implements the ex-ante efficient allocation only if there is a unique local and hence global optimum in the information acquisition stage. This efficiency result can also be generalized to environments where each agent can invest ex ante in technologies that increase their private payoffs. The ex-ante efficiency result derived for the VCG mechanism can also be extended to any ex-post efficient mechanism by the revenue equivalence theorem.

In the current model, information is acquired by all agents simultaneously. However, it is well known in statistical decision theory that a sequential decision procedure may dominate any simultaneous procedure as it economizes on the cost of information acquisition. This observation is valid in the current model as well. An important consequence of a sequential version of the VCG mechanism is that the efficient allocation is now strongly implementable as every agent acts at every node as if he were maximizing the social value function (for a more detailed argument see Bergemann and Välimäki (2000)).

The essential property that allows us to prove ex-ante efficiency with independent private values is the restriction that only agent i can (efficiently) invest in information about his own utility associated with various allocations. The logical next step is therefore to ask whether efficiency can be maintained in environments where the information of agent i is relevant to the utility calculus of agent j. We pursue this question in the context of the interdependent values model investigated recently by Dasgupta and Maskin (2000) and Jehiel and Moldovanu (2001). Before we analyze the information acquisition per se, we give a complete characterization of the ex-post efficient allocation and associated equilibrium utilities for each agent in the following section.

5. Common Values: Ex-Post Efficiency

We adopt the model of Dasgupta and Maskin (2000) to our environment with uncertainty about the true state of nature in subsection 5.1, where

---

[8]Tan (1992) makes a similar observation in the context of ex-ante R&D investments in procurement auctions.

we present necessary and sufficient conditions for efficient implementation with a direct revelation mechanism.[9] Similar results are briefly stated for a continuous allocation space in subsection 5.2.

5.1. *Finite Allocation Space*

We start by considering a set of finitely many allocations: $X = \{x^0, x^1, \ldots, x^N\}$. For a fixed α_i, player i's expected utility from an allocation x^n after observing signal s is, using independence, given by

$$u_i(x^n, s) = \sum_{\omega = \Omega} u_i(x^n, \omega) \prod_{j \in \mathscr{I}} p_j(\omega_j | s_j, \alpha_j).$$

In anticipation of the requirements for the implementability of the efficient allocation rule, we restrict our attention to an arbitrary class of one-dimensional signal realizations $S_i = [s_{-i}, \bar{s}_i] \subset \mathbb{R}$ with the result that the associated posterior beliefs $p_i(\cdot | s_i, \alpha_i)$ form a one-dimensional manifold in $\Delta(\Omega_i)$. In this section the allocation problem is analyzed exclusively at the ex-post stage. The utilities are therefore written as functions of (x, s) rather than (x, ω) and consequently $x(s)$ is *an ex-post efficient allocation rule* conditional on the signal s. We assume $u_i(x, s)$ to be continuously differentiable in s for all i.

Next we present necessary and sufficient conditions for efficient implementation in an ex-post equilibrium. By the revelation principle, we can restrict ourselves to direct mechanisms and truth telling strategies.

Definition 3: A direct revelation mechanism (x, t) permits *implementation in an ex-post equilibrium* if $\forall i, \forall s \in S$:

$$u_i(x(s), s) - t_i(s) \geq u_i(x(\hat{s}_i, s_{-i}), s) - t_i(\hat{s}_i, s_{-i}), \forall \hat{s}_i \in S_i.$$

An ex-post equilibrium, while not requiring dominant strategies, remains a Bayesian equilibrium for any prior distribution over types. For the rest of this subsection, we fix the realization of the signals s_{-i} and focus on truth telling conditions for agent i. Let the set S_i^n be defined as

---

[9]Dasgupta and Maskin (2000) actually restrict attention to multi-object auctions and achieve implementation through an indirect mechanism in which the bidders report their valuations contingent on the reports by the other bidders, but not directly their signals. Jehiel and Moldovanu (2001) present sufficient conditions in a linear model for general allocation problems with a direct revelation mechanism.

the subset of S_i for which x^n is an efficient allocation:

$$S_i^n = \{s_i \in S_i | u(x^n, s_i, s_{-i}) \geq u(x^k, s_i, s_{-i}), \ \forall x^k \neq x^n\}.$$

For any two sets S_i^k and S_i^l with a nonempty intersection, we call a point $S_i^{kl} \in S_i^k \cap S_i^l$ a *k to l change point* if there exists an $\varepsilon > 0$ such that either[10]:

$$s_i \in [s_i^{kl} - \varepsilon, s_i^{kl}) \Rightarrow s_i \in S_i^k, \quad s_i \notin S_i^l \tag{5.1}$$

or

$$\forall s_i \in (s_i^{kl}, s_i + \varepsilon] \Rightarrow s_i \notin S_i^k, \quad s_i \in S_i^l. \tag{5.2}$$

Symmetrically, we can define $s_i^{lk} \in S_i^k \cap S_i^l$ to be an *l to k change point*. By extension, let $s^{kl} \triangleq (S_i^{kl}, S_{-i})$. Every change point s^{kl} has the property that at $s = s^{kl}$:

$$\frac{\partial u(x^k, s)}{\partial s_i} \leq \frac{\partial u(x^l, s)}{\partial s_i}.$$

Consider next the ex-post truthtelling condition for agent i:

$$u_i(x(s), s) - t_i(s) \geq u_i(x(\hat{s}_i, s_{-i}), (s_i, s_{-i})) - t_i(\hat{s}_i, s_{-i}), \quad \forall \hat{s}_i \in S_i.$$

It follows that the transfer payment of agent i has to be constant conditional on the allocation $x(s) = x^n$ and we denoted it by t_i^n.

Proposition 1: *A necessary condition for ex-post implementation is that for $\forall k$, $\forall l$, at $s = s^{kl}$*

$$\frac{\partial u_i(x^k, s)}{\partial s_i} \leq \frac{\partial u_i(x^l, s)}{\partial s_i} \tag{5.3}$$

Proof: See Appendix. $\square$

The inequality (5.3) is a familiar local sorting condition and implies that the incentive compatible transfers are uniquely determined (up to a common constant) at the change point s^{kl} by

$$t_i^k - t_i^l = u_{-i}(x^l, s^{kl}) - u_{-i}(x^k, s^{kl}). \tag{5.4}$$

As the transfer payments t_i^n for every allocation x^n are necessarily determined at the change points, it follows that (generically) every pair

---

[10]The condition is written as an either/or condition as the social utility may display the same partial derivative with respect to s_i for the alternatives x^k and x^l over an interval where the social values of the alternatives x^k and x^l are equal.

of sets S_i^k and S_i^l must have an intersection that forms a connected set, as otherwise t_i^k and t_i^l would be overdetermined. The latter condition can be rephrased as follows:

Definition 4: The collection $\{S_i^n\}_{n=0}^N$ satisfies *monotonicity* if for every n:

$$s_i, s_i' \in S_i^n \Rightarrow \lambda s_i + (1 - \lambda)s_i' \in S_i^n, \quad \forall \lambda \in [0, 1].$$

A sufficient condition for monotonicity is that the social value $u(x^n, s)$ be single-crossing in (x^n, s_i). If monotonicity is satisfied, then there exists an optimal policy $x(s)$ such that x^n is chosen on a connected subset $R_i^n \subseteq S_i^n$ and nowhere else.[11] After possibly relabeling the indices, we can endow the allocation space X with the following order, denoted by $\prec$:

$$x^0 \prec x^1 \prec \cdots \prec x^N, \tag{5.5}$$

such that for all $s_i \in R_i^k$ and $s_i' \in R_i^l$, with

$$s_i < s_i' \Rightarrow x(s_i, s_{-i}) = x^k \prec x^l = (s_i', s_{-i}). \tag{5.6}$$

For the remainder of this section we continue to work with the order defined by (5.5) and (5.6).

Proposition 2: *A generically necessary condition for ex-post implementation is monotonicity.*

Proof: See Appendix. □

The class of mechanisms that implement the efficient allocation with the transfers determined by (5.4) is referred to as *the generalized Vickrey Clark Groves mechanism,* where we initialize t_i^0 by

$$t_i^0 \triangleq h_i(S_{-i}) - u_{-i}(x_0, (\underline{s}_i, s_{-i})), \tag{5.7}$$

for some arbitrary $h_i(s_{-i})$. Next we strengthen the local sorting condition to obtain sufficient conditions for ex-post implementation by extending the local to a global sorting condition.

---

[11]The socially optimal policy $x(s)$ is not unique as any (randomized) allocation over the set $\{x^k, x^l\}$ is optimal for all $s_i \in S_i^k \cap S_i^l$, and in particular at the change points. Moreover for some x^k the corresponding set R_i^k may be empty and, in consequence, the associated optimal allocation policy would use only a strict subset of the feasible allocations. Naturally, the order defined in (5.5) and (5.6) would then extend only over the subset of allocations selected by the allocation rule $x(s)$.

Proposition 3: *Sufficient conditions for ex post implementation are:*

(i) *monotonicity is satisfied for all i and s;*
(ii) *for all i, s and n,*

$$\frac{\partial u_i(x^{n-1}, s)}{\partial s_i} \leq \frac{\partial u_i(x^n, s)}{\partial s_i}. \tag{5.8}$$

Proof: See Appendix. □

Thus if the utility of every agent i displays supermodularity in (x^n, s_i) and monotonicity is satisfied, then an ex-post implementation exists. We wish to emphasize that the particular order imposed on the allocation space X may depend on i and s_{-i}, and all that is required is that for every s_{-i}, an order on X can be constructed such that the conditions above for necessity and sufficiency can be met.

It may be noted that monotonicity and supermodularity are strictly weaker than the conditions suggested by Dasgupta and Maskin (2000) in the context of a multi-unit auction. In the linear (in the signals) version of the model that is investigated by Jehiel and Moldovanu (2001), where

$$u_i(x^n) = \sum_{j=1}^{l} u_{ij}(x^n) s_j,$$

the necessary and sufficient conditions coincide. For details we refer the reader to an earlier version of the paper (Bergemann and Välimäki (2000)).

5.2. *Continuum of allocations*

The sorting and monotonicity conditions naturally extend to the case of a continuum of allocations. Let $X \subset \mathbb{R}$ be a compact interval of the real line. As before, fix the realization of signal s_{-i} and let $x(s)$ denote the efficient allocation rule. We also assume that $u_i(s, x)$ is twice continuously differentiable, and thus $x(s)$ is differentiable almost everywhere. If $x(s_i, s_{-i})$ is monotonic in s_i, we can impose a complete order, denoted by $\prec$, on the allocation space X such that the order on X mirrors the order of the signal space by requiring that for all s_i, s_i' and $x(s_i, s_{-i}) \neq x(s_i', s_{-i})$

$$s_i < s_i' \Rightarrow x(s_i, s_{-i}) \prec (s_i', s_{-i}). \tag{5.9}$$

We endow X with the complete order defined by (5.9).

Proposition 4: *Sufficient conditions for ex-post implementation are*:

(i) *monotonicity*;
(ii) *global sorting condition*:

$$\frac{u_i(x,s)}{\partial s_i \partial x} \geq 0, \quad \forall i, \forall s, \forall x.$$

Proof: See Appendix. ☐

As in the discrete case, monotonicity is generically necessary. If the global sorting condition is weakened to a local sorting condition at $x = x(s)$,

$$\frac{u_i(x(s),s)}{\partial s_i \partial x} \geq 0, \quad \forall i, \forall s,$$

then we obtain the corresponding necessary conditions for implementation. Proposition 4 generalizes an earlier proposition by Jehiel and Moldovanu (2001) from a linear to a nonlinear environment with one-dimensional signals. The transfer payments in the generalized VCG mechanism can be represented as

$$t_i(s) = -\int_{s_{-i}}^{s_i} \frac{\partial u_{-i}(x(v_i, s_{-i}), (v_i, s_{-i}))}{\partial x} \frac{\partial x(v_i, s_{-i})}{\partial v_i} dv_i + t_i(s_{-i}, s_{-i}),$$

$$(5.10)$$

where $t_i(s_{-i}, s_{-i}) = h_i(s_{-i})$ for some arbitrary $h_i(s_{-i})$.

6. Common Values: Ex-Ante Inefficiency

In this section, we analyze the implications of ex-post efficient mechanisms for the ex-ante decisions of the agents to acquire information. This problem is addressed by extending the monotone environment for a single decision-maker defined by Karlin and Rubin (1956) to a multiple agent decision environment. The monotone environment is introduced first in subsection 6.1 and the informational inefficiency is analyzed in subsection 6.2.

6.1. *Monotone environment*

We investigate the possibility of achieving efficient ex-ante decisions, while requiring efficiency in the second stage mechanism for an arbitrary choice of signals by the agents. This implementation requirement imposes certain restrictions on the posteriors and the utility functions. As we consider the ex-ante decision problem, the appropriate sorting and monotonicity conditions have to be formulated for the state space Ω, rather than the signal space S.

The first restriction concerns the implementability of the efficient allocation. Jehiel and Moldovanu (2001) show that it is generically impossible to implement the efficient allocation if the players must submit multidimensional reports. Hence we assume that for each player i, there exists a one-dimensional sub-manifold M_i, with $M_i \subset \Delta(\Omega_i)$, and a function $\lambda_i : S_i \times A_i \to M_i$ such that the ex-post efficient allocation $x(\cdot)$ can be determined by $x(\lambda(s, \alpha))$ rather than $x(p(\omega|s, \alpha))$, where

$$\lambda(s, \alpha) - (\lambda_1(s_1, \alpha_1), \dots, \lambda_I(s_I, \alpha_I)).$$

In the most straightforward case λ_i is an invertible mapping that associates to every posterior p_i in M_i exactly one signal realization s_i that generates the posterior. Then (M_i, λ_i) can be thought of as a direct dimensionality restriction on the posterior beliefs that can be generated by a family of signals. However, λ_i does not have to be invertible and then λ_i can be thought of as a sufficient statistic relative to the social allocation problem. We briefly present illustrations for this condition below.

First, observe that for a binary state space $\Omega_i = \{w_i^0, w_i^1\}$ the dimensionality assumption is satisfied for every class of signals, as the posterior on Ω_i can be represented by a single number in the unit interval, say $p_i(s_i, \alpha_i) \triangleq p_i(w_i^0|s_i, \alpha_i)$. Since $p_i(s_i, \alpha_i) \in [0, 1]$, we can take $M_i = [0, 1]$ for all i.

A second class of information structures satisfying the assumption is given by the following model, sometimes referred to as the "hard news" model in the literature. Suppose $\Omega_i \subset \mathbb{R}$. The signal realization is with probability α_i perfectly informative or $s_i \in \Omega_i$, where the conditional probability of $s_i = \omega_i$ being realized is given by the prior distribution $q_i(\omega_i)$; and with probability $1 - \alpha_i$, the signal realization is completely uninformative and hence the agent i maintains his prior as his posterior. In this class of models, the choice of α_i determines the probability of observing a completely informative signal of the state.[12] This class of models can be further extended to the case where conditional on receiving information, the signal is not perfectly informative, or $s_i = \omega_i + \varepsilon_i$, where ε_i can be

---

[12]We thank the co-editor for suggesting this class of information structures.

distributed arbitrarily and possibly dependent on ω_i or $\varepsilon_i \sim g_i(\cdot; \omega_i)$. As long as the distribution of the error term is assumed to be independent of the choice of α_i, the posterior $p(\omega_i|s_i, \alpha_i)$ is independent of $\alpha_i \in A_i$ and as long as s_i is one-dimensional, the posterior will lie on a one-dimensional submanifold of $\Delta(\Omega_i)$ and the requirement is satisfied.

A third class of models where the dimensionality condition holds for arbitrary families of information structures is one where the payoffs are linear in the states. In this case, the payoff from allocation x in state ω to agent i is given by

$$u_i(\omega, x) \triangleq \sum_{j=1}^{I} u_{ij}(x)\omega_j. \tag{6.1}$$

The linear setting is the one investigated in Jehiel and Moldovanu (2001) as well. In our model, the state ω is the primitive notion rather than the vector of signals s as in Jehiel and Moldovanu (2001), and as a result, the linearity is assumed here with respect to the state rather than the signal. The reason why the dimensionality condition holds in the linear model, is that the posterior expectation of ω_i, expressed by

$$\pi_i(s_i, \alpha_i) = \mathbb{E}[\omega_i|s_i, \alpha_i],$$

is sufficient for the determination and implementation of the efficient choice rule. Thus we let the $\lambda_i : S_i \times A_i \to M_i$ define a set of equivalent signal realizations and information acquisition decisions in the sense that each element in the equivalence class generates the same posterior expectation, or

$$\forall(s_i, \alpha_i), (s_i', \alpha_i') \in S_i, \quad \pi_i(s_i, \alpha_i) = \pi_i(s_i', \alpha_i') \Leftrightarrow \lambda_i(s_i, \alpha_i) = \lambda_i(s_i', \alpha_i').$$

As π_i is a real number, we can find a one-dimensional manifold M_i such that $\lambda_i(s_i, \alpha_i) \in M_i$ for all $(s_i, \alpha_i) \in S_i \times A_i$.[13] Thus with linear preferences, the dimensionality restriction on the signals is satisfied for every family A_i of signals. This argument also shows that the basic intuition developed in the introductory example for a binary state space extends to an arbitrary state space Ω_i.

---

[13]Observe that M_i can always be embedded in $\Delta(\Omega_i)$ so that the restriction $M_i \subset \Delta(\Omega_i)$ is satisfied as well.

With this dimensionality restriction in place, we can assume without loss of generality that every signal realization s_i leads to a fixed posterior belief $p_i(\omega_i|s_i, \alpha_i)$ independent of the choice of signal α_i. Two distinct signal choices α_i and α_i' differ in the frequency by which signal realizations s_i are observed.

Next we consider the appropriate sorting and monotonicity conditions in the state space Ω. We require that for every i, the allocation space X can be endowed with a complete order, denoted by $\prec$, such that $u_i(x, \omega_i, \omega_{-i})$ and $u(x, \omega_i, \omega_{-i})$ are supermodular in (x, ω_i) for all ω_{-i}.[14] Observe that the ranking of the allocations is allowed to vary with i as in the previous section, but here the ranking has to be invariant with respect to ω_{-i}. In addition we require that for all i, the posterior probabilities satisfy the monotone likelihood ratio property: for all $s_i' > s_i$ and $\omega_i' > \omega_i$, $p_i(\omega_i'|s_i')p_i(\omega_i|s_i) - p_i(\omega_i'|s_i)p_i(\omega_i|s_i') \geq 0$. The supermodularity of $u_i(x, \omega)$ in (x, ω_i) guarantees (globally) the sorting condition, whereas the supermodularity of $u(x, \omega_i, \omega_{-i})$ guarantees the monotonicity of the efficient allocation. The monotone likelihood ratio condition implies that supermodularity in (x, ω_i) translates into supermodularity in (x, s_i) when taking expectations with respect to the posterior beliefs based on the signal realizations.

Proposition 5: *Suppose the monotone likelihood ratio and supermodularity conditions hold for all i; then:*

(i) $u_i(x, s_i, s_{-i})$ *and* $u(x, s_i, s_{-i})$ *are supermodular in* (x, s_i);
(ii) *for all* s_i, s_i' *with* $x(s_i, s_{-i}) \neq x(s_i', s_{-i})$,

$$s_i < s_i' \Rightarrow x(s_i, s_{-i}) \prec x(s_i', s_{-i}).$$

Proof: See Appendix. $\square$

6.2. *Inefficiency*

In order to determine whether agent i has the socially correct incentives to acquire information, we must compare the returns from information acquisition for the social planner and agent i. The information provided by

---

[14]In the monotone environment of Karlin and Rubin (1956), the utility function of the decision maker was only assumed to be single-crossing in (x, ω_i). The stronger condition of supermodularity is imposed here as we consider a multi-dimensional signal space and when taking the expectations over ω_{-i}, supermodularity in (x, ω_i) is preserved while the single-crossing property is not.

a signal realization s_i affects both the valuation of any particular allocation x as well as the choice of the (socially) optimal allocation $x(s)$. By incentive compatibility, the generalized VCG mechanism guarantees that the social utility and the private utility of agent i are evaluated at $x(s)$ for every s. In the private values environment, the congruence between social and private utilities go even further since as functions of s_i they are identical up to a constant, possibly dependent on s_{-i}.

In the common value environment, in contrast, the marginal utility of s_i is in general different for the social utility and the private utility of agent i. This discrepancy is due to the fact that with common values, the utility of all agents but i, $u_{-i}(x, s)$, is responsive to the signal realization s_i. As a result, the social planner's preferences for information are in general different from the individual preferences. In order to determine how this discrepancy affects the incentives to acquire information, we make use of a characterization of the transfer function associated with the generalized VCG mechanism.

Theorem 2 (Inefficiency in Mechanisms): *Every ex-post efficient mechanism:*

(i) *leads (weakly) to underacquisition of information by agent i if $u_{-i}(x, \omega_i, \omega_{-i})$ is supermodular in (x, ω_i);*
(ii) *leads (weakly) to overacquisition of information by agent i if $u_{-i}(x, \omega_i, \omega_{-i})$ is submodular in (x, ω_i).*

Proof: See Appendix. □

We emphasize that the inefficiency result above is only a local result in the sense that we compare the decision of agent i with the planner's decision for agent i, when both take the decisions of the remaining agents as given. In particular, the theorem is not a statement about the (Nash) equilibrium decisions of the agents. The interaction between the information structures chosen by the agents may conceivably lead to an equilibrium outcome in which all agents acquire too much information relative to the social optimum even though the local prediction, based on $u_{-i}(x, \omega)$ being, say, supermodular in (x, ω_i) for all i, is that all agents acquire too little information. In this case, the theorem would still tell us that relative to the equilibrium information structure α_{-i}, the social planner would like agent i to acquire more information than i chooses to acquire in equilibrium. Observe, however, that for an important class of models, the result above is also global. This is the case where a single player has the opportunity to acquire (additional) information.

Next we give a brief outline of the proof. The key function in the proof is the difference between the social utility function and agent i's private utility function. We show that this difference (i) has a global maximum at $x(s)$ and (ii) is supermodular in (x, s_i). The first attribute holds locally in the generalized VCG mechanism and can be suitably extended to a global property. The difference between social and private utility is composed of the gross utility of all agents but i and the transfer payment of agent i. The latter is constant in s_i conditional on x in the generalized VCG mechanism and hence the second attribute follows by the hypothesis of supermodularity of $u_{-i}(x, \omega)$ in (x, ω_i) after using Proposition 5. Finally, since the difference satisfies the same supermodularity conditions as $u(x, \omega)$ and $u_i(x, \omega)$, we can order signal structures in their informativeness according to the criterion of effectiveness suggested by Lehmann (1988). As the difference is increasing in the effectiveness order, we know that the marginal value of information is larger to the social utility than to agent i's private utility.

The inefficiency results in Theorem 2 are stated simply in terms of the marginal utility of the remaining agents after excluding agent i. If the marginal preferences of the complement set to i are congruent (in their direction) with agent i, then i has insufficient incentives to acquire information. With congruent marginal preferences, the ex-post efficient mechanism induces positive informational externalities that lead agent i to underinvest in information. If the marginal utilities of agent i and all remaining agents move in opposite directions, then the resulting negative informational externality leads the agent to overinvest in information.

We conclude this section with a generalization of the single unit auction model presented earlier to a finite number of bidders and a finite state space. In a single unit auction, the feasible allocations are simply the assignments of the object to the various bidders, and we denote by x_j the assignment of the object to agent j. We restrict attention to symmetric environments where for all $\omega = (\omega_i, \omega_{-i})$ and $\omega' = (\omega_i', \omega_{-i})$ such that $x(\omega), x(\omega') \neq x_i$, we have $x(\omega) = x(\omega')$. In other words, information about ω_i is never pivotal for an allocative decision between x_j and x_k. The nature of the inefficiency in the information acquisition can then be decided on the basis of the properties of the utility function of each bidder at $x_j : u_j(x_j, \omega)$. The utility of agent j is trivially zero for $u_j(x_k, \omega)$ for all $x_k \neq x_j$.

Theorem 3 (Inefficiency in Auctions): *Every ex-post efficient single-unit auction:*

(i) *leads (weakly) to underacquisition of information by agent i if $u_j(x_j, \omega_i, \omega_{-i})$ is nonincreasing in ω_i for all $j \neq i$;*

(ii) *leads (weakly) to overacquisition of information by agent i if $u_j(x_j, \omega_i, \omega_{-i})$ is nondecreasing in ω_i for all $j \neq i$.*

Proof: See Appendix. ☐

7. Conclusion

This paper considers the efficiency of information acquisition in a mechanism design context. In the private values world, any mechanism that implements the efficient allocation, also leads to an efficient level of information acquisition by the agents ex-ante. The efficiency results with private values also extend to a setting where the information is acquired sequentially before a final social allocation is implemented.

The common value model we investigated here is one where the components ω_i of the state of the world $\omega = (\omega_1, \ldots, \omega_I)$ are distributed independently. The results in this paper can be generalized to settings including ones where the signals are single dimensional and independent conditional on the state of the world as long as we make the appropriate assumptions on utilities in terms of allocations and signals directly. If we move away from ex-post implementation to Bayesian implementation, the mechanisms suggested by Cremer and McLean (1985, 1988) can be adapted to our environment to induce efficient information acquisition in models with correlated signals.

Finally, this paper considered information acquisition with a fixed number of agents. It may be of interest to investigate the limiting model as the number of agents gets large. Intuitively, one might expect that the problem of each individual agent might be closer to the private value model. If the responsiveness of the marginal utility of all other agents to the signal of agent i declines, then the sub- or supermodularity of $u_{-i}(x, s)$ in (x, s_i) may vanish and yield efficiency in the limit.

Appendix

The appendix collects the proofs to the propositions and theorems in the main body of the text.

Proof of Theorem 1: A necessary and sufficient condition for a local social optimum α is that for all i, α_i solves

$$\alpha_i \in \underset{\alpha_i' \in A_i}{\arg\max} \left\{ \int u(x(p), p) dF^{(\alpha_i', \alpha_{-i})}(p) - c(\alpha_i' - \alpha_{-i}) \right\}. \qquad \text{(A.1)}$$

In contrast, the expected equilibrium utility of agent i under the VCG mechanism is maximized by α_i where

$$\alpha_i \in \underset{\alpha_i' \in A_i}{\arg\max} \left\{ \int u_i(x(p), p_i) + u_{-i}(x(p), p_{-i}) \right.$$

$$\left. -h(p_{-i})) dF^{(\alpha_i', \alpha_{-i})}(p) - c_i(\alpha_i') \right\},$$

or

$$\alpha_i \in \underset{\alpha_i' \in A_i}{\arg\max} \left\{ \int u(x(p), p) + dF^{(\alpha_i', \alpha_{-i})}(p) - c_i(\alpha_i') \right\}, \qquad \text{(A.2)}$$

where $h(p_{-i})$ can be omitted from the objective function using independence: $F^{(\alpha_i', \alpha_{-i})}(p) = F^{\alpha_i'}(p_i) F^{(\alpha_{-i})}(p_{-i})$. The equivalence of (A.1) and (23) follows from the additive separability of the cost function $c(\alpha)$. $\square$

Proof of Proposition 1: The argument is by contradiction. We suppose that condition (5.1) for the change point s^{kl} is met; a similar argument would apply if instead (5.2) would hold. Suppose that (5.3) doesn't hold; then there exist some $\varepsilon > 0$ such that

$$u_i(x^k, (s_i^{kl} - \epsilon, s_{-i})) - u_i(x^l, (s_i^{kl} - \epsilon, s_{-i})) < u_i(x^k, s^{kl}) - u_i(x^l, s^{kl}) \quad \text{(A.3)}$$

But at the same time we require implementation, or

$$u_i(x^k, (s_i^{kl} - \epsilon, s_{-i})) - t_i^k \geq u_i(x^l, (s_i^{kl} - \epsilon, s_{-i})) - t_i^l,$$

and

$$u_i(x^k, s^{kl}) - t_l^k \leq u_i(x^l, (s^{kl})) - t_i^l,$$

which jointly imply that

$$u_i(x^k, (s_i^{kl} - \epsilon, s_{-i})) - u_i(x^l, (s_i^{kl} - \epsilon, s_{-i})) \geq u_i(x^k, s^{kl}) - u_i(x^l, s^{kl}),$$

which leads immediately to a contradiction with (A.3). $\square$

Proof of Proposition 2: Suppose monotonicity fails to hold. Then there exists at least one set S_i^k such that for $s_i, s_i' \in S_i^k$ and for some $\lambda \in (0,1), \lambda s_i +$

$(1 - \lambda)s'_i \notin S_i^k$, but $\lambda s_i + (1 - \lambda)s'_i \in S_i^l$. By Proposition 1, the differences $t_i^k - t_i^l$ are uniquely determined by the change points. It follows that if a set S_i^k is not connected, then there are more equations (as defined by the incentive compatibility conditions at the change points) than variables, t_i^n's, and generically, in the payoffs of $u_i(x, s)$, the system of equations has no solution. $\qquad \square$

Proof of Proposition 3: By Proposition 1, the transfers are uniquely determined up to a common constant. Consider any adjacent sets R_i^{n-1} and R_i^n:

$$\forall s_i \in R_i^{n-1} : u_i(x^n, s) - u_i(x^{n-1}, s) \leq t_i^n - t_i^{n-1},$$

and

$$\forall s \in R_i^n : u_i(x^n, s) - u_i(x^{n-1}, s) \geq t_i^n - t_i^{n-1}.$$

Now consider any arbitrary pair S_i^k and S_i^m ordered so that $x_k \prec x_m$. We want to show that

$$\forall x_k \prec x_m, \forall s \in R_i^k : u_i(x^k, s) - u_i(x^m, s) \geq t_i^k - t_i^m, \qquad (A.4)$$

as well as

$$\forall x_m \succ x_k, \forall s_i \in R_i^m : u_i(x^m, s) - u_i(x^k, s) \geq t_i^m - t_i^k.$$

Consider (A.4). We can expand the difference on the right-hand side to

$$u_i(x^k, s) - u_i(x^m, s) \geq \sum_{l=k}^{m-1} t_i^l - t_l^{l+1}. \qquad (A.5)$$

Consider the uppermost element of the sum:

$$t_i^{m-1} - t_i^m = u_i(x^{m-1}, s^m) - u_i(x^m, s^m),$$

and for all $s < s^m$,

$$t_i^{m-1} - t_i^m \leq u_i(x^{m-1}, s) - u_i(x^m, s),$$

or

$$u_i(x^m, s) - t_i^m \leq u_i(x^{m-1}, s) - t_i^{m-1}, \qquad (A.6)$$

by (5.8). Replacing the left-hand side by the right-hand side of (A.6) in the inequality (A.5), the modified inequality becomes a priori harder to satisfy.

Doing so leads to

$$u_i(x^k, s) - u_i(x^{m-1}, s) \geq \sum_{l=k}^{m-2} t_i^l - t_i^{l+1},$$

and by repeatedly using the argument in (A.6), (A.5) is eventually reduced to

$$u_i(x^k, s) - u_i(x^{k+1}, s) \geq t_i^k - t_i^{k+1},$$

which is satisfied by (5.8), when the transfers are as in (5.4). □

Proof of Proposition 4: The sufficient conditions with a continuum of allocations can be obtained directly by considering the conditions of the discrete allocation model in the limit as the set of discrete allocations converges to the set of a continuum of allocations. The details are omitted. □

Proof of Proposition 5: By assumption, $u(x, \omega_i, \omega_{-i})$ is supermodular in $u(x, \omega_i)$ for every ω_{-i}. The supermodularity property is preserved under expectations:

$$u(x, \omega_i, s_{-i}) = \sum_{\Omega_{-i}} u(x, \omega_i, \omega_{-i}) \prod_{j \neq i} p_j(\omega_j | s_j),$$

and a fortiori $u(x, \omega_i, s_{-i})$ satisfies the single crossing property in (x, ω_i). By Lemma 1 of Karlin and Rubin (1956), it follows that $u(x, s_i, s_{-i})$ satisfies the single crossing property in (x, s_i). A similar argument applies to $u_i(x, \omega_i, s_{-i})$. Furthermore, by Theorem 1 of Karlin and Rubin (1956), it follows that an optimal strategy which is monotone in s_i exists. This proves the first part of the theorem.

If $u(x, \omega_i, s_{-i})$ is supermodular in (x, ω_i) for every s_{-i} then $u(x, s_i, s_{-i})$, defined as

$$u(x, s_i, s_{-i}) = \sum_{\Omega_i} u(x, \omega_i, s_{-i}) p_i(\omega_i | s_i),$$

is also supermodular in (x, s_i) by Theorem 3.10.1 in Topkis (1998) since $p_i(s_i, \omega_i)$ satisfies the monotone likelihood ratio. □

Proof of Theorem 2: The following proof is written for a continuum of allocations, but all arguments go through with the obvious notational modifications for a finite set of allocations. We start with the net utility of agent i under the generalized VCG mechanism, which is given by

$$u_i(x, s) - t_i(s),$$

where $s = (s_i, s_{-i})$ is the true signal by truthtelling under the VCG mechanism. For a fixed s_{-i}, we can rewrite the transfer $t_i(s_i, s_{-i})$ to be determined directly by x rather than (s_i, s_{-i}). This is without loss of generality as we recall that $t_i(s_i, s_{-i})$ is constant in s_i conditional on x. The net utility of agent i can now be written directly as

$$v_i(x, s) \triangleq u_i(x, s) - t_i(x) \tag{A.7}$$

The transfer function $t_i(x)$ is given by analogy with (5.10) as

$$t_i(x) = - \int_{\underline{x}}^{x} \frac{\partial u_{-i}(z, s(z))}{\partial z} dz + t_i(\underline{x}), \tag{A.8}$$

where $s(x)$ is the inverse function of $x(s)$ for a fixed s_{-i}, or $s(z) = x^{-1}(z)$. The function $s(x)$ is well-defined if $x(s)$ is strictly increasing in s_i. If x has 'flats' in s_i, then the integral would have to be modified in the obvious way, It follows directly from (A.7) that $v_i(x, s)$ is supermodular in (x, s_i) if and only if $u_i(x, s)$ is supermodular in (x, s_i), which in turn is guaranteed by the supermodularity of $u_i(x, w)$ in (x, w_i), as shown in Proposition 5.

Next we show that $u(x, s) - v_i(x, s)$ is (i) supermodular in (x, s_i) and (ii) achieves a global maximum at $x = x(s)$ for all s. The first property is guaranteed by the same argument as before if $u_{-i}(x, s)$ is supermodular in (x, s_i) as

$$u(x, s) - v_i(x, s) = u_{-i}(x, s) + t_i(x). \tag{A.9}$$

Observe next that $u_{-i}(x, s) + t_i(x)$ has a stationary point at $x = x(s)$ for all s by (A.8):

$$\frac{\partial u_i(x, s)}{\partial x} + \frac{\partial t_i(x)}{\partial x} = \frac{\partial u_{-i}(x, s)}{\partial x} - \frac{\partial u_{-i}(x, s(x))}{\partial x} = 0.$$

Notice also that locally at $x = x(s)$ the function is concave in x as the second derivative with respect to x is given by

$$\frac{\partial^2 u_{-i}(x, s)}{\partial x^2} - \frac{\partial^2 u_{-i}(x, s(x))}{\partial x^2} - \frac{\partial^2 u_{-i}(x, s(x))}{\partial x \partial s} \frac{ds(x)}{\partial x},$$

as the first two terms cancel at $s = s(x)$, and

$$\frac{\partial^2 u_{-i}(x, s(x))}{\partial x \partial s_i} \frac{ds(x)}{dx} \geq 0$$

by the supermodularity of $u_{-i}(x, s)$ and $u(x, s)$ in (x, s_i). However our standing assumptions don't allow us to conclude that the local maximum is also a global maximum. This final obstacle can be removed by modifying the objective function $u_{-i}(x, s) + t_i(x)$ through the addition of a new function $g(x, s)$ with

$$G(x, s) \triangleq u_{-i}(x, s) + t_i(x) + g(x, s),$$

such that the following properties are satisfied:

(a) $g(x(s), s) = 0$, for all s;

(b) $G(x, s)$ is supermodular in (x, s_i);

and

(c) $G(x(s), s) \geq G(x, s), \quad \forall s, x.$

If a function $g(x, s)$ exists such that $G(x, s)$ satisfies the properties (a)–(c), then $G(x, s)$ satisfies assumption (i) and (ii) of Lehmann's theorem. Moreover the expected value of $G(x, s)$ evaluated at $x = x(s)$ is equal to $u_{-i}(x, s) + t_i(x)$ evaluated at $x = x(s)$. To accomplish this define an auxiliary function $b(s)$ by

$$b(s) \triangleq u(x(s), s) - u_{-i}(x(s), s) - t_i(x(s)),$$

and define $g(x, s)$ to be

$$g(x, s) \triangleq u(x, s) - u_{-i}(x, s) - t_i(x) - b(s).$$

It is now easy to verify that $G(x, s)$ shares the supermodularity properties of $u(x, s)$, has a global maximum at $x = x(s)$ for every s, and indeed $g(x(s), s) = 0$. It remains to take expectations. We maintain s_{-i} to be fixed. We take the expectation with respect to the distribution F^{α_i} and

$F^{\alpha_i'}$ and denote

$$G(\alpha_i, s_{-i}) \triangleq \mathbb{E}_{s_i}[G(x(s_i, s_{-i}), (s_i, s_{-i}))|F^{\alpha_i}]$$

and

$$G(\alpha_i', s_{-i}) \triangleq \mathbb{E}_{s_i}[G(x(s_i, s_{-i}), (s_i, s_{-i}))|F^{\alpha_i'}].$$

It then follows by Lehmann's theorem that if α_i is more effective than α_i', we have

$$G(\alpha_i, s_{-i}) \geq G(\alpha_i', s_{-i}).$$

From (a)–(c), we can then conclude that

$$u_{-i}(\alpha_i, s_{-i}) + t_i(\alpha_i, s_{-i}) \geq u_{-i}(\alpha_i', s_{-i}) + t_i(\alpha_i', s_{-i}),$$

adopting again the notation that

$$u_{-i}(\alpha_i, s_{-i}) + t_i(\alpha_i, s_{-i}) \triangleq \mathbb{E}_{s_i}[u_{-i}(x(s_i, s_{-i}), (s_i, s_{-i})) + t_i(x(s_i, s_{-i}))|F^{\alpha_i}],$$

and similarly for α_i'. By (A.9), this is equivalent to

$$u(\alpha_i, s_{-i}) - u(\alpha_i', s_{-i}) \geq v_i(\alpha_i, s_{-i}) - v_i(\alpha_i', s_{-i}). \qquad (A.10)$$

As the inequality holds for every s_{-i}, it remains to hold after taking expectation over the realization of s_{-i}, which concludes the proof. The corresponding result for submodularity can be obtained by simply reversing the inequalities. □

Proof of theorem 3: This theorem is a special case of Theorem 2 after introducing the following ranking for the allocations. With a single unit auction, the set of allocations is simply the assignment of the object to a particular bidder. For every i, partition the set of allocations X into x_i and x_{-i} and order the assignments such that $x_i \succ x_{-i}$. (The order among the remaining bidders is irrelevant.) By definition of the single object auction

$$u_i(x_{-i}, \omega) = 0.$$

To verify the supermodularity property, it is therefore sufficient to examine the behavior of

$$u_i(x_i, \omega) - u_i(x_{-i}, \omega),$$

as a function of ω_i. Similarly for $u_{-i}(x, \omega)$. The result is now a direct consequence of Theorem 2. □

Acknowledgments

The authors thank Sandeep Baliga, Jeff Ely, Steve Matthews, Stephen Morris, Joe Ostroy, Nicola Persico, Martin Pesendorfer, Phil Retry, Bill Zame, and particularly Jon Levin for several helpful discussions. We are especially grateful for suggestions front two anonymous referees and a co-editor. Comments from seminar participants at Minnesota, U.C.L.A., and Yale are greatly appreciated. Financial support from NSF Grant SBR 9709887 and 9709340, respectively, is acknowledged.

References

AURIOL, E., AND R. J. GARY-BOBO (1999): "On the Optimal Number of Representatives," CMS- EMS Discussion Paper 1286.

BERGEMANN, D., AND J. VÄLIMÄKI (2000): "Information Acquisition and Efficient Mechanism Design," Cowles Foundation Discussion Paper No. 1248.

BULOW, J., M. HUANG, AND P. KLEMPERER (1999): "Toeholds and Takeovers," *Journal of Political Economy*, 107, 427–454.

CLARKE, E. (1971): "Multipart Pricing of Public Goods," *Public Choice*, 8, 19–33.

CREMER, J., F. KHALIL, AND J.-C. ROCHET (1998a): "Contracts and Productive Information Gathering," *Games and Economic Behavior*, 25, 174–193.

——— (1998b): "Strategic Information Gathering Before a Contract is Offered," *Journal of Economic Theory*, 81, 163–200.

CREMER, J., AND R. MCLEAN (1985): "Optimal Selling Strategies Under Uncertainty for a Discriminating Monopolist When Demands are Interdependent," *Econometrica*, 53, 345–361.

——— (1988): "Full Extraction of the Surplus in Bayesian and Dominant Strategy Auctions," *Econometrica*, 56, 1247–1258.

DASGUPTA, P., AND E. MASKIN (2000): "Efficient Auctions," *Quarterly Journal of Economics*, 115, 341–388.

D'ASPREMONT, C., S. BHATTACHARYA, AND L.-A, GERARD-VARET (2000): "Bargaining and Sharing Innovative Knowledge," *Review of Economic Studies*, 67, 255–271.

GROVES, T. (1973): "Incentives in Teams," *Econometrica*, 41, 617–631.

JEHIEL, P., AND B. MOLDOVANU (2001): "Efficient Design with Interdependent Valuations," *Econometrica*, 69, 1237–1259.

KARLIN, S., AND H. RUBIN (1956): "The Theory of Decision Procedures for Distributions with Monotone Likelihood Ratio," *Annals of Mathematical Statistics*, 27, 272–299.

LEHMANN, E. L. (1988): "Comparing Location Experiments," *Annals of Statistics*, 16, 521–533.

MASKIN, E. (1992): "Auctions and Privatization," in *Privatization: Symposium in Honor of Herbert Giersch*, ed. by H. Siebert. Tuebingen: J.C.B. Mohr, pp. 115–136.

MATTHEWS, S. (1977): "Information Acquisition in Competitive Bidding Process," California Institute of Technology.

——— (1984): "Information Acquisition in Discriminatory Auctions," in *Bayesian Models in Economic Theory*, ed. by M. Boyer and R. E. Kihlstrom. Amsterdam: North-Holland, pp. 181–207.

PERSICO, N. (2000): "Information Acquisition in Auctions," *Econometrica*, 68, 135–148.

ROGERSON, W. P. (1992): "Contractual Solutions to the Hold-Up Problem," *Review of Economic Studies*, 59, 777–794.

STEGEMAN, M. (1996): "Participation Costs and Efficient Auctions," *Journal of Economic Theory*, 71, 228–259.

TAN, G. (1992): "Entry and R and D in Procurement Contracting," *Journal of Economic Theory*, 58, 41–60.

TOPKJS, D. M. (1998): *Supermodularity and Complementarity*, Princeton: Princeton University Press.

VICKREY, W. (1961): "Counterspeculation, Auctions and Competitive Sealed Tenders," *Journal of Finance*, 16, 8–37.

Chapter 14

The Dynamic Pivot Mechanism*

Dirk Bergemann[†] and Juuso Välimäki[‡]

[†]*Department of Economics, and Cowles Foundation of Research in Economics,*
Yale University, New Haven, CT 06520-8268, USA
dirk.bergemann@yale.edu
[‡]*Department of Economics, Aalto University School of Business,*
02150 Espoo, Finland
juuso.valimaki@aalto.fi

We consider truthful implementation of the socially efficient allocation in an independent private-value environment in which agents receive private information over time. We propose a suitable generalization of the pivot mechanism, based on the marginal contribution of each agent. In the dynamic pivot mechanism, the ex-post incentive and ex-post participation constraints are satisfied for all agents after all histories. In an environment with diverse preferences it is the unique mechanism satisfying ex-post incentive, ex-post participation, and efficient exit conditions.

We develop the dynamic pivot mechanism in detail for a repeated auction of a single object in which each bidder learns over time her true valuation of the object. The dynamic pivot mechanism here is equivalent to a modified second price auction.

Keywords: Pivot mechanism, dynamic mechanism design, ex-post equilibrium, marginal contribution, multi-armed bandit, Bayesian learning.

1. Introduction

In this paper, we generalize the idea of the pivot mechanism (due to Green and Laffont (1977)) to dynamic environments with private information. We design an intertemporal sequence of transfer payments which allows

---

[1]This chapter is reproduced from *Econometrica*, **78**, 771–789, 2010.

each agent to receive her flow marginal contribution in every period. In other words, after each history, the expected transfer that each agent must pay coincides with the dynamic externality cost that she imposes on the other agents. In consequence, each agent is willing to truthfully report her information in every period.

We consider a general intertemporal model in discrete time and with a common discount factor. The private information of each agent in each period is her perception of her future payoff path conditional on the realized signals and allocations. We assume throughout that the information is statistically independent across agents. At the reporting stage of the direct mechanism, each agent reports her information. The planner then calculates the efficient allocation given the reported information. The planner also calculates for each agent i the optimal allocation when agent i is excluded from the mechanism. The total expected discounted payment of each agent is set equal to the externality cost imposed on the other agents in the model. In this manner, each agent receives as her payment her marginal contribution to the social welfare in every conceivable continuation game.

With transferable utilities, the social objective is simply to maximize the expected discounted sum of the individual utilities. Since this is essentially a dynamic programming problem, the solution is by construction time-consistent. In consequence, the dynamic pivot mechanism is time-consistent and the social choice function can be implemented by a sequential mechanism without any ex-ante commitment by the designer (apart from the commitment to the transfers promised for the current period). In contrast, in revenue-maximizing problems, it is well known that the optimal solution relies critically on the ability of the principal to commit to a contract, see Baron and Besanko (1984). Interestingly, Battaglini (2005) showed that in dynamic revenue-maximizing problems with stochastic types, the commitment problems are less severe than with constant types.

The dynamic pivot mechanism yields a positive monetary surplus for the planner in each period and, therefore, the planner does not need outside resources to achieve the efficient allocation. Finally, the dynamic pivot mechanism induces all agents to participate in the mechanism after all histories.

In the intertemporal environment there are many transfer schemes that support the same incentives as the pivot mechanism. In particular, the monetary transfers necessary to induce the efficient action in period t may become due at some later period s provided that the net present value of the transfers remains constant. We say that a mechanism supports efficient exit if an agent who ceases to affect current and future allocations also ceases to pay and receive transfers. This condition is similar to the requirement often made in the scheduling literature that the mechanism be an *online mechanism* (see Lavi and Nisan (2000)). We establish that in an environment with diverse preferences, the dynamic pivot mechanism is the only efficient mechanism that satisfies ex-post incentive compatibility, ex-post participation, and efficient exit conditions.

The basic idea of the dynamic pivot mechanism is first explored in the context of a scheduling problem where a set of privately informed bidders compete for the services of a central facility over time. This class of problems is perhaps the most natural dynamic allocation analogue to the static single-unit auction. The scheduling problem is kept deliberately simple and all the relevant private information arrives in the initial period. Subsequently, we use the dynamic pivot mechanism to derive the dynamic auction format for a model where bidders learn their valuations for a single object over time. In contrast to the scheduling problem where a static mechanism could still have implemented the efficient solution, a static mechanism now necessarily fails to support the efficient outcome as more information arrives over time. In turn, this requires a more complete understanding of the intertemporal trade-offs in the allocation process. By computing the dynamic marginal contributions, we can derive explicit and informative expressions for the intertemporal transfer prices.

In recent years, a number of papers have been written with the aim to explore various issues arising in dynamic allocation problems. Among the contributions which focus on socially efficient allocation, Cavallo, Parkes, and Singh (2006) proposed a Markovian environment for general allocation problems and analyzed two different classes of sequential incentives schemes: (i) Groves-like payments and (ii) pivot-like payments. They established that Groves-like payments, which award every agent positive monetary transfers equal to the sum of the valuation of all other agents, guarantee interim incentive compatibility and ex-post participation constraints after all histories. In contrast, pivot-like payments guarantee interim incentive compatibility and ex-ante participation constraints. Athey and Segal (2007) considered a more general dynamic model in which the current payoffs are allowed to depend on the entire past history including past signals and past actions. In addition, they also allowed for hidden action as well as hidden information. The main focus of their analysis is on incentive compatible mechanisms that are budget balanced in every period of the game. Their mechanism, called balanced team mechanism, transfers the insight from the Arrow (1979) and D'Aspremont and Gerard-Varet (1979) mechanisms into a dynamic environment. In addition, Athey and Segal (2007) presented conditions in terms of ergodic distributions over types and patients agents such that insights from repeated games can be employed to guarantee interim participation constraints. In contrast, we emphasize voluntary participation without any assumptions about the discount factor or the ergodicity of the type distributions. We also define an efficient exit condition which allows us to single out the dynamic pivot mechanism in the class of efficient mechanisms.

The focus of the current paper is on the socially efficient allocation, but a number of recent papers have analyzed the design of dynamic revenue-maximizing mechanisms, beginning with the seminal contributions by Baron and Besanko (1984) and Courty and Li (2000), who considered optimal intertemporal pricing policies with private information in a setting with two periods. Battaglini (2005) considered the revenue-maximizing long-term contract of a monopolist in a model with an infinite time horizon when the valuation of the buyer changes in a Markovian fashion over time. In particular, Battaglini (2005) showed that the optimal continuation contracts for a current high type are efficient, as his payoff is determined by the allocations for the current low type (by incentive compatibility). The net payoffs of the types then have a property related to the marginal contribution here. But as Battaglini (2005) considered revenue-maximizing contracts, the lowest type served receives zero utility, and hence the notion of marginal contribution refers only to the additional utility generated by higher types, holding the allocation constant, rather than the entire incremental social value. Most recently, Pavan, Segal, and Toikka (2008) developed a general allocation model and derived the optimal dynamic revenue-maximizing mechanism. A common thread in these papers is a suitable generalization of the notion of virtual utility to dynamic environments.

2. Model

2.1. *Uncertainty*

We consider an environment with private and independent values in a discrete-time, infinite-horizon model. The flow utility of agent $i \in \{1, 2, \ldots, I\}$ in period $t \in \mathbb{N}$ is determined by the current allocation $a_t \in A$, the current monetary transfer $p_{i,t} \in \mathbb{R}$, and a state variable $\theta_{i,t} \in \Theta_i$.

The von Neumann-Morgenstern utility function u_i of agent i is quasilinear in the monetary transfer:

$$u_i(a_t, p_{i,t}, \theta_{i,t}) \triangleq v_i(a_t, \theta_{i,t}) - p_{i,t}.$$

The current allocation $a_t \in A$ is an element of a finite set A of possible allocations. The state of the world $\theta_{i,t}$ for agent i is a general Markov process on the state space Θ_i. The aggregate state is given by the vector $\theta_t = (\theta_{1,t}, \ldots, \theta_{I,t})$ with $\Theta = \mathrm{X}_{i=1}^{I} \Theta_i$.

There is a common prior $F_i(\theta_{i,0})$ regarding the initial type $\theta_{i,0}$ of each agent i. The current state $\theta_{i,t}$ and the current action a_t define a probability distribution for next period state variables $\theta_{i,t+1}$ on Θ_i. We assume that this distribution can be represented by a stochastic kernel $F_i(\theta_{i,t+1}; \theta_{i,t}, a_t)$.

The utility functions $u_i(\cdot)$ and the probability transition functions $F_i(\cdot; a_t, \theta_{i,t})$ are common knowledge at $t = 0$. The common prior $F_i(\theta_{i,0})$ and the stochastic kernels $F_i(\theta_{i,t+1}; \theta_{i,t}, a_t)$ are assumed to be independent across agents. At the beginning of each period t, each agent i observes $\theta_{i,t}$ privately. At the end of each period, an action $a_t \in A$ is chosen and payoffs for period t are realized. The asymmetric information is therefore generated by the private observation of $\theta_{i,t}$ in each period t. We observe that by the independence of the priors and the stochastic kernels across i, the information of agent i, $\theta_{i,t+1}$, does not depend on $\theta_{j,t}$ for $j \neq i$. The expected absolute value of the flow payoff is assumed to be bounded by some $K < \infty$ for every i, a, θ and allocation plan $a' : \Theta \to A$:

$$\int |v_i(a'(\theta'), \theta_i')| \, dF(\theta'; a, \theta) < K.$$

The nature of the state space Θ depends on the application at hand. At this point, we stress that the formulation accommodates the possibility of random arrival or departure of the agents. The arrival or departure of agent i can be represented by an inactive state $\underline{\theta}_i$, where $v_i(a_t, \underline{\theta}_i) = 0$ for all $a_t \in A$ and a random time τ at which agent i privately observes her transition in or out of the inactive state.

2.2. Social efficiency

All agents discount the future with a common discount factor $\delta, 0 < \delta < 1$. The socially efficient policy is obtained by maximizing the expected discounted sum of valuations. Given the Markovian structure, the socially optimal program starting in period t at state θ_t can be written as

$$W(\theta_t) \triangleq \max_{\{a_s\}_{s=t}^{\infty}} \mathbb{E}\left[\sum_{s=t}^{\infty} \delta^{s-t} \sum_{i=1}^{I} v_i(a_s, \theta_{i,s})\right].$$

For notational ease, we omit the conditioning state in the expectation operator, when the conditioning event is obvious, as in the above, where $\mathbb{E}[\cdot] = \mathbb{E}_{\theta_t}[\cdot]$. Alternatively, we can represent the social program in its recursive form:

$$W(\theta_t) = \max_{a_t} \mathbb{E}\left[\sum_{i=1}^{I} v_i(a_t, \theta_{i,t}) + \delta \mathbb{E} W(\theta_{t+1})\right].$$

The socially efficient policy is denoted by $\mathbf{a}^* = \{a_t^*\}_{t=0}^{\infty}$. The social externality cost of agent i is determined by the social value in the absence of agent i:

$$W_{-i}(\theta_t) \triangleq \max_{\{a_s\}_{s=t}^{\infty}} \mathbb{E}\left[\sum_{s=t}^{\infty} \delta^{s-t} \sum_{j \neq i} v_j(a_s, \theta_{j,s})\right].$$

The efficient policy when agent i is excluded is denoted by $\mathbf{a}^*_{-i} = \{a^*_{-i,t}\}_{t=0}^{\infty}$. The *marginal contribution* $M_i(\theta_t)$ of agent i at signal θ_t is defined by

$$M_i(\theta_t) \triangleq W(\theta_t) - W_{-i}(\theta_t). \tag{2.1}$$

The marginal contribution of agent i is the change in the social value due to the addition of agent i.[1]

2.3. *Mechanism and equilibrium*

We focus attention on direct mechanisms which truthfully implement the socially efficient policy $\mathbf{a}^*$. A dynamic direct mechanism asks every agent i to report her state $\theta_{i,t}$ in every period t. The report $r_{i,t} \in \Theta_i$ may or may not be truthful. The public history in period t is a sequence of reports and allocations until period $t-1$, or $h_t = (r_0, a_0, r_1, a_1, \ldots, r_{t-1}, a_{t-1})$, where each $r_s = (r_{1,s}, \ldots, r_{I,s})$ is a report profile of the I agents. The set of possible public histories in period t is denoted by H_t. The sequence

---

[1] In symmetric information environments, we used the notion of marginal contribution to construct efficient equilibria in dynamic first price auctions; see Bergemann and Välimäki (2003, 2006).

of reports by the agents is part of the public history and we assume that the past reports of each agent are observable to all the agents. The private history of agent i in period t consists of the public history and the sequence of private observations until period t, or $h_{i,t} = (\theta_{i,0}, r_0, a_0, \theta_{i,1}, r_1, a_1, \ldots, \theta_{i,t-1}, r_{t-1}, a_{t-1}, \theta_{i,t})$. The set of possible private histories in period t is denoted by $H_{i,t}$. An (*efficient*) *dynamic direct mechanism* is represented by a family of allocations and monetary transfers, $\{a_t^*, p_t\}_{t=0}^{\infty} : a_t^* : \Theta \to \Delta(A)$, and $p_t : H_t \times \Theta \to \mathbb{R}^I$. With the focus on efficient mechanisms, the allocation a_t^* depends only on the current (reported) state $r_t \in \Theta$, while the transfer p_t may depend on the entire public history.

A (pure) reporting strategy for agent i in period t is a mapping from the private history into the state space: $r_{i,t} : H_{i,t} \to \Theta_i$. For a given mechanism, the expected payoff of agent i from reporting strategy $\mathbf{r}_i = \{r_{i,t}\}_{t=0}^{\infty}$ given the strategies $\mathbf{r}_{-i} = \{r_{-i,t}\}_{t=0}^{\infty}$ is

$$\mathbb{E} \sum_{t=0}^{\infty} \delta^t [v_i(a^*(r_t), \theta_{i,t}) - p_i(h_t, r_t)].$$

Given the mechanism $\{a_t^*, p_t\}_{t=0}^{\infty}$ and the reporting strategies $\mathbf{r}_{-i}$, the optimal strategy of bidder i can be stated recursively:

$$V_i(h_{i,t}) = \max_{r_{i,t} \in \Theta_i} \mathbb{E}\{v_i(a_t^*(r_{i,t}, r_{-i,t}), \theta_{i,t}) - p_i(h_t, r_{i,t}, r_{-i,t}) + \delta V_i(h_{i,t+1})\}.$$

The value function $V_i(h_{i,t})$ represents the continuation value of agent i given the current private history $h_{i,t}$. We say that a dynamic direct mechanism is *interim incentive compatible* if for every agent and every history, truthtelling is a best response given that all other agents report truthfully. We say that the dynamic direct mechanism is *periodic ex-post incentive compatible* if truthtelling is a best response regardless of the history and the current state of the other agents.

In the dynamic context, the notion of ex-post incentive compatibility is qualified by periodic, as it is ex-post with respect to all signals received in period t, but not ex-post with respect to signals arriving after period t. The periodic qualification arises in the dynamic environment, as agent i may receive information at some later time $s > t$ such that in retrospect she would wish to change the allocation choice in t and hence her report in t.

Finally we define the *periodic ex-post* participation constraints of each agent. After each history h_t, each agent i may opt out (permanently) from the mechanism. The value of the outside option is denoted $O_i(h_{i,t})$ and it is defined by the payoffs that agent i receives if the planner pursues the efficient policy $\mathbf{a}^*_{-i}$ for the remaining agents. The periodic participation constraint requires that each agent's equilibrium payoff after each history weakly exceeds $O_i(h_{i,t})$. For the remainder of the text, we say that a mechanism is *ex-post incentive compatible and individually rational* if it satisfies the periodic ex-post incentive and participation constraints.

3. Scheduling: An Example

We consider the problem of allocating time to use a central facility among competing agents. Each agent has a private valuation for the completion of a task which requires the use of the central facility. The facility has a capacity constraint and can only complete one task per period. The cost of delaying any task is given by the discount rate $\delta < 1$. The agents are competing for the right to use the facility at the earliest available time. The objective of the social planner is to sequence the tasks over time so as to maximize the sum of the discounted utilities. In an early contribution, Dolan (1978) developed a static mechanism to implement a class of related scheduling problems with private information.

An allocation policy in this setting is a sequence of choices $a_t \in \{0, 1, \ldots I\}$,, where a_t denotes the bidder chosen in period t. We allow for $a_t = 0$ and hence the possibility that no bidder is selected in t. Each agent has only one task to complete and the value $\theta_{i,0} \in \mathbb{R}_+$ of the task is constant over time and independent of the realization time (except for discounting). The transition function is then given by

$$\theta_{i,t+1} = \begin{cases} 0, & \text{if } a_t = i. \\ \theta_{i,t}, & \text{if } a_t \neq i. \end{cases}$$

For this scheduling model, we find the marginal contribution of each agent and derive the associated dynamic pivot mechanism. We determine the marginal contribution of bidder i by comparing the value of the social program with and without i. With the constant valuations over time for all i, the optimal policy is given by assigning in every period the alternative j with the highest remaining valuation. To simplify notation, we define the positive valuation $v_i \triangleq \theta_{i,0}$. We may assume without loss of generality (after relabelling) that the valuations v_i are ordered with respect to the index i: $v_i \geq \cdots \geq v_I \geq 0$. Due to the descending order of valuations, we identify each task i with the period $i + 1$ in which it is completed along the efficient path:

$$W(\theta_0) = \sum_{i=1}^{I} \delta^{t-1} v_t. \tag{3.1}$$

Similarly, the efficient program in the absence of task i assigns the tasks in ascending order, but necessarily skips task i in the assignment process:

$$W_{-i}(\theta_0) = \sum_{t=1}^{i-1} \delta^{t-1} v_t + \sum_{t=i}^{I-1} \delta^{t-1} v_{t+1}. \tag{3.2}$$

By comparing the social program with and without i, (3.1) and (3.2), respectively, we find that the assignments for agents $j < i$ remain unchanged after i is removed, but that each agent $j > i$ is allocated the slot one period earlier than in the presence of i. The marginal contribution of i from the point of view of period 0 is

$$M_i(\theta_0) = W(\theta_0) - W_{-i}(\theta_0) = \sum_{t=i}^{I} \delta^{t-1}(v_t - v_{t+1}).$$

The social externality cost of agent i is established in a straightforward manner. At time $t = i - 1$, agent i completes her task and realizes the value v_i. The immediate opportunity cost is the next highest valuation v_{i+1}. But this overstates the externality, because in the presence of i, all less valuable tasks are realized one period later. The externality cost of agent i is hence equal to the next valuable task v_{i+1} minus the improvement in future allocations due to the delay of all tasks by one period:

$$p_i(\theta_t) = v_{i+1} - \sum_{t=i+1}^{I} \delta^{t-i}(v_t - v_{t+1}) = (1-\delta)\sum_{t=i}^{I} \delta^{t-i} v_{t+1}. \tag{3.3}$$

Since we have by construction $v_t - v_{t+1} \geq 0$, the externality cost of agent i in the intertemporal framework is less than in the corresponding single allocation problem where it would be v_{i+1}. Consequently, the final expression states that the externality of agent i is the cost of delay imposed on the remaining and less valuable tasks.[2]

---

[2]In the online Supplementary Material (Bergemann and Välimäki (2010)), we show that the socially efficient scheduling can be implemented through a bidding mechanism rather than the direct revelation mechanism used here. In a recent and related contribution, Said (2008) used the dynamic pivot mechanism and a payoff equivalence result to construct bidding strategies in a sequence of ascending auctions with entry and exit of the agents.

4. The Dynamic Pivot Mechanism

We now construct the dynamic pivot mechanism for the general model described in Section 2. The marginal contribution of agent i is her contribution to the social value. In the dynamic pivot mechanism, the marginal contribution will also be the information rent that agent i can secure for herself if the planner wishes to implement the socially efficient allocation. In a dynamic setting, if agent i can secure her marginal contribution in every continuation game of the mechanism, then she should be able to receive the *flow* marginal contribution $m_i(\theta_t)$ in every period. The flow marginal contribution accrues incrementally over time and is defined recursively:

$$M_i(\theta_t) = m_i(\theta_t) + \delta \mathbb{E} M_i(\theta_{t+1}).$$

The flow marginal contribution can be expressed directly in terms of the social value functions, using the definition of the marginal contribution given in (2.1) as

$$(m_i(\theta_t) \triangleq W(\theta_t) - W_{-i}(\theta_t) - \delta \mathbb{E}[W(\theta_{t+1}) - W_{-i}(\theta_{t+1})]. \tag{4.1}$$

The continuation payoffs of the social programs with and without i, respectively, may be governed by different transition probabilities, as the respective social decisions in period t, $a_t^* \triangleq a^*(\theta_t)$ and $a_{-i,t}^* \triangleq a_{-i}^*(\theta_{-i,t})$, may differ. The continuation value of the socially optimal program, conditional on current allocation a_t and state θ_t is

$$W(\theta_{t+1}|a_t, \theta_t) \triangleq \mathbb{E}_{F(\theta_{t+1}; a_t, \theta_t)} W(\theta_{t+1}),$$

where the transition from state θ_t to state θ_{t+1} is controlled by the allocation a_t. For notational ease, we omit the expectations operator $\mathbb{E}$ from the conditional expectation. We adopt the same notation for the marginal contributions $M_i(\cdot)$ and the individual value functions $V_i(\cdot)$. The flow marginal contribution $m_i(\theta_t)$ is expressed as

$$m_i(\theta_t) = \sum_{j=1}^{I} v_j(a_t^*, \theta_{j,t}) - \sum_{j \neq i} v_j(a_{-i,t}^*, \theta_{j,t})$$
$$+ \delta[W_{-i}(\theta_{t+1}|a_t^*, \theta_t) - W_{-i}(\theta_{t+1}|a_{-i,t}^*, \theta_t)].$$

A monetary transfer $p_i^*(\theta_t)$ such that the resulting flow net utility matches the flow marginal contribution leads agent i to internalize her social externalities:

$$p_i^*(\theta_t) \triangleq v_i(a_t^*, \theta_{i,t}) - m_i(\theta_t). \tag{4.2}$$

We refer to $p_i^*(\theta_t)$ as the transfer of the dynamic pivot mechanism. The transfer $p_i^*(\theta_t)$ depends only on the current report θ_t and not on the entire public history h_t. We can express $p_i^*(\theta_t)$ in terms of the flow utilities and the social continuation values:

$$p_i^*(\theta_t) = \sum_{j \neq i} [v_j(a_{-i,t}^*, \theta_{j,t}) - v_j(a_t^*, \theta_{j,t})]$$
$$+ \delta[W_{-i}(\theta_{t+1}|a_{-i,t}^*, \theta_t) - W_{-i}(\theta_{t+1}|a_t^*, \theta_t)]. \tag{4.3}$$

The transfer $p_i^*(\theta_t)$ for agent i depends on the report of agent i only through the determination of the social allocation which is a prominent feature of the static Vickrey–Clarke–Groves mechanisms. The monetary transfers $p_i^*(\theta_t)$ are always nonnegative, as the policy $a_{-i,t}^*$ is by definition an optimal policy to maximize the social value of all agents exclusive of i. It follows that in every period t, the sum of the monetary transfers across all agents generates a weak budget surplus.

Theorem 1 (Dynamic Pivot Mechanism): *The dynamic pivot mechanism $\{a_t^*, p_t^*\}_{t=0}^\infty$ is ex-post incentive compatible and individually rational.*

Proof: By the unimprovability principle, it suffices to prove that if agent i receives as her continuation value her marginal contribution, then truthtelling is incentive compatible for agent i in period t, or

$$v_i(a^*, (\theta_t), \theta_{i,t}) - p_i^*(\theta_t) + \delta M_i(\theta_{t+1}|a_t^*, \theta_t)$$
$$\geq v_i(a^*(r_{i,t}, \theta_{-i,t}), \theta_{i,t}) - p_i^*(r_{i,t}, \theta_{-i,t})$$
$$+ \delta M_i(\theta_{t+1}|a * (r_{i,t}, \theta_{-i,t}), \theta_t) \tag{4.4}$$

for all $r_{i,t} \in \Theta_i$ and all $\theta_{-i,t} \in \Theta_{-i}$, and we recall that we denote the socially efficient allocation at the true state profile θ_t by $a_t^* \triangleq a^*(\theta_t)$. By construction of p_i^* in (4.3), the left-hand side of (4.4) represents the marginal contribution of agent i. We can express the marginal contributions $M_i(\cdot)$ in terms of the different social values to get

$$W(\theta_t) - W_{-i}(\theta_t)$$
$$\geq v_i(a^*(r_{i,t}, \theta_{-i,t}), \theta_{i,t}) - p_i^*(r_{i,t}, \theta_{-i,t})$$
$$+ \delta(W(\theta_{t+1}|a^*(r_{i,t}, \theta_{-i,t}), \theta_t) - W_{-i}(\theta_{t+1}|a^*(r_{i,t}, \theta_{-i,t}), \theta_t)). \tag{4.5}$$

We then insert the transfer price $p_i^*(r_{i,t}, \theta_{-i,t})$ (see (4.3)) into (4.5) to obtain

$$W(\theta_t) - W_{-i}(\theta_t)$$

$$\geq v_i(a^*(r_{i,t}, \theta_{-i,t}), \theta_{i,t}) - \sum_{j \neq i} v_j(a_{-i,t}^*, \theta_{j,t}) - \delta W_{-i}(\theta_{t+1} | a_{-i,t}^*, \theta_t)$$

$$+ \sum_{j \neq i} v_j(a^*(r_{i,t}, \theta_{-i,t}), \theta_{j,t}) + \delta W(\theta_{t+1} | a^*(r_{i,t}, \theta_{-i,t}), \theta_t).$$

But now we reconstitute the entire inequality in terms of the respective social values:

$$W(\theta_t) - W_{-i}(\theta_t) \geq \sum_{j=1}^{I} v_j(a^*(r_{i,t}, \theta_{-i,t}), \theta_{j,t})$$

$$+ \delta W(\theta_{t+1} | a^*(r_{i,t}, \theta_{-i,t}), \theta_t) - W_{-i}(\theta_t).$$

The above inequality holds for all $r_{i,t}$ by the social optimality of $a^*(\theta_t)$ of in state (θ_t). $\qquad\qquad\qquad\qquad\qquad\qquad\qquad\qquad\qquad\qquad\qquad\qquad\square$

The dynamic pivot mechanism specifies a unique monetary transfer after every history. It guarantees that the ex-post incentive and ex-post participation constraints are satisfied after every history. In the intertemporal environment, each agent evaluates the monetary transfers to be paid in terms of the expected discounted transfers, but is indifferent (up to discounting) over the incidence of the transfers over time. This temporal separation between allocative decisions and monetary decisions may be undesirable for many reasons. First, if the agents and the principal do not have the ability to commit to *future* transfer payments, then delays in payments become problematic. In consequence, an agent who is not pivotal should not receive or make a payment. Second, if it is costly (in a lexicographic sense) to maintain accounts of future monetary commitments, then the principal wants to close down (as early as possible) the accounts of those agents who are no longer pivotal.[3]

This motivates the following efficient exit condition. Let state θ_{τ_i} in period τ_i be such that the probability that agent i affects the efficient social decision a_t^* in period t is equal to zero for all $t \geq \tau_i$, that is, $\Pr(\{\theta_t | a_t^*(\theta_t) \neq a_{-i,t}^*(\theta_t)\} | \theta_{\tau_i}) = 0$. In this case, agent i is irrelevant for the mechanism in period τ_i, and we say that the mechanism satisfies the efficient exit

---

[3]We would like to thank an anonymous referee for the suggestion to consider the link between exit and uniqueness of the transfer rule.

condition if agents neither make nor receive transfers in periods where they are irrelevant for the mechanism.

Definition 1 (Efficient Exit): A dynamic direct mechanism satisfies the efficient exit condition if for all $i, h_{\tau_i}, \theta_{\tau_i}$,

$$p_i(h_{\tau_i}, \theta_{\tau_i}) = 0.$$

We establish the uniqueness of the dynamic pivot mechanism in an environment with diverse preferences and the efficient exit condition. The assumption of diverse preferences allows for rich preferences over the current allocations and indifference over future allocations.

Assumption 1 (Diverse Preferences):

(i) *For all i, there exists $\underline{\theta}_i \in \Theta_i$ such that for all a,*

$$v_i(a, \underline{\theta}_i) = 0 \quad and \quad F_i(\underline{\theta}_i; a, \underline{\theta}_i) = 1.$$

(ii) *For all i, a, and $x \in \mathbb{R}_+$, there exists $\theta_i^{a,x} \in \Theta_i$ such that*

$$v_i(a_t, \theta_i^{a,x}) = \begin{cases} x, & if \ a_t = a, \\ 0, & if \ a_t \neq a, \end{cases}$$

and for all a_t,

$$F_i(\underline{\theta}_i; a_t, \theta_i^{a,x}) = 1.$$

The diverse preference assumption assigns to each agent i a state, $\underline{\theta}_i$, which is an absorbing state and in which i gets no payoff from any allocation. In addition, each agent i has a state in which i has a positive valuation x for a specific current allocation a and no value for other current or any future allocations. The diverse preferences condition is similar to the rich domain conditions introduced in Green and Laffont (1977) and Moulin (1986) to establish the uniqueness of the Groves and the pivot mechanism in a static environment. Relative to their conditions, we augment the diverse (flow) preferences with the certain transition into the absorbing state $\underline{\theta}_i$. With this transition we ensure that the diverse flow preferences continue to matter in the intertemporal environment.

The assumption of diverse preference in conjunction with the efficient exit condition guarantees that in every dynamic direct mechanism there are *some* types, specifically types of the form $\theta_i^{a,x}$, that receive exactly the flow transfers they would have received in the dynamic pivot mechanism.

Lemma 1: *If $\{a_t^*, p_t\}_{t=0}^{\infty}$ is ex-post incentive compatible and individually rational, and satisfies the efficient exit condition, then*

$$p_i(h_t, \theta_t^{a,x}, \theta_{-i,t}) = p_i^*(\theta_i^{a,x}, \theta_{-i,t}) \quad \text{for all } i, a, x, \theta_{-i,t}, h_t.$$

Proof: In the dynamic pivot mechanism, if the valuation x of type $\theta_i^{a,x}$ for allocation a exceeds the social externality cost, that is,

$$x \geq W_{-i}(\theta_{-i,t}) - \sum_{j \neq i} v_j(a, \theta_{j,t}) - \delta W_{-i}(\theta_{-i,t+1}|a, \theta_{-i,t}), \quad (4.6)$$

then $p_i^*(\theta_i^{a,x}, \theta_{-i,t})$ is equal to the above social externality cost; otherwise it is zero.

We now argue by contradiction. By the ex-post incentive compatibility constraints, all types $\theta_i^{a,x}$ of agent i, where x satisfies the inequality (4.6), must pay the same transfer. To see this, suppose that for some $x, y \in \mathbb{R}_+$ satisfying (4.6), we have $p_i(h_t, \theta_i^{a,x}, \theta_{-i,t}) < p_i(h_t, \theta_i^{a,y}, \theta_{-i,t})$. Now type $\theta_i^{a,y}$ has a strict incentive to misreport $r_{i,t} = \theta_i^{a,x}$, a contradiction. We therefore denote the transfer for all x and $\theta_i^{a,x}$ satisfying (4.6) by $p_i(h_t, a, \theta_{-i,t})$, and denote the corresponding dynamic pivot transfer by $p_i^*(a, \theta_{-i,t})$.

Suppose next that $p_i(h_t, a, \theta_{-i,t}) > p_i^*(a, \theta_{-i,t})$. This implies that the ex-post participation constraint for some x with $p_i(h_t, a, \theta_{-i,t}) > x > p_i^*(a, \theta_{-i,t})$ is violated, contradicting the hypothesis of the lemma. Suppose to the contrary that $p_i(h_t, a, \theta_{-i,t}) < p_i^*(a, \theta_{-i,t})$, and consider the incentive constraints of a type $\theta_i^{a,x}$ with a valuation x such that

$$p_i(h_t, a, \theta_{-i,t}) < x < p_i^*(a, \theta_{-i,t}). \quad (4.7)$$

If the inequality (4.7) is satisfied, then it follows that $a^*(\theta_i^{a,x}, \theta_{-i,t}) = a_{-i}^*(\theta_{-i,t})$ and, in particular, that $a^*(\theta_i^{a,x}, \theta_{-i,t}) \neq a$. If the ex-post incentive constraint of type $\theta_i^{a,x}$ were satisfied, then we would have

$$v_i(a^*(\theta_i^{a,x}, \theta_{-i,t}), \theta_i^{a,x}) - p_i(h_t, \theta_i^{a,x}, \theta_{-i,t})$$
$$\geq v_i(a, \theta_i^{a,x}) - p_i(h_t, a, \theta_{-i,t}). \quad (4.8)$$

Given that $\theta_i = \theta_i^{a,x}$, we rewrite (4.8) as $0 - p_i(h_t, \theta_i^{a,x}, \theta_{-i,t}) \geq x - p_i(h_t, a, \theta_{-i,t})$. But given (4.7), this implies that $p_i(h_t, \theta_i^{a,x}, \theta_{-i,t}) < 0$. In other words, type $\theta_i^{a,x}$ receives a strictly positive subsidy even though her report is not pivotal for the social allocation as $a^*(\theta_i^{a,x}, \theta_{-i,t}) = a_{-i}^*(\theta_{-i,t})$. Now, a positive subsidy violates the ex-post incentive constraint of the absorbing type $\underline{\theta}_i$. By the efficient exit condition, type $\underline{\theta}_i$, should not receive any contemporaneous (or future) subsidies. But by misreporting her type to be

$\theta_i^{a,x}$, type $\underline{\theta}_i$, would gain access to a positive subsidy without changing the social allocation. It thus follows that $p_i(h_t, \theta_i^{a,x}, \theta_{-i,t}) = p_i^*(\theta_i^{a,x}, \theta_{-i,t})$ for all a and all x. $\qquad\square$

Given that the transfers of the dynamic pivot mechanism are part of every dynamic direct mechanism with diverse preferences, we next establish that every type $\theta_{i,0}$ in $t = 0$ has to receive the same ex-ante expected utility as in the dynamic pivot mechanism.

Lemma 2: *If $\{a_t^*, p_t\}_{t=0}^{\infty}$ is ex-post incentive compatible and individually rational, and satisfies the efficient exit condition, then for all i and all θ_0, $V_i(\theta_0) = M_i(\theta_0)$.*

Proof: The argument is by contradiction. Consider i such that $V_i(\theta_0) \neq M_i(\theta_0)$. Suppose first that $V_i(\theta_0) > M_i(\theta_0)$. Then there is a history h_τ and a state θ_τ such that $p_i(h_\tau, \theta_\tau) < p_i^*(\theta_\tau)$. We show that such a transfer $p_i(h_\tau, \theta_\tau)$ leads to a violation of the ex-post incentive constraint for some type $\theta_i^{a,x} \in \Theta_i$. Specifically consider the incentive constraint of a type $\theta_i^{a_\tau^*,x}$ with $p_i(h_\tau, \theta_\tau) < x < p_i^*(\theta_\tau)$ at a misreport $\theta_{i,\tau}$:

$$v_i(a^*(\theta_i^{a_\tau^*,x}, \theta_{-i,t}), \theta_i^{a_\tau^*,x}) - p_i(h_\tau, \theta_i^{a_\tau^*,x}, \theta_{-i,\tau})$$

$$+ \delta V_i(h_{i,\tau+1} | a^*(\theta_i^{a_\tau^*,x}, \theta_{-i,\tau}), (\theta_i^{a_\tau^*,x}, \theta_{-i,\tau}))$$

$$\geq v_i(a^*(\theta_{i,\tau}, \theta_{-i,\tau}), \theta_i^{a_\tau^*,x}) - p_i(h_\tau, \theta_\tau)$$

$$+ \delta V_i(h_{i,\tau+1} | a^*(\theta_{i,\tau}, \theta_{-i,\tau}), (\theta_i^{a_\tau^*,x}, \theta_{-i,\tau})). \qquad (4.9)$$

By hypothesis, we have $p_i(h_\tau, \theta_\tau) < x < p_i^*(\theta_\tau)$ and if $x < p_i^*(\theta_\tau)$, then we can infer from marginal contribution pricing that $a^*(\theta_i^{a_\tau^*,x}, \theta_{-i,\tau}) \neq a^*(\theta_{i,\tau}, \theta_{-i,\tau})$. But as the type $\theta_i^{a_\tau^*,x}$ has only a positive valuation for $a^*(\theta_{i,\tau}, \theta_{-i,\tau})$, it follows that the left-hand side of (4.9) is equal to zero. However, the right-hand side is equal to $v_i(a^*(\theta_{i,\tau}, \theta_{-i,\tau}), \theta_i^{a_\tau^*,x}) - p_i(h_\tau, \theta_\tau) = x - p_i(h_\tau, \theta_\tau) > 0$, leading to a contradiction.

Suppose next that for some $\varepsilon > 0$, we have

$$M_i(\theta_0) - V_i(\theta_0) > \varepsilon. \qquad (4.10)$$

By hypothesis of ex-post incentive compatibility, we have for all reports $r_{i,0}$,

$$M_i(\theta_0) - [v_i(a^*(r_{i,0}, \theta_{-i,0}), \theta_{i,0}) - p_i(h_0, r_{i,0}, \theta_{-i,0})$$

$$+ \delta V_i(h_{i,1} | a^*(r_{i,0}, \theta_{-i,0}), \theta_{i,0})] > \varepsilon.$$

Given a_0^*, we can find, by the diverse preference condition, a type $\theta_i = \theta_i^{a_0^*, x}$ such that $a_0^* = a^*(\theta_i^{a_0^*, x}, \theta_{-i,0})$. Now by Lemma 1, there exists a report $r_{i,0}$ for agent i, namely $r_{i,0} = \theta_i^{a_0^*, x}$, such that a_0^* is induced at the price $p_i^*(\theta_0)$. After inserting $r_{i,0} = \theta_i^{a_0^*, x}$ into the above inequality and observing that $v_i(a^*(r_{i,0}, \theta_{-i,0}), \theta_{i,0}) - p_i(h_0, r_{i,0}, \theta_{-i,0}) = m_i(\theta_0)$, we conclude that $M_i(\theta_1) - V_i(h_{i,1} | a_0^*(r_{i,0}, \theta_{-i,0}), \theta_{i,0}) > \varepsilon/\delta$.

Now we repeat the argument we started with (4.10) and find that there is a path of realizations $\theta_0, \ldots, \theta_t$, such that the difference between the marginal contribution and the value function of agent i grows without bound. But the marginal contribution of agent i is finite given that the expected flow utility of agent i is bounded by some $K > 0$, and thus eventually the ex-post participation constraint of the agent is violated and we obtain the desired contradiction. $\square$

The above lemma can be viewed as a revenue equivalence result of all (efficient) dynamic direct mechanisms. As we are analyzing a dynamic allocation problem with an infinite horizon, we cannot appeal to the revenue equivalence results established for static mechanisms. In particular, the statement of the standard revenue equivalence results involves a fixed utility for the lowest type. In the infinite-horizon model here, the diverse preference assumption gives us a natural candidate of a lowest type in terms of $\underline{\theta}_i$, and the efficient exit condition determines her utility. The remaining task is to argue that among all intertemporal transfers with the same expected discounted value, only the time profile of the dynamic pivot mechanism satisfies the relevant conditions. Alternative payments streams could either require an agent to pay earlier or later relative to the dynamic pivot transfers. If the payments were to occur later, payments would have to be lower in an earlier period by the above revenue equivalence result. This would open the possibility for a "short-lived" type $\theta_i^{a,x}$ to induce action a at a price below the dynamic pivot transfer and hence violate incentive compatibility. The reverse argument applies if the payments were to occur earlier relative to the dynamic pivot transfer, for example, if the agent were to be asked to post a bond at the beginning of the mechanism.

Theorem 2 (Uniqueness): *If the diverse preference condition is satisfied and if $\{a_t^*, p_t\}_{t=0}^{\infty}$ is ex-post incentive compatible and individually rational, and satisfies the efficient exit condition, then it is the dynamic pivot mechanism.*

Proof: The proof is by contradiction. Suppose not. Then by Lemma 2 there exists an agent i, a history h_τ, and an associated state $\theta_{i,\tau}$ such that $p_i(h_\tau, \theta_\tau) \neq p_i^*(\theta_\tau)$. Suppose first that $p_i(h_\tau, \theta_\tau) < p_i^*(\theta_\tau)$. We show that the current monetary transfer $p_i(h_\tau, \theta_\tau)$ violates the ex-post incentive constraint of some type $\theta_i^{a,x}$. Consider now a type $\theta_i^{a_\tau^*,x}$ with a valuation x for the allocation a_τ^* such that $x > p_i^*(\theta_\tau)$. Her ex-post incentive constraints are given by

$$v_i(a^*(\theta_i^{a,x}, \theta_{-i,t}), \theta_i^{a,x}) - p_i(h_t, \theta_i^{a,x}, \theta_{-i,t})$$

$$+ \delta V_i(h_{i,t+1} | a^*(\theta_i^{a,x}, \theta_{-i,t}), (\theta_i^{a,x}, \theta_{-i,t}))$$

$$\geq v_i(a^*(r_{i,t}, \theta_{-i,t}), \theta_{i,t}) - p_i(h_t, r_{i,t}, \theta_{-i,t})$$

$$+ \delta V_i(h_{i,t+1} | a^*(r_{i,t}, \theta_{-i,t}), (\theta_i^{a,x}, \theta_{-i,t}))$$

for all $r_{i,t} \in \Theta_i$. By the efficient exit condition, we have for all $r_{i,t}$,

$$V_i(h_{i,t+1} | a^*(\theta_i^{a,x}, \theta_{-i,t}), (\theta_i^{a,x}, \theta_{-i,t}))$$

$$= V_i(h_{i,t+1} | a^*(r_{i,t}, \theta_{-i,t}), (\theta_i^{a,x}, \theta_{-i,t})) = 0.$$

By Lemma 1, $p_i(h_t, \theta_i^{a,x}, \theta_{-i,t}) = p_i^*(\theta_i^{a,x}, \theta_{-i,t}) = p_i^*(\theta_\tau)$. Consider then the misreport $r_{i,\tau} = \theta_{i,\tau}$ by type $\theta_i^{a,x}$. The ex-post incentive constraint now reads $x - p_i^*(\theta_\tau) \geq x - p_i(h_\tau, \theta_\tau)$, which leads to a contradiction, as by hypothesis we have $p_i(h_\tau, \theta_\tau) < p_i^*(\theta_\tau)$.

Suppose next that $p_i(h_\tau, \theta_\tau) > p_i^*(\theta_\tau)$. Now by Lemma 2, it follows that the ex-ante expected payoff is equal to the value of the marginal contribution of agent i in period 0. It therefore follows from $p_i(h_\tau, \theta_\tau) > p_i^*(\theta_\tau)$ that there also exists another time τ' and state $\theta_{\tau'}$ such that $p_i(h_\tau, \theta_\tau) < p_i^*(\theta_\tau)$. By repeating the argument in the first part of the proof, we obtain a contradiction. □

We should reiterate that in the definition of the ex-post incentive and participation conditions, we required that a candidate mechanism satisfies these conditions after all possible histories of past reports. It is in the spirit of the ex-post constraints that these constraints hold for all possible states rather than strictly positive probability events. In the context of establishing the uniqueness of the mechanism, it allows us to use the diverse preference condition without making an additional assumption about the transition probability from a given state $\theta_{i,t}$ into a specific state $\theta_i^{a,x}$. We merely require the existence of these types in establishing the above result.

5. Learning and Licensing

In this section, we show how our general model can be interpreted as one where the bidders learn gradually about their preferences for an object that is auctioned repeatedly over time. We use the insights from the general pivot mechanism to deduce properties of the efficient allocation mechanism. A primary example of an economic setting that fits this model is the leasing of a resource or license over time.

In every period t, a single indivisible object can be allocated to a bidder $i \in \{1, \ldots, I\}$, and the allocation decision $a_t \in \{1, 2, \ldots, I\}$ simply determines which bidder gets the object in period t. To describe the uncertainty explicitly, we assume that the true valuation of bidder i is given by $\omega_i \in \Omega_i = [0, 1]$. Information in the model represents, therefore, the bidder's prior and posterior beliefs on ω_i. In period 0, bidder i does not know the realization of ω_i, but she has a prior distribution $\theta_{i,0}(\omega_i)$ on Ω_i. The prior and posterior distributions on Ω_i are assumed to be independent across bidders. In each subsequent period t, only the winning bidder in period $t-1$ receives additional information leading to an updated posterior distribution $\theta_{i,t}$ on Ω_i, according to Bayes' rule. If bidder i does not win in period t, we assume that she gets no information, and consequently the posterior is equal to the prior. In the dynamic direct mechanism, the bidders simply report their posteriors at each stage.

The socially optimal assignment over time is a standard multi-armed bandit problem and the optimal policy is characterized by an index policy (see Whittle (1982)). In particular, we can compute for every bidder i the index based exclusively on the information about bidder i. The index of bidder i after private history $h_{i,t}$ is the solution to the optimal stopping problem

$$\gamma_i(h_{i,t}) = \max_{\tau_i} \mathbb{E} \left\{ \frac{\sum_{l=0}^{\tau_i} \delta^l v_i(a_{t+l})}{\sum_{l=0}^{\tau_i} \delta^l} \right\},$$

where a_{t+l} is the path in which alternative i is chosen l times following a given past allocation $(a_0, \ldots, a_t)$. An important property of the index policy is that the index of alternative i can be computed independent of any information about the other alternatives. In particular, the index of bidder i remains constant if bidder i does not win the object. The socially efficient allocation policy $\mathbf{a}^* = \{a_t^*\}_{t=0}^{\infty}$ is to choose in every period a bidder i if $\gamma_i(h_{i,t}) \geq \gamma_j(h_{j,t})$ for all j.

In the dynamic direct mechanism, we construct a transfer price such that under the efficient allocation, each bidder's net payoff coincides with her flow marginal contribution $m_i(\theta_t)$. We consider first the payment of the bidder i who has the highest index in state θ_t and who should therefore receive the object in period t. To match her net payoff to her flow marginal contribution, we must have

$$m_i(\theta_t) = v_i(h_{i,t}) - p_i(\theta_t). \tag{5.1}$$

The remaining bidders, $j \neq i$, should not receive the object in period t and their transfer price must offset the flow marginal contribution: $m_j(\theta_t) = -p_j(\theta_t)$. We expand $m_i(\theta_t)$ by noting that i is the efficient assignment and that another bidder, say k, would be the efficient assignment in the absence of i:

$$m_i(\theta_t) = v_i(h_{i,t}) - v_k(h_{k,t}) - \delta(W_{-i}(\theta_{t+1}|i, \theta_t) - W_{-i}(\theta_{t+1}|k, \theta_t)).$$

The continuation value without i in $t+1$, but conditional on having assigned the object to i in period t, is simply equal to the value conditional on θ_t, or $W_{-i}(\theta_{t+1}|i, \theta_t) = W_{-i}(\theta_t)$. The additional information generated by the assignment to agent i only pertains to agent i and hence has no value for the allocation problem once i is removed. The flow marginal contribution of the winning agent i is, therefore,

$$m_i(\theta_t) = v_i(h_{i,t}) - (1 - \delta)W_{-i}(\theta_t).$$

It follows that $p_i^*(\theta_t) = (1 - \delta)W_{-i}(\theta_t)$, which is the flow externality cost of assigning the object to agent i. A similar analysis leads to the conclusion that each losing bidder makes zero payments: $p_j^*(\theta_t) = -m_j(\theta_t) = 0$.

Theorem 3 (Dynamic Second Price Auction): *The socially efficient allocation rule* **a*** *is ex-post incentive compatible in the dynamic direct mechanism with the payment rule* **p***, *where*

$$p_j^*(\theta_t) = \begin{cases} (1 - \delta)W_{-j}(\theta_t), & \text{if } a_t^* = j, \\ 0, & \text{if } a_t^* \neq j. \end{cases}$$

The incentive compatible pricing rule has a few interesting implications. First, we observe that in the case of two bidders, the formula for the dynamic second price reduces to the static solution. If we remove one bidder, the social program has no other choice but to always assign it to the remaining bidder. But then the expected value of that assignment policy is simply equal to the expected value of the object for bidder j in period t by the martingale property of the Bayesian posterior. In other words, the transfer is equal to the current expected value of the next best competitor. It should be noted, though, that the object is not necessarily assigned to the bidder with the highest current flow payoff. With more than two bidders, the flow value of the social program without bidder i is different from the flow value of any remaining alternative. Since there are at least two bidders left after excluding i, the planner has the option to abandon any chosen alternative if its value happens to fall sufficiently. This option value increases the social flow payoff and hence the transfer that the efficient bidder must pay. In consequence, the social opportunity cost is higher than the highest expected valuation among the remaining bidders.

Second, we observe that the transfer price of the winning bidder is independent of her own information about the object. This means that for all periods in which the ownership of the object does not change, the transfer price stays constant as well, even though the value of the object to the winning bidder may change.

Acknowledgments

We thank the editor and four anonymous referees for many helpful comments. The current paper is a major revision and supersedes "Dynamic Vickrey–Clarke–Groves Mechanisms" (2007). We are grateful to Larry Ausubel, Jerry Green, Paul Healy, John Ledyard, Benny Moldovanu, Michael Ostrovsky, David Parkes, Alessandro Pavan, Ilya Segal, Xianwen Shi, and Tomasz Strzalecki for many informative conversations. The authors acknowledge financial support through National Science Foundation Grants CNS 0428422 and SES 0518929, and the Yrjö Jahnsson's Foundation, respectively.

References

Arrow, K. (1979): "The Property Rights Doctrine and Demand Revelation Under Incomplete Information," in *Economics and Human Welfare: Essays in Honor of Tibor Scitovsky*, ed. by M. Boskin. New York: Academic Press, 23–39. [773]

Athey, S., and I. Segal (2007): "An Efficient Dynamic Mechanism," Discussion Paper, Harvard University and Stanford University. [773]

Baron, D., and D. Besanko (1984): "Regulation and Information in a Continuing Relationship," *Information Economics and Policy*, 1, 267–302. [772,773]

Battaglini, M. (2005): "Long-Term Contracting With Markovian Consumers," *American Economic Review*, 95, 637–658. [772,773]

Bergemann, D., and J. Välimäki (2003): "Dynamic Common Agency," *Journal of Economic Theory*, 111, 23–48. [775]

——— (2006): "Dynamic Price Competition," *Journal of Economic Theory*, 127, 232–263. [775]

——— (2010): "Supplement to 'The Dynamic Pivot Mechanism'," *Econometrica Supplemental Material*, 78, http://www.econometricsociety.org/ecta/Supmat/7260_extensions.pdf. [778]

Cavallo, R., D. Parkes, and S. Singh (2006): "Optimal Coordinated Planning Among Self-Interested Agents With Private State," in *Proceedings of the 22nd Conference on Uncertainty in Artificial Intelligence*, Cambridge. [773]

Courty, P., and H. Li (2000): "Sequential Screening," *Review of Economic Studies*, 67, 697–717. [773]

D'Aspremont, C., and L. Gerard-Varet (1979): "Incentives and Incomplete Information," *Journal of Public Economics*, 11, 25–45. [773]

Dolan, R. (1978): "Incentive Mechanisms for Priority Queuing Problems," *Bell Journal of Economics*, 9, 421–436. [777]

Green, J., and J. Laffont (1977): "Characterization of Satisfactory Mechanisms for the Revelation of the Preferences for Public Goods," *Econometrica*, 45, 427–438. [771,782]

Lavi, R., and N. Nisan (2000): "Competitive Analysis of Incentive Compatible Online Auctions," in *Proceedings of the 2nd Conference of Electronic Commerce*. New York: ACM Press, 233–241. [772]

Moulin, H. (1986): "Characterization of the Pivotal Mechanism," *Journal of Public Economics*, 31, 53–78. [782]

Pavan, A., I. Segal, and J. Toikka (2008): "Dynamic Mechanism Design: Revenue Equivalence, Profit Maximization and Information Disclosure," Discussion Paper, Northwestern University and Stanford University. [773]

Said, M. (2008): "Auctions With Dynamic Populations: Efficiency and Revenue Maximization," Discussion Paper, Yale University. [778]

Whittle, P. (1982): *Optimization Over Time*, Vol. 1. Chichester: Wiley. [786]

Appendix

SUPPLEMENT TO "THE DYNAMIC PIVOT MECHANISM"
(*Econometrica*, Vol. 78, No. 2, March 2010, 771–789)

By Dirk Bergemann and Juuso Välimäki

Dept. of Economics, Yale University, 30 Hillhouse Avenue,
New Haven, CT 06520, U.S.A.;
dirk.bergemann@yale.edu
and
Dept. of Economics, Helsinki School of Economics,
Arkadiankatu 7, 00100, Helsinki, Finland;
juuso.valimaki@hse.fi.

We show that the socially efficient solution to the scheduling problem in Section 3 of the paper can be realized through a bidding mechanism, specifically a dynamic version of the ascending price auction, rather than a direct revelation mechanism. We also give a slight modification of the example where the bidding mechanism is inefficient.

In the scheduling problem in Section 3 of the main paper, a number of bidders compete for a scare resource, namely early access to a central facility. We show here that the efficient allocation can be realized through a bidding mechanism rather than a direct revelation mechanism. We find a dynamic version of the ascending price auction where the contemporaneous use of the facility is auctioned. As a given task is completed, the number of effective bidders decreases by one. We can then use a backward induction algorithm to determine the values for the bidders starting from a final period in which only a single bidder is left without effective competition.

Consider then an ascending auction in which all tasks except that of bidder I have been completed. Along the efficient path, the final ascending auction will occur at time $t = I - 1$. Since all other bidders have vanished along the efficient path at this point, bidder I wins the final auction at a price equal to zero. By backward induction, we consider the penultimate auction in which the only bidders left are $I - 1$ and I. As agent I can anticipate to win the auction tomorrow even if she were to loose it today, she is willing to bid at most

$$b_I(v_I) = v_I - \delta(v_I - 0), \tag{S1}$$

namely the net value gained by winning the auction today rather than tomorrow. Naturally, a similar argument applies to bidder $I-1$: by dropping out of the competition today, bidder $I-1$ would get a net present discounted value of $\delta\omega_{I-1}$ and hence her maximal willingness to pay is given by

$$b_{I-1}(v_{I-1}) = v_{I-1} - \delta(v_{I-1} - 0).$$

Since $b_{I-1}(v_{I-1}) \geq b_I(v_I)$, given $v_{I-1} \geq v_I$, it follows that bidder $I - 1$ wins the ascending price auction in $t = I - 2$ and receives a net payoff

$$v_{I-1} - (1 - \delta)v_I.$$

We proceed inductively and find that the maximal bid of bidder $I - k$ in period $t = I - k - 1$ is given by

$$b_{I-k}(v_{I-k}) = v_{I-k} - \delta(v_{I-k} - b_{I-(k-1)}(v_{I-(k-1)})). \tag{S2}$$

In other words, bidder $I - k$ is willing to bid as much as to be indifferent between being selected today and being selected tomorrow, when she would be able to realize a net valuation of $v_{I-k} - b_{I-(k-1)}$, but only tomorrow, and so the net gain from being selected today rather than tomorrow is

$$v_{I-k} - \delta(v_{I-k} - b_{I-(k-1)}).$$

The maximal bid of bidder $I - (k-1)$ generates the transfer price of bidder $I - k$ and by solving (S2) recursively with the initial condition given by (S1), we find that the price in the ascending auction equals the externality cost in the direct mechanism. In this class of scheduling problems, the efficient allocation can therefore be implemented by a bidding mechanism.[4]

We end this section with a minor modification of the scheduling model to allow for multiple tasks. For this purpose, it is sufficient to consider an example with two bidders. The first bidder has an infinite series of single-period tasks, each delivering a value of v_1. The second bidder has only a single task with a value v_2. The utility function of bidder 1 is thus given by

$$v_1(a_t, \theta_{1,t}) = \begin{cases} v_1, & \text{if } a_t = 1 \text{ for all } t, \\ 0, & \text{if otherwise,} \end{cases}$$

whereas the utility function of bidder 2 is as described earlier.

The socially efficient allocation in this setting either has $a_t = 1$ for all t if $v_1 \geq v_2$ or $a_0 = 2$, $a_t = 1$ for all $t \geq 1$ if $v_1 < v_2$. For the remainder of this example, we assume that $v_1 > v_2$. Under this assumption, the efficient policy never completes the task of bidder 2. The marginal contributions of each bidder are

$$M_1(\theta_0) = (v_1 - v_2) + \frac{\delta}{1 - \delta} v_1$$

and

$$M_2(\theta_0) = 0.$$

---

[4]The nature of the recursive bidding strategies bears some similarity to the construction of the bidding strategies for multiple advertising slots in the keyword auction of Edelman, Ostrovsky, and Schwartz (2007). In the auction for search keywords, the multiple slots are differentiated by their probability of receiving a hit and hence generating a value. In the scheduling model here, the multiple slots are differentiated by the time discount associated with different access times.

Along any efficient allocation path, we have $M_i(\theta_0) = M_i(\theta_t)$ for all i and the social externality cost of agent 1, $p_1^*(\theta_t)$ for all t, is $p_1^*(\theta_t) = (1-\delta)v_2$. The externality cost is again the cost of delay imposed on the competing bidder, namely $(1-\delta)$ times the valuation of the competing bidder. This accurately represents the social externality cost of agent 1 in every period even though agent 2 never receives access to the facility.

We contrast the efficient allocation and transfer with the allocation resulting in the dynamic ascending price auction. For this purpose, suppose that the equilibrium path generated by the dynamic bidding mechanism is efficient. In this case, bidder 2 is never chosen and hence receives a net payoff of 0 along the equilibrium path. But this means that bidder 2 would be willing to bid up to v_2 in every period. In consequence, the first bidder receives a net payoff of $v_1 - v_2$ in every period and her discounted sum of payoff is then

$$\frac{1}{1-\delta}(v_1 - v_2) < M_1(\theta_0). \tag{S3}$$

But more important than the failure of the marginal contribution is the fact that the equilibrium does not support the efficient assignment policy. To see this, notice that if bidder 1 loses to bidder 2 in any single period, then the task of bidder 2 is completed and bidder 2 drops out of the auction in all future stages. Hence the continuation payoff for bidder 1 from dropping out in a given period and allowing bidder 2 to complete his task is given by

$$\frac{\delta}{1-\delta}v_1. \tag{S4}$$

If we compare the continuation payoffs (S3) and (S4), respectively, then we see that it is beneficial for bidder 1 to win the auction in all periods if and only if

$$v_1 \geq \frac{v_2}{1-\delta},$$

but the efficiency condition is simply $v_1 \geq v_2$. It follows that for a large range of valuations, the outcome in the ascending auction is inefficient and assigns the object to bidder 2 despite the inefficiency of this assignment. The reason for the inefficiency is easy to detect in this simple setting. The forward-looking bidders consider only their individual net payoffs in future periods. The planner, on the other hand, is interested in the level of gross payoffs in future periods. As a result, bidder 1 is strategically willing and able to depress the future value of bidder 2 by letting bidder 2 win today to increase the future difference in the valuations between the two bidders. But from the point of view of the planner, the differential gains for bidder 1 are immaterial and the assignment to bidder 2 represents an inefficiency. The rule of the ascending price auction, namely that the highest bidder wins, only internalizes the individual *equilibrium payoffs* but not the social payoffs.

This small extension to multiple tasks shows that the logic of the marginal contribution mechanism can account for subtle intertemporal changes in the payoffs. On the other hand, common bidding mechanisms may not resolve the dynamic allocation problem in an efficient manner. Indirectly, it suggests that suitable indirect mechanisms have yet to be devised for scheduling and other sequential allocation problems.

Reference

Edelman, B., M. Ostrovsky, and M. Schwartz (2007): "Internet Advertising and the Generalized Second Price Auction: Selling Billions of Dollars Worth of Keywords," *American Economic Review*, 97, 242–259.

Chapter 15

Dynamic Revenue Maximization: A Continuous Time Approach*

Dirk Bergemann[†] and Philipp Strack[‡]

Department of Economics, and Cowles Foundation of Research in Economics, Yale University, New Haven, CT 06520-8268, USA
[†] *dirk.bergemann@yale.edu*
[‡] *philipp.strack@yale.edu*

We characterize the revenue-maximizing mechanism for time separable allocation problems in continuous time. The willingness-to-pay of each agent is private information and changes over time.

We derive the dynamic revenue-maximizing mechanism, analyze its qualitative structure and frequently derive its closed form solution. In the leading example of repeat sales of a good or service, we establish that commonly observed contract features such as flat rates, free consumption units and two-part tariffs emerge as part of the optimal contract. We investigate in detail the environments in which the type of each agent follows an arithmetic or geometric Brownian motion or a mean-reverting process. We analyze the allocative distortions and show that depending on the nature of the private information the distortion might increase or decrease over time.

JEL classification: D44; D82; D83

Keywords: Dynamic mechanism design, repeated sales, stochastic flow, flat rates, two-part tariffs, leasing.

---

*This chapter is reproduced from *Journal of Economic Theory*, **159**, 819–853, 2015, with Philipp Strack.

1. Introduction

1.1. *Motivation*

We analyze the nature of the revenue-maximizing contract in a dynamic environment with private information at the initial time of contracting as well as in all future periods. We consider a setting in continuous time and are mostly concerned with environments where the uncertainty, and in particular the private information of the agent is described by a Brownian motion. The present work makes progress by considering allocation problems that we refer to as *weakly time separable*. Namely, (*i*) the set of available allocations at time t is independent of the history of allocations and (*ii*) the flow utility functions of the agent and the principal at time t depend only on the *initial* and the *current* private information of the agent (and hence the qualifier of weakly time separable).

With time separability, the allocation rule that maximizes the expected dynamic virtual surplus has the property that the allocation at time t is a function of the report of the agent at time 0 and time t only. As a result, at every time $t > 0$, each agent is only facing a static reporting problem since the current report is only used to determine the current allocation. A notable implication of this separability is that the incentive compatibility conditions can be decomposed completely into a time 0 problem and a sequence of static problem at all times $t > 0$. The restriction to time separable allocation problems is sufficiently mild to include many of the allocation problems explicitly analyzed in the literature so far, for example the optimal quantity provision by the monopolist as in Battaglini (2005) or the auction environment of Eső and Szentes (2007).

The specific contribution of the continuous time setting to the analysis of the optimal mechanism arises *after* establishing the necessary conditions for optimality under time separability. And in fact, we obtain the first-order conditions by using the envelope theorem using a small class of relevant deviations which is precisely the approach taken in discrete time, see for Eső and Szentes (2007) and Pavan *et al.* (2014) for the seminal contributions. The resulting dynamic version of the virtual utility accounts for the influence that the present private information has on the future state of the world (and hence future private information of the agent) through a term that Pavan *et al.* (2014) refer to as impulse response function. Now, in continuous time, the equivalent expression, which is commonly referred to as *stochastic flow*, is compact and summarizes the nature of the underlying stochastic process in an explicit formula.

We then make use of the information conveyed by the stochastic flow in three distinct ways.

First, we explicitly derive the nature of the optimal allocation policy and the associated transfer rules. We consider in some detail a number of well-known stochastic processes, in particular the arithmetic and the geometric Brownian motion. The natural starting point here is to consider the case in which the private information of the agent is the current state of the process, in particular the initial state of the Brownian motion is private information, but we also analyze the problem when either the drift or volatility of the process are private information. In Section 5 we consider the nature of the optimal mechanism for repeated sales when the type of the agent follows a geometric Brownian motion. We establish that commonly observed contract features such as flat rates, free consumption units, two-part tariffs and leasing arrangements emerge as solutions to the optimal contract design.

Second, we derive sufficient conditions for the optimality of the dynamic mechanism in terms of the primitives of the stochastic process. This is demonstrated in detail in Section 6 where we, for example, derive sufficient conditions for optimality when the private information of agent cannot be ordered by first-order stochastic dominance. In particular, we can allow the variance rather than the mean of the stochastic process to form the private information, and yet display transparent sufficient conditions for optimality. In much of the earlier literature, the types had to be assumed to be ordered according to first-order stochastic dominance in order to give rise to sufficient conditions for optimality.

Third, we systematically extend the analysis from Markovian settings where the initial private information (as well as any future private information) is the state of the stochastic process to settings in which the initial private information can present a parameter of the stochastic process, such the mean or variance of the Markov process. The subsequent private information continues to pertain to the state of the Markov process. This specification of the private information, the initial information about the parameter of the process and the ongoing information about the state of the process still conforms with our restriction to weakly time separable environments.

The initial private information may represent the drift or the volatility of the Brownian motion, or the long-run mean or the reversion rate of a mean-reverting Ornstein-Uhlenbeck process. The resulting informational term in the virtual utility, which is referred to as *generalized stochastic flow* in probability theory, still permits a compact representation that can be used for the determination of the optimal policy and/or for the sufficient conditions. With the notable exception of the recent papers by Boleslavsky and Said (2013) and Skrzypacz and Toikka (forthcoming), and a discussion in the supplementary appendix of Pavan *et al.* (2014), the earlier contributions with an infinite horizon did not allow for the possibility that the initial private information may pertain to a parameter of the stochastic process itself, such as the drift or the volatility. Interestingly, the continuous time version of the resulting generalized impulse response function is often a deterministic function of the initial state and time, whereas the corresponding discrete time process has a generalized impulse response function that depends on the realization of the entire sample path. This is shown for example in Section 5 where the initial private information is the mean of the geometric Brownian motion. The discrete time counterpart of this process, namely the multiplicative random walk, was analyzed earlier by Boleslavsky and Said (2013). Here the generalized impulse response term involves the number of *realized* upticks and downticks. In the continuous time equivalent, the generalized stochastic flow is simply the expected number of upticks or downticks which is a deterministic function of time and the initial state.

We should add that the current focus on time separable allocation problems is restrictive in that it excludes problems such as the optimal timing of a sale of a durable good, where the present decision, say a sale, naturally preempts certain future decision, say a sale, again. But our setting allows us to restrict attention to a small class of deviations, deviations that we call *consistent*. The consistent deviations, by themselves only necessary conditions, nonetheless completely describe the indirect utility of the agent in any incentive compatible mechanism. More precisely, at time zero the initial shock of the agent is drawn and the initial shock determines the probability measure of the entire future valuation process. If the agent deviates he changes the probability measure of the reported valuation process. To avoid working with the change in measures directly we restrict attention to consistent deviations. We call a deviation consistent if, after his initial misreport, say b instead of a, the agent reports his valuation as if it would follow the same Brownian motion as the one which drives his true valuation. As there is a true initial shock, namely b, which could have made these subsequent reports truthful, the principal cannot detect such a deviation and is forced to assign the allocation and transfer process of the imitated shock b. In particular, this allows us to evaluate the payoffs of the truthful and the consistently deviating agent with respect to the same expectation operator. Now, as we assume the initial shock to be one-dimensional and given that all deviations are parameterized over the time zero shock, standard mechanism design arguments deliver the smoothness of the value function of the agent.

Within the class of time separable allocation policies we can rewrite the sufficiency conditions exclusively in terms of the flow virtual utilities. By using the class of consistent deviations and allowing for time separable allocation policies, we can completely avoid the verification of the incentive compatibility conditions via backward induction methods which was the basic instrument to establish the sufficient conditions used in much of the preceding literature with dynamic adverse selection.

1.2. Related literature

The analysis of the revenue-maximizing contract in an environment where the private information may change over time appears first in Baron and Besanko (1984). They considered a two period model of a regulator facing a monopolist with unknown, but in every period, constant marginal cost. Besanko (1985) offers an extension to a finite horizon environment with a general cost function, where the unknown parameter is either i.i.d. over time or follows a first-order autoregressive process. Since these early contributions, the literature has developed rapidly. Courty and Li (2000) consider the revenue-maximizing contract in a sequential screening problem where the preferences of the buyer change over time. Battaglini (2005) considered a quantity discriminating monopolist who provides a menu of choices to a consumer whose valuation can change over time according to a commonly known Markov process. In contrast to the earlier work, he explicitly considered an infinite time horizon and showed that the distortion due to the initial private information vanishes over time. Eső and Szentes (2007) rephrased the two period sequential screening problem by showing that the additional signal arriving in period two can always be represented by a signal that is orthogonal to the signal in period one. Eső and Szentes (2014) generalize this insight in an infinite horizon environment and show that the information rent of the agent is only due to his initial information.

Pavan *et al.* (2014) consider a general environment in an infinite horizon setting and allowing for general allocation problems, encompassing the earlier literature (with continuous type spaces). They obtain general necessary conditions for incentive compatibility and present a variety of sufficient conditions for revenue-maximizing contracts for specific classes of environments. They also observed the beneficial implications of time separable environments for a tighter characterization of the optimal contract.

A feature common to almost all of the above contributions is that the private information of the agent is represented by the current state of a one-dimensional Markov process, and that the new information that the agent receives is controlled by the current state, and in turn, leads to a new state of the Markov process. Notably, Pavan *et al.* (2014), Boleslavsky and Said (2013) and Skrzypacz and Toikka (forthcoming) allowed for the possibility that the initial private information is about a parameter of the stochastic process itself.[1] For example, Boleslavsky and Said (2013) let the initial private information of the agent be the mean of a multiplicative random walk. Interestingly, this dramatically changes the impact that the initial private information has on the future allocations. In particular, the distortions in the future allocation may now increase over time rather than decline as in much of the earlier literature. The reason is that the influence of the parameter of the stochastic process, such as the drift or the variance, on the valuation may increase over time.[2] Finally, Kakade *et al.* (2013) consider a class of dynamic allocation problems, a suitable generalization of the single unit allocation problem and impose a separability condition (additive or multiplicative) on the interaction of the initial private information and all subsequent signals. The separability condition allows them to obtain an explicit characterization of the revenue-maximizing contract and derive transparent sufficient conditions for the optimal contract.

The remainder of the paper proceeds as follows. Section 2 presents the model. In Section 3 we derive the necessary and sufficient conditions for the revenue-maximizing contract. In Section 4 we analyze the implications of the revenue-maximizing contract for the structure of the intertemporal distortions. The nature of the optimal contract for repeat purchases of a product or service is analyzed in Section 5 in an environment where the type follows a geometric Brownian motion. Section 6 examines the

---

[1] This is equivalent to assuming that the private information of the agent corresponds to the state of a two-dimensional Markov process, whose first component is constant after time zero, but influences the transitions of the second component.

[2] In a recent contribution, Garrett and Pavan (2012) also exhibit the possibility of increasing distortions over time, but the source there is a trade-off in the retention decision of a known agent versus a hiring decision of new, hence less well known agent.

optimal allocation among competing bidders when the private valuation is either driven by the arithmetic Brownian motion or the mean-reverting Ornstein-Uhlenbeck process. Section 7 concludes. Appendix A contains some auxiliary proofs and additional results.

2. Model

There are n agents indexed by $i \in \{1, \ldots, n\} = N$. Time is continuous and indexed by $t \in [0, T]$, where the time horizon T can be finite or infinite. If the time horizon is infinite, then we assume a discount rate $r \in \mathbb{R}_+$ which is strictly positive, $r > 0$.

The flow preferences of agent i are represented by a quasilinear utility function:

$$v_t^i \cdot u^i(t, x_t^i) - p_t^i. \tag{2.1}$$

The function $u^i : \mathbb{R}_+ \times \mathbb{R}_+ \to [0, \bar{u}]$ is continuous and strictly increasing in x, decreasing in t and satisfies $u^i(t, 0) = 0$ for all $t \in \mathbb{R}_+$. We refer to $u^i(t, x_t^i)$ as the *valuation* of $x_t^i \in [0, \bar{x}] \subset \mathbb{R}_+$ with $0 \leq \bar{x} < \infty$. The allocation x_t^i can be interpreted as either the quantity or quality of a good that is allocated to agent i at time t. The *type* of agent i in period t is given by $v_t^i \in \mathbb{R}$ and the flow utility in period t is given by the product of the type and the valuation. The payment in period t is denoted by $p_t^i \in \mathbb{R}$.

The type v_t^i of agent i at time t depends on his *initial shock* θ^i at time $t = 0$ and the contemporaneous shock W_t^i at time t:

$$v_t^i \triangleq \phi^i(t, \theta^i, W_t^i). \tag{2.2}$$

The function $\phi^i : \mathbb{R}_+ \times \Theta \times \mathbb{R} \to \mathbb{R}$ aggregates the initial shock θ^i and the contemporaneous W_t^i of agent i into his type v_t^i. The initial private information θ^i is not restricted to be the initial type v_0^i, but might be any other characteristic determining the probability measure over paths of the types $(v_t)_{t \in \mathbb{R}_+}$. In the case of the arithmetic or geometric Brownian motion, the initial shock θ^i could constitute the initial value v_0^i, but it could also be the drift μ^i or the variance $(\sigma^i)^2$ of the Brownian motion. Similarly, in the case of a mean reverting process, the initial shock θ^i could constitute the mean reversion speed or the long run-average of the stochastic process. In any event, at time zero each agent i privately learns his initial shock $\theta^i \in (\underline{\theta}, \bar{\theta}) = \Theta \subseteq \mathbb{R}$, which is drawn from a common prior distribution $F^i : \mathbb{R} \to [0, 1]$, independently across agents.

The distribution F^i has a strictly positive density $f^i > 0$ with decreasing inverse hazard rate $(1 - F^i)/f^i$. The contemporaneous shock is given by a random process $(W_t^i)_{t \in \mathbb{R}_+}$ of agent i that changes over time as a consequence of a sequence of incremental shocks and W_t^i is assumed to be independent of W_t^j for every $j \neq i$. In Sections 5 and 6, the valuation function $u^i(t, x_t^i)$ is simply a linear function $u^i(t, x_t^i) = x_t^i$ and the type v_t^i can then be directly interpreted as the *marginal willingness to pay* of agent i.

The function ϕ^i is twice differentiable in every direction and in the following we use a small annotation for partial derivatives:

$$\phi_\theta^i(t, \theta^i, w^i) \triangleq \frac{\partial \phi^i(t, \theta^i, w^i)}{\partial \theta}. \tag{2.3}$$

If θ^i is the initial value of the process of agent i, that is $v_0^i = \theta^i$, then the derivative ϕ_θ^i is commonly referred to as the *stochastic flow*, or *generalized stochastic flow* if θ^i determines the evolution of a diffusion by influencing the drift or variance term (see for example Kunita, 1997). The stochastic flow process $(\phi_\theta^i(t, \theta, W_t^i))_{t \in \mathbb{R}_+}$ is the analogue of the impulse response functions described in the discrete time dynamic mechanism design literature (see Pavan *et al.*, 2014, Definition 3). As we will see in the examples presented later the stochastic flow is of a very simple form for many classical continuous time diffusion processes, like the arithmetic and geometric Brownian motion.

We assume that for every agent i a higher initial shock θ^i leads to a higher type, $\phi_\theta^i(t, \theta^i, w^i) \geq 0$; and an agent i who observed a higher value of the process W_t^i has a higher type, $\phi_w^i(t, \theta^i, w^i) > 0$ for every $(t, \theta^i, w^i) \in \mathbb{R}_+ \times \Theta \times \mathbb{R}$.

Assumption 1 (Decreasing influence of initial shock): The relative impact of the initial shock on the type:

$$\frac{\phi_\theta^i(t, \theta^i, w^i)}{\phi^i(t, \theta^i, w^i)} \tag{2.4}$$

is decreasing in w^i for every $(t, \theta^i, w^i) \in \mathbb{R}_+ \times \Theta \times \mathbb{R}$.

Assumption 2 (Decreasing influence of initial vs contemporaneous shock): The ratio of the marginal impact of initial and contemporaneous shocks:

$$\frac{\phi_\theta^i(t, \theta^i, w^i)}{\phi_{w^i}^i(t, \theta^i, w^i)} \tag{2.5}$$

is decreasing in θ^i for every $(t, \theta^i, w^i) \in \mathbb{R}_+ \times \Theta \times \mathbb{R}$.

The last assumption implies that the type with a large initial shock is influenced more by the contemporaneous shocks that arrive after time zero.

Assumption 3 (Finite expected impact of the initial shock): The expected influence of the initial shock on the type grows at most exponentially: there exists $C \in \mathbb{R}_+$, $q \in (0, r)$ such that $\mathbb{E}[\phi_\theta^i(t, \theta^i, W_t^i)] \leq Ce^{qt}$ for all $t \in \mathbb{R}_+$ and $\theta^i \in \Theta$.

Assumption 3 ensures that the effect of a marginal change in the agent's type on the sum of discounted expected future types is finite.

At every point in time t the principal chooses an allocation $x_t \in X$ from a compact, convex set $X \subset \mathbb{R}_+^n$, where x_t^i can be interpreted as the quantity or quality of a good that is allocated to agent i at time t. We assume that it is always possible to allocate zero to an agent:

$$x \in X \Rightarrow (x^1, \ldots, x^{i-1}, 0, x^{i+1}, \ldots, x^n) \in X.$$

To ensure that the problem is well-posed we assume that every feasible allocation process $x^i = (x_t^i)$ gives finite expected utility to agent i,

$$\mathbb{E}\left[\int_0^T e^{-rt} \mathbf{1}_{\{v_t^i \geq 0\}} v_t^i u^i(t, x_t^i) dt | \theta^i \right] < \infty,$$

for every θ^i in the support of F^i. The principal receives the sum of discounted flow payments $\sum_{i \in N} p_t^i$ minus the production costs $c(x_t)$:

$$\mathbb{E}\left[\int_0^T e^{-rt} \left(\sum_{i \in N} p_t^i - c(x_t)\right) dt \right]. \tag{2.6}$$

The cost $c : X \rightarrow \mathbb{R}_+$ is continuous and increasing in every component with $c(0) = 0$. With minor abuse of language we shall refer throughout the paper to the net revenue (or profit) maximization problem given by (2.6) as simply the revenue maximization problem.

Definition 1 (Value function): The indirect utility, or value function, $V^i(\theta^i)$ of agent i given his initial shock θ^i, his consumption process $(x_t^i)_{t \in \mathbb{R}_+}$ and his payment process $(p_t^i)_{t \in \mathbb{R}_+}$ is

$$V^i(\theta^i) = \mathbb{E}\left[\int_0^T e^{-rt}(u^i(t, x_t^i)v_t^i - p_t^i)\mathrm{d}t \Big| \theta^i\right]. \tag{2.7}$$

A contract specifies an allocation process $(x_t)_{t \in \mathbb{R}_+}$ and a payment process $(p_t)_{t \in \mathbb{R}_+}$. The allocation x_t and the payment p_t can depend on all types reported $(v_s^i)_{s \le t, i \in N}$ by the agents prior to time t. We assume that the agent has an outside option of zero and thus require the following definition:

Definition 2 (Incentive and participation constraints): A contract $(x_t, p_t)_{t \in \mathbb{R}_+}$ is acceptable if for every agent i it is individually rational to accept the contract

$$V^i(\theta^i) \ge 0 \quad \text{for all } \theta^i \in \Theta,$$

and it is optimal to report his shock θ^i and his type $(v_t^i)_{t \in \mathbb{R}_+}$ truthfully at every point in time $t \in \mathbb{R}_+$.

Given the transferable utility, we define the flow welfare function $s : \mathbb{R}_+ \times \mathbb{R}^n \times X \to \mathbb{R}$ that maps an allocation $x \in X$ and a vector of types $v \in \mathbb{R}^n$ into the associated flow of welfare

$$s(t, v, x) = \sum_{i \in N} v_t^i u^i(t, x^i) - c(x). \tag{2.8}$$

The social value of the allocation process $(x_t)_{t \in [0,T]}$ aggregates the discounted flow of social welfare over time and is given by:

$$\mathbb{E}\left[\int_0^T e^{-rt}\left(\sum_{i \in N} v_t^i u^i(t, x_t^i) - c(x_t)\right)\mathrm{d}t\right] = \mathbb{E}\left[\int_0^T e^{-rt}s(t, v_t, x_t)\mathrm{d}t\right]. \tag{2.9}$$

As the allocation x_t at time t does not influence the future evolution of types or the set of possible future allocations the problem of finding a socially efficient allocation is time-separable. We define the optimal allocation function $x^\dagger : \mathbb{R}_+ \times \mathbb{R}^n \to \mathbb{R}^n$ that maps a point in time t and a vector of types v into the set of optimal allocations

$$x^\dagger(t, v) = \arg\max_{x \in X} s(t, v, x). \tag{2.10}$$

An allocation process $(x_t)_{t \in [0,T]}$ is welfare maximizing if and only if $x_t \in x^\dagger(t, v_t)$ almost surely for every $t \in [0, T]$.

Given the essentially static character of the social allocation problem, it follows immediately that the *welfare maximizing* allocation $x^\dagger$ can be implemented via a sequence of static Vickrey–Clarke–Groves mechanisms and associated payments:

$$p_t^{\dagger i} \triangleq p^{\dagger i}(t, v_t)$$
$$= \max_{x \in X} \sum_{j \neq i} [u^j(t, x) - u^j(t, x^\dagger(t, v_t))] v_t^j - c(x) + c(x^\dagger(t, v_t)). \quad (2.11)$$

3. Revenue Maximization

In this section we derive a revenue-maximizing direct mechanism. Without loss of generality we restrict attention to direct mechanisms, where every agent i reports his initial shock θ^i and his type v_t^i truthfully. We first obtain a revenue equivalence result for incentive compatible mechanisms.

3.1. *Necessity*

We begin by establishing that the value function of the agent if he reports truthfully is Lipschitz continuous. As ϕ^i is strictly increasing in w^i we can implicitly define the function $\omega : \mathbb{R}_+ \times \Theta \times \mathbb{R} \to \mathbb{R}$ by

$$v_i = \phi^i(t, \theta^i, \omega(t, \theta^i, v^i)) \quad \text{for all } (t, \theta^i) \in \mathbb{R}_+ \times \Theta. \quad (3.1)$$

Thus ω identifies the value that the contemporaneous shock W_t^i has to have at time t to generate a contemporaneous type v^i given the initial shock θ^i. We derive a necessary condition for incentive compatibility that is based only on the robustness of the mechanism to a small class of deviations, which we refer to as *consistent deviations*.

Definition 3 (Consistent deviation): A deviation by agent i is referred to as a *consistent deviation* if agent i with type $v_0^i = \phi^i(0, a, W_0^i)$ (and associated initial shock $a \in \Theta$) misreports $\widehat{v}_0^i = \phi^i(0, b, W_0^i)$ (and associated initial shock $b \in \Theta$) at $t = 0$ and continues to misreport:

$$\widehat{v}_t^i = \phi^i(t, b, \omega(t, a, v_t^i)), \quad (3.2)$$

instead of his true type v_t^i at all future dates $t \in \mathbb{R}_+$.

Thus, an agent who misreports with a consistent deviation, continues to misreport his true type v_t^i in all future periods. More precisely, agent i's reported type $\hat{v}_t^i = \phi^i(t, b, W_t^i)$ equals the type he would have had if his initial shock would have been b instead of a. We note that the misreport generated by a consistent deviation has the property that the principal can infer from the misreport the true realized path of the contemporaneous shocks W_t^i. Now, since the allocation depends on the type v_t^i rather than the path of contemporaneous shocks W_t^i, the (inferred) truthfulness in the shocks is not of immediate use for the principal. We now show that this, one-dimensional, class of consistent deviations is sufficient to uniquely pin down the value function of the agent in any incentive compatible mechanism at time $t = 0$. The class of consistent deviations we consider here are not local deviations at one point in time, but rather represent a global deviation in the sense that the agent changes his reports at every point in time.

As $\phi^i(0, \theta^i, W_0^i)$ is strictly increasing in θ^i, it is convenient to describe the initial report directly in terms of the true initial shock a and the reported initial shock b. We thus define $V^i(a, b)$ to be the indirect utility of agent i with initial shock a but who reports shock b and misreports his type consistently as $\hat{v}_t^i = \phi^i(t, b, \omega(t, a, v_t^i))$. Note that by construction $W_t^i = \omega(t, a, v_t^i)$. Consequently the allocation agent i gets by consistently deviating and reporting b is the same allocation that he would get if his initial shock were b and he were to report it truthfully. Hence $V^i(a, b)$ is the indirect utility of an agent who has the initial shock a but reports initial shock b and misreports his type consistently and is given by:

$$V^i(a, b) = \mathbb{E}\left[\int_0^T e^{-rt}(u^i(t, x_t^i(b))\phi^i(t, a, W_t^i) - p_t^i(b))\mathrm{d}t\right].$$

Note, that when restricted to consistent deviations the mechanism design problem turns into a standard one-dimensional problem, and the Envelope theorem yields the derivative of the indirect utility function of the agent:

Proposition 1 (Regularity of value function): *The indirect utility function V^i of every agent $i \in N$ in any incentive compatible mechanism is Lipschitz continuous and has the weak derivative*

$$V_\theta^i(\theta^i) = \mathbb{E}\left[\int_0^T e^{-rt}u^i(t, x_t^i(\theta))\phi_\theta^i(t, \theta^i, W_t^i)\mathrm{d}t\right] \quad a.e. \qquad (3.3)$$

Proof: As the agent can always use consistent deviations, a necessary condition for incentive compatibility is $V(a, a) = \sup_b V(a, b)$. As ϕ^i is differentiable the derivative of V with respect to the first variable is given by

$$V_a(a,b) = \frac{\partial}{\partial a}\mathbb{E}\left[\int_0^T e^{-rt}(u^i(t,x_t^i(b))\phi^i(t,a,W_t^i)) - p_t^i(b))\mathrm{d}t\right]$$

$$= \mathbb{E}\left[\int_0^T e^{-rt}(u^i(t,x_t^i(b))\phi_\theta^i(t,a,W_t^i))\mathrm{d}t\right]$$

$$\leq \bar{u}\mathbb{E}\left[\int_0^T e^{-rt}\phi_\theta^i(t,a,W_t^i)\mathrm{d}t\right],$$

which is bounded by a constant by Assumption 3. By the Envelope theorem (see Milgrom and Segal, 2002, Theorem 1 and Theorem 2) we have that $V^i(\theta^i) = V^i(\theta^i,\theta^i)$ is absolutely continuous an the (weak) derivative is given by (3.3). As argued above (3.3) is bounded and thus V^i is Lipschitz continuous. □

We introduce a dynamic version of the virtual utility function J^i : $\mathbb{R}_+ \times \Theta \times \mathbb{R} \to \mathbb{R}$ as:

$$J^i(t,\theta^i,v_t^i) = v_t^i - \frac{1 - F^i(\theta^i)}{f^i(\theta^i)}\phi_\theta^i(t,\theta^i,\omega(t,\theta^i,v_t^i)). \tag{3.4}$$

We observe that the above virtual utility is modified relative to its static version only by the term of the stochastic flow ϕ_θ^i that multiplies the inverse hazard rate. Thus, the specific impact of the private information in the dynamic mechanism is going to arrive exclusively through the stochastic flow ϕ_θ^i (see (2.3)), the continuous time equivalent of the impulse response function. The properties of the virtual utility are summarized in the following proposition:

Proposition 2 (Monotonicity of virtual utility): *If the virtual utility* $J^i(t,\theta^i,v_t^i)$ *is positive then it is non-decreasing in* θ^i *and* v_t^i.

The proof of Proposition 2 given in Appendix A establishes the monotonicity of the virtual utility from Assumptions 1 and 2 using algebraic arguments. We observe that Proposition 2 establishes the monotonicity of the virtual utility only for the case that the virtual utility is positive. In fact, our assumptions are not strong enough to ensure the monotonicity of the virtual utility independent of its sign. The reason not to impose stronger monotonicity conditions is that for many important examples discussed later (for example the geometric Brownian motion with unknown initial value) the virtual utility is only monotone if positive.

We can now establish a revenue equivalence result that describes the revenue of the principal in any incentive compatible mechanism solely in terms of the allocation process $x = (x_t)_{t \in \mathbb{R}_+}$ and the expected time zero value the lowest type derives from the contract $V^i(\underline{\theta})$.

Theorem 1 (Revenue equivalence): *For any incentive compatible direct mechanism the expected payoff of the principal depends only on the allocation process $(x_t)_{t \in \mathbb{R}_+}$ and is given by the dynamic virtual surplus:*

$$
\mathbb{E}\left[\int_0^T e^{-rt} \left(\sum_{i \in N} p_t^i - c(x_t) \right) dt \right]
$$

$$
= \mathbb{E}\left[\int_0^T e^{-rt} \left(\sum_{i \in N} J^i(t, \theta_t^i, v_t^i) u^i(t, x_t^i) - c(x_t) \right) dt \right]
$$

$$
- \sum_{i \in N} V^i(\underline{\theta}). \tag{3.5}
$$

Proof: Partial integration gives that in any incentive compatible mechanism (x, p) the expected transfer received by the principal from agent i equals the expected virtual utility of agent i:

$$
\mathbb{E}\left[\int_0^T e^{-rt} p_t^i dt \right]
$$

$$
= \mathbb{E}\left[\int_0^T e^{-rt} u^i(t, x_t^i) v_t^i dt \right] - \int_{\underline{\theta}}^{\bar{\theta}} f(\theta^i) V^i(\theta^i) d\theta^i
$$

$$
= \mathbb{E}\left[\int_0^T e^{-rt} u^i(t, x_t^i) v_t^i dt \right] - \int_{\underline{\theta}}^{\bar{\theta}} f^i(\theta^i) \frac{1 - F^i(\theta^i)}{f^i(\theta^i)} V_\theta^i(\theta^i) d\theta^i - V^i(\underline{\theta})
$$

$$
= \mathbb{E}\left[\int_0^T e^{-rt} u^i(t, x_t^i) \left(v_t^i - \frac{1 - F^i(\theta^i)}{f^i(\theta^i)} \phi_\theta^i(t, \theta^i, W_t^i) \right) dt \right] - V^i(\underline{\theta}).
$$

Summing up the transfers of all agents and subtracting the cost gives the result.   □

As Theorem 1 provides a necessary condition for incentive compatibility it follows that if there exists an incentive compatible contract (x, p) such that the allocation process x maximizes the expected virtual surplus given by (3.5), then it maximizes the principal's surplus. Clearly, to maximize the virtual surplus it is optimal to set the transfer to the lowest initial shock equal to zero: $V^i(\underline{\theta}) = 0$ for all agents $i \in N$. We denote by $J(t, \theta, v_t) \in \mathbb{R}^n$ the vector of virtual utilities, $J(t, \theta, v_t)^i = J^i(t, \theta^i, v_t^i)$. The revenue of the principal defined by (3.5) equals the expected welfare when true utilities (types) v are replaced with virtual utilities J, hence referred to as the *dynamic virtual surplus*:

$$\mathbb{E}\left[\int_0^T e^{-rt} s(t, J(t, \theta_t, v_t), x_t)\mathrm{d}t\right] - \sum_{i \in N} V^i(\underline{\theta}), \qquad (3.6)$$

where we defined the flow social value $s(\cdot)$ earlier in (2.8). In the next step we establish that there exists a direct mechanism that maximizes the dynamic virtual surplus defined in (3.6). To do so let us first state the following result which ensures that there exists a time separable allocation that maximizes the dynamic virtual surplus:

Proposition 3 (Virtual surplus maximizing allocation): *There exists an allocation function $x^* : \mathbb{R}_+ \times \Theta \times \mathbb{R}^n \to X$ that maximizes the dynamic virtual surplus. Furthermore, the allocation $x^{*i}(t, \theta, v_t)$ of agent i is non-decreasing in his type v_t^i and his initial shock θ^i.*

Proof: For every t, θ, v_t there exists a non-empty set of allocations which maximize the flow of virtual surplus,

$$X^*(t, \theta, v_t) = \arg\max_{x \in X}\{s(t, J(t, \theta, v_t), x)\}$$

$$= \arg\max_{x \in X}\left\{\sum_{j \in N} J^j(t, \theta^j, v_t^j)u^j(t, x^j) - c(x)\right\}.$$

As u^i and c are increasing in x^i it is optimal to set the consumption of agent i to zero, $x^i = 0$, if his virtual utility $J^i(t, \theta^i, v_t^i)$ is negative. As u^i is increasing in x and J^i is increasing in θ^i and v^i by Proposition 2 it follows that the objective function of the principal $\sum_{i \in N} \max\{0, J^i(t, \theta^i, v_t^i)\}u^i(t, x^i) - c(x)$ is super-modular in (θ^i, x^i) and (v_t^i, x^i). By Topkis' theorem, there exists a quantity $x^*(t, \theta, v_t) \in X^*(t, \theta, v_t)$ that maximizes the flow virtual surplus such that the allocation $x^{*i}(t, \theta, v_t)$ of agent i is non-decreasing in θ^i and v_t^i. As the virtual surplus of the principal at time t depends only on t, the initial reports θ, and the type v_t, this flow allocation that conditions only on (t, θ, v_t) is an optimal allocation process:

$$\sup_{(x_t)} \mathbb{E} \left[\int_0^T e^{-rt} s(t, J^i(t, \theta_t^i, v_t^i), x_t) \mathrm{d}t \right]$$

$$= \mathbb{E} \left[\int_0^T e^{-rt} \sup_{x \in X} s(t, J^i(t, \theta_t^i, v_t^i), x_t) \mathrm{d}t \right].$$

$\square$

3.2. *Sufficiency*

To prove incentive compatibility of the optimal allocation process let us first establish a version of a classic result in static mechanism design.

Proposition 4 (Static implementation): *Let $X \subset \mathbb{R}$ and let $V : X \times X \to \mathbb{R}$ be absolutely continuous in the first variable with weak derivative $V_1 : X \times X \to \mathbb{R}_+$ and let V_1 be increasing in the second variable. Then the payment*

$$p(x) = V(x, x) - \int_0^x V_1(z, z) \, dz$$

ensures that truth-telling is optimal: $V(x, x) - p(x) \geq V(x, \hat{x}) - p(\hat{x})$ for all $x, \hat{x} \in X$.

Proposition 4 is similar to Lemma 1 in Pavan *et al.* (2014) and Proposition 2 in Rochet (1987) and differs only in the continuity requirements, namely absolute continuous here rather than Lipschitz continuous there.

In the first step we construct flow payments that make truthful reporting of types optimal (on and off the equilibrium path) if the virtual surplus maximizing allocation process x^* is implemented. Define the payment process $p_t \triangleq p(t, \theta, v_t)$ where the flow payment $p^i : t \times \Theta \times \mathbb{R}^n \to \mathbb{R}$ of agent i is given by:

$$p^i(t, \theta, v_t) \triangleq v_t^i u^i(t, x^{*i}(t, \theta, v_t)) - \int_0^{v_t^i} u^i(t, x^{*i}(t, \theta, (z, v_t^{-i}))) \mathrm{d}z. \quad (3.7)$$

Proposition 5 (Incentive compatible transfers): *In the contract (x^*, p) it is optimal for every agent at every $t > 0$ to report his type v_t^i truthfully, irrespective of the reported shock θ^i and past reported types $(v_s)_{s<t}$.*

Proof: As the allocation $x^*(t, \theta, v_t)$ and the payment $p(t, \theta, v_t)$ at time t are independent of all past reported types $(v_s)_{s<t}$ the reporting problem of

each agent i is time-separable. As u^i is increasing in x, and x^* is increasing in v^i by Proposition 2, we can apply Proposition 4 to

$$(v^i, \hat{v}^i) \mapsto v^i u^i(t, x^*(t, \theta, (\hat{v}^i, v^{-i}))),$$

and so guarantee that the payment scheme $p(t, \theta, v)$ makes truthful reporting of types optimal for all $t, \theta, v, \hat{v}^i$. □

It remains to augment the payments from Proposition 5 with additional payments that make it optimal for the agents to report their initial shocks θ truthfully. As the payments of Proposition 5, see (3.7) ensure truthful reporting of types even after initial misreports, we know how agents will behave even after an initial deviation. This insight transforms the time zero reporting problem into a static design problem in which the payments of Proposition 4 can be used to provide incentives at time $t = 0$.

We define the payment process for agent i as the sum of the flow incentive payment $p^i(t, \theta, v_t)$ and an *annuitized* payment $\pi^i(\theta)$ that depends only on the initial report θ:

$$p_t^{i*} \triangleq p^i(t, \theta, v_t) + \pi^i(\theta) \tag{3.8}$$

where the annuitized payment $\pi^i : \Theta \to \mathbb{R}$ of agent i is given by:

$$\pi^i(\theta) = E \left[\int_0^T \frac{re^{-rt}}{1 - e^{-rt}} \left[\int_0^{v_t^i} u^i(t, x^{*i}(t, \theta, (z, v_t^{-i}))) \mathrm{d}z \right. \right.$$

$$\left. \left. - \int_{\underline{\theta}}^{\theta^i} \phi_\theta^i(t, z, W_t^i) u^i(t, x^{*i}(t, (z, \theta^{-i}), (\phi^i(t, z, W_t^i), v^{-i}))) \mathrm{d}z \right] \mathrm{d}t \right]. \tag{3.9}$$

Theorem 2 (Revenue maximizing contract): *The virtual surplus maximizing contract* (x^*, P^*) *maximizes the revenue of the principal. In the virtual surplus maximizing contract it is optimal for every agent i to report his shock θ^i and type v_t^i truthfully for all $t \geq 0$, irrespective of the reported shocks θ^i and past reported types* $(v_s^i)_{s<t}$.

Proof: We start with the flow payments p^i of Proposition 5 given by (3.8). By construction of the payments each agent reports his type truthfully independent of his initial report θ^i. Let $\hat{V}(\theta^i, \hat{\theta}^i)$ be the agent's value if his true initial shock is θ^i but he reports $\hat{\theta}^i$ and reports truthful after time zero

$$\hat{V}(\theta^i, \hat{\theta}^i) = \mathbb{E}\left[\int_0^T e^{-rt}[v_t^i u^i(t, x^{*i}(t, (\hat{\theta}^i, \theta^{-i}), v_t)) - p(t, (\hat{\theta}^i, \theta^{-i}), v_t)]dt\right].$$

As it is optimal to report v_t^i truthfully we have that

$$\frac{\partial}{\partial v_t^i}(v_t^i u^i(t, x^{*i}(t, (\hat{\theta}^i, \theta^{-i}), v_t)) - p(t, (\hat{\theta}^i, \theta^{-i}), v_t)) = u^i(t, x^*(t, \hat{\theta}^i, \theta^{-i}), v_t)).$$

Thus, the derivative of agent i's value with respect to his initial shock is given by

$$\hat{V}_\theta(\theta^i, \hat{\theta}^i) = \mathbb{E}\left[\int_0^T e^{-rt}[\phi_\theta^i(t, \theta^i, W_t^i)u^i(t, x^{*i}, (t, (\hat{\theta}^i, \theta^{-i}), v_t))]dt\right].$$

As ϕ_θ^i is positive, u^i is increasing in x, and x^{*i} is increasing in $\hat{\theta}^i$ by Proposition 2, Proposition 4 implies that truthful reporting of θ^i is optimal for agent i if he has to make a payment of $\pi^i(\theta)(1 - e^{-rT})/r$ at time zero. As the principal can commit to payments we can transform this payment into a constant flow payment with the same discounted present value by multiplying with $r/(1 - e^{-rT})$. Note, that as the payment $\pi(\theta)$ does not depend on the types it is optimal for the agent to report his types truthfully in the contract (x^*, P^*) where $P_t^* \triangleq p(t, \theta, v_t) + \pi(\theta)$. $\qquad\square$

Theorem 2 describes a revenue-maximizing direct mechanism in which the agents report their types and the principal decides on an allocation and associated transfers at every point in time. The next result shows that in the case of a *single agent* there also exists a simple *indirect* mechanism in the form of a two-part tariff which maximizes the intertemporal revenues of the principal. In this mechanism the agent picks a specific contract at time zero and then chooses how much to consume at every point in time. The price paid by the agent at time t for his consumption x_t at time t depends only the initial contract choice through the fixed payment π and the level of consumption x_t at time t through the variable payment, and thus takes the form of a two-part tariff.

Proposition 6 (Two-part tariff): *With a single agent there exists a revenue-maximizing two-part tariff: at time zero the agent chooses an fixed payment π and then at every point in time t chooses his consumption x_t and associated price $\tilde{p}(t, \pi, x_t)$.*

Proof: Define the set of types such that a given allocation x is optimal at time t

$$V^*(t, \theta, x) = \{v \in \mathbb{R} : x = x^*(t, \theta, v)\}.$$

For every allocation x such that $V^*(t, \theta, x) \neq \emptyset$ there exists at least one type v such that the agent would receive this allocation x if he reported v in the direct mechanism of Theorem 2. The payment of the mechanism described in Theorem 2 depends only on the allocation, but not on the type v. Thus, we have that the following payment implements the virtual surplus maximizing allocation in an indirect mechanism:

$$\tilde{p}(t, \theta, x) = \begin{cases} \inf\{p(t, \theta, v) : v \in V^*(t, \theta, x)\}, & \text{if } V^*(t, \theta, x) \neq \emptyset; \\ \infty, & \text{otherwise.} \end{cases} \quad (3.10)$$

By convention, we assign an arbitrarily large payment, ∞ the choice of an allocation that is never optimal and $V^* = \emptyset$. By Theorem 2, there exists an incentive-compatible flow payment that is constant over time, $\pi(\theta)$ such that the agent reveals his true initial type, θ. Hence we can choose the allocation dependent payment $p(t, \theta, v)$ to depend on the corresponding fixed payment $\pi(\theta)$ and let $p(t, \pi(\theta), x)$ be the consumption dependent payment. If $\pi : \Theta \to \mathbb{R}$ fails to be invertible, then the agent can be offered a choice of menus across all $\tilde{p}(t, \theta', x)$ for all $\theta' \in \Theta$ such that the associated fixed payment $\pi(\theta) = \pi(\theta')$ as given by (3.9). □

The revenue-maximizing mechanism suggested by Proposition 6 is a menu over static contracts. This means that it is sufficient that the payments and allocations at time t depend only on the time t types and the time zero shocks θ.

3.3. *The relation between discrete and continuous time models*

We should emphasize that the basic proof strategy to construct the optimal dynamic mechanism in continuous time mirrors the approach taken in discrete time, see Esö and Szentes (2007) and Pavan *et al.* (2014). As in these earlier contributions, we obtain the first-order conditions by using the envelope theorem using a small class of relevant deviations. Thus, the valuable insights from discrete time carry over to continuous time. Similarly, for the sufficient conditions, we use monotonicity conditions and time separability of the allocations to guarantee that it remains optimal for the agent to report truthfully after any misreport. Here, the continuous time version of the sufficiency arguments have the advantage that they can be expressed directly in terms of the primitives of the stochastic process which we will illustrate in Section 6.

A brief, but more detailed comparison with the discrete time arguments might be instructive at this point. Eső and Szentes (2007) and Pavan *et al.* (2014) show that the additional signals arriving after the initial period can be represented as signals that are orthogonal to the past signals. In the present setting, the type v_t at every point in time is represented as a function ϕ^i of the initial shock θ^i, and an independent time t signal contribution (increment) dW_t, i.e. $v_t = \phi^i(t, \theta^i, W_t^i)$. Our use of consistent deviations is similar to the deviations used in Pavan *et al.* (2014) and Eső and Szentes (2014) where each agent reports the shock W_t after time zero truthfully to establish revenue equivalence.

We can also relate the relevant conditions that guarantee the monotonicity of the type with respect to the initial shock. Indeed, our Assumptions 1 and 2 are closely related to Assumptions 1 and 2 of Eső and Szentes (2007). In particular, we show in Appendix A that our Assumption 1 is implied by Assumption 1 in Eső and Szentes (2007) and thus weaker. Furthermore, Assumption 2 of our setup is exactly equivalent to Assumption 2 in Eső and Szentes (2007). Hence, the basic conditions on the payoffs and the shocks extend the conditions of Eső and Szentes (2007) directly to an environment with many periods and many (flow) allocation decisions.

Pavan *et al.* (2014) observed in the context of a discrete time environment that time-separability of the allocation plus monotonicity of the virtual utility in θ^i and v_t^i is sufficient to ensure strong monotonicity of the virtual surplus maximizing allocation (monotonicity in θ^i and v_t^i after every history). Furthermore, they show that strong monotonicity is sufficient for the implementability of the virtual surplus maximizing allocation (Corollary 1). In Section 5 in the supplementary appendix they use this insight to describe optimal mechanisms for discrete time situations where the private information of the agent is not the initial state of the process, but a parameter influencing the transitions.

As the allocation at time t does not change the set of possible allocations at later times our environment is time-separable. Our assumptions are similar to the assumptions made in the section discussing separable environments in Pavan *et al.* (2014) in the sense that they ensure strong monotonicity which in turn implies implementability of the virtual surplus maximizing allocation.

4. Long-run Behavior of the Distortion

In this section we analyze how the allocative distortions behave in the long-run. We are interested in the expected social welfare generated by the revenue-maximizing allocation compared to the expected welfare generated by the socially optimal allocation. We begin with the following definition and recall that the flow social welfare $s(\cdot)$ is the sum of the flow utilities over all agents, see (2.8).

Definition 4 (Vanishing distortion): The allocative distortion vanishes in the long-run if the social welfare generated by the revenue-maximizing allocation converges to the social welfare generated by the socially optimal allocation as $t \to \infty$:

$$\lim_{t\to\infty} \mathbb{E}[s(t, v_t, x(t, v_t)) - s(t, v_t, x(t, J(t, \theta, v_t)))] = 0.$$

The characterization of the long-run behavior comes in two parts. We first provide sufficient condition for the distortions to vanish in the long-run. Then we provide necessary conditions for persistence of allocative distortions in the long-run in the case of a single agent.

Proposition 7 (Long-run behavior of the distortion): *The following two statements characterize the long-run behavior of the distortions:*

(a) *The distortion vanishes in the long run if the expected type of any initial shock converges to the expected type of the lowest shock, i.e*

$$\lim_{t\to\infty} \mathbb{E}[v_t|\theta^i = x] - \mathbb{E}[v_t|\theta^i = \underline{\theta}] \to 0. \tag{4.1}$$

(b) *If $n = 1$, $u(t, x) = x$, $c(x)$ is twice continuously differentiable, strictly convex with $0 < c''(x) \leq D$ and the expected type for a (non-zero measure) set of shocks does not converge to the expected type of the lowest shock (i.e. (4.1) is not satisfied), then the allocative distortion does not vanish.*

Proof: First note that the difference in the expected type between a random and the lowest initial shock equals

$$\mathbb{E}[v_t] - \mathbb{E}[v_t|\theta^i = \underline{\theta}] = \mathbb{E}[\phi^i(t, \theta^i, W_t^i) - \phi^i(t, \underline{\theta}, W_t^i)]$$

$$= \mathbb{E}\left[\int_{\underline{\theta}}^{\bar{\theta}} \frac{1 - F^i(z)}{f^i(z)} \phi^i_{\theta}(t, z, W^i_t) f(z) \mathrm{d}z\right]$$

$$= \mathbb{E}\left[\frac{1 - F^i(\theta^i)}{f^i(\theta^i)} \phi^i_{\theta}(t, \theta^i, W^i_t)\right].$$

Part (a): We prove that the distortion vanishes if $\lim_{t \to \infty} \mathbb{E}[\frac{1 - F^i(\theta^i)}{f(\theta^i)} \phi^i_{\theta}(t, \theta^i, W_t)] = 0$. We first show that the welfare loss at a fixed point in time can be bounded by the difference between virtual utility $J \in \mathbb{R}^n$ and type $v \in \mathbb{R}^n$

$$s(t, v, x^*(t, v)) - s(t, v, x^*(t, J))$$

$$= \left(\sum_{i \in N} v^i u^i(t, x^{*i}(t, v)) - c(x^*(t, v))\right)$$

$$- \left(\sum_{i \in N} v^i u^i(t, x^{*i}(t, J)) - c(x^*(t, J))\right)$$

$$= \left(\sum_{i \in N} v^i u^i(t, x^{*i}(t, v)) - c(x^*(t, v))\right)$$

$$- \left(\sum_{i \in N} J^i u^i(t, x^{*i}(t, J)) - c(x^*(t, J))\right) - \sum_{i \in N} (v_i - j^i) u^i(t, x^{*i}(t, J))$$

$$\leq \left(\sum_{i \in N} v^i u^i(t, x^{*i}(t, v)) - c(x^*(t, v))\right)$$

$$- \left(\sum_{i \in N} J^i u^i(t, x^{*i}(t, v)) - c(x^*(t, v))\right) - \sum_{i \in N} (v_i - j^i) u^i(t, x^{*i}(t, J))$$

$$= \sum_{i \in N} (v^i - j^i)(u^i(t, x^{*i}(t, v)) - u^i(t, x^{*i}(t, J)).$$

As the set of possible allocations X is compact and u^i is continuous there exists a constant $C > 0$ such that

$$\sum_{i \in N} (v^i - j^i)(u^i(t, x^{*i}(t, v)) - u^i(t, x^{*i}(t, J)) \leq C \sum_{i \in N} (v^i - j^i).$$

Hence the welfare loss resulting from the revenue-maximizing allocation resulting from the revenue-maximizing allocation is linearly bounded by the difference between virtual utility and type. As the difference between v_t^i and J_t^i equals $\frac{1-F^i(\theta^i)}{f^i(\theta^i)}\phi_\theta^i(t,\theta^i,W_t^i)$ it follows that

$$\mathbb{E}[s(t,v_t,x^*(t,v_t)) - s(t,v_t,x^*(t,J_t))] \leq C\,\mathbb{E}\left[\sum_{i\in N}(v^i - J^i)\right]$$

$$= C\,\mathbb{E}\left[\sum_{i\in N}\frac{1-F^i(\theta^i)}{f^i(\theta^i)}\phi_\theta^i(t,\theta^i,W_t^i)\right]$$

$$= C(\mathbb{E}[v_t] - \mathbb{E}[v_t|\theta^i = \underline{\theta}]).$$

Taking the limit $t \to \infty$ gives the result.

Part (b): We prove that the distortion does not vanish in the long run if the expected type of any initial shock does not converge to the expected type of the lowest initial shock. First, we prove that the distortion changes the allocation. As $u^i(t,x) = x$ is linear and c is convex this implies that the function $x \mapsto vx - c(x)$ is concave and has an interior maximizer for every (t,v). This implies that for every point in time t and every type v

$$0 = v - c'(x^*(t,v)).$$

By the implicit function theorem

$$x_v^*(t,v) = \frac{1}{c''(x^*(t,v))} \geq \frac{1}{D}.$$

Intuitively this means that the allocation is responsive to the type v. We calculate the change in social welfare induced by the type v and the virtual valuation J

$$s(t,v,x^*(t,v)) - s(t,v,x^*(t,J))$$

$$= [vx^*(t,v) - c(x^*(t,v))] - [vx^*(t,J) - c(x^*(t,J))]$$

$$= \int_J^v x^*(t,z)\mathrm{d}z - (v-J)x^*(t,J)$$

$$= \int_J^v x^*(t,z) - x^*(t,J)\mathrm{d}z$$

$$\geq \frac{1}{D}\int_J^v (z-J)\mathrm{d}z = \frac{(v-J)^2}{2D}.$$

As the difference between type and virtual utility is given by $\frac{1-F^i(\theta^i)}{f^i(\theta^i)}$ $\phi_\theta^i(t, \theta^i, W_t^i)$ taking expectations yields

$$\mathbb{E}[s(t, v, x(v)) - s(t, v, x(J))] \geq \frac{1}{2D}\mathbb{E}\left[\left(\frac{1-F^i(\theta^i)}{f^i(\theta^i)}\phi_\theta^i(t, \theta^i, W_t^i)\right)^2\right]$$

$$\geq \frac{1}{2D}\mathbb{E}\left[\left(\frac{1-F^i(\theta^i)}{f^i(\theta^i)}\phi_\theta^i(t, \theta^i, W_t^i)\right)\right]^2$$

$$= \frac{(\mathbb{E}[v_t] - \mathbb{E}[v_t|\theta^i = \underline{\theta}])^2}{2D},$$

where the middle step follows from Jensen's inequality. As $\lim_{t \to \infty} \mathbb{E}$ $[v_t|\theta = x] - \mathbb{E}[v_t|\theta^i = \underline{\theta}] \neq 0$ for positive probability set of initial shock x it follows that $\lim_{t \to \infty} \mathbb{E}[v_t] - \mathbb{E}[v_t|\theta^i = \underline{\theta}] \neq 0$.   □

The sufficient condition for the allocative distortion to vanish requires that the conditional expectation of the type v_t at some distant horizon t converges for all initial realizations of the shock, θ, to the conditional expectation of the type v_t given the lowest initial shock $\underline{\theta}$. Clearly, in any model where the initial state θ is the current state of a recurrent Markov process, such as in Battaglini (2005), the sufficient condition will be satisfied as the influence of the initial state on the distribution of the future states of the Markov process is vanishing.

In turn, the failure of the sufficient condition is almost a necessary condition for the allocative distortion to persist. However, in addition we need to guarantee that the allocation problem is sufficiently responsive to the conditional expectation of the agent everywhere. This can be achieved by the linearity and convexity conditions in Proposition 7. We state the necessary conditions only for the problem with a single agent. With many agents, we would have to be concerned with the further complication that the distortion that each individual agent faces may be made obsolete by the distortion faced by the other agents, and thus a more stringent, and perhaps less transparent set of conditions would be required.

5. Repeated Sales

A common economic situation that gives rise to a dynamic mechanism design problem is the repeated sales problem where the buyer is unsure about his future valuation for the good. Examples of such situations are gym membership and phone contracts. At any given point in time the buyer knows how much he values making a call or going to the gym, but he might only have a probabilistic assessment on how much he values the service tomorrow or a year in the future. Usually, it is harder for the buyer to assess how much he values the good at times that are further in the future. Mathematically this uncertainty about future valuations can be captured by modeling the buyer's valuation as a stochastic process.

From the point of view of the seller the question arises whether the uncertainty of the buyer can be used to increase revenues by using a dynamic contract. A variety of dynamic contracts are used, for example for gym memberships and mobile phone contracts, as documented in DellaVigna and Malmendier (2006) or Grubb and Osborne (2015):

1. *Flat Rates* in which the buyer only pays a fixed fee regardless of his level consumption.
2. *Two-Part Tariffs* in which the buyer selects from a menu a fixed fee and a price of consumption. He pays the fixed fee independent of his level of consumption and a unit price for the realized consumption level. Tariffs with higher fixed fees feature lower unit prices of consumption.
3. *Leasing Contracts* in which the buyer selects the length of the lease term and the price charged per unit of time.

While those dynamic contracts can be observed in a wide range of situations, their theoretical properties, surprisingly, have not been widely analyzed. Using a dynamic mechanism design perspective, we can explain why and under what circumstances these specific (and other) features of dynamic contracts and consumption plans might be offered. For the purpose of this section, we consider a single buyer, and hence omit the superscript i. We assume that $u\,(t, x_t) = x_t$ for all t, and the flow utility of the agent is described by

$$v_t \cdot x_t - p_t, \qquad (5.1)$$

and hence v_t immediately represents the willingness-to-pay of the agent in period t.

In the following we describe the revenue-maximizing dynamic contract offered by a monopolistic seller. In general, dynamic contracts could have complicated features as the payments at time t could depend on all the past consumption decisions and messages sent by the agent. However we will show, using the results of the previous section, in particular Proposition 6 that offering a menu of simple static contracts is sufficient to maximize the expected intertemporal revenue.

5.1. *Unknown initial value*

We shall assume that the type $(v_t)_{t\in\mathbb{R}_+}$ of the buyer follows a geometric Brownian motion with zero drift, and possible shifted upwards by $\underline{v} \geq 0$:

$$\mathrm{d}v_t = (v_t - \underline{v})\sigma\mathrm{d}W_t, \tag{5.2}$$

where $(W_t)_{t\in\mathbb{R}_+}$ is a Brownian motion and solution to the above differential equation is given by:

$$v_t = \phi(t, \theta, W_t) = v_0 \exp\left(-\frac{\sigma^2}{2}t + \sigma W_t\right) + \underline{v}. \tag{5.3}$$

The choice of the *shifted geometric Brownian motion* as the type process ensures that the valuation v_t for the good will be greater than $\underline{v}$ at every point in time t. With zero drift, the valuation at time t is the agent's best estimate of his valuation at later times $s > t$:

$$v_t = \mathbb{E}[v_s|v_t].$$

In this subsection, the initial shock θ^i is taken to be the initial valuation of the buyer $v_0 \in (\underline{v}, \infty)$. We assume that the distribution function F is such that

$$v_0 \mapsto \frac{1 - F(v_0)}{f(v_0)v_0} \tag{5.4}$$

is decreasing — a condition that is strictly weaker than the familiar increasing hazard rate condition — and that $f(\underline{v}) \geq 1/\underline{v}$.

At every point in time t the buyer chooses an amount of consumption $x_t \in X \subseteq \mathbb{R}_+$ and pays p_t such that his overall utility equals

$$\mathbb{E}\left[\int_0^\infty e^{-rt}(v_t \cdot x_t - p_t)\mathrm{d}t\right].$$

To evaluate dynamic contracts from the sellers perspective, we assume that the seller faces continuous, non-decreasing production cost $c : X \to \mathbb{R}_+$, such that his overall payoff equals

$$\mathbb{E}\left[\int_0^\infty e^{-rt}(p_t - c(x_t))\mathrm{d}t\right].$$

We can then specialize the form of the revenue-maximizing contract obtained earlier in Theorem 2 to the specific environment of the shifted geometric Brownian motion here. The *stochastic flow* of the shifted geometric Brownian motion is simply

$$\phi_\theta(t, \theta, W_t) = \frac{v_t - \underline{v}}{v_0},$$

and thus the virtual utility, derived earlier in its general form in (3.4), can now be written as:

$$J(t, v_0, v_t) = v_t \left(1 - \frac{1 - F(v_0)}{f(v_0)v_0} \right) + \frac{1 - F(v_0)}{f(v_0)v_0} \underline{v}. \tag{5.5}$$

By Theorem 2 it is then sufficient to verify that the above virtual surplus is increasing in v_0 and v_t to guarantee that a virtual surplus-maximizing contract exists.

Proposition 8 (Virtual utility with geometric Brownian motion): *The virtual utility $J(t, v_0, v_t)$ in the environment of the geometric Brownian motion defined by (5.3)–(5.4) is increasing in $v_0 \in \mathbb{R}_+$ and $v_t \in \mathbb{R}_+$ and a virtual surplus-maximizing contract exists.*

As shown in Theorem 1 the seller aims to maximize

$$\mathbb{E}\left[\int_0^T e^{-rt} (J_t x_t - c(x_t)) \right].$$

In the case of the geometric Brownian motion, the virtual utility, and hence the virtual surplus, are simply a linear function of the type v_t of the agent. The intercept and the slope of the function are determined by the value of the initial shock $\theta = v_0$ and the lower bound $\underline{v}$. We can therefore directly identify an indirect mechanism, a pricing mechanism, that aligns the preferences of the agents with those of the principal. Let us define

$$M(v_0) \triangleq \left(1 - \frac{1 - F(v_0)}{f(v_0)v_0} \right)^{-1}, \tag{5.6}$$

and so we can express the virtual surplus of the principal as follows:

$$J(t, v_0, v_t)x_t - c(x) = \left(v_t \left(1 - \frac{1 - F(v_0)}{f(v_0)v_0} \right) + \frac{1 - F(v_0)}{f(v_0)v_0} \underline{v} \right) x - c(x)$$

$$= M(v_0)^{-1}(v_t x - M(v_0)c(x) + (M(v_0) - 1)x\underline{v}).$$

It follows that a consumption based payment $p_t(v_0, x_t)$ given by

$$p_t(v_0, x_t) \triangleq M(v_0)c(x_t) - (M(v_0) - 1)x_t\underline{v},$$

perfectly aligns the interest of the buyer and the seller, the agent and the principal at every point in time $t > 0$. After all, it leads agent and principal to solve their respective optimality conditions at the same x_t. It remains to prove that it is incentive compatible for the buyer to report his time zero type truthfully. The following results describes optimal contracts (indirect mechanism) for the seller.

Proposition 9 (Revenue maximizing indirect mechanism): *A revenue-maximizing indirect mechanism is given by a menu $(\pi, p(\pi, x_t))$ of membership fees π and consumption prices $p(\pi, x_t)$ of the form*

$$p(\pi, x_t) = M(\pi)c(x_t) - (M(\pi) - 1)\underline{v}x_t. \tag{5.7}$$

Thus, the optimal contract is of the following form: At time zero the seller offers a *menu* of static contracts each consisting of a time independent and recurrent membership fee $\pi \geq 0$, and a consumption dependent payment:

$$p(\pi, x_t) = M(\pi)c(x_t) - [M(\pi) - 1]\underline{v}x_t.$$

The consumption dependent payment p consists of a price of consumption of $M(\pi) \geq 1$, literally the *mark-up*, and a linear consumption discount $(M(\pi) - 1)\underline{v}x_t$. If the buyer accepts a contract he has to pay a recurring membership fee $\pi \geq 0$ independent of his consumption. At the same time he has to pay $p(\pi, x_t)$ depending on his consumption x_t in period t such that his overall payment at time t equals

$$p_t = \pi + p(\pi, x_t) = \pi + M(\pi)c(x_t) - [M(\pi) - 1]\underline{v}x_t. \tag{5.8}$$

The optimal fixed fee $\pi(v_0)$ that is chosen by the agent at the beginning of the contracts depends on the agent's initial valuation v_0. It will be such that

$$M(\pi(v_0)) = \left(1 - \frac{1 - F(v_0)}{f(v_0)v_0}\right)^{-1}.$$

With the general characterization of the optimal contract given by Proposition 9 we next establish under what conditions on the nature of the private information and the cost of delivering the service $c(x)$ the above mentioned contract features will arise as a part of an optimal contract.

5.1.1. *Flat rate contracts*

In a flat rate contract the payment $p_t = \pi$ is constant over time and independent of the level of consumption chosen by the buyer. Suppose that the set of possible allocation is given by $X = [0,1]$, and that the minimal valuation $\underline{v}$ equals zero. Assume that the cost of production $c(x)$ is constant and normalized to zero, $c(x) = 0$. As the buyer's utility given by (5.1) increases linearly in the level of the consumption, he will always want to consume the good at the maximal possible level if he faces a flat rate. A direct consequence of the transfers described in (5.8) is the following result characterizing an optimal mechanism with zero (marginal) cost of production: The optimal mechanism is a flat rate where every agent who accepts the contract at time zero, makes a constant flow payment, independent of his consumption, and consumes the maximal possible amount: $x_t = 1$.

Now, if his current valuation v_t is below the flat rate $p_t = \pi$, not only is his current flow of utility negative, but so is his expected continuation utility of the contract:

$$\mathbb{E}\left[\int_t^\infty e^{-rs}(v_s - p_s)dt\,|\,v_t\right] = \frac{v_t - \pi}{r}. \tag{5.9}$$

As a consequence of condition (5.9) only the agent with an initial valuation $v_0 \geq \pi$ will accept the contract. All agents with an initial valuation $v_0 < \pi$ reject the contract and never consume the good no matter how high the consumption utility is at times $t > 0$.

5.1.2. *Two-part tariffs*

With constant cost of production, the optimal contract leads to flat rate tariffs. We next consider the case of increasing and convex costs. We maintain the assumption that the minimal valuation $\underline{v}$ equals zero and assume that the cost function $c(x)$ is strictly increasing and convex for $x \in \mathbb{R}_+$.[3] In particular, we allow for a constant but positive marginal cost of production. By condition (5.8) a two-part tariff where the agent pays π independent of his consumption and $M(\pi)c(x)$ depending on his consumption x is a revenue-maximizing contract for the principal. It is

---

[3]As we established the revenue equivalence theorem under the assumption that the quantity x is bounded we understand the model with unbounded quantities as the limit of optimal mechanisms when the bound on x converges to infinity.

worth emphasizing that a simple menu of static two-part tariffs can hence maximize the expected dynamic revenue of the principal.

We illustrate the structure of the two-part tariff with the following quadratic cost function $c(x) = x^2/2$ and assume that the initial valuation v_0 is exponentially distributed with mean μ:

$$F(v_0) = 1 - \exp\left(-\frac{v_0}{\mu}\right),$$

which will allow us to explicitly compute the terms of the contract. By Proposition 9, we know that if the agent decided on a contract $(\pi, M(\pi))$ then the optimal consumption of the agent at time t is given by

$$\{x_t\} = \arg\max_{x \geq 0} \left(x\, v_t - M(\pi)\frac{x^2}{2}\right) = \frac{v_t}{M(\pi)}. \tag{5.10}$$

Hence, the agent's expected time zero utility from the contract $(\pi, M(\pi))$ is

$$\max_{(x_t)t \in \mathbb{R}_+} \mathbb{E}\left[\int_0^\infty e^{-rt}(v_t x_t - \pi - M(\pi)c(x_t))\right]$$

$$= \mathbb{E}\left[\int_0^\infty e^{-rt}\left(\frac{v_t^2}{2M(\pi)} - \pi\right)\right]$$

$$= \frac{v_0}{2M(\pi)(r - \sigma)} - \frac{\pi}{r}. \tag{5.11}$$

Hence, the agent chooses his optimal contract at time 0 by maximizing his expected net utility (5.11) over π to select a contract $(\pi, M(\pi))$ based only on his time zero valuation v_0. Let us denote by $\pi(v_0)$ the fixed fee chosen by the agent of initial valuation v_0. By Proposition 8, in the optimal contract the mark-up $M(\pi(v_0))$ computed earlier in its general form as (5.6) is given as a function of the initial valuation v_0 by:

$$M(\pi(v_0)) = \begin{cases} \dfrac{v_0}{v_0 - \mu}, & \text{if } v_0 \geq \mu, \\[2mm] \infty, & \text{otherwise.} \end{cases} \tag{5.12}$$

Hence every buyer who has initially a valuation v_0 below the average time zero valuation μ will be excluded and never consume the good no matter how high his future valuation is. The mark-up decreases and converges to one, and hence the socially efficient allocation as $v_0 \to \infty$. Since the mark-up in the incentive compatible indirect mechanism has to satisfy

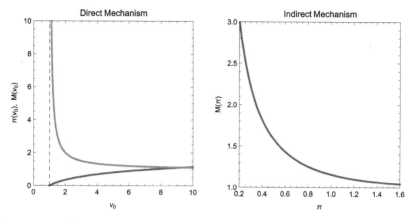

Figure 15.1.   The two-part tariff in the direct mechanism (left) and in the indirect mechanism (right). The markup $M(v_0)$ is in red and the membership fee $\pi(v_0)$ in blue. (For interpretation of the colors in this figure, the reader is referred to the web version of this article.)

(5.12) we can compute the membership fee π to be paid by the buyer with initial valuation v_0 from the incentive compatibility at time $t = 0$ based on (5.11) and get:

$$\pi(v_0) = \frac{r\mu}{2(r-\sigma)} \left(\ln \frac{v_0}{\mu} \right).$$

We illustrate the nature of the optimal tariff in Fig. 15.1. The left panel expresses the membership fee π and the mark-up $M(\pi)$ in the two-part tariff as a function of the initial type (blue and red curve respectively). The right panel incentive compatible choice describes the equilibrium trade-off between the level of the membership π and the mark-up $M(\pi)$ which is given by:

$$M(\pi) = \left[1 - \exp\left(\frac{-2\pi(r-\sigma)}{r\mu} \right) \right]^{-1}.$$

It follows that a lower mark-up $M(\pi)$ is purchased at the expense of higher membership fee π. In the optimal contract, an agent with a higher initial valuation is willing to purchase the rights to lower mark-up by means of higher membership fee. In consequence, an agent with a higher membership fee faces less distortion with respect to his flow consumption decision.

In Fig. 15.2 we illustrate how the intertemporal distortions induced by the optimal contract influence the consumption choices over time. The solid lines are two path realizations of the geometric Brownian motion without

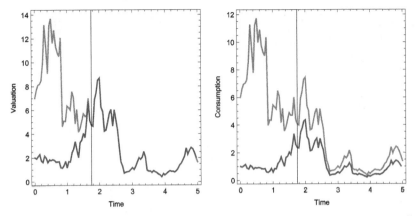

Figure 15.2. Consumption path and valuation if the initial valuation v_0 is exponentially distributed with mean one and the valuation evolves as a geometric Brownian motion without drift. (For interpretation of the colors in this figure, the reader is referred to the web version of this article.)

drift. One path starts at an initial valuation of $v_0 = 2$ (blue) and one at $v_0 = 7$ (red). It so happens that both of these paths coincide after time $t = 1.75$. Now the dashed lines represent the consumption levels in the revenue-maximizing contract. Note, that even after the valuations coincide in the sample path, the consumption levels associated with different initial valuations differ. In particular, the optimal consumption level react with differential intensity to changes in the valuations. By contrast, the consumption of the agent in the social welfare maximizing allocation would exactly equal his valuation.

5.1.3. *Free minute contract*

We now consider the case in which the minimal type $\underline{v}$ of the agent is strictly positive and that the density at the minimal valuation is bounded away from zero, or $f(\underline{v}) > 1/\underline{v}$. In addition we assume that the marginal cost of providing the good vanishes for small quantities, i.e. $c'(0) = 0$. When the agent decides how much to consume at time t he solves the maximization problem:

$$\max_x \{xv_t - (\pi + M(\pi)c(x) - (M(\pi) - 1)\underline{v}x)\}.$$

This leads to the first-order condition:

$$0 = v_t - M(\pi)c'(x) + (M(\pi) - 1)\underline{v} \Leftrightarrow c'(x) = \underline{v} + \frac{(v_t - \underline{v})}{M(\pi)}.$$

As the marginal cost of providing the good vanishes if the quantity goes to zero it follows that the consumption of the agent is bounded from below at every point in time by $c'^{-1}(\underline{v})$. Hence we can interpret the amount $c'^{-1}(\underline{v})$ as a quantity provided to the agent for free. This is a feature that is common in mobile phone contracts. In such a contract the agent can consume a certain number of minutes for free and only has to pay for the consumption exceeding this amount.

5.2. *Unknown drift*

We now return to the cost structure that lead to the flat rate contract earlier, namely, the cost of production is constant and normalized to zero and $x_t \in [0,1]$. Different from the preceding analysis we now assume that the initial private information θ of the agent is about the drift of the geometric Brownian motion, or $\mu = \theta$, rather than the initial value of the geometric Brownian motion v_0 which we now assume to be public information. We set the lower bound on valuations $\underline{v}$ to zero. In consequence, the valuation v_t now evolves according to:

$$\mathrm{d}v_t = v_t(\theta \mathrm{d}t + \sigma \mathrm{d}W_t),$$

and the solution to the above differential equation is given by:

$$v_t = \phi(t, \theta, W_t) = v_0 \exp\left(\left(\theta - \frac{\sigma^2}{2}\right)t + \sigma W_t\right). \tag{5.13}$$

The derivative of the type $v_t = \phi(t, \theta, W_t)$ with respect to θ, the *generalized stochastic flow*, now equals:

$$\phi\theta = \phi t,$$

and the virtual utility is now given by:

$$J(t, \theta, v_t) = v_t\left(1 - \frac{1 - F(\theta)}{f(\theta)}t\right). \tag{5.14}$$

Interestingly, the distortion is still formed on the basis of a multiplicative handicap, but now the handicap factor is increasing linearly in time as expressed by the second term of the virtual utility. It follows that in contrast to the above cases of unknown initial value, the distortion is now growing over time. As v_t is positive, it follows that the virtual valuation is strictly positive until a deterministic time T is reached which is precisely given by the hazard rate:

$$L(\theta) = \frac{f(\theta)}{1 - F(\theta)},$$

and thereafter the virtual utility turns negative. Thus, the allocation of the object to agent i ends with probability one at time $L(\theta)$. The optimal contract can now be implemented by a constant leasing payment $p(\theta)$ the agent makes at every time $t \in [0, L(\theta)]$.

Proposition 10 (Leasing contract): *The revenue-maximizing mechanism allocates the object to the agent with initial shock θ if and only if $t \in [0, L(\theta)]$ and requires a payment of*

$$p(t,\theta) = \begin{cases} \dfrac{r}{1 - e^{-rL(\theta)}} \left(v_0 \dfrac{e^{(\theta-r)L(\theta)} - 1}{\theta - r} \right. \\[2ex] \qquad \left. - \displaystyle\int_0^\theta \dfrac{e^{(z-r)L(z)}[L(z)(z-r) - 1]}{(z-r)^2} dz \right) & \text{if } t \in [0, L(\theta)]; \\[2ex] 0, & \text{otherwise.} \end{cases}$$

To establish the above formula for the payments we calculate the expected value that the agent with initial shock θ derives from getting the object until time $L(\theta)$. By the envelope theorem the payment equals this value minus the integral over the marginal value of those types with a lower initial shock.

In a recent paper, Boleslavsky and Said (2013) derive the revenue-maximizing contract in a discrete time setting where the private information of a single agent is the uptick probability of a multiplicative random walk. As it is well known, the geometric Brownian motion can be viewed as the continuous time limit of the discrete time multiplicative random walk stochastic process. Thus, it is interesting to compare their results to the implications following our analysis. In terms of the private information of the agent, the unknown drift in the geometric Brownian motion here represents the unknown uptick probability analyzed in Boleslavsky and Said (2013). As the general convergence result of the stochastic process itself would suggest, we can also establish, see Appendix A for the details, that the continuous time limit of the virtual valuation derived in Boleslavsky and Said (2013) is the virtual utility derived above by (5.14). However, in the continuous time limit the expression for the virtual utility, see

(5.14), becomes notably easier to express and to interpret. The analysis in Boleslavsky and Said (2013) explicitly verifies the validity of the incentive constraints in the case of a single agent. With the general approach taken here, we can obtain sufficient conditions for the revenue optimal contract and associated allocations even in the presence of many agents. In fact, the next section considers such an allocation problem, namely the allocation of a single unit among competing bidders. This second class of allocation problems is notably more restrictive in terms of the cost of providing the service, namely constant for a single unit, but allow us to obtain some novel insight regarding the structure of the intertemporal distortion with many agents.

6. Sequential Auctions and Distortions

We illustrate the impact that the structure of the private information has on the intertemporal policies and the allocative distortion within the context of a sequential auction model. The allocation problem is as follows. At every point in time t, the owner of a single unit of a, possibly divisible, object wishes to allocate it among the competing bidders, $i = 1, \ldots, n$. The allocation space is given at every instant t by $x_t^i \in [0, 1]$ and $\sum_{i=1}^n x_t^i \leq 1$. The marginal cost of providing the object is constant and normalized to zero. The flow utility of each agent i is given by $v_t^i \cdot x_t^i - p_t^i$.

We can interpret the allocation process as a process of intertemporal licensing where the current use of the object is determined on the basis of the past and current reports of the agents. In particular, the assignment of the object can move back and forth between the competing agents. Alternatively, the description of the valuation could be rephrased as a description of the marginal cost of producing a single good, and the associated allocation process is the solution to a long-term procurement contract with competing producers. As in the static theory of optimal procurement, the virtual utility would then be replaced by the virtual cost, but the structure of the allocation process would remain intact.

6.1. *Arithmetic Brownian motion*

In the previous section we represented the valuation process by a geometric Brownian motion, now we consider the arithmetic Brownian motion, thus indicating the versatility of the current approach. The arithmetic Brownian motion v_t^i is completely described by its initial value v_0^i, the drift μ^i and the variance σ^i of the diffusion process W_t. The willingness to pay of agent

i evolves according to:

$$dv_t^i = \mu^i dt + \sigma^i dW_t^i,$$

and the willingness-to-pay of agent i, v_t^i, is:

$$v_t^i = v_0^i + \mu^i t + \sigma^i W_t^i. \tag{6.1}$$

We analyze the incentive problem when either one of the three determinants of the Brownian motion, the initial value, the drift or the variance is unknown, whereas the remaining two are commonly known. Surprisingly, we find that even though we consider the same stochastic process, the nature of the private information, i.e. about which aspect of the process the agent is privately informed, has a substantial impact on the optimal allocation. In particular, we find that the distortion is either constant, increasing or random (and increasing in expectation) depending on the precise nature of the private information.

Unknown initial value — constant distortion. We begin with the case where the initial value of the Brownian motion, $v_0^i = \theta^i$, is private information to agent i, as are all future realizations v_t^i. In contrast, the drift μ^i and the variance σ^i of the Brownian motion are assumed to be commonly known. Given the representation of the Brownian motion (6.1), we have

$$v_t^i = \phi^i(t, \theta^i, W_t^i) = \theta^i + \mu^i t + \sigma^i W_t^i. \tag{6.2}$$

The partial derivative of ϕ^i with respect to θ^i is given by $\phi_\theta^i = 1$ and thus the virtual utility is given by:

$$J^i(t, \theta^i, v_t^i) = v_t^i - \frac{1 - F^i(\theta^i)}{f^i(\theta^i)}, \tag{6.3}$$

and the distortion imposed by the revenue-maximizing mechanism is *constant over time*. In every period, the object is allocated to the agent i_t^* with the highest virtual utility, provided that the valuation is positive. Thus, the allocation proceeds by finding the bidder with the highest valuation, after taking into account a handicap, that is determined once and for all through the report of the initial shock.

Earlier, we gave a general description of the payments decomposed into an annualized upfront payment π and a flow payment p_t. In the present auction environment, we can give an explicit description of the flow payments in terms of the virtual utility of the agents. The associated flow transfer of the bidders, p_t^i follows directly from the logic of the second price auction:

$$p_t^i = \begin{cases} \max_{j \neq i} \left\{ v_t^j - \dfrac{1 - F^j(\theta^j)}{f^j(\theta^j)} \right\} + \dfrac{1 - F^i(\theta^i)}{f^i(\theta^i)}, & \text{if } i = i_t^*; \\ 0, & \text{if } i \neq i_t^*. \end{cases} \qquad (6.4)$$

Thus, it is only the winning bidder who incurs a flow payment. By rewriting (6.4), we find that the winning bidder has to pay his valuation, but receives a discount, namely his information rent, which is exactly equal to the difference in the virtual utility between the winning bidder and the next highest bidder, i.e.

$$p_t^{i*} = v_t^{i*} - \left(J^{i*}(t, \theta^{i*}, v_t^{i*}) - \max_{j \neq i*}\{J^j(t, \theta^j, v_t^j)\} \right). \qquad (6.5)$$

By construction of the transfer function, the flow net utility of the bidder is positive whenever he is assigned the object, as

$$v_t^{i*} \geq v_t^j - \dfrac{1 - F^j(\theta^j)}{f^j(\theta^j)} + \dfrac{1 - F^i(\theta^{i*})}{f^i(\theta^{i*})}, \qquad (6.6)$$

and thus, the flow allocation proceeds as a "handicap" second price auction, where the price of the winner is determined by the current value of the second highest bidder, as measured by the virtual utility. The "handicap" is computed as the difference between the constant handicap of the current winner and the current second highest bidder. The above version of the handicap auction appeared in Eső and Szentes (2007) in a two period model of a single unit auction. Similarly, Board (2007) develops a handicap auction in a discrete time, infinite horizon model, but one in which the object is allocated only once — at an optimal stopping time. There the handicap is represented — as it is here — by the constant terms, $(1 - F^j(\theta^j))/f^j(\theta^j)$ and $(1 - F^i(\theta^i))/f^i(\theta^i)$, but the second highest value is computed as the continuation value of the remaining bidders, as in Bergemann and Välimäki (2010).

Unknown drift — increasing distortion. We now consider the case where the initial private information is the drift of the Brownian motion. Let $v_t^i \in \mathbb{R}_+$ be an arithmetic Brownian motion with drift θ^i and known variance σ^i and known initial value, v_0^i:

$$v_t^i = \phi^i(t, \theta^i, W_t^i) = v_0^i + \theta^i t + \sigma^i W_t^i. \tag{6.7}$$

The derivative of the valuation ϕ^i with respect to the initial private information θ^i, which is now the drift of the Brownian motion, is given by $\phi_\theta^i = t$. Thus the virtual utility is now:

$$J^i(t, \theta^i, v_t^i) = v_t^i - \frac{1 - F^i(\theta^i)}{f^i(\theta^i)} t. \tag{6.8}$$

The flow payment is of exactly the same form as (6.5), and the virtual utility function is given by (6.8). The distortion is still formed on the basis of the handicap, by the inverse hazard rate $(1 - F(\theta^i))/f(\theta^i)$, but now the handicap is *increasing linearly in time*. In contrast to the case of the unknown starting value, the distortion is *growing* deterministically over time, rather than vanishing over time. Since v_t^i might be growing as well, the deterministic increase in the distortion does not allow us to conclude that the assignment of the object is terminated with probability one at some finite time T, a conclusion that we arrived earlier in Section 5 where we considered the geometric Brownian motion.

Unknown variance — random distortion. We conclude the analysis with the case of unknown variance and the valuation v_t^i then evolves according to:

$$v_t^i = \phi^i(t, \theta^i, W_t^i) = v_0^i + \mu^i t + \theta^i W_t^i. \tag{6.9}$$

Now, the initial private information θ^i represents the volatility of the Brownian motion. The derivative of the valuation ϕ^i with respect to the initial private information θ^i now takes the form:

$$\phi_\theta^i = \frac{\phi^i - v_0^i - \mu^i t}{\theta^i} = W_t^i$$

In consequence the virtual utility of agent i can be expressed as:

$$\begin{aligned}
J^i(t, \theta^i, v_t^i) &= v_t^i - \frac{1 - F^i(\theta^i)}{f^i(\theta^i)} W_t^i \\
&= v_t^i - \frac{1 - F^i(\theta^i)}{f^i(\theta^i)} \frac{v_t^i - v_0^i - \mu^i t}{\theta^i} \\
&= v_t^i \left(1 - \frac{1 - F^i(\theta^i)}{f^i(\theta^i)\theta^i}\right) + \frac{1 - F^i(\theta^i)}{f^i(\theta^i)\theta^i}(v_0^i + \mu^i t). \tag{6.10}
\end{aligned}$$

We observe from the first line of the above virtual utility that the expected virtual utility equals the initial value v_0^i for any time zero shock θ^i. The variance of the Brownian motion does not lend itself to an ordering along the first-order stochastic dominance criterion, rather it is ordered by second-order stochastic dominance. Formally, in the case of unknown variance ϕ^i is not increasing in θ_i and fails Assumption 2. But as those assumptions were only used to establish that the virtual utility is increasing in θ^i, v_t^i if it takes positive values, we can dispense with them here as we can ensure monotonicity here by requiring that μ^i, $v_0^i \leq 0$.

The basic proof idea is to use the convexity of the objective function to guarantee that an increase in variance leads to an increase in the expected (virtual) valuation. After all, if the virtual utility turns negative, the seller does not want to assign the object to the buyer, thus the revenue is flat and equal to zero. It therefore follows that the revenue of the seller has a convex like property. But in contrast to the utility of the buyer, which is linear in v_t^i, and hence strictly convex if truncated below by zero, the virtual surplus of the seller has additional terms, as displayed by (6.10) which need to be controlled to guarantee the monotonicity of the virtual utility. From the expression of the virtual utility function we can immediately derive sufficient conditions for the monotonicity. Thus if we assume that the initial value v_0^i is negative, $v_0^i \leq 0$, and the arithmetic Brownian motion has a negative drift $\mu^i \leq 0$, then we are guaranteed that the convexity argument is sufficiently strong.

Formally, let $\hat{\theta}^i$ be the solution to $\hat{\theta}^i - \frac{1-F^i(\hat{\theta}^i)}{f^i(\hat{\theta}^i)} = 0$. As

$$J^i(t, \theta^i, v_t^i) \leq v_t^i \left(1 - \frac{1 - F^i(\theta^i)}{f^i(\theta^i)\theta^i}\right)$$

the virtual utility $J^i(t, \theta^i, v_t^i)$ is only positive if the valuation v_t^i is negative, for all $\theta^i < \hat{\theta}^i$. But this implies that the gross expected utility of all agents with initial shock $\theta^i < \hat{\theta}^i$ is negative, and hence they cannot generate a nonnegative revenue due to the ex-ante participation constraint, and hence, it can never be optimal to allocate to an agent with variance $\theta^i < \hat{\theta}^i$. Thus, we ignore agents with low variance $\theta^i < \hat{\theta}^i$ and never allocate the object to them. As $\frac{1-F^i(\theta^i)}{f^i(\theta^i)}$ is decreasing we have that $1 - \frac{1-F^i(\theta^i)}{f^i(\theta^i)\theta^i} > 0$ for all $\theta^i > \hat{\theta}$ and hence $J^i(t, \theta^i, v_t^i)$ is increasing in v_t^i and θ^i for all $v_t^i > 0, \theta^i > \hat{\theta}^i$. Hence, by the argument of Proposition 5, there exists a payment such that truthful reporting of valuations becomes optimal irrespective of the reported types.

As the virtual utility

$$J^i(t, \theta^i, \phi^i(t, \theta^i, W_t^i)) = W_t^i \left(\theta^i - \frac{1 - F^i(\theta^i)}{f^i(\theta^i)} \right) + \mu^i t + v_0^i$$

is increasing in θ^i whenever $W_t^i > 0$ and decreasing whenever $W_t^i < 0$ it follows that the product

$$W_t^i u^i(t, x^{*i}(t, (\hat{\theta}^i, \theta^{-1}), v_t^i))$$

is increasing in the reported shock $\hat{\theta}^i$. The derivative of the agents utility with respect to his initial shock simplifies to

$$\mathbb{E} \left[\int_0^T e^{-rt} [W_t^i u^i(t, x^{*i}(t, (\hat{\theta}^i, \theta^{-i}), v_t^i))] dt \right]$$

and thus, by the argument of Theorem 2, the virtual surplus maximizing allocation for the shocks $\theta^i > \hat{\theta}^i$ is incentive compatible.

The last two examples emphasize that our approach can accommodate not only private information about the initial state of a random process, but also private information about a parameter of the stochastic process, such as the mean or the variance of the process.

6.2. *Ornstein–Uhlenbeck process*

Finally, we describe the implications for the revenue-maximizing allocation if the stochastic process is given by the Ornstein–Uhlenbeck process, which is the continuous-time analogue of the discrete-time AR(1) process. This example is closely connected to the discrete time literature. Besanko (1985) showed that the distortions induced by the discrete-time AR(1) process vanish for privately known initial values of the process if and only if the process is mean-reverting. Furthermore, the AR(1) process, was the leading example in the analysis of the impulse response function in Pavan *et al.* (2014).

The Ornstein–Uhlenbeck process v_t^i is completely described by its initial value v_0^i, the mean reversion level μ, the mean reversion speed $M \geq 0$ and the variance $\sigma \geq 0$ of the diffusion process B_t. The willingness to pay of agent i evolves according to the stochastic differential equation:

$$dv_t^i = m(\mu - v_t) dt + \sigma dB_t^i,$$

where B_t is a standard Brownian motion. The Ornstein–Uhlenbeck process can be represented using a distinct Brownian motion $\tilde{B}$ as:

$$v_t = v_0 e^{-mt} + \mu(1 - e^{-mt}) + \frac{\sigma e^{-mt}}{\sqrt{2m}} \tilde{B}_{2mt-1}. \tag{6.11}$$

Hence we can define the process W as a time-changed Brownian motion by

$$W_t^m = \frac{e^{-mt}}{\sqrt{2m}} \tilde{B}_{2mt-1}.$$

Using W we can represent the valuation of the agent as

$$v_t = v_0 e^{-mt} + \mu(1 - e^{-mt}) + \sigma W_t^m.$$

Unknown initial value. Consider the case where the valuation process is an Ornstein–Uhlenbeck process and the initial valuation is private information, i.e. $v_0^i = \theta^i$. Given the representation (6.11) it follows that

$$\frac{\partial \phi^i(t, \theta^i, W_T^i)}{\partial \theta^i} = e^{-mt}.$$

Thus, Assumptions 1 and 2 are satisfied. The virtual utility J^i equals

$$J^i(t, \theta^i, v_t^i) = v_t^i - \frac{1 - F^i(\theta^i)}{f^i(\theta^i)} e^{-mt}.$$

Hence the optimal mechanism is a handicap mechanism with a deterministic handicap that is exponentially decreasing over time. As the Ornstein–Uhlenbeck process converges to a stationary distribution which is independent of the starting value θ^i, Proposition 7 applies and the distortion vanishes in the long run. Intuitively the initial valuation does not change the expected valuation in the long run.

Unknown long run average. We can also take a parameter of the stochastic process to be the private information of the agent, that is we can take the expected long run average of the process to be the private information of agent i, i.e. $\mu = \theta^i$. Given the representation (6.11) it follows that

$$\frac{\partial \phi_t^i}{\partial \theta^i} = 1 - e^{-mt}.$$

Thus, Assumptions 1 and 2 are satisfied. The virtual utility J^i equals

$$J^i(t, \theta^i, v_t^i) = v_t^i - \frac{1 - F^i(\theta^i)}{f^i(\theta^i)} (1 - e^{-mt}).$$

Hence the optimal mechanism is a handicap mechanism with a deterministic handicap that is increasing over time. As the Ornstein–Uhlenbeck process converges in the long run to a stationary distribution which depends on the long run average θ^i the distortion increases in the long run. Intuitively, the expected valuation converges to the long run average θ^i, and so does the virtual utility, it converges to the long rune average of the virtual utility as well. In a notable recent contribution, Skrzypacz and Toikka (forthcoming) consider dynamic mechanisms for repeated trade under private information. In particular they analyze the discrete time version of the mean-reverting process in which the persistence of the stochastic process is private information, the equivalent of the mean-reversion speed m here. They establish that the allocative distortion is increasing over time rather than decreasing as it is when the initial state of the stochastic process constitutes the private information.

7. Conclusion

We analyzed a class of dynamic allocation problems with private information in continuous time. In contrast to much of the received literature in dynamic mechanism design, the private information of each agent was not restricted to the current state of the Markov process. In particular, the initial private information was allowed to pertain to a one-dimensional parameter of the stochastic process such as the drift of the arithmetic or geometric Brownian motion, or the speed of the mean-reverting process. By allowing for a richer class of private information structures, we gained a better understanding about the nature of the distortion due the private information. In contrast to the settings where the private information always pertains to the state of the Markov process and where the distortions induced by the revenue-maximizing allocation are typically vanishing over time, we have shown that the distortion can be constant, increasing or decreasing over time. The analysis of the private information in terms of the stochastic flow, the equivalent of the impulse response functions in continuous time, allowed us directly link the nature of the private information to the nature of the intertemporal distortion.

A distinct advantage of the continuous and time-separable approach taken here is that we could offer explicit solutions, in terms of the optimal allocation, the level of distortion and the transfer payments. We highlighted this advantage in the analysis of the repeated sales environment in which we gave complete, explicit and surprisingly simple solutions to a class of sales or licensing problems. In particular, we showed that we can implement the dynamic optimal contract by means of an essentially static contract, a membership contract, that displayed such common empirical features as flat rates, free consumption units and two-part tariffs.

Appendix A

Proof of Proposition 2. As there is no risk of confusing agents we drop the upper indices in the proof and denote by (θ, v) the type and the type of agent i. Assume that the virtual utility is positive $J(t, \theta, v) > 0$. We first prove the monotonicity in v and than in θ.

Part 1: $J(t, \theta, v) > 0 \Rightarrow J_v(t, \theta, v) \geq 0$. Note that

$$
J(t, \theta, v) = v - \frac{1 - F(\theta)}{f(\theta)} \phi_\theta^i(t, \theta, w(t, \theta, v))
$$

$$
= v \left(1 - \frac{1 - F(\theta)}{f(\theta)} \frac{\phi_\theta^i(t, \theta, w(t, \theta, v))}{\phi^i(t, \theta, w(t, \theta, v))} \right).
$$

As $\phi_\theta^i > 0$ it follows that $J(t, \theta, v) \leq v$ and hence $v \geq 0$. Consequently the second term needs to be positive as well. Clearly, $v \mapsto v$ is non-decreasing. As ϕ_θ^i / ϕ^i is decreasing in w by (2.4) and $w(t, \theta, v)$ is increasing in v, so the second term is increasing in v.

Part 2: $J(t, \theta, v) > 0 \Rightarrow J_\theta(t, \theta, v) \geq 0$. It remains to prove that the virtual utility $J(t, \theta, v) = v - \frac{1 - F(\theta)}{f(\theta)} \phi_\theta^i(t, \theta, w(t, \theta, v))$ is non-decreasing in θ. First, note that $\frac{1 - F(\theta)}{f(\theta)}$ is decreasing in θ by assumption. Second, note that $0 = \phi_\theta^i + \phi_w^i w_\theta$ and hence

$$
\frac{\partial}{\partial \theta} \phi_\theta^i(t, \theta, w(t, \theta, v))
$$

$$
= \phi_{\theta\theta}^i(t, \theta, w(t, \theta, v)) + \phi_{\theta w}^i(t, \theta, w(t, \theta, v)) w_\theta(t, \theta, v)
$$

$$
= \phi_{\theta\theta}^i(t, \theta, w(t, \theta, v)) - \phi_{\theta w}^i(t, \theta, w(t, \theta, v)) \frac{\phi_\theta^i(t, \theta, w(t, \theta, v))}{\phi_w^i(t, \theta, w(t, \theta, v))}.
$$

Now we replace $w(t, \theta, v)$ by w and prove that the derivative is negative for any $w \in \mathbb{R}$:

$$
= \phi_\theta^i(t, \theta, w) \left(\frac{\phi_{\theta\theta}^i(t, \theta, w)}{\phi_\theta^i(t, \theta, w)} - \frac{\phi_{\theta w}^i(t, \theta, w)}{\phi_w^i(t, \theta, w)} \right)
$$

$$= \phi_\theta^i(t,\theta,w)\left(\frac{\partial}{\partial\theta}\log(\phi_\theta^i(t,\theta,w)) - \frac{\partial}{\partial\theta}\log(\phi_w^i(t,\theta,w))\right)$$

$$= \phi_\theta^i(t,\theta,w)\frac{\partial}{\partial\theta}\log\left(\frac{(\phi_\theta^i(t,\theta,w)}{(\phi_w^i(t,\theta,w)}\right)$$

$$\leq 0.$$

The last step follows as $\frac{\phi_\theta^i(t,\theta,w)}{\phi_w^i(t,\theta,w)}$ is decreasing in θ by (2.5), and so the logarithm is decreasing as well. □

Proof of Proposition 4. We have that

$$V(x,\hat{x}) - p(\hat{x}) = V(x,\hat{x}) - V(\hat{x},\hat{x}) + \int_0^{\hat{x}} V_1(z,z)\mathrm{d}z$$

$$= \int_{\hat{x}}^x V_1(z,\hat{x})\mathrm{d}z + \int_0^{\hat{x}} V_1(z,z)\mathrm{d}z$$

$$= \int_{\hat{x}}^x V_1(z,\hat{x}) - V_1(z,z)\mathrm{d}z + \int_0^{\hat{x}} V_1(z,z)\mathrm{d}z$$

$$\leq \int_0^x V_1(z,z)\mathrm{d}z = V_1(x,x) - p(x). \qquad\square$$

Proof of Proposition 8. We begin with the case where $\underline{v} = 0$. Note that in this case Assumptions 1 and 2 are satisfied and thus Proposition 2 yields the monotonicity of the virtual valuation $J(t,v_0,v_t)$ in v_0 and v_t conditional on $J_t \geq 0$.

If $\underline{v}$ is greater than zero it follows from $f(\underline{v}) > 1/\underline{v}$ and the monotonicity of $\frac{1-F(v_0)}{f(v_0)v_0}$ that for all $v_0 \geq \underline{v}$

$$1 - \frac{1 - F(v_0)}{f(v_0)v_0} > 0.$$

Hence, the virtual utility defined in (5.5) is increasing in v_t and v_0. The proof of Theorem 2 shows that this is sufficient for the existence of a payment that makes it incentive compatible to report the time zero valuation truthfully. □

Proof of Proposition 10. We can explicitly calculate the time zero expected utility the agent derives from consuming the good when she reported a shock $\hat{\theta}$ if her true shock equals θ

$$
\hat{V}(\theta, \hat{\theta}) = \mathbb{E}\left[\int_0^{L(\hat{\theta})} e^{-rt} v_t \mathrm{d}t\right] = \int_0^{L(\hat{\theta})} e^{-rt} \mathbb{E}[v_t] \mathrm{d}t = \int_0^{L(\hat{\theta})} e^{-rt} e^{\theta t} v_0 \mathrm{d}t
$$

$$
= v_0 \left[\frac{e^{(\theta-r)t}}{\theta - r}\right]_{t=0}^{t=L(\hat{\theta})} = v_0 \frac{e^{(\theta-r)L(\hat{\theta})} - 1}{\theta - r}.
$$

Thus, time zero transfers that make this allocation incentive compatible are given by

$$
\hat{V}(\theta, \theta) - \int_0^\theta \frac{\partial V}{\partial \hat{\theta}}(z, z) \mathrm{d}z = v_0 \frac{e^{(\theta-r)L(\theta)} - 1}{\theta - r}
$$
$$
- \int_0^\theta \frac{e^{(z-r)L(z)}[L(z)(z-r) - 1]}{(z-r)^2} \mathrm{d}z.
$$

If payment is made as a flow transfer on the time interval $[0, L(\theta)]$ we need to adjust it by multiplying with $r(1 - e^{-rL(\theta)})^{-1}$.   □

A.1. *Relationship to Eső and Szentes (2007)*

In Lemma 2 Eső and Szentes (2007) show that their Assumption 1 is equivalent to (in our notation)

$$
\phi_{\theta w}^i(t, \theta, w) \le 0, \tag{A.1}
$$

and their Assumption 2 is equivalent to (in our notation)

$$
\frac{\phi_{\theta\theta}^i(t, \theta, w)}{\phi_\theta^i(t, \theta, w)} \le \frac{\phi_{\theta w}^i(t, \theta, w)}{\phi_w^i(t, \theta, w)}. \tag{A.2}
$$

As

$$
\frac{\partial}{\partial w} \frac{\phi_\theta^i}{\phi^i} = \frac{\phi_{\theta w}^i \phi^i - \phi_\theta^i \phi_w^i}{\phi^{i2}},
$$

Assumption 1 of Eső and Szentes (2007) implies our Assumption 1 and is thus stronger. As

$$
\frac{\partial}{\partial \theta} \frac{\phi_\theta^i}{\phi_w^i} = \frac{\phi_{\theta\theta}^i \phi_w^i - \phi_\theta^i \phi_{\theta w}^i}{\phi_w^{i2}} = \frac{\phi_\theta^i}{\phi_w^i}\left(\frac{\phi_{\theta\theta}^i}{\phi_\theta^i} - \frac{\phi_{\theta w}^i}{\phi_w^i}\right).
$$

Hence, Assumption 2 of our setup is exactly equivalent to Assumption 2 in Eső and Szentes (2007).

A.2. *Relationship to Boleslavsky and Said (2013)*

We briefly establish the relationship between the multiplicative random walk in the discrete time environment of Boleslavsky and Said (2013) and the geometric Brownian motion analyzed here. Let $(X_k)_{k \in \mathbb{N}}$ be a multiplicative random walk, i.e.

$$X_{k+1} = \begin{cases} u\, X_k, & \text{with probability } \theta, \\ d\, X_k, & \text{with probability } 1 - \theta; \end{cases}$$

for some $d < 1 < u$ and let the uptick probability $\theta \in (0,1)$ be the private information. Boleslavsky and Said (2013) show, see page 11, equation (2.7), that the virtual utility in period k equals[4]

$$v_k^i \left(1 - \sum_{s \le k} \mathbf{1}_{\{X_s = dX_{s-1}\}} \frac{u - d}{d(1 - \theta)} \frac{1 - F^i(\theta)}{f^i(\theta)} \right).$$

In the next step we let the period length Δ go to zero. To do so let $d \equiv d^\Delta$, $u \equiv u^\Delta$ and $t \equiv \Delta k \in \mathbb{N}$. The virtual utility at the physical time t thus equals

$$v_t^i \left(1 - \sum_{s \le \frac{t}{\Delta}} \mathbf{1}_{\{X_s = dX_{s-1}\}} \left(\left(\frac{u}{d}\right)^\Delta - 1 \right) \frac{1 - F^i(\theta)}{f^i(\theta)(1 - \theta)} \right).$$

Note that $\sum_{s \le \frac{t}{\Delta}} \mathbf{1}\{X_s = dX_{s-1}\}$ is Binomial distributed and converges to its expectation for $\Delta \to 0$, i.e.

$$\lim_{\Delta \to 0} \sum_{s \le \frac{t}{\Delta}} \mathbf{1}_{\{X_s = dX_{s-1}\}} = \mathbb{E} \left[\sum_{s \le \frac{t}{\Delta}} \mathbf{1}_{\{X_s = dX_{s-1}\}} \right] = (1 - \theta) \frac{t}{\Delta}.$$

As $\lim_{\Delta \to 0} \frac{1}{\Delta} \left(\left(\frac{u}{d}\right)^\Delta - 1 \right) = 1$ we have that the virtual utility goes to:

$$v_t^i \left(1 - (1 - \theta) \frac{t}{\Delta} \left(\left(\frac{u}{d}\right)^\Delta - 1 \right) \frac{1 - F^i(\theta)}{f^i(\theta)(1 - \theta)} \right)$$

---

[4]For convenience we translated their result into our notation. We use k for the period to clearly differentiate between periods and physical time.

$$= v_t^i \left(1 - t \frac{1}{\Delta} \left(\left(\frac{u}{d} \right)^\Delta - 1 \right) \frac{1 - F^i(\theta)}{f^i(\theta)} \right)$$

$$= v_t^i \left(1 - \frac{1 - F^i(\theta)}{f^i(\theta)} t \right),$$

which establishes the convergence to the virtual utility derived earlier in (5.14).

Acknowledgments

The first author acknowledges financial support through NSF Grants SES 0851200 and ICES 1215808. We are grateful to the Symposium Editor, Alessandro Pavan, the Associate Editor and two anonymous referees for many valuable suggestions. We thank Juuso Toikka for very helpful conversations. We thank Heng Liu, Preston McAfee, Balázs Szentes and seminar audiences at the University of Chicago, INFORMS 2012, and Microsoft Research for many helpful comments.

References

Baron, D., Besanko, D., 1984. Regulation and information in a continuing relationship. Inf. Econ. Policy 1, 267–302.

Battaglini, M., 2005. Long-term contracting with Markovian consumers. Am. Econ. Rev. 95, 637–658.

Bergemann, D., Välimäki, J., 2010. The dynamic pivot mechanism. Econometrica 78, 771–790.

Besanko, D., 1985. Multi-period contracts between principal and agent with adverse selection. Econ. Lett. 17, 33–37.

Board, S., 2007. Selling options. J. Econ. Theory 136, 324–340.

Boleslavsky, R., Said, M., 2013. Progressive screening: long-term contracting with a privately known stochastic process. Rev. Econ. Stud. 80, 1–34.

Courty, P., Li, H., 2000. Sequential screening. Rev. Econ. Stud. 67, 697–717.

DellaVigna, S., Malmendier, U., 2006. Paying not to go to the gym. Am. Econ. Rev. 96, 694–719.

Eső, P., Szentes, B., 2007. Optimal information disclosure in auctions. Rev. Econ. Stud. 74, 705–731.

Eső, P., Szentes, B., 2014. Dynamic Contracting: An Irrelevance Result. Oxford University and LSE.

Garrett, D., Pavan, A., 2012. Managerial turnover in a changing world. J. Polit. Econ. 120, 879–925.

Grubb, M.D., Osborne, M., 2015. Cellular service demand: biased beliefs, learning and bill shock. Am. Econ. Rev. 105 (1), 234–271.

Kakade, S., Lobel, I., Nazerzadeh, H., 2013. Optimal dynamic mechanism design and the virtual pivot mechanism. Oper. Res. 61, 837–854.

Kunita, H., 1997. Stochastic Flows and Stochastic Differential Equations. Cambridge University Press, Cambridge.

Milgrom, P., Segal, I., 2002. Envelope theorem for arbitrary choice sets. Econometrica 70, 583–601.

Pavan, A., Segal, I., Toikka, J., 2014. Dynamic mechanism design: a Myersonian approach. Econometrica 82, 601–653.

Rochet, J.-C., 1987. A necessary and sufficient condition for rationalizability in a quasi-linear context. J. Math. Econ. 16, 191–200.

Skrzypacz, A., Toikka, J., forthcoming. Mechanisms for repeated trade. Am. Econ. J. Microecon.

Chapter 16

Dynamic Mechanism Design: An Introduction[*]

Dirk Bergemann[†] and Juuso Välimäki[‡]

[†]*Department of Economics, and Cowles Foundation of Research in Economics, Yale University, New Haven, CT 06520-8268, USA*
dirk.bergemann@yale.edu
[‡]*Department of Economics, Aalto University School of Business, 02150 Espoo, Finland*
juuso.valimaki@aalto.fi

We provide an introduction to the recent developments in dynamic mechanism design, with a primary focus on the quasilinear case. First, we describe socially optimal (or efficient) dynamic mechanisms. These mechanisms extend the well-known Vickrey-Clark-Groves and D'Aspremont-Gérard-Varet mechanisms to a dynamic environment. Second, we discuss revenue optimal mechanisms. We cover models of sequential screening and revenue maximizing auctions with dynamically changing bidder types. We also discuss models of information management where the mechanism designer can control (at least partially) the stochastic process governing the agents' types. Third, we consider models with changing populations of agents over time. After discussing related models with risk-averse agents and limited liability, we conclude with a number of open questions and challenges that remain for the theory of dynamic mechanism design.

Keywords: Dynamic mechanism design, sequential screening, dynamic pivot mechanism, bandit auctions, information management, dynamic pricing.

JEL Classification: D44, D82, D83.

---

[*]This chapter is reproduced from *Journal of Economic Literature*, 2019.

1. Introduction

In the analysis of economic environments in which information is dispersed amongst agents, the paradigm of mechanism design has been developed to analyze questions of optimal information collection and resource allocation. The aim of these models is to come up with a framework that is sufficiently flexible to treat applications in various fields of economics yet precise enough to yield concrete insights and predictions. Over the last decade, the mechanism design approach has been applied to a variety of dynamic settings. In this survey, we review the basic questions and modeling issues that arise when trying to extend the static paradigm to dynamic settings. We do not aim at maximal generality of the results that we present but we try to bring out the main ideas in the most natural settings where they arise.

By far the best-understood setting for mechanism design is the one with independent private values and quasilinear payoffs. Applications of this model include negotiations, auctions, regulation of public utilities, public goods provision, nonlinear pricing, and labor market contracting, to name just a few. In this survey, we concentrate for the most part on this simplest setting.

It is well-known that in dynamic principal-agent models, private information held by the agent requires the optimal contract to a long-term arrangement, one that cannot be replicated by a sequence of short-term contracts, this is due the "ratchet effect". We follow the literature on static mechanism design by allowing the principal to commit at the beginning of the game to a contract that covers the entire length of the relationship.

The leading example for this survey is the problem of selling a number of goods–possibly limited–over time as the demand for the goods evolves. The dynamics that arise in such problems pertain both to the evolution of the willingness-to-pay as well as to the set of feasible allocations over time–through a variety of natural channels:

(i) The sales problem may be non-stationary because the goods have either a fixed supply or an expiry date like airline tickets for a particular flight date and time. The key feature here is that the opportunity cost of selling a unit of the good today is determined by the opportunities for future sales. Markets where such concerns are important, such as for airline tickets, have recently witnessed a number of new pricing practices: frequently changing prices and options for buyers to reserve a certain price for a given period of time.

A vast literature under the heading *revenue management* in operations research tackles applied problems of this sort.

(*ii*) Realized sales today help predict future sales if there is uncertainty about the rate at which buyers enter the market. Professional and college sports teams base their prices for remaining tickets on the sales to date. This form of dynamic pricing is used for concert tickets, hotel booking and transportation services, such as the surge pricing of Uber.

(*iii*) The valuations of the buyers may evolve over time as they learn more about the product by using it or by observing others. Cheap trial periods for online services are a particular form of intertemporal price discrimination in this setting.

(*iv*) The cost of serving the market may change over time due to exogenous improvement in technology or through learning by doing.

The general model that we consider will encompass all of these different trading environments. We cover the optimal timing of a single sale as well as repeated sales over the time horizon. In all of the above applications, the types of some agents and/or the set of allocations available change in a non-trivial manner across periods. For us, this is the distinguishing feature of dynamic mechanism design.

The techniques of dynamic mechanism design have witnessed a rapid increase in use in many market places over the past decade, often under the term "dynamic pricing". The essence of dynamic pricing is to frequently adjust the price of the object over time in response to changes in the estimated demand. The optimal price is commonly adjusted through an algorithm that responds to temporal supply and demand conditions, time, competing prices and customer attributes and behavior. The adoption of dynamic pricing strategies (particularly in e-commerce, e.g. Amazon) is facilitated by the rapid increase in real-time data on market conditions and customer behavior, which are used to condition the price and allocation policies.

Beyond the dynamic pricing of individual items, sophisticated dynamic contracts appear to be increasing in use as well. Airlines now frequently offer option contracts to allow customers to secure a certain fare for a fixed time period before they purchase the product, a "fare lock". Many subscriptions services offer a trial period with a low price before the price resets at a higher level.

A different class of applications of dynamic mechanism design arises in such common situations as the pricing of memberships, such as fitness clubs, or long-term contracts, such as mobile phone contracts or equipment service contracts. At any given point in time, the potential buyer knows how much she values the service, but is uncertain about how future valuations for the service may evolve. From the point of view of the service provider, the question is then how to attract (and sort) the buyers with different current and future valuations for his services. The menu of possible contracts presumably has to allow the buyers to express their private current willingness-to-pay as well as their expectations over future willingness-to-pay. A variety of dynamic contracts are empirically documented, for example in gym memberships and mobile phone contracts, as described in Della Vigna and Malmendier (2006) and Grubb and Osborne (2015), respectively. These include (*i*) flat rates, in which the buyer only pays a fixed fee regardless of her consumption; (*ii*) two-part-tariffs in which the buyer selects from a menu of fixed fees and variable price per unit of consumption; and (*iii*) leasing contracts where the length of the lease term is the object of choice for the consumer. We will highlight in Section 5 how these and other features of observed contract varieties may arise as solutions to dynamic mechanism design problems.

Since we insist on full commitment power throughout this survey, we bypass the vast literature on Coasian bargaining that has the lack of commitment at its heart. Similarly, we do not consider contract dynamics resulting from renegotiation. Since both the seller and the buyer commit to the mechanism, we also restrict the type of participation constraints that we allow. In particular, we do not try to give full analysis of models where the buyer's outside option changes over time as in Harris and Holmstrom (1982). Dynamic games where the players engage in interactions while their payoff relevant types are subject to stochastic changes are closely related to the material in this survey.[1] Since the analysis of such games requires sequential rationality on part of all players, our focus on the optimal commitment solutions for the designer rules them out of the scope of this survey. This survey will also not include a comprehensive review of the recent work on dynamic taxation and dynamic public finance, a line of research that has a strong focus on strictly concave rather than quasilinear payoffs. However in Section 7, we shall provide connections between theses two classes of payoff environments and comment on the similarities and differences in the analysis and the results.

---

[1] In repeated and dynamic games, the vector of continuation payoffs plays the role of monetary transfers in mechanism design problems.

Our aim in writing this survey is twofold. We want to give an overview of the tools and techniques used in dynamic mechanism design problems in order to give the reader an understanding of the scope of applications that can be tackled within this framework. We give a number of examples where the optimal solution can be fully characterized. For many interesting dynamic contracting problems, finding optimal incentive compatible mechanisms is beyond the scope of the currently available techniques. In such cases, one must look for partial solutions or approximate solutions. Second, we want to present and discuss the recent literature in this area.

We begin with mechanisms that achieve a *socially efficient* allocation. A dynamic version of the *pivot mechanism* gives each agent a private payoff equal to her marginal contribution to the utilitarian social surplus. A mechanism that has this property is attractive since it gives each agent the societally optimal incentives to make private investments. In a dynamic context, such investments could generate more accurate valuations or reduce future costs. We give a simple formula for the periodic payments that support the efficient allocation rule. In contrast to optimal or revenue maximizing mechanisms, the dynamic pivot mechanism does not rely on strong assumptions about commitment or constant outside options.

For the case of *revenue maximizing* mechanisms, the central trade-off in the static case is between social surplus and information rent going to the agent. We investigate how far the well-known results from the static model extend. In particular, we try to stay as close as possible to the best-understood model where a principal offers a contract to a privately informed agent with a single-dimensional type and with supermodular preferences over allocations and types. We see conditions where the usual results of no distortions for the highest type agent and downward distortions for all other agents hold. But sometimes the direction of the distortions may be reversed due to the type dynamics.

On the more dynamic implications, one might guess that the part of private information held by the bidders at the moment of contracting is the only source of information rent. The rest of the stochastic type process is uncertain to both the seller and the buyer, and after the initial report the two parties share a common probability distribution on future types. We discuss a way of formalizing this line of thought and we will see the extent to which this result holds.

One of the key implications for the revenue maximizing allocation stems from this intuition. For most stochastic processes (e.g. ergodic and strongly mixing processes), knowing the value of the process in period t tells us little about the value of the process in period $t + k$ for k large. Hence one might conjecture that the private information θ_0 held by the agent at the moment of signing the contract provides little private information about the valuation θ_t for large t. As a result, distortions from the efficient allocation path should vanish as t becomes large. This property of the optimal allocation path is in fact quite robust. Whereas most of the analysis that we present relies on arguments based on the envelope theorem (i.e. arguments depending on local optimality of truthful reporting in the mechanism), the property of vanishing distortions holds also for a much wider class of models where the so called first-order approach fails.

The third part considers models with changing populations of agents over time. Obviously this part has no counterpart on the static side. It allows us to ask new questions relating to the properties of the payment rules. For example, with changing populations, it makes sense to require that agents receive or make transfers only in the periods when they are alive. These restrictions lead to interesting new findings about the settings where efficient outcomes can be achieved. Another novel finding in this literature is that having forward-looking buyers may sometimes be good for the revenue maximization. This is very much in contrast with the typical Coasian reasoning and also represents a novel finding relative to the literature on revenue management.

In the last substantive section of this survey, we briefly consider related models from public finance and financial economics. The key departure in these models is the lack of quasi-linearity. The models in dynamic public finance are primarily concerned with consumption smoothing over risky outcomes. Hence the models feature agents with strictly concave utilities in consumption and leisure. In addition to the possibility of having risk-averse decision makers, the models in financial economics often feature a limited liability constraint on the transfer rules: owners can pay the managers but managers cannot be asked to make (arbitrarily large) payments to the owners. We discuss the similarities in the analysis and contrast the results of these models with the models under quasilinear utility. Finally, we make some connections to the rapidly growing computer science literature on mechanism design. Rather than concentrating on the properties of the optimal mechanism for a fixed stochastic model, this literature seeks mechanisms that guarantee a good payoff across a variety of different stochastic models.

The interested reader will find complementary material and more technical detail in the recent textbooks by Börgers (2015) and Gershkov and Moldovanu (2014). We have also included a few more technical observations in the Appendix. The earlier survey by Bergemann and Said (2011) focuses on dynamic auctions, and the more recent survey by Pavan (2017) focuses on issues of robustness and endogenous types. Bergemann and Pavan (2015) provide an introduction to recent research in dynamic mechanism design collected in a symposium issue of the *Journal of Economic Theory*. The textbook by Talluri and van Ryzin (2004) is a classic introduction into revenue management from the operations research perspective.

2. The Dynamic Allocation Problem

2.1. *Allocations, preferences and types*

In this section, we present a dynamic and stochastic payoff environment that is general enough to cover all the later sections. We consider a discounted discrete-time model with a finite or infinite ending date T. Each agent $i \in \{1, 2, \ldots, I\}$ receives in each period $t \leq T$ a payoff that depends on the current physical allocation $x_t \in X_t$, the current monetary *payment* (or transfer) $p_{i,t} \in \mathbb{R}$, and the private information

$$\theta_t = (\theta_{i,t}, \theta_{-i,t}) \in \times_{i=1}^{I} \Theta_i = \Theta \subset \mathbb{R}^I.$$

Throughout this survey, we assume private values and quasilinear utilities. As a result, the Bernoulli utility function u_i of agent i takes the form:

$$u_i(x_t, p_t, \theta_t) \triangleq v_i(x_t, \theta_{i,t}) - p_{i,t}.$$

We assume that the type $\theta_{i,t}$ of agent i follows a controlled Markov process on the state space Θ_i. The flow payoffs of the social planner are defined by:

$$u_0(x_t, p_t, \theta_t) \triangleq v_0(x_t) + \sum_{i=1}^{I} p_{i,t}.$$

The set X_t of feasible allocations in period t may depend on the vector of past allocations

$$x^{t-1} \triangleq (x_0, \ldots, x_{t-1}) \in X^{t-1}.$$

For example, the seller may only have K units of the object for sale, and a sale today diminishes the number of available objects tomorrow.

The dependence of the set of feasible allocations tomorrow on the current feasible set and current allocation is denoted by a transition function g:

$$X_{t+1} \triangleq g\left(X_t, x_t\right).$$

There is a common prior c.d.f. $F_{i,0}\left(\theta_{i,0}\right)$ regarding the initial type $\theta_{i,0}$ of each agent i. The current type $\theta_{i,t}$ and the current action x_t determine the distribution of the type $\theta_{i,t+1}$ in the next period. We assume that this distribution can be represented by a Markovian transition function (or stochastic kernel)

$$F_i(\theta_{i,t+1}|\theta_{i,t}, x_t).$$

The utility functions u_i and the transition functions F_i are all common knowledge at $t = 0$. At the beginning of each period t, each agent i observes $\theta_{i,t}$ privately. At the end of each period, an allocation $x_t \in X_t$ is chosen by the principal and payoffs for period t are realized. The asymmetric information is therefore generated by the private observation of $\theta_{i,t}$ in each period t. To ensure that all the expectations in the model are well-defined and finite, we assume that

$$|v_i(x, \theta_i)| < K,$$

for some $K < \infty$ for all i, x and θ_i.

2.2. *Possible interpretations of types*

Up to now, we have been very general about the interpretation of $\theta_{i,t}$. There are at least three separate cases that deserve mention here. In the first, all agents are present in all periods of the game, and their types evolve according to an exogenous stochastic process on Θ_i. In the second, all agents are present in all periods, but their future types depend endogenously on current allocations. In the third case, not all agents are present in all periods.

The first case seems appropriate for procuring goods over time from firms whose privately known costs follow a stochastic process $F(\theta_{i,t}|\theta_{i,t-1})$. For example, we could take $\theta_{i,t} \in \mathbb{R}$, with

$$\theta_{i,t+1} = \gamma\theta_{i,t} + \varepsilon_{i,t+1},$$

where the $\varepsilon_{i,t}$ are i.i.d. shocks.

For an example of the second class of models, consider an employer i who learns privately about the (firm-specific) productivity ω_i of a given worker which is constant over time. In this case, a risk neutral employer would compute the posterior distribution on i's productivity:

$$\theta_{i,t} \triangleq \mathbb{E}[\omega_i + h_{i,t}],$$

where $h_{i,t}$ is the information set of firm i at time t. It makes sense to assume now that $\theta_{i,t+1}$ depends on $\theta_{i,t}$ and the allocation x_t, in particular whether the worker was employed by firm i in period t or not. Hence the type evolution is endogenous to the allocation problem.

At the cost of some notational inconvenience, we could have allowed the payoffs and the transitions to depend on the full history of allocations: $x^t = (x, \ldots, x_t)$. It will become clear that none of the results would change as a result of this more general formulation. Hence we can accommodate other endogenous models such as learning by doing where the production cost of a firm decreases stochastically in its past cumulative production, or habit formation and preference for variability over time.

The case where not all agents are present at all times requires a bit more discussion. An agent may for example decide to wait in order to get a better deal on a purchase. If all agents are present from the start, this case is covered by the previous two specifications. If agents arrive stochastically over time, they can enter contracts only after arrival. We may then assume that the arrivals are either private information to the agent or publicly observed. For the first case, we assume that each agent i can have a particular type, the null type, $0 \in \Theta_i$ for all i that we interpret as indicating that the agent is not present. We assume that $v_i(x_t, 0) = 0$ for all i, x_t, and that agents with type 0 cannot make or receive any transfers. The interpretation is that i is born at the first time $t = \tau$ where $\theta_{i,t} \neq 0$ and hence his arrival time τ is private information to the agent. Alternatively, we can assume each agent's arrival time τ is publicly observable. In Section 6, we analyze and contrast such models.

2.3. *Dynamic direct mechanisms*

In a dynamic direct mechanism every agent i is asked to report her type $\theta_{i,t}$ in every period t. We say that the dynamic direct mechanism is truthful if the reported type $r_{i,t} \in \Theta_i$ coincides with the true type for all i, t after all histories of realized and reported types. The dynamic revelation principle as first stated in Myerson (1986), and recently extended by Sugaya and Wolitzky (2017), argued that there is no loss of generality in restricting attention to dynamic direct mechanisms where the agents report their information truthfully.

The mechanism designer chooses how much of the information in the reports and allocations to disclose to the players. In this survey, we assume that the physical allocation $x_t \in X_t$ is publicly observed.[2] It is clear that restricting the information available to agent i makes it easier to satisfy the incentive compatibility constraints for that player. Hence it might be easier to achieve incentive compatibility when other agents' past reports are kept secret. In all of the applications covered in this survey, we can find an optimal mechanism where all past reports and all past allocations are made public. In what follows we consider dynamic direct mechanisms that have this feature.

Let $r_{i,t} \in \Theta_i$ denote the report of agent i in period t and let $r_t = (r_{i,t}, r_{-i,t})$ be the vector of reported types in period t. We denote the public history $(x_s, r_s)_{s<t}$ at period t by $h_t \in H_t$. When agent i chooses her report in period t, she knows her own type $\theta_{i,t}$, and all her past realized types. In the Markovian setting, the only payoff relevant private information is her current type and hence we let the private history of agent i be $h_{i,t} = (\theta_{i,t}, h_t) \in H_{i,t}$. A reporting strategy $\mathbf{r}_i = (r_{i,t})_{t=1}^T$ for agent i is given by

$$r_{i,t} : \Theta_i \times H_t \to \Theta_i.$$

A dynamic direct mechanism $(\mathbf{x}, \mathbf{p}) = (x_t, p_t)_{t=1}^T$ assigns physical allocations and transfer payments to the agents as a function of their current reports and the public history:

$$x_t : \Theta \times H_t \to X_t,$$

$$p_t : \Theta \times H_t \to \mathbb{R}^I.$$

Notice that the reporting strategy $\mathbf{r}$ and the allocation process $\mathbf{x}$ induce a stochastic process for the sequence of types through the transition probability $F(\theta_{t+1}|\theta_t, x_t(r_t(\theta_t)))$. We shall be particularly interested in *truthful reporting* strategies. For this purpose, let $\hat{\mathbf{r}}$ denote the reporting strategy profile where

$$r_{i,t}(h_{i,t}) = \theta_{i,t}.$$

for all $i, t, h_{i,t}$.

---

[2]In some cases such as allocating a fixed capacity over time, each player is only informed of her own allocation and the number of remaining units may be kept secret.

The physical allocation rule $\mathbf{x}$, and any vector $\mathbf{r}_{-i}$ of reporting strategies for agents other than i induce a Markovian decision problem for agent i with the dynamic programming formulation:

$$V_{i,t}(\theta_{i,t}, h_t; \mathbf{r}_{-i}) = \max_{r_{i,t}} \mathbb{E}[v_i(x_t(r_{i,t}, r_{-i,t}(\tilde{\theta}_{-i,t}, h_t), h_t), \theta_{i,t})$$

$$- p_{i,t}(r_{i,t}, r_{-i,t}(\tilde{\theta}_{-i,t}, h_t), h_t)$$

$$+ \delta V_{i,t+1}(\tilde{\theta}_{i,t+1}, \tilde{h}_{t+1}; \mathbf{r}_{-i}) | \theta_{i,t}, h_t].$$

We designate a random variables, such as $\tilde{\theta}_{i,t}$ by tilde. The expectation above is taken with respect to the stochastic process $\{\tilde{\theta}_t\}$ induced by the transition probability F, by the allocation rule x and by the reporting strategy $\mathbf{r}$. For the remainder of the paper we shall suppress the time index when the conditioning variable implicitly defines the time index for the relevant function. Thus for example, we will write $V_i(\theta_{i,t}, h_t; \mathbf{r}_{-i})$ rather than $V_{i,t}(\theta_{i,t}, h_t; \mathbf{r}_{-i})$, and $x(r_{i,t}, r_{-i,t}(\tilde{\theta}_{-i,t}, h_t), h_t)$ rather than $x_t(r_{i,t}, r_{-i,t}(\tilde{\theta}_{-i,t}, h_t), h_t)$.

We define

$$V_i(\theta_{i,t}, h_t) \triangleq V_i(\theta_{i,t}, h_t; \hat{\mathbf{r}}_{-i})$$

to be the value function of agent i under truthful reporting by the other agents. We say that $(\mathbf{x}, \mathbf{p})$ is Bayes-Nash implementable if for all i, t, h_t,

$$\theta_{i,t} \in \arg_{r_{i,t}} \max \mathbb{E}[v_i(x(r_{i,t}, \tilde{\theta}_{-i,t}, h_t), \theta_{i,t}) - p_i(r_{i,t}, \tilde{\theta}_{-i,t}, h_t)$$

$$+ \delta V_i(\tilde{\theta}_{i,t+1}, \tilde{h}_{t+1}) | \theta_{i,t}, h_t].$$

Thus, taking the expectation over the other agents' true type realization truth-telling in every period and after every history is an optimal strategy for agent i.

We will sometimes refer to a stronger notion of implementability, called *periodic ex post* implementability. To define this notion, we let

$$V_i(\theta_t, h_t) \triangleq \max_{r_{i,t}} \{v_i(x(r_{i,t}, \theta_{-i,t}, h_t), \theta_{i,t}) - p_i(r_{i,t}, \theta_{-i,t}, h_t)$$

$$+ \delta \mathbb{E}[V_i(\tilde{\theta}_{t+1}, \tilde{h}_{t+1}) | \theta_t, h_t]\}.$$

Observe that the value function is now defined on the entire set of type vectors, θ_t, and public histories, rather than just the individual type $\theta_{i,t}$.

A mechanism $(\mathbf{x}, \mathbf{p})$ is periodic ex-post implementable if for all i, t, θ_t, h_t:

$$\theta_{i,t} \in \arg_{r_{i,t}} \max\{v_i(x(r_{i,t}, \theta_{-i,t}, h_t), \theta_{i,t}) - p_i(r_{i,t}, \theta_{-i,t}, h_t)$$
$$+ \delta\mathbb{E}[V_i(\tilde{\theta}_{t+1}, h_{t+1})|\theta_t, h_t]\}.$$

Whenever $(\mathbf{x}, \mathbf{p})$ is periodic ex-post implementable, no agent wants to change her report after learning the contemporaneous reports of the other agents. This means that as in the static setting, ex-post implementation is a solution concept that is stronger than Bayes Nash implementation but weaker than dominant strategy implementability.

3. Efficient Dynamic Mechanisms

We begin the analysis by describing three dynamic mechanisms that attain the intertemporally efficient allocation in the presence of private information arriving over time: (i) the team mechanism, (ii) the dynamic pivot mechanism and (iii) the dynamic AGV mechanism. We illustrate these mechanisms by considering allocating a fixed number of objects over time.

3.1. *The team mechanism*

We start by constructing a simple mechanism that makes truthful reporting incentive compatible in the sense of periodic ex-post incentive compatibility. In this mechanism, called the team mechanism, the agents will have the right incentives to report their types truthfully as their payoff is the entire social surplus generated in the allocation problem. Hence the right place to start the construction is the social planner's optimal allocation problem with publicly observable types. The utilitarian welfare maximization, including all agents and the principal, solves the following program:

$$W(\theta_0, X_0) \triangleq \max_{\{x_t \in X_t\}_{t=0}^T} \mathbb{E}\left\{\sum_{t=0}^T \delta^t \sum_{i=0}^I v_i(x_t, \theta_{i,t})\right\},$$

where the expectation is taken with respect to $F(\theta_{t+1}|\theta_t, x_t)$ and the feasibility condition $x_t \in X_t$ for all t. Notice, that we allow the value function in period 0 (and all future periods) to depend explicitly on the set of feasible allocations. We define the social flow payoff as:

$$w(x_t, \theta_t) \triangleq \sum_{i=0}^{I} v_i(x_t, \theta_{i,t}).$$

We can write this in terms of a dynamic program:

$$W(\theta_t, X_t) = \max_{x_t \in X_t} \left\{ w(x_t, \theta_t) + \delta \mathbb{E} W(\theta_{t+1}, X_{t+1}) \right\},$$

subject to the transition function of the state

$$\theta_{t+1} \sim F(\cdot | \theta_t, x_t)$$

and the feasibility constraint:

$$X_{t+1} = g(X_t, x_t).$$

Let $x_t^*(\theta_t, X_t)$ denote an optimal policy for this program.

As in the static setting, periodic ex-post incentive compatibility follows if we give each agent the entire social surplus. By the one-shot deviation principle, it is sufficient to set for all i and all θ_t:

$$p_i^*(r_{i,t}, r_{-i,t}) = - \sum_{j \neq i} v_j(x^*(r_{i,t}, r_{-i,t}), r_{j,t}) \triangleq -w_{-i}(r_t).$$

In other words, each agent is paid in each period the efficient gross surplus that the other agents receive at the efficient allocation.

Up to now, we have allowed correlated types, and in fact, $(\mathbf{x}^*, \mathbf{p}^*)$ is periodic ex-post incentive compatible for correlated as well as for independent types. Strengthening the notion of incentive compatibility to dominant strategies is unfortunately not possible. The easiest way to see this is to notice that opponents' future reports depend on past allocations (and possibly also on other agents' past reports). To ensure the implementability of efficient allocations in all future periods, a Vickrey-Clark-Groves (VCG) term $w_{-i}(r_s)$ must be included in all future transfers.

We emphasize that we only ensure periodic ex-post incentive compatibility, but not dominant strategy incentive compatibility. We illustrate by means of an example in Section 3.3 that the notion of dominant strategy incentive compatibility is typically too demanding in dynamic settings.

For periodic ex-post incentive compatibility, we can also allow for interdependent types between the players as long as the payoff consequences to i resulting from information of player j become observable at some later point in time.[3] Unfortunately, the team mechanism results in a deficit of size $(I - 1)W_0(\theta_0, X_0)$. In the next two subsections, we consider efficient mechanisms that reduce and sometimes eliminate this deficit. We shall see that modified versions of the pivot mechanism and the D'Aspremont-Gérard-Varet (AGV) mechanism in the static setting perform well within the dynamic environment.

3.2. *Leading example: Sequential allocation of fixed capacity*

Throughout this survey, we illustrate the results with the following sequential allocation problem. There is a fixed supply of goods given at $t = 0$ and there is uncertainty about the demand which is realized stochastically over time.

There is supply of K identical units of the good and I potential bidders, each with unit demand, over two periods, $t \in \{1, 2\}$. For each $i \in \mathcal{I} \triangleq \{1, \ldots, I\}$, we let $x_{i,t} = 1$ denote the event that a good is allocated to agent i in period t, otherwise $x_{i,t} = 0$. The allocation is once and for all: after one of the goods has been allocated to i, it cannot be taken away and allocated to another bidder. By x_t, we denote the vector of allocation decisions in period t. The capacity constraint states that $\sum_t \sum_i x_{it} \leq K$.

At the beginning of period 1, each bidder observes his type $\theta_{i,1} \in [0, 1]$. The type $\theta_{i,2} \in [0, 1]$ in period $t = 2$ depends on the vector of realized types in period 1, θ_1. In a dynamic direct mechanism, the agents report their realized types in each period. The allocation rule depends on both of these announcements. The payoff of agent i in the mechanism is $\sum_t \delta^{t-1}(\theta_{i,t} x_{i,t} - p_{i,t})$.

We denote the number of unallocated goods in period $t = 2$ by $K_2 = K - \sum_i x_{i1}$. Efficient allocation in $t = 2$ requires allocating the K_2 units to those bidders that have the highest $\theta_{i,2}$ amongst the ones with $x_{i,1} = 0$. Solving for the efficient allocation in $t = 1$ is not trivial even in this simple allocation problem. Let $\mathcal{I}_1$ denote the set of bidders that receive the good in period 1 and let K_1 denote the number of goods allocated in $t = 1$. Let $\mathcal{I}_2 = \mathcal{I} \backslash \mathcal{I}_1$, and we can write the value function for the efficient period 2 allocation as:

$$W_2(\theta_2, \mathcal{I}_2, K_2) = \max_{\{x_{i,2}\}} \sum_{i \in \mathcal{I}_2} \theta_{i,2} x_{i,2},$$

---

[3]In this way, the team mechanism can be extended to cover the mechanisms first displayed in Mezzetti (2004).

subject to $\sum_i x_{i,2} \leq K_2$. Similarly, for the first period, we can write

$$W_1(\theta_1, \mathcal{I}, K) = \max_{\{x_{i,1}\}} \mathbb{E}\left[\sum_{i \in \mathcal{I}} \theta_{i,1} x_{i,1} + \delta W_2(\mathcal{I}_2, K - K_1) | \theta_1\right],$$

subject to $\sum_i x_{i,1} = K_1$.

The first period decision fixed first the number K_1 to be allocated in $t = 1$ and then asks how to optimally choose the identities of the K_1 agents to receive the goods in $t = 1$. The next step optimizes over K_1. The hard step is obviously in determining the set of agents to receive the goods in the first period. Unless we specify the model further, little can be said in general about the features of the optimal allocation decision. The first period decision incorporates a few dynamic considerations. Bidder i may be present only in $t = 1$, in which case $\theta_{i,2} = 0$, or alternatively she could only arrive in $t = 2$, in which case we could take $\theta_{i,1} = 0$. Bidder i's true valuation may be learned in $t = 2$, in which case, we could set $\theta_{i,2} \sim F_i(\cdot | \theta_{i,1})$. Nevertheless, since the allocation choices are finite in the above problem, an optimal allocation policy $\mathbf{x}^*$ exists and it is easy to see that by setting

$$p_{i,t}^*(r_{i,t}, r_{-i,t}) \triangleq -w_{-i}(r_t),$$

the mechanism $(\mathbf{x}^*, \mathbf{p}^*)$ is periodic ex-post incentive compatible.

3.3. *Impossibility of dominant strategy implementation*

We first illustrate the impossibility of implementing the efficient allocation rule in dominant strategies by specifying the above example even further.

We consider two bidders, thus $I = 2$, who compete for a single object, thus $K = 1$, in a two-period model. We suppose further that bidder 1 draws a valuation θ_1 uniformly from $[0, 1]$ in period 1. Her valuation for the good remains unchanged (but is discounted) in period 2. Bidder 2 is active only in period 2 in the sense that her valuation for allocation in $t = 1$ is 0 with probability 1. Her valuation θ_2 in period $t = 2$ is independent of the valuation of bidder 1 and also drawn from the uniform distribution on $[0, 1]$. The allocation decision is non-trivial in $t = 2$ only if the good was not allocated in $t = 1$, i.e. if $K_2 = 1$. In this case, $x_{1,2}^*(r) = 1$ and $x_{2,2}^*(r) = 0$ if $r_{1,1} \geq r_{2,2}$, otherwise $x_{1,2}^*(r) = 0$ and $x_{2,2}^*(r) = 1$.

The efficient allocation in $t = 1$ gives:

$$x^*(\theta_1) = 1 \Leftrightarrow \theta_1 \geq \delta\mathbb{E}[\max\{\theta_1, \tilde{\theta}_2\}],$$

which results in a threshold value θ^* to allocate the object to the first bidder in the first period if:

$$\theta_1 > \theta^* = \frac{1}{\delta} - \sqrt{\frac{1}{\delta^2} - 1}.$$

The threshold θ^* is strictly increasing in the discount factor δ as the future realization of θ_2 provides a greater option value with a higher discount factor.

Green and Laffont (1977) and Holmstrom (1979) show that the transfer rules for efficient mechanisms are uniquely pinned down (up to a constant) when the type sets are path-connected. Combining this with the logic of VCG mechanisms allows us to conclude that in $t = 1$, the expected transfer of bidder 1 depends on her reported type $r_{1,1}$ only through its impact on the efficient allocation:

$$\mathbb{E}[p_1(r_{1,1}, r_{2,2})] = -\mathbb{E}[\delta r_{2,2} x^*_{2,2}(r_{1,1}, r_{2,2}) + \phi_1(r_{2,2})] \qquad (3.1)$$

where the (arbitrary) component function $\phi_1(r_{2,2})$ that enters the price p_1 depends only on the report of bidder 2.

The payment p_1 of bidder 1 displays no interaction between his report and the report of bidder 2 beyond their joint effect on the payoff to bidder 2 in the efficient allocation rule, the first term on the RHS of (3.1). Now, in order to secure the efficient decision in period $t = 1$, the transfer in equation (3.1) must be calculated using the truthful reporting strategy for bidder 2 (and hence the true distribution of θ_2) in $t = 2$.

But now we can show that these transfers can't possibly implement truthful reporting as a dominant strategy for player 1. To see this, suppose that bidder 1 expects that bidder 2 does not report truthfully, but rather reports $r_{2,2} = 0$ for all θ_2, and thus

$$\mathbb{E}[r_{2,2} x^*_{2,2}(r_{1,1}, r_{2,2})] = 0.$$

Now truthful reporting does not constitute an optimal report in period $t = 1$. Under this candidate reporting strategy of bidder 2 and given the payment rule, bidder 1 would optimally exaggerate her type in period $t = 1$ and report $r_{1,1} > \theta^*$ for all valuation $\theta_1 > 0$ since she would prefer an early allocation of the object, or

$$\theta_1 > \delta\theta_1,$$

and given her expectation about the reporting strategy of bidder 2, she would not forego any compensation she might have received if the object were to be allocated to bidder 2 in period 2. Thus, dominant strategy implementation is impossible to guarantee even in this elementary two period allocation problem.

Notice that a similar impossibility argument would emerge if bidder 1 would expect bidder 2 to report $r_{2,2} = 1$ for all θ_2. In this case, bidder 1 would not report truthfully but rather downward misreport in order to obtain the compensatory payments due to the high report of bidder 2 since now bidder 1 would expect to get $\mathbb{E}[r_{2,2}x_{2,2}^*(r_{1,1}, r_{2,2})] = 1$.

Thus, in dynamic mechanism design problems incentive compatibility in dominant strategies is too demanding. This example also shows that implementing the efficient allocation rule will typically not be detail-free, and in particular will depend on the common prior regarding the distribution of the types. In the current example, the efficient allocation decisions in the initial period (the computation of θ^*) depended on the distribution of future types, and as a result, the optimal mechanism–in particular the transfer function–also reflects the informational details of the valuation process.[4]

3.4. *The dynamic pivot mechanism*

For the remainder of this section, we concentrate on two particular efficient mechanisms that have further desirable properties. We begin with the dynamic pivot mechanism, introduced in Bergemann and Välimäki (2010), which ensures that each agent's payoff in the mechanism corresponds to her marginal contribution to the societal welfare after all histories. In the dynamic pivot mechanism, all agents have the correct societal incentives to engage in private investments in e.g. increasing their own payoffs through cost-reducing investments. In the next subsection, we consider the dynamic counterpart of the AGV mechanism where the focus shifts towards budget

---

[4]The failure of dominant strategy implementability in dynamic mechanism is hence comparable to the failure of static AGV mechanism to be dominant strategy incentive compatible. In both instances, it is critical to use the same expectations regarding the behavior of the other agents in the determination of the efficient rule and in the computation of expected payments for an individual agent.

balance. For dynamic bargaining processes and dynamic problems of public goods provision, these considerations are of obvious importance just as they are in the static case.

We now construct the dynamic pivot mechanism for the general model described in Section 2 under the assumption of independent private values. We give an example in the Appendix showing that dynamic pivot mechanisms do not always exist if the values are correlated. We recall that in the static pivot mechanism–introduced by Green and Laffont (1977)–the transfers are constructed as follows:

$$p_i(\theta) = -\sum_{j \neq i} v_j(x^*(\theta), \theta_j) + \sum_{j \neq i} v_j(x^*_{-i}(\theta_{-i}), \theta_j), \qquad (3.2)$$

where $x^*_{-i}(\theta_{-i})$ is the optimal allocation if agent i is not participating, thus

$$x^*_{-i}(\theta_{-i}) \in \arg \max_{x_t \in X_t} \sum_{j \neq i} v_j(x_t, \theta_{j,t}).$$

The idea behind the pivotal transfers is to equate each agent's expected payoff to her expected contribution to the social value. At state (θ_t, X_t), we compute the (dynamic) marginal contribution of agent i:

$$M_i(\theta_t, X_t) \triangleq W(\theta_t, X_t) - W_{-i}(\theta_t, X_t),$$

where W and W_{-i} are the value functions of social surpluses with and without i in the society (in all future periods), respectively.

In the dynamic pivot mechanism, we show that the marginal contribution will also be equal to the equilibrium payoff that agent i can secure for herself along the socially efficient allocation. If agent i receives her marginal contribution in every continuation game of the mechanism, then she should receive the *flow* marginal contribution $m_i(\theta_t, X_t)$ in each period. The flow marginal contribution accrues incrementally over time and is defined recursively:

$$M_i(\theta_t, X_t) = m_i(\theta_t, X_t) + \delta \mathbb{E} M_i(\theta_{t+1}, X_{t+1}). \qquad (3.3)$$

A monetary transfer $p^*_i(\theta_t, X_t)$ such that the resulting flow net utility matches the flow marginal contribution leads agent i to internalize her social externalities:

$$p^*_{i,t}(\theta_t, X_t) \triangleq v_i(x^*_t, \theta_{i,t}) - m_{i,t}(\theta_t, X_t). \qquad (3.4)$$

We refer to $p_i^*(\theta_t, X_t)$ as the transfer of the dynamic pivot mechanism. Notice that in contrast to the static transfer payment, the reported type of agent i has also an indirect effect through $\delta \mathbb{E} W_{-i}(\theta_{t+1}, X_{t+1})$.[5] This reflects the intertemporal internalization of future externalities that is necessary for aligning the incentives with the planner's dynamic optimum. Given that we started our construction from the requirement that each agent receives her full marginal contribution $W(\theta_t, X_t) - W_{-i}(\theta_t, X_t)$, we are obviously in the realm of (dynamic) VCG mechanisms.

Theorem 1 (Dynamic Pivot Mechanism): *The dynamic pivot mechanism* $\{x_t^*, p_t^*\}_{t=0}^{\infty}$ *is ex-post incentive compatible and individually rational.*

It should be noted that as in any dynamic context, it is hard to pin down the exact timing of payments. Making a payment of p in period t has the same payoff consequences as making a payment of p/δ in $t + 1$. In Bergemann and Välimäki (2010), we give sufficient conditions for the uniqueness of the above payment rule. Similar to the static case (Moulin (1986)), these conditions require a rich domain of possible preferences. In a dynamic context, this also requires an assumption that amounts to allowing the agents to leave the mechanism stochastically. By this we mean that after leaving, no more transfers can be enacted.

The dynamic pivot mechanism has properties that other VCG schemes do not necessarily have. All payments are online in the sense that once an agent is irrelevant for future allocations, she is not asked to make any payments. Furthermore the property of equating equilibrium payoffs with marginal contributions gives the individual agents the socially correct incentives to engage in privately costly investments in θ_i. For a class of dynamic auctions, Mierendorff (2013) develops a dynamic Vickrey auction that satisfies a strong ex-post individual rationality requirement.

We illustrate how the payments in the dynamic pivot mechanism are computed for the leading example of Section 3.2.

Example 1 (Dynamic Pivot Mechanism for Fixed Capacity Allocation): We first compute the marginal contributions of the agents. If agent 1 is not present, then the expected social surplus is the discounted expected value of the good to bidder 2, i.e. $\delta/2$. The expected social surplus

---

[5] Since $W_{-i}(\theta_t, X_t) = \max_{x_t \in X_t} \{w_{-i}(x_t, \theta_t) + \delta \mathbb{E} W_{-i}(\theta_{t+1}, X_{t+1})\}$.

at any moment in time when bidder 2 is not present is simply θ_1. If the good has already been allocated, then there is no social surplus for the continuation problem. Without loss of generality, we can restrict attention to integer allocations, $x \in \{0, 1\}$. The marginal contribution of bidder 1 in period 1 and 2 are then:

$$M_1(\theta_1, 1) = \begin{cases} \theta_1 - \delta/2, & \text{if } x_{1,1}^*(\theta_1) = 1, \\ \delta \mathbb{E} \max\{\theta_1 - \theta_2, 0\}, & \text{otherwise;} \end{cases}$$

$$M_1(\theta_1, \theta_2, 1) = \max\{\theta_1 - \theta_2, 0\};$$

and similarly for bidder 2:

$$M_2(\theta_{1,}, 1) = \begin{cases} 0, & \text{if } x_{1,1}^*(\theta_1) = 1, \\ \delta \mathbb{E} \max\{\theta_2 - \theta_1, 0\}, & \text{otherwise;} \end{cases}$$

and

$$M_2(\theta_1, \theta_2, 1) = \max\{\theta_2 - \theta_1, 0\}.^6$$

Using the implicit definition of the flow marginal contribution given by (3.3), we get

$$m_1(\theta_1, 1) = \begin{cases} \theta_1 - \delta/2, & \text{if } x_1^*(\theta_1) = 1, \\ 0, & \text{otherwise.} \end{cases}$$

By equating

$$p_1(\theta_1, 1) = v_1(x_{i,t}^*, \theta_{i,t}) - m_{i,1}(\theta_{1,1}, 1),$$

we get

$$p_1(\theta_1, 1) = \begin{cases} \delta/2, & \text{if } x_1^*(\theta_1) = 1, \\ 0, & \text{otherwise.} \end{cases}$$

Similarly,

$$p_1(\theta_1, \theta_2, 1) = \theta_2, \text{if } \theta_1 \geq \theta_2, \quad p_1(\theta_1, \theta_2, 1) = 0 \text{ otherwise;}$$

$$p_2(\theta_1, \theta_2, 1) = \theta_1, \text{if } \theta_2 \geq \theta_1, \quad p_1(\theta_1, \theta_2, 1) = 0 \text{ otherwise.}$$

This suggests a simple indirect implementation of the efficient two-period auction. Bidder 1 is given the option of purchasing the good at the opportunity cost $\delta/2$ of allocating the good in $t = 1$. If he does not exercise the option, then the good is sold in a second price auction without reserve prices in period 2. It should be noted that finding the right price for this indirect implementation is remarkable easy in comparison to finding the efficient threshold type θ^*.

More generally, it is quite easy to compute the direct version of the dynamic pivot mechanism on the basis of the dynamic social surpluses using dynamic programming techniques. For example, Bergemann and Välimäki (2003), (2006), use the construction of the dynamic marginal contribution to solve for the equilibrium of dynamic common agency problems and dynamic competition problems, respectively. These earlier contributions considered symmetric but imperfect information environment; Bergemann and Välimäki (2010) establish that the underlying principles extend to asymmetric information environments as well, and then can solve priority queuing problems as analyzed by Dolan (1978).

Since we have assumed independent types, additional assumptions on the connectedness of the type spaces and payoff functions guarantee a dynamic payoff equivalence result via the envelope theorem of Milgrom and Segal (2002). By imposing an individual rationality or participation constraint for the agents, it is often possible to show that similar to the static setting, the dynamic pivot mechanism maximizes the expected transfers from the agents among all efficient mechanisms. A negative surplus in the dynamic pivot mechanism then implies an impossibility result mirroring the static Myerson-Satterthwaite theorem on budget balanced efficient dynamic mechanisms that satisfy incentive compatibility and individually rationality. Skrzypacz and Toikka (2015) consider a model of repeated bilateral monopoly with varying degrees of persistence for the buyer's valuations and the seller's costs. With perfectly persistent types, the Myerson-Satterthwaite theorem applies and efficient trade is impossible. With independent types, the expected gains from future trades can be used to relax the participation constraint and efficient trading may become possible. Different levels of persistence then determine different sets of efficient budget balanced trading rules for the problem.

3.5. *Balancing the budget*

The static VCG mechanisms is defined by $(x^*(\theta), p(\theta))$, and for all i and all θ, we have

$$p_i(r) = -w_{-i}(r) + \phi_i(r_{-i}),$$

where the second component of the transfer function, $\phi_i(r_{-i})$ of agent i is an arbitrary function that only depends on the reports of the other agents, r_{-i}. In other words, the transfer of agent i depends on her own announcement only through its impact on the other players' payoffs via the efficient allocation rule. The static VCG mechanisms are dominant strategy incentive compatible, i.e. they induce truthtelling as a best response against any reported type vector of other agents. As a result, a modified mechanism where

$$p_i(r) = -\mathbb{E}_{\theta-i}[w_{-i}(r_i, \theta_{-i}) + \phi_i(\theta_{-i})],$$

and the expectation is taken with respect the marginal distribution on the other agents' types (recall that we have assumed independence here) is incentive compatible as long as the other agents announce their types truthfully. Budget balance is obtained by specifying (with the understanding that $I + 1 = 1$):

$$\phi_{i+1}(r_{-(i+1)}) \triangleq \mathbb{E}_{r-i}[w_{-i}(x^*(r_i, r_{-i}), r_{-i})].$$

Notice that we have to give up on dominant strategy incentive compatibility here. If the other players lie about their type, the first term in the transfer does not equate the bidder's payments to the social surplus and hence it may well be that lying is optimal for i too. Observe the similarity of this reasoning to the failure of dominant strategy incentive compatibility in the dynamic example above.

For the dynamic mechanism, the transfer payments must be constructed in such a way that similar problems do not arise due to the dynamic nature of the announcements. Supposing that the incentive payments are made as above based on the expectations over other players' types, but that the realizations of other players' types can be inferred from the allocations prior to one's own announcement, the simple AGV-mechanism is no longer incentive compatible. In order to secure incentive compatibility, Athey and Segal (2013) modify the transfers to overcome this problem by aligning agent i's incentive pay with the change in the expected externality on the other agents resulting from i's report.

The resulting balanced budget mechanisms can be quite complicated and it may not be easy to find natural indirect mechanisms for their implementation. One instance where this can be done is in the context of dynamically allocating the capacity shares in a joint project when private information about future profits arrives over time. In a dynamic sharing problem, Kurikbo, Lewis, Liu, and Song (2019) find a version of the dynamic mechanism with budget balance that can also handle individual rationality constraints.

3.6. *Interdependent values and correlation*

In a dynamic setting, it is possible to use the intertemporal correlation of the reports of the agents and this allows for new types of implementations of the efficient allocation path. To see this, consider the following simple version of the famous lemons problem with common values.

Example 2 (Common Values vs Correlated Private Values):
    Two agents decide the allocation of an indivisible object in a two-period model. The allocation $x_t \in \{1,2\}$ for $t \in \{1,2\}$ records who gets the object in period t. Agent 1 is privately informed of the quality of the object, i.e. $\theta_1 \in \{0,2\}$, and her value from consuming the good in $t = 1$ is given by her private type θ_1. If the good is allocated to agent 2 in $t = 1$, then agent 2 receives her gross utility $\theta_2 = 2\theta_1 - 1$ after consuming the good in period $t = 1$. In $t = 2$, agent 2 just reports her type θ_2 and transfers are made.

If all transactions and transfers take place in $t = 1$, the efficient allocation rule where agent 1 consumes the good if and only if $\theta_1 \leq 1$ is not monotone and hence not incentive compatible in the static mechanism. By allowing agent 2 to learn her value after consuming the good, the common values lemons model becomes a dynamic private values model since conditional on knowing θ_2, player 2 does not care about θ_1. As a result, we can compute the efficient team mechanism to support trade here. The reported θ_1 determines the allocation: there is trade if and only if $\theta_1 \geq 1$ and the reported θ_2 determines the payment $t_1 = \theta_2$ that agent 1 receives in $t = 1$.

The idea that incentive constraints can be relaxed by using future realizations of correlated signals is elaborated further in Mezzetti (2004), (2007) and Deb and Mishra (2014). For many applications, experienced utilities after trade are natural signals of this type. It should also be noted that there are often quite natural implementations of the efficient mechanisms. Contracts with money back clauses can be used to facilitate trade that would otherwise be limited by the lemons problem.

Dynamic VCG mechanisms have been generalized to cover the case of correlated and interdependent values in Liu (2018). Correlation across agents allows for the use of dynamic versions of mechanisms in the style

of Cremer and McLean (1985), (1988). Liu (2018) also covers the case of
interdependent values but independently distributed signals and develops a
dynamic version of the generalized VCG mechanism along the lines of the
static version of Dasgupta and Maskin (2000).

4. Optimal Dynamic Mechanisms

We now shift our attention from socially efficient mechanism to revenue-
maximizing, or (revenue) optimal mechanisms. In static environments,
the key economic insight is the resulting trade-off between efficiency
and information rent left to the agents. In socially efficient mechanisms,
this trade-off is absent since with quasi-linear preferences. After all, the
utilitarian solution does not preclude information rents. By contrast, if a
seller tries to maximize her sales revenue or if a regulator does weigh profits
higher (or lower) than the consumer surplus in her objective function, then
this trade-off emerges. Deviations from the surplus-maximizing allocations
are generally optimal since the reductions in information rent to the
privately informed parties more than compensate for the losses in the social
surplus. For dynamic models of mechanism design, the key issue is then how
information rent accrues to privately informed agents over the contracting
horizon.

Consider a model where a privately informed agent contracts with an
uninformed principal at the beginning of a dynamic allocation problem.
If the agent knows all her future information types, the model is a
multi-dimensional mechanism design problem and as it is well-known, it
is very hard to characterize the optimal contract in such environments.
Fortunately, it is often quite reasonable that the agent does not know her
future types. Of course, her current type allows her to predict her future
types more accurately than the principal (except in the less interesting
case of i.i.d. types). The main analytical challenge in optimal dynamic
contracting problems is to characterize how the initial private information
affects the future information rents and how the optimal allocation trades
off these effects relative to the social surplus maximizing allocation.

Optimal dynamic contracting in an environment where the agent's
private information may change over time appears first in Baron and
Besanko (1984). They consider a two-period model of a regulator facing
a monopolist with a privately known marginal cost in the first period
and where the second-period marginal cost is unknown to both parties
(but may depend stochastically on the first period cost). Within a similar

model, Riordan and Sappington (1987) analyze the optimal task assignment between the agent and the principal across the two periods. Besanko (1985) covers a finite-horizon with a general cost function, where the unknown cost parameter is either i.i.d. over time or follows a first-order autoregressive process.

In the past decade, the literature has developed considerably beyond these early contributions. Much of the early literature was focused on the case where the allocation problem itself is assumed to be time-invariant in the sense that the set of feasible choices in t is independent of allocation decisions for $s < t$. These papers also assumed that the distribution of future types is independent of current allocation decisions. The first assumption is violated in any dynamic problem of capacity allocation and also in models with a fixed decision date but a dynamic flow of private information prior to the decision. The second assumption is violated in models with endogenous learning about the payoffs from a fixed set of alternatives. Examples of this type include dynamic assignment of workers to tasks or dynamic sales of experience goods.

The first analysis of the revenue-maximizing dynamic sales problem with a fixed selling date appears in Courty and Li (2000) under the name of *sequential screening*. Board (2007) extended the analysis to the case where the sales date itself is also endogenously chosen. Battaglini (2005) considers a nonlinear pricing model (with variable quantity or quality) in which the buyer's valuation changes over time according to a commonly known Markov process with two states. In contrast to the earlier work, he explicitly considers an infinite time horizon and shows that the distortion due to the initial private information vanishes over time.

Pavan, Segal, and Toikka (2014) present a general infinite-horizon model that allows for general allocation problems and endogenous type processes. Their model encompasses the earlier literature with continuous type spaces and emphasizes the connections to static allocation problems of the Myersonian type. They obtain necessary conditions for incentive compatibility and present a variety of sufficient conditions for revenue-maximizing contracts for specific classes of environments.

Our goal here is to use the tools from static mechanism design as much as possible to understand the basic analytics of the dynamic problem. Hence we start with a brief review of the main results in static mechanism design. Then we connect the static and dynamic formulations by transforming the original dynamic problem into an equivalent one where it is easier to separate the initial private information and future information that can be taken to be independent of the initial type as proposed in Eső and Szentes (2007). After formulating the optimization problem of the principal, we discuss some examples with explicit solutions to the optimal contracting problem.

4.1. *Preliminaries from static mechanism design*

Not surprisingly, the assumptions needed for tractability in static mechanism design problems are also needed in the dynamic case. We assume that all payoff functions are linear in transfers, the agents' types are intervals of the real line, $\Theta_i = [\underline{\theta}_i, \overline{\theta}_i] \subset \mathbb{R}, \widetilde{\theta}_i$ is independent of $\widetilde{\theta}_j$, and that the agents' payoff functions $v_i(x, \theta)$ are strictly supermodular in (x, θ_i). To make the connection to the dynamic setting more immediate, we allow the allocation decision $x \in X$ to be multidimensional. In the presentation below, we concentrate on the case with a single agent and therefore we omit the identity subscripts.[6] A direct mechanism $(x(\theta), p(\theta))$ is incentive compatible if for all types $\theta \in \Theta$ and all reports $r \in \Theta$, we have

$$U(\theta; \theta) \triangleq v(x(\theta), \theta) - p(\theta) \geq v(x(r), \theta) - p(r) \triangleq U(\theta; r),$$

and let

$$V(\theta) \triangleq U(\theta; \theta).$$

The envelope theorem of Milgrom and Segal (2002) gives the following necessary condition for incentive compatibility, their Theorem 2 and 3, respectively. We denote by $v_2(x, \theta)$ the partial derivative with respect to the second argument, here θ.

Theorem 2 (Payoff Equivalence Theorem):

Assume that $v(x, \cdot)$ is differentiable for all $x \in X$ and that there exists a $K < \infty$ such that for all $x \in X$ and all θ,

$$|v_2(x, \theta)| \leq K.$$

Then $V(\theta)$ is absolutely continuous, $V'(\theta) = v_\theta(x, (\theta), \theta)$ for almost every θ, and therefore

$$V(\theta) = V(\underline{\theta}) + \int_{\underline{\theta}}^{\theta} v_2(x(s), s)ds. \tag{4.1}$$

---

[6]Since types are independent, incentive compatibility reduces to individual incentive compatibility by taking expectations over the other agents' types.

This result is called the payoff equivalence theorem because we can now pin down the transfers by just determining the physical allocation $x(\theta)$ and the additive constant $V(\underline{\theta})$:

$$p(\theta) = v(x(\theta), \theta) - V(\underline{\theta}) - \int_{\underline{\theta}}^{\theta} v_2(x(s), s)ds. \qquad (4.2)$$

With the help of this necessary condition for implementability, we can rewrite the full incentive compatibility requirement as the following *integral monotonicity* condition:

$$\int_{r}^{\theta} (v_2(x(s), s) - v_2(x(r), s))ds \geq 0. \qquad (4.3)$$

As an intermediate step towards the dynamic analysis, consider for a moment the case where the principal and the agent can write contracts based on a publicly observable sequence of random variables $\{\tilde{\varepsilon}_t\}_{t=1}^{T}$ so that the allocation x_0 is determined by the initial private information θ and x_t is determined by θ and the sequence of realizations $\varepsilon^t \triangleq (\varepsilon_1, \ldots, \varepsilon_t)$. Assume also that the following separability requirement is satisfied:

$$v_t(x, \theta, \varepsilon^T) = v_t(x_t, \theta, \varepsilon^t).$$

Incentive compatibility is then equivalent to requiring that for all $\theta, r \in \Theta$,

$$\int_{r}^{\theta} \mathbb{E}_{\varepsilon}[\sum_{t=0}^{\infty} \delta^t \frac{\partial v_t}{\partial \theta}(x_t(s, \tilde{\varepsilon}^t), s) - \sum_{t=0}^{\infty} \delta^t \frac{\partial v_t}{\partial \theta}(x_t(s, \tilde{\varepsilon}^t), r)]ds \geq 0 \qquad (4.4)$$

In the static setting, single-dimensional types and single-dimensional allocations with supermodular payoff functions yield a full characterization of incentive compatible allocation rules: a mechanism is incentive compatible if and only if the physical allocation is monotone and the transfers are pinned down by the payoff equivalence theorem. The static problem is then solved by maximizing the designer's objective function over the feasible set of monotone allocation rules.

Unfortunately it is not possible to find an equally attractive characterization for incentive compatibility in the dynamic model. In some sense, this is not too surprising. The static formula for contingent allocations in equation (4.4) requires monotonicity on average when taking expectations over $\{\tilde{\varepsilon}_t\}_{t=1}^{T}$. If $x_t(\theta, \tilde{\varepsilon}_t)$ is monotonic in θ for all realizations ε^t, then incentive compatibility follows. This sufficient condition is obviously not a necessary condition since the sum of non-monotonic functions may well be monotonic.

In the remainder of this section, we express the process of types θ_t in terms of the initial type θ_0 and a sequence of independent (uniform) random variables $\{\tilde{\varepsilon}_t\}_{t=1}^T$. The main analysis considers the case where the $\{\tilde{\varepsilon}_t\}_{t=1}^T$ are privately observed by the agent, but we also discuss the connections to the static case above where all future information $\{\tilde{\varepsilon}_t\}_{t=1}^T$ is publicly observed.

4.2. *Orthogonalized information*

Eső and Szentes (2007) emphasize the benefits from distinguishing between private information at the time of contracting (captured in θ_0) and subsequent independent private information. With Markovian types, each $\tilde{\theta}_t$ is statistically dependent on $\tilde{\theta}_{t-1}$ and as a result, all future types are also influenced by the initial type $\tilde{\theta}_0$. We present below an equivalent formulation for the type process $\{\tilde{\theta}_t\}_{t=0}^T$ that allows us to distinguish in a transparent manner between the initial and the future private information despite the statistical dependency.

Consider an arbitrary random variable $\tilde{x}$ with distribution function F. Then the random variable $\tilde{y}$, with

$$\tilde{y} \triangleq F(\tilde{x}),$$

is by construction uniformly distributed on $[0, 1]$. Building on this simple observation, consider next a random variable $\tilde{\theta}_1$ with a conditional distribution $F(\cdot|\theta_0)$ dependent on some realization θ_0. Then

$$\tilde{\varepsilon}_1 \triangleq F(\tilde{\theta}_1|\theta_0)$$

is uniformly distributed for all θ_0. As a result $\tilde{\varepsilon}_1$ is independent of θ_0 by construction. We can view ε_1 as the realized percentile in the conditional distribution $F(\cdot|\theta_0)$. Since $F(\cdot|\theta_0)$ is an increasing function, knowledge of ε_1 and θ_0 allows for solving $\theta_1 = F^{-1}(\varepsilon_1|\theta_0)$.[7] Hence the information content of $(\tilde{\theta}_0, \tilde{\theta}_1)$ is the same as the information content of $(\tilde{\theta}_0, \tilde{\varepsilon}_1)$.

As a final preparatory step, let us consider the effect of $\tilde{\theta}_0$ on $\tilde{\theta}_1$. Letting $\theta_1 = F^{-1}(\varepsilon_1|\theta_0)$, we can evaluate the effect of a change in θ_0 on θ_1 for a

---

[7]If $F(\cdot|\theta_0)$ is constant for some interval or if it has upward jumps, we can define $F^{-1}(\varepsilon_1|\theta_0) = \inf\{\theta_1|F(\cdot|\theta_0) \geq \varepsilon_1\}$.

fixed ε_1.[8]

$$\frac{\partial F^{-1}(\varepsilon_1|\theta_0)}{\partial \theta_0} = \frac{\dfrac{\partial F(\theta_1|\theta_0)}{\partial \theta_0}}{f(\theta_1|\theta_0)} \triangleq I_1(\theta_0, \theta_1).$$

The function $I_1(\cdot)$ is called the *impulse response function* in Pavan, Segal, and Toikka (2014) and it will play a key role in the following analysis. Since $\tilde{\theta}_0$ is independent of $\tilde{\varepsilon}_1$, the distribution of $\tilde{\varepsilon}_1$ does not vary as θ_0 changes. This fact implies that a characterization of the information rent of the buyer follows by the envelope theorem if $\partial F(\theta_1|\theta_0)/\partial \theta_0$ and $f(\theta_1|\theta_0)$ are sufficiently well-behaved. We refer to

$$\left(\tilde{\theta}_0, \tilde{\varepsilon}_1, F^{-1}(\varepsilon_1|\theta_0) \right)$$

as the canonical (orthogonal) representation of the original information.

Since $\tilde{\theta}_0$ is independent of $\tilde{\varepsilon}_1$ by construction, we may view $\tilde{\theta}_0$ as the true private information to the agent and $\tilde{\varepsilon}_1$ as information not available at the moment of contracting. If $\tilde{\varepsilon}_1$ were publicly observable and contractible, then we would indeed be dealing with the static problem in the previous subsection. If the solution of the mechanism design problem with observable and unobservable $\tilde{\varepsilon}_1$ coincide, then the principal does not have to give any information rent to the agent in excess of that contained in $\tilde{\theta}_0$. We will discuss below when this conclusion holds and when it does not.

As long as $\tilde{\theta}_1$ is first-order stochastically increasing in $\tilde{\theta}_0$, we have $I_1 \geq 0$. In the Appendix, we provide a construction of the t-period impulse responses for a general Markov process $\{\tilde{\theta}_t\}_{t=0}^T$ and we record here that:

$$I_t(\theta^t, x^{t-1}) \triangleq -\prod_{k=1}^{t} \frac{\dfrac{\partial F(\theta_k|\theta_{k-1}, x_{k-1})}{\partial \theta_{k-1}}}{f(\theta_k|\theta_{k-1}, x_{k-1})},$$

where we allow the transition function to depend on the current state and the current allocation, thus a controlled stochastic process.

A very rough protocol for solving dynamic mechanism design problems can now be given as follows. First, find the dynamic equivalent of the envelope formula (4.1) in the payoff equivalence theorem to compute the

---

[8]Since

$$\varepsilon_1 \triangleq F(F^{-1}(\varepsilon_1|\theta_0)|\theta_0),$$

the second equality in the formula follows by total differentiation with respect to θ_0.

transfers as a function of the allocation process. Second, consider the relaxed principal's problem where her payoff is maximized subject to the constraint that the transfer is computed from (4.2). Third, verify that the obtained solution satisfies the dynamic equivalent of the full incentive compatibility requirement (4.3). We pursue this program in the next few subsections.

4.3. *Dynamic payoff equivalence*

Since the dynamic mechanism design problem inherits the trade-off between efficiency and information rent from the static problem, we must find a characterization for the information rent in terms of the allocation as in the static payoff equivalence theorem. This is where the above orthogonalized model becomes useful. In a model of perfect commitment, the mechanism designer maximizes her payoff from the perspective of $t = 0$. The orthogonalization gives a tractable solution for the agent's equilibrium payoff in $t = 0$ (and hence also her expected transfers).

Recall that in a dynamic direct mechanism $(\mathbf{x}, \mathbf{p})$ the agent reports her type θ_t at each t. Any allocation rule x under truthful reporting induces a stochastic process whose transitions are given by:

$$\theta_{t+1} \sim F(\cdot|\theta_t, x_t),$$

and we denote this process by $\lambda[\mathbf{x}]$. By the dynamic revelation principle Myerson (1986), it is without loss of generality to consider a dynamic direct mechanism where the buyer reports her type θ_t truthfully in each t, and any such mechanism is said to be incentive compatible.

We want to compute the equilibrium payoff to the agent with initial private information θ_0:

$$V(\theta_0) = \mathbb{E}^{\lambda[\mathbf{x}]}\left[\sum_{t=0}^{T} \delta^t (v(x_t(\theta_t, h_t), \theta_t) - p_t(\theta_t, h_t))\right],$$

where the expectation is taken with respect to $\lambda[\mathbf{x}]$. The following theorem is a special case of the characterization of local incentive compatibility in Theorem 1 of Pavan, Segal, and Toikka (2014).[9]

---

[9]Modulo changing the notation for a different starting date and starting history, the same characterization holds for $\partial V_s(\theta_s, \theta^{s-1})/\partial\theta_s$ for all s and all $\theta^s \in \Theta^s$.

Theorem 3 (Dynamic Payoff Equivalence): *If* $(\mathbf{x}, \mathbf{p})$ *is incentive compatible, then* $V_0(\theta_0)$ *is Lipschitz continuous and has almost everywhere the derivative:*

$$V'(\theta_0) = \mathbb{E}^{\lambda[x]} \left[\sum_{t=0}^{T} I_t(\theta^t, x^{t-1}) \delta^t \frac{\partial v(x_t(\theta_t, h_t), \theta_t)}{\partial \theta_t} \right]. \tag{4.5}$$

The reason for using the canonical state representation is that it allows for an application of the Milgrom-Segal envelope theorem more easily than the original formulation. The representation also shows that the equilibrium payoff to the agent from truthful reporting is the same in the original game and the game where the $\widetilde{\varepsilon}_t$ are publicly observed. This indeed follows immediately from the assumption of truthtelling and the envelope formula. We will return to this issue, but it should be noted already here that there are instances where truthtelling is not optimal in a mechanism with privately observed $\widetilde{\varepsilon}_t$ but where the allocation rule can be implemented with observable $\widetilde{\varepsilon}_t$.

As in the static case, the payoff equivalence theorem shows that the (dynamic) allocation pins down the agent's payoff and therefore her transfers up to a constant. To interpret the result, let

$$U(x, \theta) \triangleq \sum_{t=0}^{T} \delta^t v(x_t, \theta_t)$$

so that

$$\frac{\partial U}{\partial \theta_t} = \delta^t \frac{\partial v}{\partial \theta_t}.$$

The derivative of the indirect utility (4.5) then becomes:

$$V_0'(\theta_0) = \mathbb{E} \sum_{t=0}^{T} I_t(\theta^t, x^{t-1}) \delta^t \frac{\partial v(x_t, \theta_t)}{\partial \theta_t}.$$

The impulse response function measures the effect of a small change in θ_s on θ_t and $\partial v / \partial \theta_t$ measures the induced change in period t utility. All the other effects across periods depend on the reported types, not the true types. The transfers that support the indirect utility of the agent can then be derived just as in the static model.

4.4. *Dynamic incentive compatibility*

Characterizing incentive compatible dynamic mechanisms is not hard at a very abstract level. A counterpart for the static integral monotonicity condition in equation (4.3) can be given as follows. A mechanism $(\mathbf{x}, \mathbf{p})$ is incentive compatible if the transfers are computed from formula (4.5) and if for each s, θ^s and and report m_s in period s,

$$\int_{m_s}^{\theta_s} \left[\frac{\partial V_s\left(q, \theta^{s-1}\right)}{\partial \theta_s} - \frac{\partial V_s^{m_s}\left(q, \theta^{s-1}\right)}{\partial \theta_s} \right] dq \geq 0, \qquad (4.6)$$

where $\partial V_s^{m_s}(q, \theta^{s-1})/\partial \theta_s$ is the derivative of the continuation payoff with respect to the true type q given the reports $\theta^s = (m_s, \theta^{s-1})$ and given that all future reports are truthful.

Even though a condition for full incentive compatibility can be expressed in a concise form, it does not yield an easy characterization of the feasible mechanisms for the principal. Pavan, Segal, and Toikka (2014) offer stronger sufficient conditions for implementability that are easier to verify. A very strong sufficient condition is that the payoff functions are supermodular in the allocation and type and that the allocation (at all histories) is non-decreasing in all types.

It should also be noted that while the canonical representation is convenient for proving the payoff equivalence theorem, it is not helpful for analyzing full incentive compatibility. When one writes the payoff functions in terms of the canonical representation $(\tilde{\theta}_0, \{\tilde{\varepsilon}_t\}_{t=1}^T, \{Z_t\}_{t=1}^T)$ as defined in the Appendix using the orthogonal shocks $\tilde{\varepsilon}_t$, one obtains:

$$v_t(x_t, \theta_t) = v(x_t, Z_t(\theta_0, \varepsilon^t, x^{t-1})) \triangleq \hat{v}_t(x_t, \varepsilon^t, x^{t-1}).$$

In other words, the agent's utility in period t depends on variables that are determined by past and current reported types and the whole sequence of realized ε_t. The first dependence is present in the original model as well, but the second type breaks the Markovian nature of the agents's problem. This adds new difficulties into checking full incentive compatibility since checking for optimal behavior after non-truthful messages is no longer easy.[10]

---

[10]Within the Markovian setting of the original model, the future incentives for truthtelling are independent of past types. Hence if a report is optimal on the equilibrium path for type θ_t, it will also be optimal following non-truthful reports.

4.5. *Optimal dynamic mechanism*

We proceed to describe the solution of the optimal dynamic mechanism design. Similar to the static case, we start by considering a relaxed problem where the only constraint for the problem is that the transfers are calculated from the payoff equivalence theorem.

Relaxed Problem. We denote the dynamic payoff to the mechanism designer by

$$\sum_{t=0}^{T} \delta^t (p_t - c_t(x_t)).$$

She designs a mechanism to maximize her own payoff. As always, we can write the designer's payoff as the difference between the social surplus and the agent's information rent. Using formula (4.5) to substitute for the payments gives after the usual integration by parts the following program for maximizing the designer's payoff from period $t = 0$ perspective:

$$\max_{(x,p)} \mathbb{E}^{\lambda[x]} \sum_{t=0}^{T} \delta^t (v(x(\theta_t, h_t), \theta_t) - c(x(\theta_t, h_t))) \tag{4.7}$$

$$- \mathbb{E}^{\lambda[x]} \frac{1 - F_0(\theta_0)}{f(\theta_0)} \left[\sum_{t=0}^{T} \delta^t I(\theta^t, x^{t-1}) \frac{\partial v(x(\theta_t, h_t), \theta_t)}{\partial \theta_t} \right] - V(\underline{\theta}_0).$$

The maximization is also subject to period 0 participation constraints:

$$V(\theta_0) \geq 0.$$

We denote the first line in the objective function by $\mathbb{E}^{\lambda[x]}[S(x, \theta)]$ to represent the social surplus. We have built the local incentive compatibility conditions into the objective function by using the envelope formula to represent the buyer's information rent. If the stage payoff functions of the agent are supermodular and if $\widetilde{\theta}_{t+1}$ is first-order stochastically increasing in $\widetilde{\theta}_t$, then the individual participation constraint typically bind at the optimum for the lowest type and thus $V(\underline{\theta}_0) = 0$.

Even though the relaxed problem can be written rather concisely, solving

$$\max_{x} \mathbb{E}^{\lambda[x]} \left[S(x, \theta) - \frac{1 - F(\theta_0)}{f(\theta_0)} \sum_{t=0}^{T} \delta^t I_t \frac{\partial v_t}{\partial \theta_t} \right] \tag{4.8}$$

involves dynamic programming and is not easy in general. When discussing the applications below, we shall see some instances where more or less explicit solutions to the problem exist.

Properties of the Solution to the Relaxed Problem. If the process of $\{\widetilde{\theta}_t\}_{t=1}^T$ does not depend on the allocations x^t and if there are no intertemporal restrictions on x_t, then a pointwise solution is possible as in the static case. Examples of this setting were covered already in Baron and Besanko (1984) and Besanko (1985). In this case, we can deduce some immediate properties of equation (4.8).

First, if the type is perfectly persistent, then $I_t(\theta^t) = 1$ for all t and θ^t. This implies that the optimal pointwise solution collapses to the static solution to

$$\max_x \left\{ s_t(x, \theta) - \frac{1 - F(\theta_0)}{f(\theta_0)} \frac{\partial v_t}{\partial \theta_t} \right\}. \tag{4.9}$$

Notice that here the distortions in the allocation rule remain over time.

Second, if the $\widetilde{\theta}_t$ are independent of $\widetilde{\theta}_0$ for $t > 0$, then x_t maximizes the social surplus for all $t > 0$. In this case, initial private information has no effect on future types.

Third, if the type process follows an AR(3.1) process:

$$\theta_t = \lambda \theta_{t-1} + \varepsilon_t,$$

with $\theta_{-1} = 0$, then one finds from the moving average representation:

$$\theta_t = \sum_{s=0}^t \lambda^{t-s} \varepsilon_s$$

that $I_t(\theta^t) = \lambda^t$. Since we require that $|I_t(\theta^t)| < \infty$, we must have $\lambda \leq 1$. If we have a persistent random walk, i.e. $\lambda = 1$, then the solution is as with persistent types. If $\lambda < 1$, then the distortions from the efficient allocation vanish as $t \to \infty$. This simply reflects the fact that as time goes on, the effect of the initial shock $\theta_0 = \varepsilon_0$ on θ_t vanishes and at the moment of contracting, the principal and the agent have almost identical beliefs about $\widetilde{\theta}_t$ for large t.

Full Solution. If the relaxed problem allows for an explicit solution, one can check if the sufficient conditions for full incentive compatibility are satisfied. For the examples that we describe below, the solution of the relaxed problem can be characterized in sufficient detail to allow us to verify sufficient conditions for full incentive compatibility.

The optimal solution in the original mechanism design problem coincides with the optimal solution in the modified problem of the canonical representation where the orthogonal shocks $\widetilde{\varepsilon}_t$ are publicly observable. By the payoff equivalence theorem, this implies that the agent gets the same expected payoff in the two problems. Hence it is reasonable to say that the agent does not benefit from the additional private information that she gets during the game.

At the same time, if the solution to the relaxed problem is not fully incentive compatible, this is no longer true. In the appendix, we present an example showing that the solution of the relaxed problem may not be incentive compatible even though the solution is incentive compatible when the $\widetilde{\varepsilon}_t$ are publicly observable.

Implementing the Solution. Unfortunately there is no general recipe along the lines of the taxation principle for natural indirect implementations of the optimal direct mechanism in the dynamic setting. In some cases, the solution to the optimal contracting problem is suggestive of natural ways to implement the solution. For example, the sequential screening problem and the dynamic auction formats discussed below have solutions that can be implemented through option contracts and through various types of handicapped auctions.

5. Leading Applications

We shall now discuss how the general insights translate into specific solutions in a number of important economic applications.

5.1. *Sequential screening*

Starting with Courty and Li (2000), the simplest model of bilateral trading with a dynamic flow of information has been called the *sequential screening* model. The canonical model extends over two periods $t \in \{0, 1\}$ with trade taking place only in period $t = 1$. An uninformed seller proposes a mechanism to an informed buyer with type $\theta_0 \in [\underline{\theta}, \overline{\theta}]$ in $t = 0$. Her second period type $\theta_1 \in [\underline{\theta}, \overline{\theta}]$ is unknown to both parties at the moment of contracting, but it is common knowledge that its conditional distribution is given by $F(\theta_1|\theta_0)$. The prior on θ_0 is denoted by $F_0(\theta_0)$.

The key economic question for the model is whether the seller can use the dynamic nature of information arrival to increase her expected revenue. Obviously she can sell using an optimal Myerson mechanism based on either

$\widetilde{\theta}_0$ or on $\widetilde{\theta}_1$. The main observation in sequential screening models is that she can do strictly better by offering an option contract. In $t = 0$, the seller offers a menu of strike prices for period $t = 1$. Based on different θ_0, the buyers choose different strike prices (obviously at different up-front payments) and this improves the seller's payoff.

Different interpretations are possible for θ_0. It can be thought as a prior mean for θ_0 or alternatively it can be thought of as a measure of the precision of the agent's prior information about θ_1. In the first case, it would be natural to assume that $\widetilde{\theta}_1$ is first-order stochastically increasing in $\widetilde{\theta}_0$ while in the second, one would expect $\widetilde{\theta}_1$ to be second-order stochastically increasing in $\widetilde{\theta}_0$.

Since trading takes place only in period $t = 1$, there is no loss of generality in assuming that θ_1 is the value of the buyer in $t = 1$. We also assume that the good is indivisible (or alternatively the payoffs are linear in quantities) and that the seller has no value for the object herself. This leads to the payoffs:

$$u_S(\theta_1, x, p) = p,$$

$$u_B(\theta_1, x, p) = \theta_1 x - p,$$

for the seller and the buyer respectively, where x is the probability of trading and p is the transfer from the buyer to the seller. We also assume that the outside option for the buyer yields payoff 0.

We can use the general result from the previous section to see rather quickly how to arrive at this solution. A direct dynamic mechanism is now a pair of functions:

$$x : \Theta_0 \times \Theta_1 \to [0, 1],$$

$$p : \Theta_0 \times \Theta_1 \to \mathbb{R}_+.$$

We recall that the single period impulse response function can be written as:

$$I_1(\theta_0, \theta_1) = -\frac{\dfrac{\partial F(\theta_1|\theta_0)}{\partial \theta_0}}{f(\theta_1|\theta_0)}.$$

Together with equation (4.5), this gives:

$$V'(\theta_0) = -\int_{\Theta_1} x(\theta_0, \theta_1) \frac{\partial F(\theta_1|\theta_0)}{\partial \theta_0} d\theta_1.$$

Solving for the expected transfer (i.e. the seller's expected payoff), the relaxed problem becomes (after an integration by parts and using the individual rationality constraint $V(\underline{\theta}) = 0$):

$$\max_{x} \int_{\Theta_0} \int_{\Theta_1} x[\theta_1 \frac{1 - F_0(\theta_0)}{f_0(\theta_0)} I_1(\theta_0, \theta_1)] f(\theta_1|\theta_0) f(\theta_0) d\theta_1 d\theta_0.$$

But this is a linear problem in x and hence the relaxed solution is easy to find.

Define a modified virtual value $\psi(\theta_0, \theta_1)$ by

$$\psi(\theta_0, \theta_1) \triangleq \theta_1 - \frac{1 - F_0(\theta_0)}{f_0(\theta_0)} I_1(\theta_0, \theta_1).$$

This modifies the classic Myersonian virtual value by multiplying the information rent component $(1 - F_0(\theta_0))/f_0(\theta_0)$ by the impulse response I_1.

Since the value of the integral is linear in x, it is clearly optimal to set $x(\theta_0, \theta_1) = 1$ whenever $\psi(\theta_0, \theta_1) \geq 0$ in the relaxed program. If $\psi(\theta_0, \theta_1)$ is strictly increasing in both components, then this solution solves the revenue maximization problem. Hence we assume from now on that ψ is increasing in both arguments. To complete the description of the optimal mechanism, define the following function

$$q(\theta_0) = \min\{\theta_1 \in \Theta_1 | \psi(\theta_0, \theta_1) \geq 0\}.$$

Since ψ is increasing, $q(\cdot)$ is well-defined. With the help of this function, we can characterize the optimal selling mechanisms.

Theorem 4 (Optimal Screening Mechanism): *If $\psi(\theta_0, \theta_1)$ is increasing in both arguments, then a direct dynamic mechanism (x, t) maximizes the seller's expected profit in the class of incentive compatible mechanisms if and only if*

$$x(\theta_0, \theta_1) = \mathbb{I}_{\{\theta_1 \geq q(\theta_0)\}},$$

and the transfer is computed from the envelope formula.

As anticipated, in the optimal mechanism the buyer pays an up-front fee $p(\theta_0)$ for the option of purchasing the good at strike price $q(\theta_0)$. Hence the mechanism seems to bear some relation to contracts that are actually observed in situations where uncertainty is gradually resolved and revealed about the value of the alternatives.

Courty and Li (2000) show that in the case where $\widetilde{\theta}_1$ is second-order stochastically increasing in $\widetilde{\theta}_0$, the standard Myersonian downward distortions may be reversed. If $\widetilde{\theta}_1$ is first-order stochastically increasing in $\widetilde{\theta}_0$, this is not possible. This result can be understood in terms of the sign of the impulse response function in the two cases. Under first-order stochastic dominance (FOSD), $I_1(\theta_0, \theta_1)$ is always positive. For the case of second-order stochastic dominance (SOSD), it may well be negative, thus leading to a reversal in the direction of the distortions.

Eső and Szentes (2007) extend the model to allow for multiple bidders for the good (otherwise the model is identical to the model above). They find an optimal auction-called the *handicap auction*-where the bidders can make up-front payments in the first period to influence the allocation rule determining the second period allocation (the handicaps for the final auction). In order to analyze the model, Eső and Szentes (2007) introduced the orthogonalization process described in section (4.2). They compare the revenue to the seller under two scenarios: one where she releases the orthogonal signals to the buyers and one where she does not. They conclude that the seller is always better off releasing the information.

Bergemann and Wambach (2015) and Li and Shi (2017) offer extensions of the sequential screening model that incorporate information *and* mechanism design. Li and Shi (2017) show that even though the seller always wants to release all of the orthogonalized information to the buyer, she may prefer to send garbled information based on the original (not orthogonalized) type θ_1. The question of what types of disclosure policies are optimal in this setting is still open.

5.2. *Selling options*

The second illustrative example is a stopping problem rather than a recurrent allocation problem. Suppose we would like to allocate a single object among N bidders, but we can allocate it only once and for all. Thus, the seller faces a stopping problem, and at the moment of stopping must decide to whom to allocate the object. Suppose the evolution of the willingness to pay by bidder i is given by:

$$\theta_{i,t} = \gamma \theta_{i,t-1} + \varepsilon_{i,t},$$

with $\theta_{i,0} \sim G_i(\theta_{i,0}), \varepsilon_{i,t} \sim H_i(\cdot)$, i.i.d. If we set $\gamma = 1$, we are essentially dealing with the model of Board (2007).

We can now compute the indirect utility function in the familiar way,

$$V_{i,0}(\theta_{i,0}) = \mathbb{E} \sum_{t+0}^{T} \delta^t \frac{1 - G_i(\theta_{i,0})}{g_i(\theta_{i,0})} \gamma^t x_{i,t}(\theta),$$

and find that the expected revenue to the seller is

$$\mathbb{E} \sum_{t=0}^{T} \sum_{i=1}^{N} \delta^t \left[\theta_{i,t} - \frac{1 - G_i(\theta_{i,0})}{g_i(\theta_{i,0})} \gamma^t \right] x_{i,t}(\theta),$$

The seller's problem is thus an optimal stopping problem, and her decision in period t is whether to stop the process and collect

$$\max_i \left\{ \theta_{i,t} - \frac{1 - G_i(\theta_{i,0})}{g_i(\theta_{i,0})} \gamma^t \right\},$$

or to continue until $t+1$ and draw a new valuation vector $\theta_{t+1} = \gamma \theta_t + \varepsilon_t$ for the bidders. As time progresses and t increases, the distortion relative to the planner's solution in the allocation diminishes.

5.3. *Bandit auctions*

A single indivisible object is allocated in each period amongst n possible bidders who learn about their true valuation for the good. The type of bidder i changes only in periods t where she is allocated the good: if $x_{i,t} = 0$, then $\theta_{i,t+1} = \theta_{i,t}$, if $x_{i,t} = 1$, then

$$\theta_{i,t+1} = \theta_{i,t} + \varepsilon_i(n_i(t)) \tag{5.1}$$

where ε_i is a random variable whose distribution depends on the number of periods up to $t, n_i(t)$, in which the good has been allocated to i. For some stochastic processes such as the normal learning process outlined in Section 2.2, the number of observations from the process (here $n_i(t)$) and the current posterior mean (here $\theta_{i,t}$) form a sufficient statistic. We can interpret the allocation process as intertemporal licensing where the current use of the object is determined by the past and current reports of the bidders. Notably, the assignment of the object can move back and forth between the bidders as a function of their reports. Pavan, Segal, and Toikka (2014) and Bergemann and Strack (2015) consider a revenue maximizing auction for the special case of the multi-armed bandit model in discrete or continuous time, respectively. The efficient allocation policy under private information was analyzed earlier in Bergemann and Välimäki (2010).

A useful aspect of the bandit model with the additive noise model is the easily verified property that:

$$\prod_{t=r}^{S} \left(-\frac{\frac{\partial F_i(\theta_{i,t+1}|\theta_{i,t})}{\partial \theta_{i,t}}}{f_i(\theta_{i,t+1}|\theta_{i,t})} \right) = 1. \tag{5.2}$$

Hence the revenue maximization problem is now turned (again using the usual steps) into a modified bandit problem where the seller maximizes

$$\max_{x \in X} \mathbb{E} \sum_{t=0}^{T} \sum_{i=1}^{N} \delta^t \left[\theta_{i,t} - \frac{1 - F_i(\theta_{i,0})}{F_i(\theta_{i,0})} \right] x_i(\theta_{i,t}),$$

where $X = \{(x_1, \ldots, x_N) \in \mathbb{R}_+^N | \sum_i x_i = 1\}$. Stated in this form, the problem can be solved using the dynamic allocation index, the Gittins index. Pavan, Segal, and Toikka (2014) verify that the solution satisfies the average monotonicity condition and is hence implementable. Thus, the resulting dynamic optimal auction proceeds by finding the bidder with the highest valuation after taking into account the handicap, which is determined exclusively by the initial private information $\theta_{i,0}$. Moreover, by (5.2), the impulse response function, and hence the handicap is constant in time and determined only by the initial shock.

Kakade, Lobel, and Nazerzadeh (2013) consider a class of dynamic allocation problems that includes the above bandit problem. By imposing a separability condition (additive or multiplicative) on the interaction of the initial private information and all subsequent signals, they obtain an explicit characterization of the revenue-maximizing contract and derive transparent sufficient conditions for the optimal contract.

5.4. *Repeated sales*

A common economic setting where long-term contracts govern the inter-action between buyer and seller is the repeated sales problem. The buyer anticipates that he might purchase a good or a service repeatedly over time, but is uncertain about his future valuation of the good. At any point in time his willingness-to-pay is private information, and the current willingness-to-pay is a good prediction of the future willingness to pay. A variety of dynamic contracts are used to support the provision of services, as documented by Della Vigna and Malmendier (2006), Grubb (2009) and

Eliaz and Spiegler (2008) for gym memberships, mobile phone contracts and may other services.

These allocation problems can be viewed has being separable across periods in two important aspects: (i) the set of feasible allocations at time t is independent of the history of the allocations, and (ii) the flow utility functions depends only on current type. This class of models is particularly tractable since a pointwise solutions to the relaxed problem is quite easily obtained and the conditions for full incentive compatibility can be directly checked. In fact, the earliest contributions to the dynamic mechanism design literature, Baron and Besanko (1984) and Besanko (1985) restricted attention to time-separable problems of this form.

Bergemann and Strack (2015) consider time-separable allocation problems in continuous time. They leverage the structure of the continuous-time setting to obtain closed-form solutions of the optimal contract. In the leading example of repeat sales of a good or service, they establish that many commonly observed contract features such as flat rates, free consumption units and two-part tariffs can emerge naturally as part of the optimal contract.

In their setting, the flow value is given by $v_t x_t - p_t$, v_t is the willingness-to-pay in time t and x_t the quantity or quality assigned to buyer in period t. The willingness to pay is assumed to be a function:

$$v_t = \phi(t, \theta_0, W_t),$$

that is weakly increasing in the initial type θ_0 and the value of a Brownian motion W_t in period t. With the time separability of the allocation across periods, the virtual utility in period t is simply given by

$$v_t - \frac{1 - F(\theta_0)}{f(\theta_0)} \frac{\partial \phi(t, \theta_0, W_t)}{\partial \theta_0}. \tag{5.3}$$

This is simply the continuous-time analogue to the relaxed problem that we derived earlier in (4.9), where the derivative

$$\frac{\partial \phi(t, \theta_0, W_t)}{\partial \theta_0},$$

often referred to as the *stochastic flow* replaces the product of the marginal flow value times the impulse response function. The nature of the initial information θ_0 together with the shape of the stochastic process now determine how the stochastic flow, and ultimately the optimal allocation vary over time.

Bergemann and Strack (2015) then analyze how the optimal contract depends on the nature of initial private information and the structure of the stochastic process. In their leading case the valuation evolves as a geometric Brownian motion

$$dv_t = (v_t - \underline{v})\sigma dW_t,$$

where $\underline{v} \geq 0$ is a lower bound on the flow utility, and W_t is a Brownian motion. If the initial private information is simply the initial value of the process, $\theta_0 - v_0$, then the stochastic flow is simply

$$\frac{\partial \phi(t, \theta_0, W_t)}{\partial \theta_0} = \frac{v_t - \underline{v}}{v_0}.$$

Thus, the corresponding expression from discrete time, the impulse response function, reduces to a simple expression. They can consequently show that a menu of either flat rate contracts, or different two-part tariffs, or different free minute contracts can arise as optimal solutions depending on the value of the lower bound $\underline{v}$ and the flow cost of providing the service, given by $c(x)$.

By contrast, if the initial private information θ_0 is the drift of the geometric Brownian motion, thus

$$dv_t = v_t(\theta_0 dt + \sigma dW_t),$$

then the stochastic flow can be computed to be

$$\frac{\partial \phi(t, \theta_0, W_t)}{\partial \theta_0} = v_t t.$$

Now the optimal contract is a menu of leasing contracts with deterministic deadlines as the flow virtual utility takes the form:

$$v_t \left(1 - \frac{1 - F(\theta_0)}{f(\theta_0)} t \right).$$

Interestingly, the distortion is linearly increasing in time. It follows that in contrast to much of the models analyzed so far, the allocative distortion is now increasing over time rather than decreasing over time. A noteworthy aspect of this last example is that the initial private information of the agent is not the initial value of the stochastic process, but rather a parameter of the stochastic process itself.

5.5. *Private information about the stochastic process*

In fact, a number of recent contributions have considered the possibility that the initial private information is about a parameter of the stochastic process itself, such as the drift or the variance of the process. For example, Boleslavsky and Said (2013) let the initial private information of the agent be the mean of a multiplicative random walk. This changes the impact that the initial private information has on the future allocations. The distortions in the future allocation may now increase over time rather than decline as in much of the earlier literature. The reason is that the influence of the parameter of the stochastic process on the valuation may increase over time. Pavan, Segal, and Toikka (2014) and Skrzypacz and Toikka (2015) report similar findings.[11]

The impulse response function in Boleslavsky and Said (2013) involves the number of past realized upticks and downticks of the binary random walk. Bergemann and Strack (2015) consider the continuous-time version of the multiplicative random walk, the geometric Brownian motion. Interestingly, in the continuous-time version, the impulse response function is simply the expected number of upticks or downticks, which is a deterministic function of time and the initial state. This implies that the factor modifying the standard Myersonian virtual valuation is increasing linearly over time, and the optimal contract prescribes a deterministic time at which the trade ends, thus suggesting a leasing contract with fixed term length. More generally, Bergemann and Strack (2015) allow the valuation process of the buyer to be either the arithmetic, geometric, or mean-reverting Brownian motion. Across these classes of models, they show that the allocative distortion of the revenue-maximizing contract can be constant, decreasing, increasing, or even random over time depending on the precise nature of the private information.

5.6. *Beyond the first-order approach*

The method of analysis for the dynamic contracting problem above relies heavily on the payoff equivalence theorem, also known as the first-order approach. For this approach to be successful, the models must be such that the solutions to the relaxed problem are incentive compatible. Battaglini

---

[11]This is equivalent to assuming that the private information of the agent corresponds to the state of a two-dimensional Markov process, whose first component is constant after time zero, but influences the transitions of the second component.

and Lamba (2017) show that in models with discrete types, the first-order approach becomes problematic if the agents interact frequently. In particular, the solution to the first-order problem is no longer monotonic if types are highly persistent. They propose and analyze optimal contracts in the class of strongly monotonic allocation functions and show that these contracts are approximately efficient in the class of all incentive compatible contracts.

Garrett, Pavan, and Toikka (2017) take a different approach to the problem. They characterize necessary properties of the optimal contract by a relatively simple perturbation argument. They show that regardless of whether the first-order approach is applicable or not, the optimal contract must have vanishing distortions as long as the underlying process on types is sufficiently mixing in the sense that the impact of initial information on future types vanishes. Hence this paper confirms for a larger class of models one of the key findings in Battaglini (2005) derived for models with binary types.

6. Dynamic populations

In this section, we consider mechanism design problems where the population of privately informed agents changes over time. To fix ideas, we return to our leading example of a seller who has a fixed capacity K of indivisible goods to sell by a (possibly infinite) deadline T. Potential buyers arrive according to a stochastic process and the seller wants to extract as much revenue as possible from them. Variants of this problem have been studied in the literature on *revenue management* in management science and in operations research.

Important economic examples fit this description very nicely. By far the most important and the most analyzed example is the pricing of airline tickets. As airlines customers have noticed. prices for identical tickets on a given flight vary over time. The airlines industry uses various dynamic pricing and allocation methods for the seats. They use time-varying posted prices that may depend on the query data for the flight in question, and sometimes also more complicated mechanisms allowing for the possibility of securing a future price by paying an up-front fee. These feature are important for potential buyers as well. Forward looking buyers should time their purchases optimally given their expectation of the price path. In fact, services such as Kayak have been developed to alert buyers to particularly good moments to purchase tickets.

Optimal timing of the purchases is a natural element in any dynamic mechanism design problem of this type. The findings in this literature have a clear substantive message. In a wide class of models, the literature shows that sellers cannot be made worse off if the buyers are forward looking rather than myopic. Moreover, the analysis of models with forward looking buyers guides practical implementations for the revenue maximizing scheme.

A key modeling decision with dynamic populations is whether the arrival of a buyer is publicly observable or not. We start with the simpler models where observability is assumed. We discuss also models where arrival is private information to the buyer. In this case, the buyer's type has two dimensions: age and valuation, but the model has enough structure that the analysis remains tractable.

6.1. *Observable arrival of short-lived buyers*

The first approach in the revenue management literature was to assume away problems of asymmetric information, i.e. assume that the seller observes the valuations and that the buyers are short lived (i.e. they disappear if they are not allocated the good).[12] In this framework, the seller's problem coincides with the problem of surplus maximization and the analysis is a standard (but not analytically simple) exercise in dynamic programming. Within the revenue management literature, the key analytical aspect of the problem is to find a characterization for the optimal allocation rule for the goods as a function of remaining objects k at any point in time, and remaining time t to the deadline. The key finding in this literature is that the optimal allocation rule is often given by a cut-off rule in the set of types or valuations: allocate at (k, t) if and only if $v_t \geq g(k, t)$ for some function g that is typically decreasing in k and increasing in t.[13]

The mechanism design approach to this problem emphasizes the effects of incentive compatibility when the buyers's types are not observable to the seller. When buyers are short-lived and their process of arrivals is observable to the seller, then we are specializing the general model to the case described at the end of Subsection 2.2, where the agent's type in period t is $\theta_{i,\tau_i} \in [0, \overline{\theta}]$ if $t = \tau_i$ for some (possibly random) publicly observed arrival period τ_i and

---

[12]The classical references are Derman, Lieberman, and Ross (1972) for the case of known distribution of buyer valuations and Albright (1977) for the case where the seller learns about the distribution based on the observed types.
[13]With an infinite deadline, the problem becomes stationary if the arrival process of buyers is stationary and the solution of the process simplifies considerably.

$\theta_{i,\tau} \in \theta_0$ if $t \neq \tau_i$ and her payoff from allocation x in a (possibly random) period τ_i is $v_i(x_{i,\tau_i}, \theta_{i,\tau_i})$.

In the simplest case, the buyers have unit demands for the object and they have independent valuations. By the payoff equivalence theorem, incomplete information about the type of the buyer transforms the maximization of total expected revenue to the relaxed problem of maximizing expected virtual surplus. For notational convenience, we assume that the distribution of the realized type does not depend on the arrival time τ_i, but this could be easily accommodated in the model as well.

For the case of identical objects and with $v_i(x_{i,\tau_i}, \theta_{i,\tau_i}) = \theta_{i,\tau_i} x_{i,\tau_i}$, where $x_{i,\tau_i} \in \{0,1\}$ indicates whether the object is allocated or not to i in τ_i, we can write the expected revenue of the seller in terms of the expected virtual utility:

$$\mathbb{E}_\theta \sum_{i-1}^{N} \delta^{\tau_i} \left[\theta_i - \frac{1 - F(\theta_i)}{F(\theta_i)} \right] x_{i,\tau i}$$

$$\text{such that } \sum_{i=1}^{N} x_{i,\tau}(\theta_i) \leq K \text{ for all } \theta_i \in \Theta_i,$$

where the expectation is taken over the vector θ of type processes and the allocation decisions depend only on the realized part θ^{τ_i} of the process at τ_i.

Full incentive compatibility typically boils down to an appropriate monotonicity requirement for the allocation rule in type $\theta_{i,t}$. For the case of identical objects, monotonicity is equivalent to a cut-off characterization of the allocation rule. Hence a sequence of posted prices can always implement the optimal allocation. Versions of this problem has been analyzed in a sequence of papers by Gershkov and Moldovanu (2009a), (2009b), and additional results collected in Gershkov and Moldovanu (2014). The revenue maximization problem in Gershkov and Moldovanu (2009a) allows for the possibility that the K objects to be allocated have different (vertical) qualities. Gershkov and Moldovanu (2009b) shows that learning or correlation in types may cause problems for the monotonicity even for the socially efficient allocation rule. Thus, the difficulty of obtaining a monotone allocation rule does not arise solely due to non-monotonic virtual surpluses. This observation can be illustrated nicely within our leading example.

Example 3 (Learning and Incentive Compatibility): A single indivisible good is to be allocated efficiently to one of two bidders $i \in \{1, 2\}$, both with a strictly positive valuation for the object. It is commonly known that $\tau_i = i$ and therefore the relevant social allocation decisions are given by $x \in \{1, 2\}$, where the first choice indicates allocating the object to bidder 1 in period 1 and the second indicates allocating to bidder 2 in period 2. The planner's objective is to maximize the social surplus and she has a discount factor δ.

The valuation $\theta_1 \in \{0, 1\}$ of bidder 1 is known at the outset of the game. Bidder 2 learns her value in period 2. The valuations can come from one of two possible distributions: θ_i is uniformly distributed on either $[0, \frac{1}{2}]$ or $(\frac{1}{2}, 1]$. The prior probability of each of these distributions is identical.

With observable types, the planner's optimal solution is immediate: allocate to agent 1 in period 1 iff

$$\theta_1 \in \left[\frac{1}{2}\delta, \frac{1}{2}\right] \cup \left[\frac{3}{4}\delta, 1\right].$$

As long as $\delta > 2/3$, this allocation rule is not monotone and since the bidders' payoffs are supermodular, it fails to be incentive compatible for the case of unobserved types. If a bidder can make a payment only in the period when he receives the good, we see that there is no way of implementing the efficient decision rule in the model with incomplete information. This problem does not arise if we can condition payments on the reports of both types. The team mechanism derived in Section 3 works nicely here if such contingent payments are allowed. Hence the example points out the problems that arise as a consequence of the (often quite realistic) requirement that monetary transfers occur only in conjunction with physical allocation decisions. This requirement sometimes goes under the name of "online" payments.

6.2. *Unobservable arrival of long-lived buyers*

With unobservable arrivals, the buyers will have an incentive to time their purchases strategically. If prices decrease over time, they will delay reporting to the mechanism in order to get a better deal later on. Any incentive compatible mechanism must take this possibility into account. By contrast, if the arrivals are publicly observed then this is not a concern. The seller may simply commit not to offer any contracting opportunities except in the period of arrival. If the arrivals are not observed, the seller cannot distinguish between new arrivals in any period t from those that arrived earlier and waited with their announcement.

Board and Skrzypacz (2015) consider the sales problem of K identical indivisible units to a population of arriving buyers when the arrivals are private information. In their model, the statistical properties of the arrivals and valuations are common knowledge at the beginning of the game and arrivals and types satisfy independence across buyers and across periods. They show that the optimal selling mechanism is surprisingly simple: it is a deterministic sequence of posted prices depending on (k, t). Interestingly this pattern leads to waiting by the buyers along the equilibrium path. Even if the seller knew the past realized arrivals, this would not change the solution. If the demand is decreasing over time, they find a surprisingly explicit analytical solution for the problem. The other main substantive finding is that under the optimal mechanism, the seller gets a higher revenue than she would get if the buyers were short-lived. This happens even though sales are more back-loaded in the case with forward-looking buyers and prices are falling. If the modeling assumptions are relaxed, each of these two main predictions may fail and complete solutions seem difficult to obtain. See for example the analysis in Mierendorff (2016) for the case where the agents are forward looking but may disappear or have different discount factors.

Gershkov, Moldovanu, and Strack (2018) extend the model to cover the case where the buyers' arrival process is initially not known. This relaxes the assumption of independence in arrival times. They show that even though the optimal allocation is no longer implementable through anonymous posted prices, a simple name-your-own-price mechanism can be used as an indirect mechanism that achieves the maximal revenue. Further results in the paper show that the seller does not benefit from hiding information, say, about the existing stock of k units and that forward-looking buyers still benefit the seller as in Board and Skrzypacz (2015).

Since the models of dynamic populations where the buyers' types, the willingness-to-pay, are fixed over time lead to considerations of strategic timing, it is natural to ask how the case where buyers' types change over time would change the problem. In this case, there are two reasons for optimizing over the purchasing time: the price may be more favorable in the future or the type may change to one with a higher information rent.

A related issue is the timing of the contractual agreement between principal and agents. Much of the current analysis assumes that the arrival of the agents is known to the principal and that the principal can make a single, take-it-or-leave-it offer at the moment of the agent's arrival. This constraint, while natural in a static setting, is much less plausible in dynamic settings. In particular, it explicitly excludes the possibility for the agent to postpone and delay the acceptance decision to a later time when he may have additional information about the value of the contract offered to him.

Garrett (2017) considers a model of sales of a non-durable good where the buyer appears at a random future time, the arrival is private information, and her privately known valuation changes over time. He shows that in the optimal mechanism, the principal commits to punishing the agent for late arrival by inducing more inefficiencies to diminish information rents from manipulating the entry time.

A more ambitious attempt in this direction appears in Garrett (2016) where generations of new buyers are arriving over time to contract with a seller. A full mechanism design approach is not tractable in this case, and the paper restricts attention to optimal time-dependent sequence of posted prices. Using anonymous posted prices implies that old and new buyers with the same valuation type have the same incentives for all purchases. In this sense, explicit penalties for late arrivals are not possible. In an otherwise stationary environment, the optimal posted price fluctuates. This comes as a surprise after the well known result in Conlisk, Gerstner, and Sobel (1984) showing that stationary prices are optimal because the forward-looking buyers with high values anticipate lower prices and therefore are reluctant to buy at high prices. In Garrett (2016), high valuation buyers are more keen to buy immediately since they understand that their type may decrease in the future.

Bergemann and Strack (2017) analyze a dynamic revenue-maximizing problem in continuous time when the arrival time of the agent is uncertain and unobservable to the seller. The valuation of the agent is private information as well and changes over time. They derive the optimal dynamic mechanism, characterize its qualitative structure and derive a closed form solution. As the arrival time of the agent is private information, the optimal dynamic mechanism has to be stationary to guarantee truth-telling. The truth-telling constraint regarding the arrival time can be represented as an optimal stopping problem. They show that the ability to postpone the acceptance of an offer to a future period can increase the value of the buyer and can lead to a more efficient allocation resulting in equilibrium.

7. Connections to Nearby Models

In this section, we briefly discuss two classes of dynamic contracting models that do not assume quasi-linear payoffs. Since Rogerson (1985), models of dynamic moral hazard have discussed the smoothing of dynamic risks in models with incentive problems. In dynamic settings, the distinction between dynamic moral hazard and adverse selection is almost impossible to make and many models that share the informational structure with our general dynamic model have been discussed under the name of dynamic moral hazard. The key difference between these models and those discussed in the previous sections is that with risk averse preferences, the trade-off between efficient physical allocation and efficient risk allocation emerges. Whether private information exists at the moment of contracting or not is not that important for this literature since the incentives-insurance trade-off emerges in any case as private information is generated.

In models of financial economics, a key assumption is that the privately informed managers may be risk-neutral but that they do not have sufficient funds to buy the entire enterprise. This is typically formalized through a limited liability constraint stating that the manager (the agent) cannot make payoffs to the owner (the principal). Recent work starting with Clementi and Hopenhayn (2006), DeMarzo and Sannikov (2006), DeMarzo and Fishman (2007), and Biais, Mariotti, Plantin, and Rochet (2007) has analyzed the problem of incentivizing a manager who privately observes the cash flow of a firm.

7.1. *Risk-averse agent*

In most mechanism design problems, the key problem for the designer can be formulated as follows: what is the most advantageous way of providing the agent with a fixed level of utility u_0 (e.g. to satisfy a participation constraint). With risk-averse agents and a risk-neutral principal, optimal contracts provide some amount of insurance, but incentive compatibility precludes the possibility of full insurance. This problem has attracted a large amount of attention starting with Green (1987) and Thomas and Worrall (1990).[14] Our goal here is not to assess this literature but merely point out how it connects to the models in the previous sections of this survey.

---

[14] Many of the issues also arise in dynamic incentive provision models with hidden actions starting with Rogerson (1985).

The consumer derives utility $v(x_t, \theta_t)$ in period t from allocation x_t if her type is θ_t. We consider incentive compatible dynamic direct mechanism. The agent's problem is typically formulated as a dynamic programming problem (induced by the mechanism):

$$V(\theta_t) = \max_{r_t} \left\{ v(x(r_t, h_t), \theta_t) + \mathbb{E}\left[V(\widetilde{\theta}_{t+1})|\theta_t\right] \right\}.$$

Under sufficient regularity conditions, value functions V_t satisfying these equations exist and are sufficiently well behaved for an application of the envelope theorem. Indeed, if one assumes that the set of possible types is a connected interval and that the process of types has full support, then an application of the envelope theorem yields:

$$V'(\theta_t) = \frac{\partial v(x_t, \theta_t)}{\partial \theta_t} + \int_{\underline{\theta}}^{\overline{\theta}} V(\theta_{t+1}) \frac{\partial f(\theta_{t+1}|\theta_t)}{\partial \theta_t} d\theta_{t+1}.$$

Integration by parts gives:

$$V'(\theta_t) = \frac{\partial v(x_t, \theta_t)}{\partial \theta_t} + \mathbb{E}\left[\frac{-\frac{\partial F(\theta_{t+1}|\theta_t)}{\partial \theta_t}}{f(\theta_{t+1}|\theta_t)} V'(\theta_{t+1}) \right].$$

Hence by iterating this formula forwards, one gets as before:

$$V'(\theta_0) = \mathbb{E}\left[\sum_{t=0}^{\infty} I_t \delta^t \frac{\partial v(x_t, \theta_t)}{\partial \theta_t} \right].$$

In other words, a similar envelope theorem characterization for the agent's utility is still possible in this model. Many papers in the new public finance literature adopt this approach. For example Farhi and Werning (2007) study dynamic insurance schemes from this perspective.

But following with this first-order approach to dynamic problems, the next step of substituting the agent's payoff into the principal's objective unfortunately fails because the utility is not quasi-linear. As a result, solving the model is in general more difficult than in the quasi-linear case and numerical methods are typically needed. This also implies that checking full incentive compatibility becomes much harder in this class of models.

7.2. *Managerial contracts and hidden actions*

Garrett and Pavan (2012) consider a model where a risk-neutral principal contracts with a risk-neutral manager whose type (productivity) changes over time. The manager (the agent) has to be incentivized to take the optimal action at each point in time and the distortions now refer to the dynamic distortions relative to the model where incentives are provided in a setting with no private information. The paper shows that as long as the impulse response functions in the model are positive (for the privately observed productivity of the manager), then the distortions to the incentives diminish over time and incentives become more high-powered.[15]

Limited liability protection on the part of the agent implies an upper bound on the transfers that can be made from the agent to the principal. Often this constraint takes the form that all transfers must be from the principal to the agent. This prevents the principal from selling the enterprise to the agent at the outset even when there is no initial asymmetric information and hence no losses due to information rent left with the agent. A canonical model for this literature is one where the agent reports a privately observed i.i.d. cash flow to the principal in each period. The mechanism determines the transfers to the agent and a probability of continuing the project as functions of the (history of) reported cash flows. A key finding in this literature is that over time, the contract becomes more efficient, i.e. the probability of inefficient liquidation decreases over time. It should be noted that the intuition for this finding is very different compared to the models surveyed above. With limited liability, the optimal contract effectively saves funds for the agent so that she can buy the enterprise at a later time. Since recent surveys of this large literature exist (see for example Biais, Mariotti, and Rochet (2013)), we do not survey the topic here.

8. Concluding Remarks

It was our objective to give a broad and synthetic introduction to the recent work on dynamic mechanism design. We hope we have conveyed the scope and the progress that has been made in the past decade. Still, many interesting questions remain wide open. We shall describe some of them in these final remarks.

---

[15] In contrast, Garrett and Pavan (2015) consider the case where a risk-neutral prinicpal provides incentives for a risk-averse manager and shows that the power of incentives vanishes over time to reduce the overall riskiness in the contract.

The intertemporal allocations and commitments that resulted from the dynamic mechanism balanced trade-offs over time. These trade-offs were based on the expectations of the agents and the principal over the future states. In this sense, all of the mechanisms considered were Bayesian solutions and relied on a shared and common prior of all participating players. Yet, this clearly is a strong assumption and a natural question would be to what extent weaker informational assumptions, and corresponding solution concepts, could provide new insights into the format of dynamic mechanisms. For example, the sponsored search auctions, which provide much of the revenue for the search engines on the web, are clearly repeated and dynamic allocations with private information, yet, most of the allocations and transfer are determined by spot markets or short-term arrangements rather than long-term contracts. An important question then is why not more transactions are governed by long-term arrangements that could presumably share the efficiency gains from less distortionary allocations between the buyers and the seller. An important friction to long-term arrangements is presumably the diversity in expectations about future events between buyer and seller. In a recent paper, Mirrokni, Leme, Tang, and Zuo (2017) provide lower bounds for a revenue maximizing mechanism in which the players do not have to agree on their future expectations. The mechanism that achieves the lower bound in fact satisfies the interim participation and incentive constraints for all possible realizations of future states. This approach reflects the recent interest of theoretical computer science in dynamic mechanism design, see for example Papadimitriou, Pierrakos, Psomas, and Rubinstein (2016). But in contrast to the Bayesian approach most commonly taken by economic theorists who explicitly identify and design the optimal mechanism, theoretical computer scientists often describe achievable performance guarantees. The bounds are frequently achieved by mechanisms that have computational advantages in terms of computational complexity relative to the, possible unknown, exact optimal mechanism.

As an important friction to long-term arrangements is presumably the diversity in expectations about future events among the players, it is natural to ask to what extent the relevant insights from static mechanism design can be transferred to dynamic settings. Mookherjee and Reichelstein (1992) establish that in static environments the revenue maximizing allocation can frequently be implemented by dominant rather than Bayesian incentive compatible strategies. Similarly, Bergemann and Morris (2005) present conditions for static social choice functions under which an allocation can be implemented for all possible interim beliefs that the agents may hold. The robustness to private information is arguably an even more important consideration in dynamic environments.

The central problem that the literature of dynamic mechanism has addressed is how to provide incentives to report the sequentially arriving private information. Thus, the central constraints on the design are given by the sequence of interim incentive compatibility conditions. The participation constraints on the other hand have–somewhat surprisingly for a dynamic perspective–received much less attention. A dynamic mechanism requires voluntary participation at the ex-ante and interim stages via interim or periodic ex-post constraints. The dynamic pivot mechanism that governed the dynamically efficient allocation provided such an instance.

By contrast, the dynamic revenue maximization contract only imposed the participation constraint in the initial period. Interim participation constraints can be handled by allowing the agents to post bonds at the initial stage. In general, the mechanism does not provide any guarantees about ex-post participation constraints.[16] In fact, Krähmer and Strausz (2015) show that sequential screening frequently reduces to a static screening solution if the seller has to meet the ex-post rather than the ex-ante participation constraints of the buyers. More generally, if the dynamic mechanism improves upon a static mechanism in the sequential screening model, then the ex-post participation constraint severely limits the ability of the seller to extract surplus through option contracts as shown in Bergemann, Castro, and Weintraub (2017).

An interesting set of issues arise when the mechanism itself can govern only some of the relevant economic transaction. A specific setting where this is occurs are markets with resale. Here, the design of the optimal mechanism in the initially stage of the game is affect by the interaction in the resale market, see for example Calzolari and Pavan (2006), Dworczak (2016), Carroll and Segal (2016) and Bergemann, Brooks, and Morris (2017). In particular, the information that is generated by the mechanism may affect the nature of the subsequent interaction, and thus the tools from information design and mechanism design may jointly yield interesting new insights.

---

[16]Many commonly observed dynamic contracts do in fact violate ex-post participation constraints. For example, an insurance company does not return the premium in case of no accident.

9. Appendix

9.1. *Dynamic pivot mechanism and independence*

To see why the restriction to independent values is necessary, recall the transfer rule for agent i in the static pivot mechanism:

$$p_i(\theta) = -\sum_{j \neq i} u_j(x^*(\theta), \theta_j) + \sum_{j \neq i} u_j(x^*_{-i}(\theta_{-i}), \theta_j),$$

where $x^*_{-i}(\theta_{-i})$ is the optimal allocation for agents different from i. In the static case, x^*_{-i} depends only on the vector θ_{-i} by the assumption of private values regardless of any statistical dependencies between the agents' types. In the dynamic case, with correlated values, $\theta_{i,t}$ might have an effect on $x^*_{-i,t}$ even when fixing $\theta_{-i,t}$. As a result, both sums on the right hand side of (3.2) depend on $\theta_{i,t}$ and this distorts the incentives for truthful reporting. The following example illustrates this point.

Example 4 (Capacity Allocation and Correlated Types): Three agents $i \in \{1,2,3\}$ are bidding for a single indivisible object over three periods. Let $x_t \in \{1,2,3\}$ denote the possible allocations to i in period t. If the good is allocated in period s, then $\theta_{i,t} = 0$ for all i and all $t > s$ (say because it is not worthwhile to pay a cost to learn the valuation for an object that was already sold). Assume also that $\theta_{i,t} = 0$ if $i \neq t$ to indicate that agent i is active at most in period $t = i$. The payoff to agent i in period t is then $\Pr\{x_t = i\}\theta_{i,t}$.

Assume that $\theta_{1,1} \in \{3, 3-\varepsilon\}, \theta_{2,2} = 1$ if $x_0 = N$ and $\theta_{3,3} = 0$ otherwise, $\theta_{3,3} \in \{2, \varepsilon\}$ if $x_1 = x_2 = N$ and $\theta_{1,1} = 0$ otherwise. Let π denote the prior probability that $\theta_{2,2} = 2$ given that $x_0 = x_1 = N$, and assume that $2\pi\delta > 1$. Then it is optimal to have $x_1 = N, x_2 = 2$ conditional on $x_0 = N$. As long as $\varepsilon < 1$, the efficient decision is to have $x_0 = 0$. But it is clear that this is not incentive compatible if the transfers are calculated using the pivotal rule. To minimize the payments, agent 0 should always report type $\theta_{0,0} = 3 - \varepsilon$.

9.2. *The canonical representation of a Markov process*

Consider an arbitrary Markov process $\{\tilde{\theta}_t\}_{t=0}^{T}$ with the prior distribution $F_0(\theta_0)$ and the transition kernel $\tilde{\theta}_{t+1} \sim F(\cdot|\theta_t)$ on Θ. Since $\tilde{\varepsilon}_{t+1} = F(\tilde{\theta}_{t+1}|\theta_t)$ is uniformly distributed for all θ_t, we can deduce θ_{t+1} from $(\theta_0, \varepsilon_1, \ldots, \varepsilon_t)$ by using the recursive formula:

$$\theta_t = F^{-1}(\varepsilon_t|\theta_{t-1}) \triangleq Z_t(\theta_0, \varepsilon^t). \tag{9.1}$$

Since the functions $Z_t(.,.)$ are common knowledge at the start of the game, $(\tilde{\theta}_0, \ldots, \tilde{\theta}_t)$ contains the same information as $(\tilde{\theta}_0, \tilde{\varepsilon}_1, \ldots, \tilde{\varepsilon}_t)$. Pavan, Segal, and Toikka (2014) call the collection $(\tilde{\theta}_0, \{\tilde{\varepsilon}_t\}_{t=1}^{T}, \{Z_t\}_{t=1}^{T})$ the *canonical representation* of the process $\{\tilde{\theta}_t\}_{t=0}^{T}$.

In computing the dynamic payoff equivalence formula, we need to evaluate the impact of the initial private information on future types. Using the chain rule, we can compute from equation (9.1) the impact $\hat{I}_t(\varepsilon^t)$ of the initial private information θ_0 on θ_t for any fixed sequence of orthogonalized future types ε^t:

$$\hat{I}_t(\varepsilon^t) \prod_{k=1}^{t} \frac{\partial F^{-1}(\varepsilon_k|\theta_{k-1})}{\partial \theta_{k-1}} = -\prod_{k=1}^{t} \frac{\dfrac{\partial F(\theta_k|\theta_{k-1})}{\partial \theta_{k-1}}}{f(\theta_k|\theta_{k-1})} \triangleq I_t(\theta^t). \qquad (9.2)$$

The function $I_t(\cdot)$ that expresses this impact in terms of the original type formulation is called the *impulse response function* in Pavan, Segal, and Toikka (2014) and it plays a key role the characterization of the agent's information rent.[17] Notice that as long as $\tilde{\theta}_t$ is first-order stochastically increasing in $\tilde{\theta}_{t-1}$, we can show that $I(\theta^t) \geq 0$. From now on, we assume that $|I_t(\theta^t)| < K$ for some $K < \infty$ so that the formula for determining $I_t(\theta^t)$ makes sense, and we let $I_0(\theta_0) = 1$ for all θ_0.

The canonical representation is by no means a unique representation of the original model in terms of initial information and subsequent independent information. The next Appendix gives an example where a non-canonical representation allows us to overcome a differentiability problem in the canonical representation.

We have presented the construction here for homogeneous Markov processes, but the same procedure can be used to obtain a canonical representation for the more general processes $F(\theta_{t+1}|\theta_t, x^t)$. In this more general case, we denote the impulse response functions by $I_t(\theta^t, x^{t-1})$.

---

[17]Since
$$\varepsilon_k \equiv F(F^{-1}(\varepsilon_k|\theta_{k-1})|\theta_{k-1}),$$
the second equality in the formula follows by total differentiation with respect to θ_{k-1}
$$\frac{\partial F^{-1}(\varepsilon_k|\theta_{k-1})}{\partial \theta_{k-1}} = -\frac{\dfrac{\partial F(\theta_k|\theta_{k-1})}{\partial \theta_{k-1}}}{f(\theta_k|\theta_{k-1})}.$$

9.3. *Implementability in the orthogonalized model*

This example communicated to us by Juuso Toikka illustrates that the solution of the relaxed problem may not be (fully) incentive compatible, but where it can be implemented if the orthogonalized information were publicly observable.

A seller (mechanism designer) with cost function $c_t(x_1) = \frac{x_1^2}{2}$ sells to a privately informed buyer in period $t = 1$. In period $t = 0$, there is no trade and hence the allocation decision for that period is trivial. The buyer's (the agent's) type in $t = 0$ is uniformly distributed, $\theta_0 \sim U[0, 1]$, and in period $t = 1$, it remains unchanged with probability q. With probability $(1 - q)$, the type is drawn independently of θ_0. The canonical representation of the model is $(\tilde{\theta}_0, \tilde{\varepsilon}_1, Z_1(\theta_0, \varepsilon_1))$, where

$$
Z_1(\theta_0, \varepsilon_1) = \begin{cases} \dfrac{\varepsilon_1}{1 - q} & \text{if } 0 \le \varepsilon_1 \le (1 - q)\theta_0, \\[2mm] \theta_0 & \text{if } (1 - q)\theta_0 \le \varepsilon_1 \le (1 - q)\theta_0 + q, \\[2mm] \dfrac{\varepsilon_1 - q}{1 - q} & \text{if } (1 - q)\theta_0 + q \le \varepsilon_1 \le 1. \end{cases}
$$

Notice that $Z_1(\theta_0, \varepsilon_1)$ is not differentiable in θ_0 at $\theta_0 \in \left\{ \frac{\varepsilon_1}{1-q}, \frac{\varepsilon_1-q}{1-q} \right\}$ and hence the envelope theorem is not directly applicable.[18] This problem can however be overcome by selecting a different (non-canonical) orthogonal representation with a two dimensional $\eta_1 = (\eta_{11}, \eta_{12})$ where η_{11} is a Bernoulli random variable with $\Pr\{\eta_{11} = 1\} = 1 - \Pr\{\eta_{11} = 0\} = q$ and η_{12} is an independent uniform random variable. Then we can write

$$
\theta_1 = \hat{Z}_1(\theta_0, \eta_{11}, \eta_{12}) = \eta_{11}\theta_0 + (1 - \eta_{11})\eta_{12}.
$$

Now we see that $\hat{Z}_1$ is differentiable in θ_0 and

$$
I_1(\theta_0, \theta_1) = \mathbf{1}_{\{\theta_1 = \theta_0\}} \in \{0, 1\}.
$$

Hence the envelope theorem is applicable and we can write equation (4.8) for the relaxed problem as

$$
\max_{x_1 \ge 0} \left\{ x_1(\theta_1 - (1 - \theta_0)\mathbf{1}_{\{\theta_1 = \theta_0\}}) - \frac{x_1^2}{2} \right\}.
$$

---

[18]A similar example can be constructed where the canonical representation is well behaved by smoothing the distribution of θ_1 around θ_0.

The solution to this problem is

$$
x_1(\theta_0, \theta_1) = \begin{cases} \theta_1 & \text{if } \theta_1 \neq \theta_0, \\ \max\{0, 2\theta_1 - 1\} & \text{if } \theta_1 = \theta_0. \end{cases}
$$

The allocation rule is non-monotone in θ_1 for each $\theta_0 < 1$ and as a result, it cannot be implemented if η_1 is private information. On the other hand, it can be implemented if η_1 is observed by the seller. In this case, the seller knows θ_1 if $\eta_{11} = 0$ and any rule is trivially implementable. If $\eta_{11} = 1$, the solution to the problem is the usual Mussa-Rosen rule. The second part of the allocation rule x_1 above is the optimal scheme for this case. Hence we conclude that x_1 is implementable with publicly observed η_1.

Acknowledgment

We acknowledge financial support through NSF Grant 1459899 and a Google Faculty Fellowship. We would like to thank the editor, Steven Durlauf, and four anonymous referees for their extraordinarily helpful and detailed suggestions. We would like to thank Juuso Toikka and Philipp Strack for many helpful conversations. We would like to thank our students in the Advanced Economic Theory Class at Yale University who saw early versions of this survey and Ian Ball and Jaehee Song for valuable research assistance.

References

ALBRIGHT, C. (1977): "A Bayesian Approach to Generalized House Selling Problem," *Management Science*, 24, 432–440.

ATHEY, S., AND I. SEGAL (2013): "An Efficient Dynamic Mechanism," *Econometrica*, 81, 2463–2485.

BARON, D., AND D. Besanko (1984): "Regulation and Information in a Continuing Relationship," *Information Economics and Policy*, 1, 267–302.

BATTAGLINI, M. (2005): "Long-Term Contracting with Markovian Consumers," *American Economic Review*, 95, 637–658.

BATTAGLINI, M., AND R. LAMBA (2017): "Optimal Dynamic Contracting: The First-Order Approach and Beyond," Discussion paper, Cornell University and Penn State University.

BERGEMANN, D., B. BROOKS, and S. MORRIS (2017): "Selling to Intermediaries: Optimal Auction Design in a Common Value Model," Discussion paper, Yale University, University of Chicago and Princeton University.

BERGEMANN, D., F. CASTRO, AND G. WEINTRAUB (2017): "The Scope of Sequential Screening with Ex-Post Participation Constraints," in *18th ACM Conference on Economics and Computation.*

BERGEMANN, D., AND S. MORRIS (2005): "Robust Mechanism Design," *Econometrica*, 73, 1771–1813.

BERGEMANN, D., AND A. PAVAN (2015): "Introduction to Symposium on Dynamic Contracts and Mechanism Design," *Journal of Economic Theory*, 159, 679 = 7−1.

BERGEMANN, D., AND M. SAID (2011): "Dynamic Auctions: A Survey," in *Wiley Encyclopedia of Operations Research and Management Science*, ed. by J. Cochran, no. 1757, pp. 1511–1522. Wiley, New York.

BERGEMANN, D., AND P. STRACK (2015): "Dynamic Revenue Maximization: A Continuous Time Approach," *Journal of Economic Theory* 159, 819–853.

———(2017): "Stationary Dynamic Contracts," Discussion paper, Yale University and UC Berkeley.

BERGEMANN, D., AND J. VÄLIMÄKI (2003): "Dynamic Common Agency," *Journal of Economic Theory*, 111, 23–48.

———(2006): "Dynamic Price Competition," *Journal of Economic Theory*, 127, 232–263.

———(2010): "The Dynamic Pivot Mechanism," *Econometrica*, 78, 771–790.

BERGEMANN, D., AND A. WAMBACH (2015): "Sequential Information Disclosure in Auctions," *Journal of Economic Theory*, 159, 1074–1095.

BESANKO, D. (1985): "Multi-Period Contracts Between Principal and Agent with Adverse Selection," *Economics Letters* 17, 33–37.

BIAIS, B., T. MARIOTTI, G. PLANTIN, AND J. ROCHET (2007): "Dynamic Security Design: Convergence to Continuous Time and Asset Pricing Implications," *Review of Economic Studies*, 74, 345–390.

BIAIS, B., T. MARIOTTI, and J.-C. ROCHET (2013): "Dynamic Financial Contracting," in *Advances in Economics and Econometrics, Tenth World Congress*, ed. by D. Acemoglu, M. Arellano, and E. Dekel, vol. 1. Cambridge University Press.

BOARD, S. (2007): "Selling Options," *Journal of Economic Theory*, 136, 324–340.

BOARD, S., AND A. SKRZYPACZ (2015): "Revenue Management with Forward-Looking Buyers," *Journal of Political Economy*, 124, 1046–1087.

BOLESLAVSKY, R., AND M. SAID (2013): "Progressive Screening: Long-Term Contracting with a Privately Known Stochastic Process," *Review of Economic Studies*, 80, 1–34.

BÖRGERS, T. (2015): *An Introduction to the Theory of Mechanism Design*. Oxford University Press, Oxford.

CALZOLARI, G., AND A. PAVAN (2006): "Monopoly with Resale," *RAND Journal of Economics*, 37, 362–375.

CARROLL, G., AND I. SEGAL (2016): "Robustly Optimal Auctions with Unknown Resale Opportunities," Discussion paper, Stanford University.

CLEMENTI, G., AND H. HOPENHAYN (2006): "A Theory of Financing Constraints and Firm Dynamics," *Quarterly Journal of Economics*, 121, 229–265.

CONLISK, J., E. GERSTNER, AND J. SOBEL (1984): "Cyclic Pricing by a Durable Goods Monopolist," *Quarterly Journal of Economics*, 99, 489–505.

COURTY, P., AND H. LI (2000): "Sequential Screening," *Review of Economic Studies*, 67, 697–717.

CREMER, J., AND R. MCLEAN (1985): "Optimal Selling Strategies Under Uncertainty for a Discriminating Monopolist When Demands are Interdependent," *Econometrica*, 53, 345–361.

―――― (1988): "Full Extraction of the Surplus in Bayesian and Dominant Strategy Auctions," *Econometrica*, 56, 1247–1258.

DASGUPTA, P., AND E. MASKIN (2000): "Efficient Auctions," *Quarterly Journal of Economics*, 115, 341–388.

DEB, R., AND D. MISHRA (2014): "Implementation With Contingent Contracts," *Econometrica*, 82(6), 2371–2393.

DELLA VIGNA, S., AND U. MALMENDIER (2006): "Paying Not To Go To The Gym," *American Economic Review*, 96, 694–719.

DEMARZO, P., AND M. FISHMAN (2007): "Optimal Long-Term Financial Contracting," *Review of Financial Studies*, 20, 2079–2128.

DEMARZO, P., AND Y. SANNIKOV (2006): "Optimal Security Design and Dynamic Capital Structure in a Continuous-Time Agency Model," *Journal of Finance*, 61, 2681–2724.

DERMAN, C., G. LIEBERMAN, and S. ROSS (1972): "A Sequential Stochastic Assignment Problem," *Management Science*, 18 , 349–355.

DOLAN, R. (1978): "Incentive Mechanisms for Priority Queuing Problems," *Bell Journal of Economics*, 9, 421–436.

DWORCZAK, P. (2016): "Mechanism Design with Aftermarkets: Cutoff Mechanisms," Discussion paper, Stanford University.

ELIAZ, K., AND R. SPIEGLER (2008): "Consumer Optimism and Price Discrimination," *Theoretical Economics*, 3, 459–497.

ESÖ, P., AND B. SZENTES (2007): "Optimal Information Disclosure in Auctions," *Review of Economic Studies*, 74, 705–731.

FARHI, E., AND I. WERNING (2007): "Inequality and Social Discounting," *Journal of Political Economy*, 115, 365–402.

GARRETT, D. (2016): "Intertemporal Price Discrimination: Dynamic Arrivals and Changing Values," *American Economic Review*, 106, 3275–3299.

―――― (2017): "Dynamic Mechanism Design: Dynamic Arrivals and Changing Values," *Games and Economic Behavior*, 104, 595–612.

GARRETT, D., AND A. PAVAN (2012): "Managerial Turnover in a Changing World," *Journal of Political Economy*, 120, 879–925.

―――― (2015): "Dynamic Managerial Compensation: A Variational Approach," *Journal of Economic Theory*, 159B, 775–818.

GARRETT, D., A. PAVAN, and J. TOIKKA (2017): "Robust Predictions in Dynamic Screening," Discussion paper, Northwestern University and University of Toulouse and MIT.

GERSHKOV, A., AND B. MOLDOVANU (2009a): "Dynamic Revenue Maximization with Heterogenous Objects: A Mechanism Design Approach," *American Economic Journal: Microeconomics*, 2, 98–168.

———(2009b): "Learning about the Future and Dynamic Efficiency," *American Economic Review*, 99, 1576–1588.

———(2014): *Dynamic Allocation and Pricing: A Mechanism Design Approach.* MIT Press.

GERSHKOV, A., B. MOLDOVANU, and P. STRACK (2018): "Revenue Maximizing Mechanisms with Strategic Customers and Unknown, Markovian Demand," *Management Science*, 64, 2031–2046.

GREEN, E. (1987): "Lending and the Smoothing of Uninsurable Income," in *Contractual Arragnements for Intertemporal Trade*, ed. by E. Prescott, and N. Wallace, pp. 3–25. University of Minnesota Press.

GREEN, J., AND J. LAFFONT (1977): "Characterization of Satisfactory Mechanisms for the Revelation of the Preferences for Public Goods," *Econometrica*, 45, 427–438.

GRUBB, M. (2009): "Selling to Overconfident Consumers," *American Economic Review*, 99, 1770–1807.

GRUBB, M., AND M. OSBORNE (2015): "Cellular Service Demand: Biased Beliefs, Learning and Bill Shock," *American Economic Review*, forthcoming.

HARRIS, M., AND B. HOLMSTROM (1982): "A Theory of Wage Dynamics," *Review of Economic Studies*, 49, 315–333.

HOLMSTROM, B. (1979): "Groves Scheme on Restricted Domains," *Econometrica*, 47, 1137–1144.

KAKADE, S., I. LOBEL, AND H. NAZERZADEH (2013): "Optimal Dynamic Mechanism Design and the Virtual Pivot Mechanism," *Operations Research*, 61, 837–854.

KRÄHMER, D., AND R. STRAUSZ (2015): "Optimal Sales Contracts with Withdrawal Rights," *Review of Economic Studies*, 82, 762–790.

KURIKBO, N., T. LEWIS, F. LIU, AND J. SONG (2019): "Long-Term Partnership for Achieving Efficient Capacity Allocation", *Operations Research*, 67, 984–1001.

LI, H., AND X. SHI (2017): "Discriminatory Information Disclosure," *American Economic Review*, 107, 3363–3385.

LIU, H. (2018): "Efficient dynamic mechanisms in environments with interdependent valuations: the role of contingent transfers", *Theoretical Economics*, 13, 795–830.

MEZZETTI, C. (2004): "Mechanism Design with Interdependent Valuations: Effciency," *Econometrica*, 72, 1617–1626.

———(2007): "Mechanism Design with Interdependent Valuations: Surplus Extraction," *Economic Theory*, 31, 473–488.

MIERENDORFF, K. (2013): "The Dynamic Vickrey Auction," *Games and Economic Behavior*, 82, 192–204.

MIERENDORFF, K. (2016): "Optimal Dynamic Mechanism Design with Deadlines," *Journal of Economic Theory*, 161, 190–222.

MILGROM, P., AND I. SEGAL (2002): "Envelope Theorem for Arbitrary Choice Sets," *Econometrica*, 70, 583–601.

MIRROKNI, V., R. P. LEME, P. TANG, and S. ZUO (2017): "Non-Clairvoyant Dynamic Mechanism Design," Discussion paper, Google Research.

MOOKHERJEE, D., AND S. REICHELSTEIN (1992): "Dominant Strategy Implementation of Bayesian Incentive Compatible Allocation Rules," *Journal of Economic Theory*, 56, 378–399.

MOULIN, H. (1986): "Characterization of the Pivotal Mechanism," *Journal of Public Economics*, 31, 53–78.

MYERSON, R. (1986): "Multistage Games with Communication," *Econometrica*, 54, 323–358.

PAPADIMITRIOU, C., G. PIERRAKOS, C. PSOMAS, and A. RUBINSTEIN (2016): "The Intractability of Dynamic Mechanism Design," in *SODA*, pp. 1458–1475.

PAVAN, A. (2017): "Dynamic Mechanism Design: Robustness and Endogenous Types," in *Advances in Economics and Econometrics: 11th World Congress*, ed. by M. P. B. Honore, A. Pakes, and L. Samuelson. Cambridge University Press.

PAVAN, A., I. SEGAL, AND J. TOIKKA (2014): "Dynamic Mechanism Design: A Myersonian Approach," *Econometrica*, 82, 601–653.

RIORDAN, M., AND D. SAPPINGTON (1987): "Information, Incentives and Organizational Mode," *Quarterly Journal of Economics*, 102, 243–264.

ROGERSON, W. (1985): "The First-Oder Approach to Principal-Agent Problems," *Econometrica*, 51, 357–367.

SKRZYPACZ, A., AND J. TOIKKA (2015): "Mechanisms for Repeated Trade," *American Economic Journal: Microeconomics*, 7, 252–293.

SUGAYA, T., AND A. WOLITZKY (2017): "Revelation Principles in Multistage Games," Discussion paper, Stanford University and MIT.

TALLURI, K., AND G. VAN RYZIN (2004): *The Theory and Practice of Revenue Management*. Springer.

THOMAS, J., AND T. WORRALL (1990): "Income Fluctuation and Asymmetric Information: An Example of a Repeated Principal-Agent Problem," *Journal of Economic Theory*, 51, 367–390.

Author Index

Subject Index

CPSIA information can be obtained
at www.ICGtesting.com
Printed in the USA
BVHW040922290420
578090BV00004B/7